Indicators of milk and beef quality

The EAAP series is published under the direction of Dr. P. Rafai

EAAP – European Association for Animal Production

The European Association for Animal Production wishes to express its appreciation to the *Ministero per le Politiche Agricole e Forestali* and the *Associazione Italiana Allevatori* for their valuable support of its activities

Indicators of milk and beef quality

EAAP publication No. 112

Editors:

J.F. Hocquette and S. Gigli

CIP-data Koninklijke Bibliotheek, Den Haag

ISBN 9076998485 paperback
ISSN 0071-2477

Subject headings:
Muscle and meat
Dairy products
Cattle

First published, 2005

Wageningen Academic Publishers
The Netherlands, 2005

Contents

Review papers

The challenge of quality

J.F. Hocquette[1] and S. Gigli[2]*
[1]INRA, Herbivore Research Unit, Muscle Growth and Metabolism Group, Clermont-Ferrand/Theix Research Center, Theix, 63122 Saint-Genès Champanelle, France, [2]CRA (Council Research in Agriculture), Animal Production Research Institute, Meat Department, Via Salaria 31, 00016 Monterotondo (Roma), Italy
**Correspondence: hocquet@clermont.inra.fr*

Summary

The great change in our Society is that the farming and agri-food sectors are faced with a general saturation of food markets in Europe and with an increasing demand by consumers for high-quality products. The major current questions are thus how to define quality, and how to increase the quality of animal products to satisfy the new consumer requirements.

To achieve the above objectives, the producers and sellers of animal products now need indicators for quality. This is the reason why the Cattle Commission of the European Association for Animal Production (EAAP) had a specific session on this topic in its Annual Meeting, held in Bled (Slovenia) in September 2004.

All the review papers and the short communications presented in this session are in this EAAP publication entitled "Indicators of milk and beef quality". Additional papers from invited authors were written in 2004 and then added to complete the publication although not all topics related to quality indicators are covered in the present publication. Review papers aim to describe the state of the art in different disciplines including genetics, physiology, nutrition, biochemistry as well as social and economic aspects. Short communications are recent and new results given as examples of the research done in these different areas. All the papers were peer-reviewed by international experts.

This introduction to our EAAP publication aims to remind readers of the history of Cattle in the World, the milk and beef production features in the European Union, and finally the definition of quality for food products and especially for beef, milk and cheese. Parts of this Introduction are based on previous discussions during the EAAP Round Tables, which were held over the last three years during EAAP meetings.

Keywords: cattle, milk, cheese, beef, quality

The history of cattle products

Humankind began domestication of animals and plants at least 10,000 years ago and has been modifying species and agricultural practises ever since. Cattle were domesticated from urus (B. primigenius) and aurochs 9,000 years ago. The domestication of cattle was simply for meat production at first. Then, about 6,000 years ago, humans started milking their cows. Thereafter, around 5,000 years ago, nomadism was developed. In this new way of life, humans and animals used to migrate together and humans used to depend on cow milk and not on meat (for review, see Tanabe, 2001).

The major objective of agriculture was initially to satisfy the needs of humanity quantitatively with respect to different food products. This objective was achieved at least in developed countries.

Cattle have indeed contributed to society over the millennia by providing food. However, out of a total world population of 6 billions, about 800 million people are nowadays malnourished, mainly in developing countries (EAAP Round Table, 2002).

The increasing development of trades and exchanges within and between countries during Antiquity, the Middle Ages and especially in our modern Society progressively increased the relative importance of economical factors in agriculture, and livestock management. Therefore, the production structure has changed significantly with the development of the market economy, especially during the last couple of centuries.

International trade issues are nowadays becoming increasingly important, as well as concentration of the companies. Globalisation is going ahead anyway and we cannot stop it. More recently, the huge increase in world-wide trade over the last ten years is due to Asia, which is now the new economic player in the World. In addition, imports of developed countries are increasing more than exports, which has increased the gap between developing and developed countries (EAAP Round Table, 2002).

As side effects, the real international prices of agricultural products, including beef and milk, have dramatically decreased during the last twenty years, although the prices at consumer level have not decreased in the same way (EAAP Round Table, 2002). In addition, intensive systems have been developed to a high degree in some countries. This has contributed to rural depopulation. Indeed the FAO reported in 2003 that about 4% of the population works in agriculture in the European Union as compared to more than 50% in Asia and Africa, 20% in the European new members of 2004 and only 2.2% in the USA. In addition, the increase in food supply and in fat content of the human diet has induced obesity in epidemic proportions in developed nations.

Thus, in today's society in developed countries, changing lifestyles, increasing consumer demands, globalisation, opening of international markets and obvious differences in prices in different markets are forcing producers and sellers of animal products to rethink their business. The major change is that the farming and agri-food sectors are faced with a general saturation of food markets in Europe and with an increasing demand by consumers for high-quality products, especially in terms of healthiness.

The future of livestock in Europe depends mainly on economical and social factors. From an economical point of view, livestock production has to face the reality of free world trade and the progressive removal of import barriers. The question is how to limit the negative effects of globalisation, especially in developing countries, and how to increase its benefits. Producers and sellers must also take into account the consequences of the Common Agricultural Policy (CAP) reform within European Countries, and the enlargement of the European Union, which rose to 25 members in 2004. From a social point of view, environmental degradation is sometimes blamed on livestock. However, livestock provide means of managing our landscapes and of maintaining biodiversity. We must thus recognise the wider role in society of livestock (not only cattle) in providing food products to human beings, in providing products of high quality and in satisfying new consumer expectations in terms of healthiness, environment management, animal welfare, etc. This led Professor Margaret Gill in the Plenary Session of the EAAP Meeting in Bled (September 2004) to suggest that scientists have not been addressing the right questions in their research on livestock. We can also argue that globalisation must not be reduced to trade between countries, but must also include globalisation of environment, human poverty, sustainable agriculture, etc. (EAAP Round Table, 2002). This is the reason why improved sustainability in livestock farming systems became a priority, as integrated research, for scientists of the EAAP (Gibon *et al.*, 1999a, 1999b).

Among the new challenges for scientists is also the necessity to better define quality of food products (including social aspects), and how to increase the quality of animal products to satisfy the new consumer requirements. To achieve this goal, the producers and sellers of animal products nowadays need quality indicators. This is the reason why a specific session of the EAAP meeting dealt with that subject.

Milk and beef production in Europe

The economic value of livestock production in the European Union has been recently reviewed by Liinamo and Neeteson-van Nieuwenhoven (2003), Chatellier *et al.* (2003) and OFIVAL (2004) from FAO data before the joining of 10 new countries in 2004. The major points are described below.

The majority of cattle in the World are in Asia (35% in 2002), South America (23%), Africa (17%) and North and Central America (12%). About 6% of cattle were found in the 15 countries of the European Union in 2002 and 4% in the rest of Europe. Among the 1,359 million head of cattle in the World in 2003, more than 160 million were in Brazil, 110-130 in China, 96 in the USA and about 80 in the 15 members of the European Union.

Milk production and consumption in Europe

The production of milk comes mainly from cattle in most countries (84% on average in the World, 97% in the European Union) except in India (42%) and Pakistan (31%) (FAO 2003 data). The European Union is currently the first bovine milk producer followed by the USA. They each represent 28% and 15% of the total World bovine milk production. Whereas the average milk yield per cow in the World in 2002 was 2,237 kg, it varied from 487 kg in Africa to 5,903 kg in Europe. Within Europe, Germany and then France and the United Kingdom are the largest producers of milk (24%, 20% and 13% respectively of total bovine milk production in the 15 countries of the European Union in 2003). The number of dairy cows on European farms in 2003 was on average 36 (ranging from 10 in Austria to 100 in the United Kingdom).

The European citizen consumes an average of 323 kg of dairy products (in kg of milk equivalent) per person (data 1999, 2000 and 2002). This is more than the USA and Australia, but less than New Zealand and Switzerland (FAO 1998 data). The European Union is self-sufficient for milk. The main exporters of dairy products in the World are the European Union, Australia and New Zealand.

Beef production and consumption in Europe

In 2003, the European Union was the second largest beef producer in the World (about 12%) after the USA (about 20%) and in front of Brazil (less than 12%) and China (less than 10%). The USA is one of the major exporters of beef, more than Brazil and than the European Union. Australia is, however, the country that exports the most beef on the World market. Within the European Union, France and then Germany and Italy are the largest producers of beef (22%, 19% and 16% respectively).

The highest consumption of beef is in Argentina (57.4 kg per inhabitant in 2003), followed by the USA, Australia, Brazil, Canada, Uruguay and Canada (32 to 42.6 kg) and then the European Union (20 kg per inhabitant). The European Union is self-sufficient for beef. However, in 2003, the

European Union imported about 500,000 tonnes equivalent carcass and exported (mainly to Russia) about 410,000 tonnes equivalent carcass.

In Europe, France is the first consumer of beef (27.1 kg per inhabitant per year). France is also the first exporter of beef especially to Italy (50.9%), Greece (12.4%) and Spain (10.1%). On the other hand, Italy is the first importer of beef, especially from France.

The challenge for the future

It is quite clear that world demographic development predicts more than 8 billion inhabitants in total for the year 2020. Longer-term forecasts are predicting increasing demand of food mainly from developing countries. Despite the increase in total population, the major cause will be growth in per capita consumption. Beef and milk consumption are forecast to increase by 2.1 to 2.3 fold respectively between 1993 and 2020 in developing countries. The increase in beef and milk consumption will be, however, only 7 to 9% in developed countries (Delgado *et al.*, 1999). Satisfaction of this international demand will require a 46% increase in world grain production. It was thus argued by Corbett (2001) that it would be prudent to enhance the contributions of grazing livestock to meet the increasing demand for products from cattle. Indeed, it will be important not to depend on crop-livestock and intensive systems of production to meet these new human population requirements for food products.

The European Union is, however, self-sufficient for most animal products, except for sheep and goat meat. In addition, different food crises (BSE, foot and mouth, etc.), with their effects amplified by the media, have decreased consumer confidence in animal products. Some consumers even do not accept some of the current agricultural practices. Therefore, European consumers will demand animal products of high quality and Scientists must develop socially acceptable food production methods (Gibon *et al.*, 1999a). The development of Quality Assurance Schemes, by addressing management issues from conception of the animals to the consumers' plate (from farm to fork) is thus an important challenge to meet. In addition, land is increasingly a limited factor in populated areas of developed countries. Therefore, the management of natural resources and environment will remain an important issue. In developed countries, the majority of citizens is also urban and relatively well-educated. But many of them have little knowledge of agricultural principles, and increasingly ask for more accurate information on food products. So consumers are paying more attention to guaranteed quality, based for instance on environmental and animal-friendly production. It is thus essential to increase consumer awareness of information on agriculture principles and on properties of food and, when needed, to correct any consumer myths. Lastly, urban consumers ask also for convenience (Issanchou, 1996) and this is another important feature to integrate when considering the future of livestock.

Because Europe does not live in isolation, it is also clear that European Agriculture has to remain cost, price and quality competitive, while facing strong price pressures in the international market. In addition, European consumers spend only 16% of their income on food, and this proportion is decreasing. Customers are indeed used to paying low prices for food. But the extra-costs to guarantee the safety and the quality of animal products must be included in the agro-food economy (EAAP Round Table, 2003). Economists have drawn up several scenarios describing the Evolution of Livestock Production in the World and their consequences in Europe. The options depend on the strength of the World economy and the relative importance of Europe. High quality products and regional specialities may be good choices if the relative importance of cost prices does not increase (for review, see Liinamo and Neeteson-van Nieuwenhoven, 2003). Others think that parts of the market involved in the production of cheap food will move outside Europe, whereas Europe will focus on the quality market with high quality standards (EAAP round table, 2001). It is indeed likely that the relatively high levels of income enjoyed by European citizens will increase the demands for branded products rather than consumption per capita. Labelling is, however, viewed as a trade restriction by some countries, whereas others, especially in the South of Europe, view labelling requirements as an opportunity to brand high-quality products. The success of branded products is often based on the assumption that many consumers, though not all, would be willing to pay for transparency, traceability and quality assurance in food products.

The future of livestock is thus uncertain, but, to summarise, it is likely that quality of food products, the evolution of consumers' expectations and perceptions, as well as economical factors, will play a major role in its future.

What does quality mean for bovine products?

The concept of quality

The concept of quality has been defined in many ways, and differs between countries. For northern European countries, quality refers to health and hygiene aspects and to any other public norms. Private companies think indeed that a product is of good quality if it follows pre-defined norms. In the southern European countries, the concept of quality is much wider, including sensorial traits, the geographical and human environment, any link to a specific region or to any specific method of production. We will here consider quality as all the characteristics of food products valued by the consumer. It implies stated and implicit needs of consumers. It is therefore a holistic and multifaceted concept integrating safety, sensorial and nutritional traits, traceability, or social considerations (public interest in environment management, animal welfare, etc.). Safety and Health are however the two major expectations of consumers in developed countries (EAAP Round Table, 2003).

Safety is the absence of any contamination in food products. It is not linked to the characteristics of the product itself, except in some specific cases (BSE for instance). It is often linked to the introduction of external pathogens at any stage of the food chain. It is therefore not always included in the concept of quality and will not be discussed in this publication despite its increasing importance for consumers.

Nutritional value or healthiness of food depends more on the product chemical composition. It includes the type and amount of protein, carbohydrate, lipid, vitamins and minerals in food products. Consumers in industrialised countries have displayed an aversion to a fatty diet and therefore to food products with high fat contents. On the other hand, contents in proteins, vitamins, and micro-nutrients may be positive indicators of the nutritional value of food.

Sensory quality can be defined as texture, flavour, taste and visual aspect. These traits depend directly on the product characteristics in terms of structure and constitution.

Social considerations arise from changes in consumers' attitudes towards food and ethical views of animal and environmental management. This is especially true for cattle, which are supposed to be reared at pasture in the simplistic view of most consumers. In addition, unlike monogastrics and human beings, ruminants are able to convert renewable resources (uncultivated land, crop residues, by-products) into humanly-edible food. Therefore, ruminants play a key role in sustainable livestock production systems (Oltjen and Beckett, 1996). The perception of those phenomena by consumers is also important.

Traceability is the ability to trace the food products back to the farm or animal of origin. In other words, the farming and agri-food sector is able to guarantee consumers the origin of the food products, including in terms of animal breed, farm where animals were reared, nutrition programme, etc. This concept differs from that of transparency (e.g. knowing that food products are produced without growth promoters, in respect of animal welfare, etc.). Traceability or transparency often mean quality for at least some consumers. This is not true because the origin of the product or its production conditions do not necessarily mean that it is of high quality (for instance, highly healthful, of good flavour, etc.). The confusion between these concepts is based on consumers' subjective considerations or social aspects. For instance, consumers are sometimes convinced of possible adverse health effects of foods produced using intensive farming methods. This has reinforced the interest in the benefits of extensive-based systems, for instance at pasture. We will, however, discuss in this publication some aspects of traceability because it is often included as a quality criterion in some quality conventions between producers and consumers.

We must also keep in mind the fact that consumers' wishes are not always logically consistent. For instance, consumers' demands for low prices conflict with demands for high-quality products. In beef, demands for flavour tend to conflict with demands for wholesomeness and low fat. These apparent contradictions may be explained by an increasing diversity of expectations from consumers. This may reinforce segmentation of the market and the consumers' demands for specifications. It is important to underline that the consumer's expectations (including the different quality traits listed above) vary with consumer type and also with time. It varies depending on incomes, country, culture, consumer's age, habits, etc. For instance, the price of beef has not declined as much as those for other meats during the last two decades; consequently, beef consumption is more and more correlated to incomes or consumed during festive meals in Europe (EAAP Round Table, 2003).

The concept of quality, and how it will be incorporated in the beef and milk industries, will however change depending on the future, which is itself subject to controversial issues. What will be the weight of the public attitude and wishes? What will be the nature of these wishes? What impact might international companies promoting new technologies have? How will Society meet the challenges of environmental issues and sustainable development? What is the future of local breeds and organic agriculture? None of these questions is simple to answer and therefore any forecasting is hazardous. The future of the livestock sector is thus not secure (EAAP Round Table, 2003). One point, which is however obvious, is that consumers' needs and expectations must be satisfied.

How can quality traits be optimised and guaranteed to consumers?

Quality traits depend first of all on factors relative to live animals (Figure 1). These factors affect the quality of raw products (muscle for beef and milk for dairy products). They include breeding and rearing strategies, especially nutritional management.

Genetic improvement is quite effective because it is permanent and cumulative: indeed improvements made in one generation are passed to the next ones. Emphasis is nowadays put on genetics and genomics due to the sequencing of genomes from different species, the bovine one being available in the very short-term future (Lewin, 2003). This will have direct applications in genetic improvement and traceability with the help of DNA-based techniques.

On the other hand, functional genomics is likely to impact our knowledge of ruminant physiology. Genomics is indeed a new Science which aims to better understand how biological traits are determined from genes. Genomics is changing our scientific paradigm, because the global expression of genes in cells and tissues will generate new biological hypotheses. We will thus move from hypothesis-driven research (where scientists test the relevance of biological hypotheses) to hypothesis-generation research (in which new biological hypotheses will appear from the gene expression profiles).

Many papers in this publication will thus deal with genetics and genomics. However, the advent of genomics will probably increase the gap between research and what the consumer can understand. In addition, this approach should not be confounded with genetic engineering, which is the science which involves deliberate modification of the genetic material of plants or animals (Uzogara, 2000). In fact, genetic engineering of food is developing rapidly, but it is more or less accepted. It is in fact rejected by most European consumers.

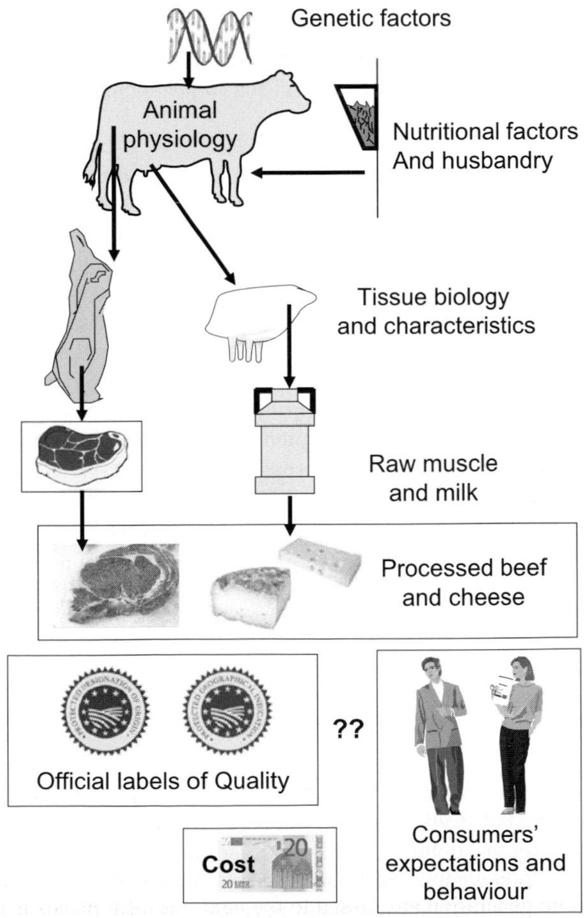

Figure 1. Quality traits of food products depend first of all on animal physiology and tissue characteristics. They also depend, on the other hand, on the process of muscle and milk transformation into beef and cheese respectively. These characteristics together with the economical and social context may help in the preparation of quality conventions. Consumers decide to purchase the food products depending on their expectations, the quality traits they recognise and the cost of the products.

Husbandry may also affect the quality of the products depending on growth paths, growth rate, nutritional level and the nature of food given to the animals. During the last decade, emphasis has been put on nutritional factors, due especially to the BSE crisis which developed due to the inappropriate type of food given to high-producing dairy cows. Nowadays, consumers are thus increasingly involved in how cattle are fed. Besides, the development of genomic tools now provides a powerful way to understand how production systems interact with Genetics.

Quality also depends on raw product transformation processes (the conversion of muscle into meat, the transformation of milk into cheese). In the case of beef, the conditions of animal slaughtering and carcass chilling play a major role in the determination of tenderness. Methods of muscle stretching have therefore been given increasing attention (Sorheim and Hildrum, 2002). More generally, it is important to develop control systems and techniques for monitoring processes for high quality added-value products. The idea is indeed to anticipate the final properties of food products by any method based on physics, biochemistry or anything else. In this way, emerging technologies (for instance, fluorescence, infrared spectroscopy, etc.), and their on-line applications, are expected to play a key role in predicting quality, and in providing the development of large-scale industrial routine methods.

Because consumers cannot spontaneously assess the quality of food products, the value of those products is recognised through "quality signals". Brand name is the archetype of such signals. We have now Protected Denomination of Origin (PDO), Protected Geographic Identification (PGI), Traditional Speciality Guaranteed (TSG) or issued from organic farming. The credibility of those quality signals is based on promises for the future quality of food products. This is more and more difficult due to the increasing complexity of the concept of quality as discussed previously (Maze *et al.*, 2001). Nevertheless, this implies the definition of a quality convention, which should specify rules to characterise the products. Various forms of quality conventions exist: they may include quality traits based on origin or animal breeds, on production methods, or on specific supply-chain structure. Quality conventions may also involve typicity (connection with a territory) especially in France and Italy (Sylvander *et al.*, 2003). The success of these conventions can be assessed with different parameters including the market size, the price differential for producers or for consumers, or simply by raising consumer expectations in terms of typicity or quality. The market attractiveness, the economical factors and good co-operation within a production chain are important parameters to ensure the practical success of such conventions. Furthermore, signalling and guaranteeing the quality of a product to consumers require the information to be transmitted from the producers to the sellers and finally to the consumers. This implies good co-operation between all operators of the food chain (Maze *et al.*, 2001).

Conclusion

Quality of food products is a very complex concept, which needs to be better defined. The consumers' expectations and their evolution with time (including apparent contradiction between them) have to be the basis for this definition. We must also face various types of consumers with different types of demands and we must distinguish perceived quality prior to purchase (mainly determined by beliefs and attitudes), at the time of purchase (mainly determined by the product characteristics in interaction with its price) and upon consumption. The different scientific disciplines in Animal Science (Genetics, Genomics, Physiology, Nutrition, Husbandry, etc.) and the expertise in various fields of technical food processing (Physics, Biochemistry, etc.) should help to better define the different dimensions of quality. They are also essential in predicting quality, i.e. in anticipating the different quality traits of food products. Research has thus to be more interdisciplinary and practically oriented to satisfy the consumers' demands, while fundamental research needs to be maintained for basic knowledge production. The economical and social aspects remain key factors in the end to meet consumers' expectations. Scientists have also to participate more in the social debates since consumers are asking for more "transparency". To conclude, we have to stress the advantages of economic approaches and of consumer-oriented approaches in addition to biological and technical disciplines for the control and the improvement of food product quality. The future of dairy and beef production in Europe depends on this evolution.

Acknowledgements

All the papers in this EAAP Publication have been reviewed by one or two external experts from European countries (Belgium, France, Germany, Ireland, Italy, Poland, Portugal, UK, Sweden, Switzerland) as well as Australia, Japan and USA and we would like to thank the referees very much for the good work they did.

References

Chatellier, V., H. Guyomard and K. Le Bris, 2003. The World Trade organization rounds on agriculture: prospects for animal production in the European Union. INRA Prod. Anim. 16: p. 301-316.

Corbett, J.L., 2001. The challenging demand for increases in meat and milk production: enhancing the contributions of grazing livestock. Asian-Aust. J. Anim. Sci. 14, Special Issue: p. 1-12.

Delgado, C., M. Rosegrant, H. Steinfeld, S Ehui and C. Courbois. 1999. Livestock to 2020. The next Food Revolution. Food, Agricultural and the Environment Discussion Paper n°28, IFPRI/FAO/ILRI. International Food Policy Research Institute, Washington D.C.

EAAP round table, 2001. Flamant, J.C. (coordinator), The future of the livestock sector in the light of the recent crises in Europe. Budapest 26th August 2001. http://www.eaap.org/round_table_budapest.htm

EAAP round table, 2002. Flamant, J.C., and P. Cunningham (coordinators), Globalisation and the livestock sector: who benefits? Cairo, 2nd September 2002. http://www.eaap.org/round_table_cairo_english.htm

EAAP round table, 2003. Flamant, J.C., and R. Chizzolini (coordinators), Changing consumers ... changing animal production? Roma, 31th August 2003. http://www.eaap.org/round_table_rome_english.htm

Food and Agriculture Organization (FAO) statistics. http://www.fao.org.

Gibon, A., A.R. Sibbald and C. Thomas, 1999a. Improved sustainability in livestock systems, a challenge for animal production science. Livest. Prod. Sci. 61: p. 107-110.

Gibon, A., A.R. Sibbald, J.C. Flamant, P. Lhoste, R. Revilla, R. Rubino and J.T. Sorensen, 1999b. Livestock farming systems research in Europe and its potential contribution for managing towards sustainability in livestock farming. Livest. Prod. Sci. 61: p. 121-137.

Issanchou, S., 1996. Consumer expectations and perceptions of meat and meat product quality. Meat Sci. 43, p. S5-S19.

Lewin, H.A., 2003. The future of cattle genome research: the beef is here. Cytogenet. Genome Res. 102: 10-15

Liinamo, A.E. and A.M. Neeteson-van Nieuwenhoven, 2003. The economic value of livestock production in the European Union. Farm Animal Industrial Platform (FAIP), AnNe Publishers, The Netherlands, 32 p.

Maze, A., S. Polin, E. Raynaud, L. Sauvee and E. Valceschini, 2001. Quality signals and governace structures within agro-food chains: a new institutional economics approach. "Economics of Contract in Agriculture and the Food Chain Supply", 78th EAAE Seminar, Copenhagen, June 15-16, 2001.

Office National Interprofessionnel des Viandes, de l'Elevage et de l'Aviculture (OFIVAL). http://www.ofival.fr

Oltjen, J.W. and J.L. Beckett, 1996. Role of ruminant livestock in sustainable agricultural systems. J. Anim. Sci. 74: p. 1406-1409.

Sorheim, O. and K.I. Hildrum, 2002. Muscle stretching techniques for improving meat tenderness. Trends Food Sci. Technol. 13: p. 127-125.

Sylvander, B., G. Belletti, A. Marescotti, E. Thévenod-Mottet, 2003. Establishing a quality convention, certifying and promoting the typicité of animal products: the beef case. International Livestock Farming System Symposium, Benevento, 26-29 August, 2003.
Tanabe, Y. 2001. The roles of domesticated animals in the cultural history of the humans. Asian-Aust. J. Anim. Sci. 14, Special Issue: p. 13-18.
Uzogara, S.G., 2000. The impact of genetic modification of human foods in the 21st century: a review. Biotechnol. Adv. 18: p. 179-206.

Genetic markers for beef quality

Ch. Kühn[1], H. Leveziel[2], G. Renand[3], T. Goldammer[1], M. Schwerin[1] and J. Williams[4]*
[1]FB Molekularbiologie, Forschungsinstitut für die Biologie landwirtschaftlicher Nutztiere, 18196 Dummerstorf, Germany, [2]Unité de Génétique Moléculaire Animale, UMR1061; INRA-Université de Limoges, Faculté des Sciences et Techniques, 87060 Limoges cedex, France, [3]Station de Génétique Quantitative et Appliquée, INRA, 78352 Jouy-en-Josas cedex, France, [4]Roslin Institute, Roslin, Midlothian, Scotland, EH25 9PS, United Kingdom
**Correspondence: kuehn@fbn-dummerstorf.de*

Summary

Improvement of beef quality by marker-assisted selection requires a set of genetic markers thoroughly characterised for their effects in the target population. Presently, however, only a limited number of genetic markers with confirmed effects is available. These comprise markers in linkage equilibrium with the desired traits that have been derived from QTL mapping experiments as well as linkage disequilibrium and direct markers mostly within functional candidate genes. The genetic markers presently available that are associated with beef quality, as well as current developments including novel techniques in molecular biology to generate new markers relevant for European beef cattle production, are described.

Keywords: genetic marker, beef, marbling, tenderness

Introduction

In this review beef quality is defined by the properties of a piece of beef meat that are appreciated during consumption. Thus, in this definition beef quality is directly related to the consumers preferences, which are primarily focused on meat tenderness and marbling. However, presently the nutritional value of beef is an increasing concern for the consumer. Focusing on this definition of beef quality, only very limited information on genetic markers is available compared to a wealth of literature on genetic markers for traits associated with growth and the composition of the whole carcass. Although beef quality is influenced by environmental effects, e.g., nutrition and by *post-mortem* carcass chilling and meat processing, the heritability of the meat quality traits is between 0.15 - 0.35 (e.g., Wheeler *et al.*, 2004). This is a substantial level of genetically controlled variation, and should enable genetic markers for these traits to be detected and used to increase beef quality through marker-assisted selection (MAS) (Dekkers, 2004).

Dekkers (2004) defined three types of genetic markers for MAS dependent on the information they provided regarding the functional mutation underlying variations of the target trait: i) direct markers coding for the functional mutation, ii) LD markers in population-wide linkage disequilibrium (LD) with the functional mutation and iii) LE markers in population-wide linkage equilibrium (LE) with the functional mutation, but more or less closely linked. Direct and LD markers can be used population-wide for MAS, regardless of family background. In contrast LE markers are limited to application within families and require confirmation of the "phase" of alleles at marker and functional mutation in each generation. This means increased costs and decreased accuracy for MAS using LE markers compared with using direct markers. However, often LE markers are the first markers that are available, because QTL mapping experiments identify markers linked to a trait that subsequently can be used for MAS. Later identification of the functional trait genes themselves and the direct markers within them requires more sophisticated and expensive investigations.

QTL for beef quality

Most of the published QTL mapping experiments on beef quality traits have been performed in the USA or Canada under North-American feedlot production conditions regarding pre- and postweaning nutrition and housing. Therefore from a European perspective care should be taken, when considering the application of the information. Several further studies on beef quality have been performed e.g. in Japan or Australia. However, the information produced has not been made freely available to the scientific community. Thus, this review considers twelve reports on QTL mapping for beef quality traits only. Among these studies, several investigated meat quality in B. taurus x B. indicus crosses, where there is known to be very large difference in meat quality traits, particularly toughness. Genetic differences in trait expression detected between indicine and taurine marker alleles, however, are of limited use for the MAS in Europe, because B. indicus cattle are not used in European beef production. The investigations summarized in Table 1 show that the published reports mainly focus on tenderness, and on the amount and the composition of intramuscular fat. The myostatin gene, which is localised in the centromeric region of BTA2 and underlies the double muscling phenotype (Grobet *et al.*, 1997) is not included in Table 1. Double muscling itself is not primarily a beef quality trait, although variations in the gene are associated with variations in marbling (Casas *et al.*, 1998) and significant differences in collagen content and solubility (Boccard, 1981).

Intramuscular fat

Several studies reported QTL for marbling as an indicator of the amount of intramuscular fat (see Table 1). Loci affecting this trait have been found on BTA2 in several different studies. Casas *et al.* (1998) directly attributed the QTL for marbling detected on BTA2 to the genotype for myostatin that is mapped to this chromosome. However, it seems unlikely that this gene is involved in the variation seen in the other studies, which either did not include breeds known to be carriers of double muscling alleles or where the most likely QTL position was on the opposite end of the chromosome. Other QTL for marbling for which confirming evidence is provided by the location being reported in more than one study were identified on BTA3 by three independent studies and on BTA27 in two separate experiments. As QTL affecting marbling on BTA2, 3, and 27 have been confirmed, MAS for intramuscular fat will probably focus on these QTL, whereas other reported QTL on BTA5, 8, 9, 10, 14, 16, 17, 23, and 29 are as yet unconfirmed. In contrast to the multitude of studies investigating the amount of intramuscular fat, there is only one report describing loci with impact on the composition of the intramuscular fat (Taylor *et al.*, 1998). This study was, however, restricted to the investigation of a single chromosome (BTA19) and to the comparison of B. taurus and B. indicus alleles. As there is an increasing consumer focus on the nutritional value of beef, further whole-genome studies looking at the composition of the intramuscular fat are needed, especially for the relevant European beef production conditions.

Tenderness

In various QTL studies, tenderness has been measured by three different approaches: either as shear force values, by taste panel analysis or by merging these data into principal components. Several studies independently identified a locus on BTA29 with effect on tenderness in the middle to telomeric region of the chromosome (Table 1). The QTL was identified in crosses between B. taurus and B. indicus as well as in crosses between B. taurus breeds. Page *et al.* (2002) suggested that genetic variants of the calpain 1 (*CAPN1*) gene are the functional background of this QTL. Other QTL with impact on beef tenderness traits identified on BTA4, 5, 9, 11, 15, and 20 have not been confirmed in independent studies nor is there evidence for a gene within the QTL region that could be considered as a strong candidate that has a physiological function associated with

Table 1. QTL for beef quality.

Trait	BTA	cM[*]	Resource population
Marbling	2[i]	10	(Piedmontese x Angus) + (Belgian Blue x MARCIII) BC[v]
Marbling	2[k]	38	(Brahman x Hereford) x composite B. taurus
Marbling	2[l]	54	(Brahman x Angus) x composite B. taurus
Marbling	2[m]	~70	Bos taurus beef breed crosses
Marbling	2[n]	120	Hereford x composite B. taurus double BC
Marbling	3[o]	28	(Brahman x Hereford) x composite B. taurus
Marbling	3[l]	56	(Brahman x Angus) x composite B. taurus
Marbling	3[p]	65	(Belgian Blue x MARC III) BC to MARCIII
Marbling	5[k]	n. i.	(Brahman x Hereford) x composite B. taurus
Marbling	5[o]	50	(Brahman x Hereford) x composite B. taurus
Marbling	8[p]	9	(Belgian Blue x MARC III) BC to MARCIII
Marbling	9[o]	71	(Brahman x Hereford) x composite B. taurus
Marbling	10[p]	59	(Belgian Blue x MARC) III BC to MARCIII
Marbling	10[o]	4	(Brahman x Hereford) x composite B. taurus
Marbling	14[o]	47	(Brahman x Hereford) x composite B. taurus
Marbling	16[l]	44	(Brahman x Angus) x composite B. taurus
Marbling	17[q]	21	(Belgian Blue x MARC III)BC to MARCIII
Marbling	23[o]	30	(Brahman x Hereford) x composite B. taurus
Marbling	27[o]	29	(Brahman x Hereford) x composite B. taurus
Marbling	27[q]	60	(Belgian Blue x MARC III BC) to MARCIII
Marbling	29[n]	2	Hereford x composite B. taurus double BC
% EEF[a]	19[r]	71	Brahman x Angus BC and F2
% oleic acid	19[r]	21	Brahman x Angus BC and F2
% ULCFA[b]	19[r]	18	Brahman x Angus BC and F2
% stearic acid	19[r]	18	Brahman x Angus BC and F2
UFA/SFA[c]	19[r]	18	Brahman x Angus BC and F2
Beef flavour	21[s]	n. i.	Hereford x composite B. taurus double BC
WBS 14 d[d]	5[q]	62-72	(Piedmontese x Angus) x MARC III
WBS 14 d[d]	9[p]	26	(Belgian Blue x MARC III) BC to MARCIII
WBS 14 d[d]	15[t]	22-28	(Brahman x Hereford) x composite B. taurus
WBS 14 d[d]	20[o]	72	(Brahman x Hereford) x composite B. taurus
WBS 14 d[d]	29[o]	54	(Brahman x Hereford) x composite B. taurus
WBS 14 d[d]	29[q]	56-65	(Piedmontese x Angus) x MARC III
WBS 3 d[e]	4[p]	19	(Belgian Blue x MARC III) BC to MARCIII
WBS 3 d[e]	20[o]	66	(Brahman x Hereford) x composite B. taurus
WBS 3 d[e]	29[q]	56-65	(Piedmontese x Angus) x MARC III
Tenderness PC[f]	11[s]	24-28	Hereford x composite B. taurus double BC
Tenderness (taste)[g]	29[u]	n. i.	Bos taurus beef breed crosses
WBS[h]	29[u]	n. i.	Bos taurus beef breed crosses

BTA: Bos taurus chromosome; [*] position of the maximum of the test statistic along the chromosome; [a] % EEF: % ether extractable fat; [b] % ULCFA: proportion unsaturated long chain fatty acids; [c] UFA/SFA: unsaturated / saturated fatty acids; [d] WBS 14d: Warner Bratzler shear force value 14 days after slaughter; [e] WBS 3d: Warner Bratzler shear force value 3 days after slaughter; [f] Tenderness PC: principal component value for tenderness; [g] Tenderness (taste): Tenderness as determined by a taste panel; [h] Warner Bratzler shear force value, time not indicated; [i] Casas *et al.*, 1998; [k] Stone *et al.*, 1999; [l] Casas *et al.*, 2004 ; [m] Schimpf *et al.*, 2000 ; [n] MacNeil and Grosz, 2002; [o] Casas *et al.*, 2003 ; [p] Casas *et al.*, 2001; [q] Casas *et al.*, 2000 ; [r] Taylor *et al.*, 1998 ; [s] MacNeil *et al.*, 2003; [t] Keele *et al.*, 1999; [u] Schmutz *et al.*, 2000; [v] BC: backcross; n.i. : not indicated.

variations in muscle structure or *post-mortem* tenderisation. Thus, in the absence of further information these QTL remain speculative and MAS for tenderness has to depend on LE markers.

Other traits

A single report on a QTL with impact on beef flavour on BTA21 is described in a Hereford x composite double backcross population, however, without indication on the precise QTL position or the underlying gene (MacNeil *et al.*, 2003).

Direct and LD genetic markers for beef quality

Before whole genome scan experiments were used to detect QTLs, studies looking for genetic markers associated with beef quality were performed by investigating polymorphisms in candidate genes, which were chosen because of their central role in biological processes with impact on meat quality. Undoubtedly, the recent description of QTL regions has further encouraged the search for candidate genes that may underlie the variations seen in the traits. Today, the selection of the functional and positional candidate genes can benefit of the huge volume of data arising from genome sequencing and annotation programmes conduced in mammalian species. However, at the present time, just a few markers and/or genes have been identified, which have a confirmed association with meat tenderness or marbling.

Tenderness

The *CAPN1* gene (Smith *et al.*, 2000) has been genetically mapped on BTA29. It encodes the Calpain 1 or μ-Calpain, which is one of the proteases implicated in the proteolysis of muscular proteins during meat ageing, and whose localisation followed the detection of a QTL for tenderness on chromosome 29 (Table 2). Among 38 SNPs identified in the *CAPN1* gene by Page *et al.* (2002), two corresponding to mutations in *CAPN1* exons 9 and 14 have been associated with variations in tenderness measured by a shear force test on *Longissimus dorsi*. It should be noted that these 2 polymorphisms correspond to amino-acid substitutions (A316G and I530V) and that three combinations (alleles or haplotypes) have been described. The results obtained in two separate populations suggest that variation in *CAPN1* is responsible for tenderness differences in beef.

Much research has also been focused on calpastatin, which is the inhibitor of the calpains. Its role in meat ageing has been demonstrated (Koohmaraie *et al.*, 1995), especially because of strong

Table 2. QTL mapping confirmation for direct and LD genetic markers for beef quality.

Gene/Marker	BTA	Trait	QTL detection
CAPN1	29	Shear force value	56-65 cM[a], 54 cM[b]
CAST	7	Shear force value	not identified
LOX	7	Shear force value	not identified
TG	14	Marbling score	47 cM[b]
CSSM34/ETH10	5	Marbling score	50 cM[a]; position not indicated[c]
DGAT1	14	Intramuscular fat content	16 cM[d]
RORC	3	Marbling score	65 cM[e]
SCD	26	MUFA content	not identified

[a] Casas *et al.*, 2000; [b] Casas *et al.*, 2003; [c] Stone *et al.*, 1999; [d] Barendse, 1999; [e] Casas *et al.*, 2001

genetic correlations between calpastatin activity and tenderness measured by Warner-Bratzler shear force (Shackelford *et al.*, 1994). Three polymorphisms within the *CAST* gene that encodes calpastatin, have been associated with an effect on tenderness measured by a shear force test on *Longissimus dorsi* (Barendse, 2002): two SNPs, located in the 3'UTR region of the gene, and a microsatellite, located in the 5' region of the gene that had been described before (Nonneman *et al.*, 1999). When haplotypes of the SNPs and the linked microsatellite locus are taken into consideration, only two haplotypes have been shown to be associated with improved tenderness despite the increased number of possible haplotypic combinations. In summary, these results suggest that the presently known markers are in LD with a causative mutation that has not yet been identified.

A patent on the *CAST* polymorphisms for genetically testing for animals that should produce tender meat was submitted by Barendse (2002), which also indicates that the *LOX* gene (Lysyl oxidase) which codes for an enzyme involved in the formation of links between collagen fibers during the early stage of its synthesis, is associated with variations in tenderness. There are 2 alleles of the *LOX* gene that haven been associated with variation in tenderness measured by a compression force test on the *Semitendinosus* muscle. It should be noted that *CAST* and *LOX* are both located on cattle chromosome 7 and that no QTL for tenderness has been reported in that genomic region.

Marbling

Two marker systems associated with intramuscular fat deposition or marbling are described in a second patent filed by Barendse (1999): the thyroglobulin gene (*TG*) located on chromosome 14 (BTA14) and two microsatellites (*CSSM34* and *ETH10*) located on chromosome 5 (BTA5).

Thyroglobulin is the precursor of two thryroid hormones, which are released by proteolytic cleavage and are involved in the development of adipocytes. A study of marbling scores in Angus, Shorthorn, and Wagyu described association to the polymorphism of the *CSSM66* microsatellite marker, which is close to the *TG* gene (7cM). Subsequently, an association between an allele of the *TG* gene and enhanced intramuscular fat content was identified (Barendse, 1999).

Interestingly, a QTL for marbling was detected in the centromeric region of BTA14, where both the *TG* gene and the *DGAT1* gene are located. The *DGAT1* gene encoding diacylglycerol-O-acyltransferase, which is involved in the last stages of fat synthesis, was identified as the gene underlying a QTL for milk fat synthesis. An Ala232Lys polymorphism of the *DGAT1* gene has also been shown to have an effect on intramuscular fat deposition in German Holstein and Charolais (Thaller *et al.*, 2003). Both genes, *TG* and *DGAT1*, seem to have independent effects, because the alleles of the two polymorphisms showed no statistically significant linkage disequilibrium.

On BTA3, where QTL mapping indicated a QTL for marbling, Barendse (2004) identified an association between intronic and exonic alleles within the retinoid related orphan receptor C (gamma) (*RORC*) gene and marbling scores with haplotypes of the *RORC* alleles displaying the strongest effects. On BTA5, the polymorphic microsatellite loci *CSSM34* and *ETH10*, which are 20 cM apart, are associated with marbling scores in the Angus, Shorthorn, and Wagyu breeds.

Stearoyl-CoA desaturase encoded by the *SCD* gene is the enzyme responsible for conversion of saturated fatty acids into mono-unsaturated fatty acids (MUFA). The fatty acid composition of beef has an impact on the softness of the fat and also on flavour. In Wagyu cattle, Taniguchi *et al.*

(2004) identified an association between a polymorphism in the *SCD* gene and the MUFA content as well as the melting point of intramuscular fat.

Critical comments

It is important to underline that most, if not all, of the studies published need to be confirmed and enlarged before the markers reported should be used for MAS, because the associations have been observed in a limited number of breeds and breeding systems (Renand *et al.*, 2003). The *DGAT1, CAPN1,* and *SCD* polymorphisms very likely represent direct markers, where the polymorphism directly explains the biological effect. In the other reports the markers that were found probably correspond to LD markers. In addition to the results reviewed above, which present the most convincing data, other publications describe possible effects of candidate gene polymorphisms on meat quality. For instance, Di Stasio *et al.* (2003) described the influence of a Growth Hormone gene allele on cooking losses and tenderness at day 11 in Piedmontese.

New prospects for the development of genetic markers for beef quality

The availability of new molecular tools for functional genome analysis and of comprehensive genomic information (DNA sequence, ESTs, genes) that is emerging for cattle enables new approaches for identifying functional polymorphisms within candidate genes with an effect on beef quality (Eggen and Hocquette, 2004).

One novel experimental approach is based on the comparative analysis of tissue-specific gene expression activity in cattle of different phenotypes (e.g., German Holstein and Charolais for intramuscular fat content; Bellmann *et al.*, 2004). In an example of this strategy, Dorroch *et al.* (2001) performed Differential Display-RT-PCR and Real-Time-RT-PCR gene expression studies in liver and intestine of German Holstein and Charolais individuals identifying 277 differentially expressed DNA sequences between the breeds. Both tissues are known to play an active role in nutrient assimilation, transformation and secretion. While 52 % of the expressed sequence tags (ESTs) are homologous to expressed sequences within public domain databases, interestingly, 17% showed no homology to any database entry, indicating a substantial proportion of potentially new sequences identified by this approach.

The differentially expressed ESTs have been used to construct a bovine transcript map (Dorroch *et al.*, 2001; Goldammer *et al.*, 2002) using the tools of comparative genome analysis as *in silico* mapping, somatic hybrid cell mapping, radiation hybrid mapping and in-situ hybridisation. All together, 146 ESTs were assigned on bovine chromosomes identifying candidate genes and genome regions potentially associated with intramuscular fat content. We have tested whether the observed ESTs mapping onto a bovine chromosome corresponded to the expected number of ESTs there given the gene content of the homologous regions of the human genome and assuming a respective distribution of cattle ESTs. The expected value was derived by assigning homologous chromosomal regions between cattle and human referring to the latest whole genome radiation hybrid map (Everts-van der Wind *et al.*, 2004) and to chromosome painting data (Hayes *et al.*, 2003). The gene content of the identified homologous human chromosomal regions was assessed in the latest version of the Human genome map (http://www.ncbi.nlm.nih.gov/mapview/; BUILD 34.3, accessed July, 8th, 2004). Figure 1 displays the results for the first six bovine chromosomes and gives a strong indication that there is a clustering of differentially expressed ESTs. Comparison of these clusters of ESTs differentially expressed in individuals with differences in intramuscular fat content with the QTL mapping results for beef quality traits suggests that one cluster overlaps with a QTL for marbling on BTA3 and most prominently with the genomic region harbouring the *RORC* gene. Correlating the chromosome regions identified in QTL analyses and

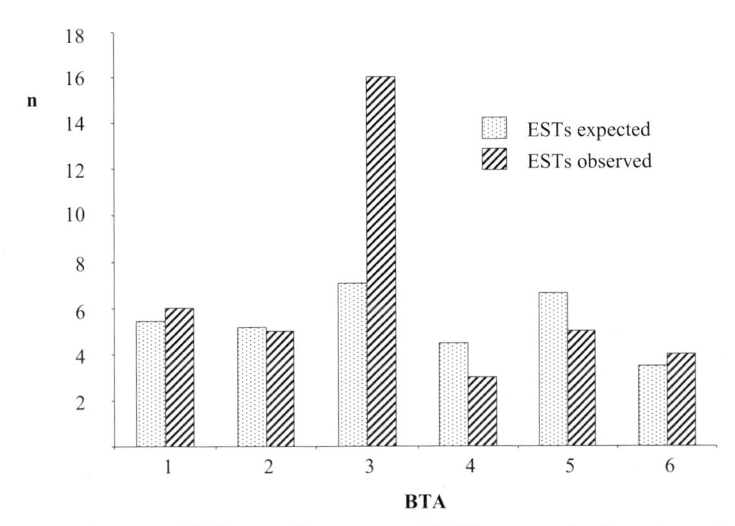

Figure 1. Distribution of differentially expressed ESTs across the first six cattle chromosomes.

the locations of genes involved in relevant metabolic pathways may allow additional candidate genes for intramuscular fat content to be identified.

Further experiments on the differentially expressed ESTs identified single nucleotide polymorphisms (SNPs) and revealed differences in the allele frequencies of the SNPs between groups of German Holstein and Charolais bulls, which differed significantly regarding intramuscular fat deposition (Schwerin *et al.*, submitted). It has to be remembered that these preliminary results, like unconfirmed QTLs, should be treated with caution. However, the ESTs and SNPs identified represent a starting point providing candidate genes, which can be explored further, e.g., in detailed investigations on a F_2 resource population between Charolais and German Holstein (Kühn *et al.*, 2002) as well as in further commercial populations on variations in intramuscular fat content. Other studies on the functional analysis of muscle tissue are also under way (e.g., Sudre *et al.*, 2003), which will provide further genes and insights into regulatory processes, which can be scanned for genetic variation responsible for differences in beef quality.

Another approach to exploring possible effects of genes on meat quality traits is the identification of SNPs in a large set of candidate genes and testing these across a wide range of genetically divergent breeds on which meat quality data is available. Within the context of an EC funded project (GeMQual, www.gemqual.org), a list of about 500 candidate genes that may be expected to have an influence on muscle development, composition, metabolism or meat ageing and hence affect the quality of meat, has been established from knowledge of their physiological role. Coding and non-coding regions from about 400 of these candidates have been sequenced to reveal polymorphisms (Levéziel *et al.*, 2003). So far a total of about 700 SNPs has been identified in 196 genes (ranging from 1 to 27 SNPs for a given gene), and 86 of these SNPs are associated with amino-acids substitutions in the proteins they encode. The SNPs are being genotyped in 450 bulls, 30 from 15 different breeds that have been measured for meat characteristics in the project to detect meat quality associated differences in allele frequencies, which will provide an indication of genes that may have an effect on the meat quality traits and so be targets for further studies.

The research for direct or LD genetic markers for beef quality will further benefit in the near future from the information arising from QTL validation programmes running in Australia, France and USA, and from new results obtained from QTL detection experiments conduced in several countries. For instance, in Europe, Roslin Institute in UK, INRA in France, and FBN in Germany have undertaken crossbred (Holstein x Charolais) or purebreed (Charolais) protocols, respectively, since several years.

It is essential that, in the future, efforts are made to collect phenotypic data on large numbers of animals and including traits, which are not currently measured routinely. The collection of such high quality data to describe beef quality by scientists, breeder associations and meat producers will be essential to unravelling the genetic control of the traits involved. Large consortia are being organized within national programmes, like the GenAnimal in France or the FUGATO initiative in Germany. Ideally these activities would be carried out at an international level e.g., under the auspice of international bodies like the European Union.

Undoubtedly, further progress will be made in the future when the entire bovine genome is sequenced. It is predicted that the complete bovine genome sequence will be available mid 2005. Part of the International bovine sequencing project is identifying SNP markers at high density by sample sequencing 10 additional individuals from 10 different cattle breeds. Commercial companies, such as MetaMorphix and Cargill in the USA, have carried out first whole genome association studies in beef cattle using SNP markers distributed throughout the bovine genome. As the availability of SNP markers increases and the genotyping costs decrease it is likely that such approaches will be adopted more widely.

With a set of genetic markers established for the relevant European production schemes and consumers interests, beef quality will be open to improvement by marker-assisted-selection. However, to achieve the underpinning information there is the requirements for good infrastructure to carry out the work. This infrastructure will include the coordinated trait recording, genotyping, databases, statistical methods, most of which has still to be set up.

References

Barendse, W., 1999. Assessing lipid metabolism. Patent WO9923248.

Barendse, W., 2002. DNA markers for meat tenderness. Patent WO02064820.

Barendse, W., 2004. DNA markers for marbling. Patent WO 2004/070055.

Bellmann, O., J. Wegner, F. Teuscher, F. Schneider and K. Ender, 2004. Muscle characteristics and corresponding hormone concentrations in different types of cattle. Livest. Prod. Sci. 85: p. 45-57.

Boccard, R., 1981. Facts and reflections on muscular hypertrophy in cattle: double muscling or culard. In: Developments in Meat Science-2, R. Lawrie (editor), Applied Science Publisher, Barking, England, p. 1-28.

Casas, E., J.W. Keele, S.D. Shackelford, M. Koohmaraie, T. S. Sonstegard, T.P.L. Smith, S.M. Kappes and R.T. Stone, 1998. Association of the muscle hypertrophy locus with carcass traits in beef cattle. J. Anim. Sci. 76: p. 468-473.

Casas, E., J.W. Keele, S.D. Shackelford, M. Koohmaraie and R.T. Stone, 2004. Identification of quantitative trait loci for growth and carcass composition in cattle. Anim. Genet. 35: p. 2-6.

Casas, E., S.D. Shackelford, J.W. Keele, M. Koohmaraie, T.P.L. Smith and R.T. Stone, 2003. Detection of quantitative trait loci for growth and carcass composition in cattle. J. Anim. Sci. 81: p. 2976-2983.

Casas, E., S.D. Shackelford, J.W. Keele, R.T. Stone, S.M. Kappes and M. Koohmaraie, 2000. Quantitative trait loci affecting growth and carcass composition of cattle segregating alternate forms of myostatin. J. Anim. Sci. 78: p. 560-569.

Casas, E., R.T. Stone, J.W. Keele, S.D. Shackelford, S.M. Kappes and M. Koohmaraie, 2001. A comprehensive search for quantitative trait loci affecting growth and carcass composition of cattle segregating alternative forms of the myostatin gene. J. Anim. Sci. 79: p. 854-860.

Dekkers, J. C. M., 2004. Commercial application of marker- and gene-assisted selection in livestock: Strategies and lessons. J. Anim. Sci. 82 (E. Suppl.): p. E313-E328.

Di Stasio, L., A. Brugiapaglia, G. Destefanis, A. Albera and S. Sartore, 2003. GH1 as candidate for variability of meat production traits in Piemontese cattle. J. Anim. Breed. Genet. 120: p. 358-361.

Dorroch, U., T. Goldammer, R.M. Brunner, S.R. Kata, C. Kühn, J.E. Womack and M. Schwerin, 2001. Isolation and characterization of hepatic and intestinal expressed sequence tags potentially involved in trait differentiation between cows of different metabolic type. Mamm. Genome 12: p. 528-537.

Eggen, A. and J.F. Hocquette, 2004. Genomic approaches to economic trait loci and tissue expression profiling: application to muscle biochemistry and beef quality. Meat Sci. 66: p. 1-9.

Everts-van der Wind, A., S.R. Kata, M.R. Band, M. Rebeiz, D.M. Larkin, R.E. Everts, C.A. Green, L. Liu, S. Natarajan, T. Goldammer, J.H. Lee, S. McKay, J.E. Womack and H.A. Lewin, 2004. A 1463 gene cattle-human comparative map with anchor points defined by human genome sequence coordinates. Genome Res. 14: p. 1424-1437.

Goldammer, T., U. Dorroch, R.M. Brunner, S.R. Kata, J.E. Womack and M. Schwerin, 2002. Identification and chromosome assignment of 23 genes expressed in meat and dairy cattle. Chromosome Res. 10: p. 411-418.

Grobet, L., L.J.R. Martin, D. Poncelet, D. Pirottin, B. Brouwers, J. Riquet, A. Schoeberlein, S. Dunner, F. Menissier, J. Masabanda, R. Fries, R. Hanset and M. Georges, 1997. A deletion in the bovine myostatin gene causes the double-muscled phenotype in cattle. Nat. Genet. 17: p. 71-74.

Hayes, H., C. Elduque, M. Gautier, L. Schibler, E. Cribiu and A. Eggen, 2003. Mapping of 195 genes in cattle and updated comparative map with man, mouse, rat and pig. Cytogenetic and Genome Res. 102: p. 16-24.

Keele, J. W., S.D. Shackelford, S.M. Kappes, M. Koohmaraie and R.T. Stone, 1999. A region on bovine chromosome 15 influences beef longissimus tenderness in steers. J. Anim. Sci. 77: p. 1364-1371.

Koohmaraie, M., J. Killefer, M.D. Bishop, S.D. Shackelford, T.L. Wheeler and J.R. Arbona, 1995. Calpastatin-based method for predicting meat tenderness. In : Expression of Tissue Proteinases and Regulation of Protein Degradation as related to Meat Quality, A. Ouali, D. Demeyer and F. Smulders (editors) ; ECCEAMST, Utrecht, The Netherlands, p. 395-410.

Kühn, C., O. Bellmann, J. Voigt, J. Wegner, V. Guiard and K. Ender, 2002. An experimental approach for studying the genetic and physiological background of nutrient transformation in cattle with respect to nutrient secretion and accretion type. Arch. Tierzucht 45: p. 317-330.

Levéziel, H., V. Amarger, M.L. Checa, A. Crisà, D. Delourme, S. Dunner, F. Grandjean, C. Marchitelli, M.E. Miranda, N. Razzaq, A. Valentini and J.L. Williams, 2003. Identification of SNPs in candidate genes which may affect Meat Quality in cattle. In: Book of Abstracts of the 54th Annual Meeting of the EAAP, Roma, Italy, Y. van der Honig (editor-in-chief), p. 94.

MacNeil, M. D. and M.D. Grosz, 2002. Genome-wide scans for QTL affecting carcass traits in Hereford x composite double backcross populations. J. Anim. Sci. 80: p. 2316-2324.

MacNeil, M. D., R.K. Miller and M.D. Grosz, 2003. Genome-wide scan for quantitative traits loci affecting palatability traits of beef. Plant and Animal Genomes XI conference, San Diego, USA

Nonneman, D., S.M. Kappes and M. Koohmaraie, 1999. Rapid communication: a polymorphic microsatellite in the promoter region of the bovine calpastatin gene. J. Anim. Sci. 77: p. 3114-3115.

Page, B. T., E. Casas, M.P. Heaton, N.G. Cullen, DL. Hyndman, C.A. Morris, A. M. Crawford, T.L. Wheeler, M. Koohmaraie, J.W. Keele and T.P.L. Smith, 2002. Evaluation of single-nucleotide polymorphisms in CAPN1 for association with meat tenderness in cattle. J. Anim. Sci. 80: p. 3077-3085.

Renand, G., C. Larzul, E. Le Bihan-Duval and P. Leroy, 2003. L'amélioration génétique de la qualité de la viande dans les différentes espèces : situation actuelle et perspectives à court et moyen terme. INRA Prod. Anim. 16: p. 159-173.

Schimpf, R. J., D.C. Winkelman-Sim, F.C. Buchanan, J.L. Aalhus, Y. Plante and S.M. Schmutz, 2000. QTL for marbling maps to cattle chromosome 2. 27th International Conference on Animal Genetics, Minneapolis, USA

Schmutz, S. M., F.C. Buchanan, Y. Plante, D.C. Winkelman-Sim, J.L. Aalhus, J.A. Boles and J.S. Moker, 2000. Mapping collagenase and a QTL to beef tenderness to cattle chromosome 29. Plant and Animal Genome VIII conference, San Diego, USA

Schwerin, M., T. Goldammer, S.R. Kata, Ch. Kühn and J. E. Womack. Genetic predisposition of nutrient utilization - Metabolic type specific hepatic and intestine gene expression and genetic variants in cattle. Genomics, submitted

Shackelford, S.D., M. Koohmaraie, L.V. Cundiff, K.E. Gregory, G.A. Rohrer and J.W. Savell, 1994. Heritabilities and phenotypic and genetic correlations for bovine postrigor calpastatin activity, intramuscular fat content, Warner-Bratzler shear force, retail product yield, and growth rate. J. Anim. Sci. 72: p. 857-863.

Smith, T.P.L., E. Casas, C.E. Rexroad III, S.M. Kappes and J.W. Keele, 2000. Bovine CAPN1 maps to a region of BTA29 containing a quantitative trait locus for meat tenderness. J. Anim. Sci., 78: p. 2589-2594.

Stone, R. T., J.W. Keele, S.D. Shackelford, S.M. Kappes and M. Koohmaraie, 1999. A primary screen of the bovine genome for quantitative trait loci affecting carcass and growth traits. J. Anim. Sci. 77: p. 1379-1384.

Sudre, K., C. Leroux, G. Pietu, I. Cassar-Malek, E. Petit, A. Listrat, C. Auffray, B. Picard, P. Martin, P. and J.F. Hocquette, 2003. Transcriptome analysis of two bovine muscles during ontogenesis. J. Biochem. 133: p. 745-756.

Taniguchi, M., T. Utsugi, K. Oyama, H. Mannen, M. Kobayashi, Y. Tanabe, A. Ogino and S. Tsuji, 2004. Genotype of stearoyl-CoA desaturase is associated with fatty acid composition in Japanese Black cattle. Mamm. Genome 15: p. 142-148.

Taylor, J. F., L.L. Coutinho, K.L. Herring, D.S. Gallagher, R.A. Brenneman, N. Burney, J.O. Sanders, R.V. Turner, S.B. Smith, R.K. Miller, J.W. Savell and S.K. Davis, 1998. Candidate gene analysis of GH1 for effects on growth and carcass composition of cattle. Anim. Genet. 29: p. 194-201.

Thaller, G., C. Kühn, A. Winter, G. Ewald, O. Bellmann, J. Wegner, H. Zühlke and R. Fries, 2003. DGAT1, a new positional and functional candidate gene for intramuscular fat deposition in cattle. Anim. Genet. 34: p. 354-357.

Wheeler, T. L., L.V. Cundiff, S.D. Shackelford and M. Koohmaraie, 2004. Characterization of biological types of cattle (Cycle VI): Carcass, yield, and longissimus palatability traits. J. Anim. Sci. 82: p. 1177-1189.

Genetic markers of milk quality in cows

A. Lundén
Department of Animal Breeding and Genetics, Swedish University of Agricultural Sciences, SLU
P.O. Box 7023, SE-750 07Uppsala, Sweden
Correspondence: Anne.Lunden@hgen.slu.se

Summary

During the last decades the market for dairy products has been internationalized, accompanied by a fierce competition between dairy companies. Today's consumers have a large variety of dairy products from different countries to choose between, and product price but also milk quality aspects like flavour and texture have become of increasing importance for consumers' choice of product, and thereby also for the dairy companies' profitability.

Most traits related to milk quality can be expected to have a genetic basis and should consequently be possible to improve by breeding, had only the genes controlling the traits been known. QTL mapping strategies based on information from the progeny testing of bulls have successfully revealed a major gene for milk fat content, the *DGAT1* gene, and there is an ongoing QTL study to identify the genes behind non-coagulation of milk. In general, QTL-based approaches are not suited to detect genes controlling milk quality traits like flavour, texture, and processing properties. This is in part due to the lack of cost effective, large scale methods for analysis on individual cow's milk, like those used in the milk recording scheme for fat and protein content. We have, however, already information on some genes with major effects on milk quality, like the milk protein genes. Variation in milk characteristics has been found that can be attributed to both qualitative, structural differences between allele products of the milk protein genes as well as quantitative differences, i.e. in amount of synthesized gene product. Another major gene for milk quality is the *FMO3* gene which in its mutated form causes fishy off-flavour in milk.

To improve milk quality we need to develop strategies to identify additional genes controlling these traits. The lack of suitable animal materials for QTL studies could partly be compensated by more information on the physio-chemical mechanisms behind the processing properties of milk and detailed knowledge of the biological pathways controlling the quality traits of interest. These strategies would hopefully provide us with candidate genes that can be tested in association studies performed in experimental herds under controlled conditions.

Keywords: major gene, milk quality, milk protein, milk flavour

Introduction

Until recently animal production was focused almost exclusively on providing large food quantities to a low cost. Today there is a growing interest also in the quality of farm products. During the last decades the dairy market has shifted from a situation where the dairy companies controlled what products should be marketed on what was then a local market. Today, many nationally produced dairy products are competing with corresponding ones produced in other countries and, consequently, the consumers have a large variety of products to choose between. Thus, the consumers' choice controls the dairies' product assortment, rather than the opposite. As a consequence, milk quality is of increasing importance, where aspects like flavour, texture, nutrition, and health attract consumer interest. From a dairy industry perspective, processing properties of the raw milk and high concentrations of valuable milk components, like protein and fat, are of major importance. There is also an ethical aspect to quality reflected by consumers' demand for production systems that consider both health and well-being of the animals and minimize negative environmental impact. In this paper only physically perceived quality aspects will be addressed.

Breeding strategies to improve milk quality

Progeny testing

Traditional breeding strategies have been very successful in producing high yielding dairy cows. In Sweden, an increase in milk yield of 1000 kg that took 45 years to achieve between the years 1900 and 1945, takes today only seven years (year 2000). Currently, the dairy cattle breeding is based on daughter group performance of potential breeding bulls. To achieve a sufficient accuracy of estimated breeding values a large number of progenies need to be tested. As a consequence, genetic improvement can only be achieved in traits that are possible to measure on a large scale, like milk yield and yield and concentrations of fat and protein measured in the routine milk recording system.

Mapping QTL for marker assisted selection

Many QTL studies have been performed on traits of importance for milk production (see the review by Geldermann *et al.* (2005) in a separate article within this publication) for use in marker assisted selection but they have so far been limited to traits that can be measured on a large scale, due to the extensive family materials that are required. The first study where progeny testing data were used in a "grand daughter design" (Weller *et al.*, 1990) to map QTL for milk production was published by Georges *et al.* (1995). Up to now, the only QTL study on milk traits that has resulted in the identification of the causal mutation behind the observed trait variation is the successful mapping of a QTL for milk fat concentration (Grisart *et al.*, 2002; 2004; Winter *et al.*, 2002). The causative mutation has convincingly been shown to be a lysine to alanine substitution in exon VIII of the *DGAT1* gene (diacylglycerol acyltransferase 1; Grisart *et al.*, 2004), coding for an enzyme that catalyzes the last step in the triacylglycerol synthesis (Cases *et al.*, 1998).

Genetic improvement of milk quality - inherent problems

So far no QTL studies have been performed regarding traits being more directly associated with milk quality, with the notable exception of the ongoing effort to map the genes behind non-coagulation of milk (see the separate article by Ojala *et al.* (2005) within this publication). This is partly due to the aforementioned lack of suitable, large scale methods for analysis on individual cow's milk. An additional complication is that the majority of milk characteristics of importance for perceived product quality and processing properties are not apparent until the milk has reached the dairy plant and the processing has started. By that time the milks from several farms have been pooled and it has thus become impossible to trace back to the individual cow any inherent milk quality defects. It is therefore important to develop measures that can be applied on individual milk samples that reflect well the processes taking place in full scale on the dairy plant.

Major genes for improved milk quality

Some quality characteristics of milk are probably controlled by relatively few genes. When we have accumulated detailed knowledge of the mechanisms and biological pathways controlling the quality traits of interest, we will possibly have a good idea of the genes involved, so called candidate genes. We have already information on several genes with major effects on milk quality, of which some will be addressed in the following text.

Major genes for cheese yield and milk processing properties

In EU, the proportion of the milk used for production of cheese is expected to increase by approximately 4% over the next ten years (http://www.fapri.iastate.edu/Outlook2004/). High concentrations of the two major milk components protein and fat, which are found in roughly equal proportions in dairy products like hard cheese, are thus of growing economical importance to the dairy sector.

Of particular relevance in the manufacturing of cheese and yoghurt products are the milk proteins which consist predominantly of caseins and whey proteins. The caseins that show most polymorphisms are the β-casein and κ-casein, whereas the most polymorphic whey protein is β-lactoglobulin. The different milk proteins with their genetically determined variants are obvious candidates when searching for genes controlling the variation in yield and manufacturing properties of dairy products. Consequently, these genes have been subjected to numerous studies, beginning with the detection of β-lactoglobulin protein polymorphism and its genetic basis by Aschaffenburg and Drewry (1955; 1957).

Effects of β-lactoglobulin polymorphism

Qualitative effects: The β-lactoglobulin protein comes in two common variants, A and B (Aschaffenburg and Drewry, 1955; 1957), and several rare variants (see review by Formaggioni *et al.*, 1999). The amino acid substitutions that discriminate the two common variants of the β-lactoglobulin protein was described by Piez *et al.* (1961), whereas the causative nucleotide substitutions were published by Alexander, *et al.* (1993). Variation in β-lactoglobulin of importance for milk quality has been found that can be attributed to both qualitative, structural differences between allele products as well as quantitative differences, i.e. in amount of synthesized β-lactoglobulin of the A and B variants (see reviews by Mercier and Vilotte, 1993; Martin *et al.*, 2002). There are numerous studies on the functional implications of the subtle differences in amino acid composition between the two major variants, A and B (Jakob and Puhan, 1992; Ng-Kwai-Hang and Grosclaude, 1992; Hill *et al.*, 1997b; Goddard, 2001). An example of the qualitative effects is the observed variation in fouling, i.e. undesired adsorption of proteins during heat treatment, inside heat exchangers in dairy plants between milk with different β-lactoglobulin variants, with the B variant being the favourable allele (Elofsson *et al.*, 1996; Hill *et al.*, 1997a). There are several reports indicating that the amino acid substitutions affect the thermal stability of the resulting protein (Imafidon *et al.*, 1991; Qin *et al.*, 1999; Boye *et al.*, 2004).

Quantitative effects: That the A variant of β-lactoglobulin is associated with higher concentrations of the protein was reported already by Aschaffenburg and Drewry (1955). In a study by Ng-Kwai-Hang and Kim (1996) on 2663 heterozygous AB cows of four breeds the ratio between the amounts of A and B β-lactoglobulin in milk was been reported to vary between 1.31 to 1.49, whereas in a similar study by Graml *et al.* (1989) on 2631 heterozygote cows of two breeds the corresponding values were 1.56 and 1.50. An extreme ratio of 4.02 was, however, observed by Kim *et al.* (1996) among heterozygous cows of the Swiss Brown breed and their crosses with Brown Swiss cattle. The high ratio was a result of a lower content of the B variant whereas the A variant was found in normal amounts. About 75% of the 20 cows with extreme ratios were off-spring of the same bull, which indicates that this extreme phenotype to a certain extent is genetically determined.

Mechanisms behind the allelic difference in gene expression: To try to elucidate the genetics behind the difference in expression of the β-lactoglobulin A and B alleles the 5'-flanking region of the gene has been searched for polymorphisms (Wagner *et al.*, 1994; Ehrmann *et al.*, 1997; Lum *et al.*, 1997). The linkage disequilibrium between the observed polymorphisms upstream the transcription start point and the polymorphisms in the coding region is very strong leading to the A and B alleles having there own promoter variants. In this region Lum *et al.* (1997) detected a nucleotide substitution in a consensus binding site for activator protein-2 (AP-2). We genotyped 361 cows for polymorphisms in the AP-2 binding site and the coding region and observed a complete linkage disequilibrium between the nucleotide variant (G), which causes a complete match with the consensus binding site for AP-2, and the more expressed A allele (Gustafsson and Lundén, 2003). The studies by Lum *et al.* (1997) and Kuss *et al.* (2003) both showed a positive association between the G nucleotide in the AP-2 consensus binding site and high expression of β-lactoglobulin.

Effect on casein number: The higher β-lactoglobulin concentration in AA milk does not seem to be accompanied by a higher concentration of total milk protein, but rather to be balanced by a lower casein content, resulting in a similar total milk protein content for all three genotype combinations of the β-lactoglobulin A and B alleles but varying casein number, i.e. ratio of casein to total protein (Bobe, *et al.*, 1999; reviews by Jakob, 1994; Hill *et al.*, 1997b). There are, however, conflicting results showing an increased milk protein concentration in association with the A allele (Ng-Kwai-Hang *et al.*, 1984; 1990). Lundén *et al.* (1997) reported a difference between the AA and BB homozygotes in casein number corresponding to 1.5 phenotypic standard deviations, making the β-lactoglobulin gene an obvious major gene for this trait.

Effect on cheese yield: Whereas several groups have reported positive associations between the B variant of β-lactoglobulin and cheese yield (Graham *et al.*, 1984; Schaar *et al.*, 1985; Aleandri *et al.*, 1990; Hill *et al.*, 1997b), Marziali and Ng-Kwai-Hang (1986) found no effect of β-lactoglobulin genotype. The idea of favouring the B allele to increase cheese yield per kg of milk protein was put into practice in something that has been refered to as "the Kaikoura experience" (Boland and Hill, 2001). A small dairy co-operative in Kaikoura, New Zealand decided to change all cows in the herds delivering milk to the dairy plant into β-lactoglobulin BB homozygotes. The company merged, however, with Fonterra Co-operative Group Ltd before the concept could be properly evaluated.

Effects of κ-casein polymorphism

Unfavourable effects of the E variant on milk coagulation: Of the four caseins in bovine milk κ-casein is the most studied (see reviews by Jakob and Puhan, 1992; Ng-Kwai-Hang and Grosclaude, 1992; Goddard, 2001; Martin *et al.*, 2002). The protein exists in several genetic variants of which A and B show the highest frequencies in most breeds (see review by Formaggioni *et al.*, 1999). However, in the Ayrshire breeds in Finland and Sweden the E allele is rather common, with frequencies of 0.31 in a material of 20 990 Finnish Ayrshire cows (Ikonen *et al.*, 1996) and 0.22 in a group of 300 young bulls and breeding bulls of the Swedish Ayrshire breed (Lundén and af Forselles, 2000). The comparatively high frequency of this allele in Finland and Sweden is likely due to the frequent use of a single Finnish Ayrshire carrier bull with high genetic merits for milk production (Velmala *et al.*, 1995), which is reflected by the fact that the E allele is almost exclusively found in a single haplotype, combined with the β-casein A1 allele (Velmala *et al.*, 1995; Ikonen *et al.*, 2001). The E variant is also relatively frequent in Italian Holstein (0.11; Leone *et al.*, 1998) and US Holstein (0.16; John R. Woollard, University of Wisconsin; personal communication). This variant was first identified by Erhardt (1989) and differs from the A and B variants in that serine is replaced by glycin in codon 155 of exon IV. Because serine is known to bind calcium phosphate and thereby adding to stabilize the casein micellar structure, it is tempting to speculate that this lost serine could potentially affect the protein's characteristics. In Finland extensive studies have been performed regarding the effects of the different κ-casein variants (Ojala *et al.*, 1997; Ikonen *et al.*, 1997; 1999a,b,c; 2001; Ruottinen *et al.*, 2004; see also the separate article by Ojala *et al.*, 2005) which indicate that the E variant is associated with poor milk coagulation properties (Oloffs *et al.*, 1992; Lodes *et al.*, 1996; Ikonen *et al.*, 1997; Ikonen *et al.*, 1999a). Detailed studies of the coagulation process after addition of chymosin to purified κ-casein samples of the A, B, and E variants revealed a delayed onset of coagulation as well as low viscosity associated with the E variant, compared to A and B, at equal concentrations of κ-casein (Allmere *et al.*, 2002). The E allele was previously mixed up with the A variant when running isoelectric focusing gels, due to similar electric charge, and may therefore in populations where A and E coexist have given the A variant an undeservedly bad reputation in cheese production. Similar to β-lactoglobulin, the expression of the A and B alleles have been shown to differ in heterozygous cows, with the B variant having a higher level of expression (van Eenennaam and Medrano, 1991; Debeljak *et al.*, 2000).

Non-coagulating milk: In addition to poorly coagulating milk, being associated with the κ-casein E variant, there is also milk that does not coagulate, i.e. no curd formation within a set time of 30 minutes after addition of milk-clotting enzyme, so called non-coagulating milk. This has previously been observed by Okigbo *et al.* (1985) and Davoli *et al.* (1990) in Holstein and Friesian cows, respectively, and the phenomenon has since then been more thoroughly addressed by Ikonen *et al.* (1997, 1999a,c) and Tyrisevä *et al.* (2003, 2004). Out of 4600 Finnish Ayrshire cows, 13% produced non-coagulating milk (Tyrisevä *et al.*, 2003). Although not as prevalent, it has been observed also in Holstein-Friesian cows (Tyrisevä *et al.*, 2004). Non-coagulation of milk has also been reported to occur in Sweden, mainly in milk from Swedish Ayrshire (SRB) cows but also in milk samples from Swedish Holstein cows (Elin Hallén, Swedish University of Agricultural Sciences, personal communication).

The κ-casein paradox: Milk coagulation begins when milk-clotting enzymes cleave the κ-casein protein and release κ-casein macropeptide (amino acid residues 106-169), thereby initiating the coagulation of para-casein (residues 1-105) that constitute the cheese curd. The fact that the polymorphic part of the κ-casein molecule which resides in the macropeptide that is cleaved off by the renneting enzyme, and thereby no longer participates in the cheese making process, still seems to affect the cheese making process has puzzled the dairy research community. Horne *et al.* (1996) speculate that the allelic variation in protein structure is mainly controlling the rate of coagulation, whereas the curd firmness is rather a quantitative effect of the amount of κ-casein associated with the different alleles.

Methods to analyse effects of casein polymorphism: There have been numerous publications throughout the years on associations between κ-casein variants and milk composition as well as processing properties, many of them reporting contradictory results. Already in 1992, in a comprehensive review on the effects of milk protein polymorphisms on technological properties of milk by Jakob and Puhan (1992), it was stressed that due to extensive linkage disequilibrium between alleles at the various casein loci the effects of casein polymorphisms should preferentially be analysed as composite genotypes or haplotypes. Lien *et al.* (1995) managed to derive the casein haplotypes of 13 sires and the corresponding paternally inherited haplotypes in 250 sons of these sires, which they included in a granddaughter design for analysis of effects of casein polymorphisms on milk production traits. By considering haplotype 'blocks' of the casein loci, instead of treating each locus separately, the analysis will cover the whole stretch of DNA within the casein gene complex, thereby also accounting for polymorphisms in regulatory sequencies between the coding regions. Also, in situations were relationships between animals are expected to be greater within milk protein genotype than across genotype, it is appropriate to use an animal model in the statistical analyses in order to get unbiased estimates of the genotype effects (Kennedy *et al.*, 1992; Famula and Medrano, 1994). This strategy, in combination with composite casein genotypes or haplotypes has since then been adopted in several studies (e.g. Lundén *et al.*, 1997; Ojala *et al.*, 1997; Ikonen *et al.*, 1997; 1999a,b; 2001; 2004; Ruottinen *et al.*, 2004; Tyrisevä *et al.*, 2003; 2004).

A major gene for milk fat content

Fat is one of the major components in the milk. The important aspects of milk fat from a dairy point of view do not only concern functional properties like butterfat spreadability and whipping properties of cream but also the milk fat associated off-flavours. The characteristics of milk fat are to a large degree determined by the proportion of different fatty acids. These are either synthesised by the udder or by the rumen microorganisms, or originates from body fat or fat in feed. Thus, environmental factors play an important role for the fatty acid composition of milk but it is also affected by the cow's genotype.

Milk fat of ruminants has an unique composition

The major part of the milk fat consists of triacylglycerols (triglycerides) that are composed of a glycerol molecule (glycerol-3-phosphate) that has three fatty acids attached to it. Characteristic of milk fat of cows and other ruminants is the presence of the short-chained, saturated fatty acids butyric acid with four carbons (C4) and capronic acid (C6). The synthesis of triacylglycerols starts by various enzymes attaching fatty acids to positions 1 and 2 of the glycerol molecule, thereby forming diacylglycerol. When a fatty acid should be attached to position 3 the enzyme diacylglycerol *O*-acyltransferase (DGAT) is required to catalyse the reaction. This is the single role of the enzyme and it is, furthermore, the only enzyme in the glycerolipid pathway that is unique to triacylglycerol synthesis. The identification of the gene (*DGAT1*) that encodes this enzyme (Cases *et al.,* 1998) constituted an important first step towards understanding the genetic mechanisms in the triacylglycerol synthesis (see review by Farese *et al.*, 2000). The enzyme has a broad specificity, i.e. it binds to all fatty acids irrespective of chain length (see Bell and Coleman, 1980). The C4 and C6 fatty acids are unique not only in the sense that they are exclusively found in ruminants, but also that they are not found in any other fat synthesising tissue than milk, not even in ruminants. Furthermore, they are almost only found in position 3 of the glycerol molecule whereas other fatty acids can be found in all three positions. The reason behind the unique presence of these short-chained fatty acids in ruminant's milk is not known although there have been alternative suggestions (Marshall and Knudsen, 1979; Marshall and Knudsen, 1980; Hansen *et al.,* 1984).

DGAT1 - a 'major gene' for milk fat content

Strong evidence of the central role of the *DGAT1* gene in the triacylglycerol synthesis were provided by Smith *et al.* (2000) who showed that the milk synthesis was completely blocked in mice homozygous for a knock-out event in the gene. Compared to mice with functional copies of the gene, these *DGAT* $^{-/-}$ mice resisted diet-induced obesity through increased energy expenditure.

As far as it is known, the *DGAT1* gene exists in two functionally separate variants (Grisart *et al.,* 2002; Winter *et al.,* 2002). In addition, the same authors report a number of silent mutations in non-coding regions. The two 'functional' variants result from a point mutation in exon VIII of the gene which has lead to a substitution of a lysine (K) by an alanine (A) in amino acid position 232. The proposed 'ancestral' variant, the K allele, is associated with increased concentrations of both milk fat and protein and also fat yield whereas it leads to reduced yields of milk and protein. In a material consisting of 1818 Dutch breeding bulls of the Holstein breed (Grisart *et al.*, 2002), in which the frequency of the *K* allele was 0.37, the *K232A* accounted for 51 % of the phenotypic variation in milk fat content. With the present breeding objectives in the two countries that participated in the study by Grisart *et al.* (2002), i.e. The Netherlands and New Zealand, where focus since the 1950's has been on fat yield rather than on total milk produced, the K and A alleles have similar economic values which could explain the intermediate frequencies of the two alleles in the countries' dairy populations (Grisart *et al.*, 2002).

Kaupe *et al.* (2004) performed a screening on an extensive number of breeds of different origins, and with varying breeding objectives, for the prevalence of the two *DGAT1* alleles (*K232A*). The beef breeds tended to have a high frequency of the A allele whereas dairy breeds showed a larger variation in A allele frequency, e.g. 0.98 in Ayrshire cattle as compared to 0.31 in Jerseys.

In addition to the *K232A* polymorphism, Kühn *et al.* (2004) report the existence of genetic variability in the promoter region of the *DGAT1* gene, consisting of a variable number of tandem repeats (VNTR) of an element consisting of 18 nucleotides. This 18 mer contains a motif that is a potential binding site of the transcription factor SP1. The number of SP1 binding sites within a cluster has previously been shown to influence the expression of the associated coding sequence (Yang *et al.*, 1995) which indicates that this promoter VNTR in the *DGAT1* gene may explain some of the observed variation in milk fat content attributed to the *K232A* polymorphism. Furthermore, Kühn *et al.* (2004) found that several paternal halfsib families homozygous for the A allele still segregated for a putative, additional QTL for fat content located at the *DGAT1* locus. Another interesting observation from this study was that the K allele was found to be in almost complete linkage disequilibrium with one of the five VNTR alleles whereas the A allele occurred in haplotype combination with all five VNTR alleles. The implications of these findings for the present breed distribution of *DGAT1* variants are discussed by the authors.

The *DGAT1* polymorphism creates exciting possibilities to accomplish a relatively rapid change in milk composition but the complex nature of its effects on yield traits and contents of fat and protein raises the question if, and how, this information should be used in selection of breeding bulls. In this context it should be mentioned that the association of perceived quality and processing properties on one hand with concentration of fat and protein is still unclear. For example milk coagulation properties showed hardly any correlation to neither concentration of total milk protein nor casein (Ikonen *et al.*, 2004).

Genetic components behind off-flavour in milk

Similar to other milk quality parameters, milk flavour is difficult to improve by traditional breeding methods. Moreover, only a few countries have the necessary testing routines to control the milk flavour on farm level. Therefore molecular genetic approaches, utilizing markers or major genes, offer an attractive way to achieve tasty milk.

Common off-flavour in cow's milk associated with milk fatty acid composition

Spontaneous oxidised flavour is one of the more frequently occurring off-flavours in cow's milk. It results from the oxidation of the double bounds that characterise unsaturated fatty acids. The process is initiated by prooxidants like copper (Neimann-Sörensen *et al.*, 1973; Timmons *et al.*, 2001) of which the concentration in milk has been shown to be partly genetically determined (Neimann-Sörensen *et al.*, 1973). Also variation in the capacity of cows to excrete the antioxidant α-tocopherol (vitamin E) in milk has a large genetic component (Jensen *et al.*, 1999). Another factor of importance for the occurrence of spontaneous oxidised flavour is milk fatty acid composition where a high proportion of polyunsaturated fatty acids (C18:2 and C18:3) is predisposing (Barrefors *et al.*, 1995; Timmons *et al.*, 2001). Karijord *et al.* (1982) estimated the heritabilities of lactation averages of individual milk fatty acids to range from 0.2 to 0.6.

We have genotyped two selection lines in the experimental dairy herd at the Swedish University of Agricultural Sciences for the *K232A* mutation in *DGAT1* (Näslund *et al.*, 2004). This herd includes cows of the Swedish Ayrshire (SRB) and the Swedish Holstein breeds. Since 1985 the Ayrshire cows have been selected for either high (HF) or low (LF) fat content, the two lines having an equal and high total milk energy production. In total, 276 cows were genotyped for the *DGAT1* polymorphism, 167 Swedish Ayrshire cows, 103 Swedish Holstein cows, and 6 Jersey cows. The frequency of the K allele was 0.18 among the 76 cows in the HF line compared to 0.01 among the 91 cows in the LF line. As a comparison, the Swedish Holstein cows showed a frequency of the K allele of 0.12 whereas five of the six Jersey cows were homozygous for the K allele and one for the A allele. The information on *DGAT1* genotype will be utilized in a study aiming at identifying the factors behind the variation in occurrence of spontaneous oxidised flavour in milk.

Genetic factors underlying fishy off-flavour in cow's milk

Fishy off-flavour in milk is a recently observed quality defect in Sweden, characterized by a taste and smell of rotting fish. Milk from a few affected cows in a herd is sufficient to impart a fishy off-flavour to the whole bulk milk. The off-flavour has been shown to be recurrent in the affected cows, regardless of the diet (Lundén *et al.*, 2003).

Trimethylamine related fishy odour in other species: Fishy body odour in association with trimethylamine (TMA) has been described in human ('Fish odour syndrome' or 'Trimethylaminuria'; OMIM #602079, http://www.ncbi.nlm.nih.gov) and chicken, predominantly in egg yolk (Hobson-Frohock *et al.* 1973). The TMA is derived from food/feed components rich in TMA or its precursors. Under normal conditions the TMA is oxidised to the odourless compound TMA-oxide by the liver enzyme flavin-containing monooxygenase (FMO; Hlavica and Kehl, 1977) and is thereafter excreted in the urine (Al Waiz *et al.*, 1987). The fishy odour is a consequence of impaired oxidation of TMA (Spellacy *et al.*, 1979) which has been shown in human to be due to mutations or deletions in the *FMO3* gene (Dolphin *et al.*, 1997; Treacy *et al.*, 1998; Akerman *et al.*, 1999; Basarab *et al.*, 1999; Forrest *et al.*, 2001) and the odour shows a recessive mode of inheritance.

A nonsense mutation behind fishy off-flavour in cow's milk: A mutation in the bovine *FMO3* orthologue has resulted in an amino acid substitution from an arginine (R) to a stop codon (X) at position 238 in the predicted polypeptide sequence, the mutation thus denoted R238X (Lundén *et al.*, 2002b). Cows producing milk with a strong fishy flavour has been shown to carry the same homozygous *FMO3* genotype (X/X) whereas cows producing milk with a normal flavour are homozygous for the 'normal' allele, i.e. carrying the R/R genotype. Due to premature termination of the gene translation that eliminates more than 50% of the complete amino acid sequence of the FMO3 enzyme, cows carrying the X/X genotype most likely lack the capacity to synthesize the FMO enzyme necessary for the oxidation of TMA, which would explain the fishy flavour of the milk. Milk from five of the homozygous cows had also been analyzed for TMA content using dynamic headspace gas chromatography (Lundén *et al.*, 2002a). The milk samples from these X/X cows all had high concentrations of TMA whereas the milk from the R/R cows showed non-detectable TMA concentrations. Similar to the human 'Fish odour syndrome', our results suggest a recessive mode of inheritance of the defect (Lundén *et al.*, 2002b).

The olfactory threshold for detecting TMA in milk has been reported to be 1-2 ppm (Metha *et al.*, 1974; von Gunten *et al.*, 1976), a figure that is confirmed by Ampuero *et al.* (2002). The TMA concentration in milk from cows homozygous for the *FMO3* nonsense mutation has been shown to vary between 1 and 16 mg/kg of milk (Ampuero *et al.*, 2002).

Fishy off-flavour is recurrent in affected cows: The degree of recurrence of fishy off-flavour and the simultaneous variation in TMA concentration in the milk during the course of lactation was studied on cows in the experimental herd at the Swedish University of Agricultural Sciences (Lundén *et al.*, 2003). Monthly milk samples were collected from one cow known to produce milk with fishy flavour and homozygous for the R238X mutation, and from one 'control' cow producing normal milk and homozygous 'normal' (R238). Altogether 26 milk samples were tested by a sensory panel and analysed for TMA concentration by both Solid-phase microextraction (SPME) and dynamic head space chromatography with mass selective (MSD), flame ionisation (FID), nitrogen/phosphor (NPD) as well as olfactometric (O) detections. Furthermore, the TMA concentration was also analysed using an MS based electronic nose (Smart Nose) in combination with numerical principal component analysis (PCA) (Ampuero *et al.*, 2002). At the first two sampling occasions, 9 and 15 days after calving, no fishy off-flavour was detected in the milk from the R238X homozygous cow although there were detectable amounts of TMA (1 mg/kg milk). From the third until the 37th lactation week, however, the milk samples from this cow were judged to have a strong fishy flavour, with a variation in TMA concentration between 1 and 16 mg/kg milk. The control samples were always judged as having a normal flavour, with non-detectable levels of TMA.

Our results provide strong evidence that the observed nonsense mutation in the bovine *FMO3* gene cause fishy off-flavour in milk. This is one of the very few identified genes influencing a production trait of economic importance. This genetic defect can be eliminated in carrier breeds by genotyping of breeding animals, something that is currently being done in Sweden. Cows with high TMA levels in their milk can now be detected using an MS based electronic nose (Smart Nose; Ampuero *et al.*, 2002).

Occurrence of fishy off-flavour in different breeds: In a screening of the four dairy breeds in Sweden the R238X mutation was only found in the Swedish Ayrshire breed (SRB), and at a relatively high frequency, 0.155 (Lundén *et al.*, 2002b). The mutation has also been found in Finnish Ayrshire bulls (http://www.svavel.se/databas/rak_tjur.asp). Pedigrees of affected cows indicate that the mutation may exist in cattle populations of Ayrshire origin also in other countries. A fishy flavour in milk that was caused by high concentration of TMA in the milk has also been reported to occur in cows of the Holstein breed (e.g. Johnson *et al.*, 1973; von Gunten *et al.*, 1976).

How do we detect additional milk quality genes?

Although we know a few genes with major effects on milk quality we can expect that there are a lot more waiting to be found and explored. But how do we go about to identify these genes? QTL strategies like the one that provided us with a major gene for milk fat content (Grisart *et al.*, 2002; 2004) would only be applicable for traits that, for various reasons, are only feasible to measure in large paternal half-sib groups.

Possibly, the lack of suitable animal materials for QTL studies could partly be compensated by increasing knowledge of the physio-chemical mechanisms behind the quality parameters and of the detailed biological pathways controlling the quality traits of interest. Strategies like these would hopefully provide us with candidate genes that can be tested in association studies. Moreover, in experimental herds with cows kept under conditions where the environmental factors are under control it will be possible to perform detailed analyses on traits directly related to milk quality and to discern the genetic factors behind the observed variation. In this context it is important to stress the need to develop measures that can be applied on individual milk samples and that in a small scale reflect the processes taking place in full scale on the dairy plants.

References

Al-Waiz, M., S.C. Mitchell, J.R. Idle and R.L. Smith, 1987. The metabolism of ^{14}C-labelled trimethylamine and its *N*-oxide in man. Xenobiotica 17: p. 551-558.

Akerman, B.R., S. Forrest, L. Chow, R. Youil, M. Knight and E.P. Treacy, 1999. Two novel mutations of the FMO3 gene in a proband with trimethylaminuria. Hum. Mutat. 13: p. 376-379.

Aleandri, R., L.G. Buttazzoni and J.C. Schneider, 1990. The effects of milk protein polymorphisms on milk components and cheese-producing ability. J. Dairy Sci. 73: p. 241-255.

Alexander, L.J., G. Hayes, W. Bawden, A.F. Stewart and A.G. Mackinlay, 1993. Complete nucleotide sequence of the bovine β-lactoglobuline gene. Anim. Biotechnol. 4: p. 1-10.

Allmere, T., A. Lundén and A. Andrén, 2002. Influence of kappa-casein E on the technological properties of milk. In: "Proceedings of the 26th IDF World Dairy Congress", 24-27 September 2002, Paris, France. p. B4-79.

Ampuero, S., T. Zesiger, A. Lundén and J.O. Bosset, 2002. Determination of trimethylamine in milk using an MS based electronic nose. Eur. Food Res. Technol. 214: p. 163-167.

Aschaffenburg, R. and J. Drewry, 1955. Occurrence of different beta-lactoglobulins in cow's milk. Nature 176: p. 218-219.

Aschaffenburg, R. and J. Drewry, 1957. Genetics of the β-lactoglobulins of cow's milk. Nature 180: p. 376-378.

Barrefors, P., K. Granelli, L.-Å. Appelqvist and L. Björk, 1995. Chemical characterisation of raw milk with and without oxidative off-flavour. J. Dairy Sci. 78: p. 2691-2699.

Basarab, T., G.H. Ashton, H. du Menagé and J.A. McGrath, 1999. Sequence variations in the flavin-containing mono-oxygenase 3 gene (FMO3) in fish odour syndrome. Br. J. Dermatol. 140: p. 164-167.

Bell, R.M. and R.A. Coleman, 1980. Enzymes of glycerolipid synthesis in eukaryotes. Annu. Rev. Biochem. 49: p. 459-487.

Bobe, G., D.C. Beitz, A.E. Freeman and G.L. Lindberg, 1999. Effect of milk protein genotypes on milk protein composition and its genetic parameter estimates. J. Dairy Sci. 82: 2797-2804

Boland, M. and J.P. Hill, 2001. Genetic selection to increase cheese yield - the Kaikoura experience. Aust. J. Dairy Technol. 56: p. 171-176.

Boye, J.I., C.Y. Ma and A. Ismail, 2004. Thermal stability of beta-lactoglobulins A and B: effect of SDS, urea, cysteine and N-ethylmaleimide. J. Dairy Res. 71: p. 207-215.

Cases, S., S.J. Smith, Y.W. Zheng, H.M. Myers, S.R. Lear, E. Sande, S. Novak, C. Collins, C.B. Welch, A.J. Lusis, S.K. Erickson and R.V. Farese Jr, 1998. Identification of a gene encoding an acyl CoA:diacylglycerol acyltransferase, a key enzyme in triacylglycerol synthesis. Proc. Natl. Acad. Sci. USA. 95: p. 13018-13023.

Davoli, R., S. Dall`Olio and V. Russo, 1990. Effect of λ-casein genotype on the coagulation properties of milk. J. Anim. Breed. Genet. 107: p. 458-464.

Debeljak, M., S. Susnik, R. Marinsek-Logar, J.F. Medrano and P. Dovc, 2000. Allelic differences in bovine kappa-CN gene which may regulate gene expression. Pflugers Arch. 439: p. R4-6.

Dolphin, C.T., A. Janmohamed, R.L. Smith, E.A. Shephard and I.R. Phillips, 1997a. Missense mutation in flavin-containing monooxygenase 3 gene, FMO3, underlies fish-odour syndrome. Nat. Genet. 17: p. 491-494.

Ehrmann, S., H. Bartenschlager and H. Geldermann, 1997. Polymorphism in the 5' flanking region of the bovine lactoglobulin encoding gene and its association with beta-lactoglobulin in the milk. J. Anim. Breed. Genet. 114: 49-53

Elofsson, U.M., M.A. Paulsson, P. Sellers and T. Arnebrant, 1996. Adsorption during heat treatment related to the thermal unfolding/aggregation of beta-lactoglobulins A and B. J. Colloid Interf. Sci. 183: p. 408-415.

Erhardt, G., 1989. K-kasein in Rindermilch - Nachweis eines weiteren Allels (λ-Cn[E]) in verschiedenen Rassen. J. Anim. Breed. Genet. 106: p. 225-231.

Famula, T.R. and J.F. Medrano, 1994. Estimation of genotype effects for milk proteins with animal and sire transmitting ability models. J. Dairy Sci. 77: p. 3153-3162.

Farese Jr, R.V., S. Cases and S.J. Smith, 2000. Triglyceride synthesis: insights from the cloning of diacylglycerol acyltransferase. Curr. Opin. Lipidol. 11: p. 229-234. Review.

Formaggioni, P., M. Summer, M. Malacarne and P. Mariani, 1999. Milk protein polymorphism: detection and diffusion of the genetic variants in bos genus. Annali della Facolta di Medicina Veterinaria Universita di Parma 19: p. 127-165, http://www.unipr.it/arpa/facvet/annali/1999/formaggioni/formaggioni.htm

Forrest, S.M., M. Knight, B.R. Akerman, J.R. Cashman and E.P. Treacy, 2001. A novel deletion in the flavin-containing monooxygenase gene (FMO3) in a Greek patient with trimethylaminuria. Pharmacogenetics 2: p. 169-174.

Geldermann, H., A. Kuss and J. Gogol, 2005. Genes as indicators for milk composition. In: Indicators of milk and beef quality, J.F. Hocquette and S. Gigli (editors), Wageningen Academic Publishers, Wageningen, The Netherlands, this book.

Georges, M., D. Nielsen, M. Mackinnon, A. Mishra, R. Okimoto, A.T. Pasquino, L.S. Sargeant, A. Sorensen, M.R. Steele and X. Zhao, 1995. Mapping quantitative trait loci controlling milk production in dairy cattle by exploiting progeny testing. Genetics 139: p. 907-920.

Goddard, M., 2001. Genetics to improve milk quality. Aust. J. Dairy Technol. 56: p. 166-170.

Graham, E.R.B., D.M. McLean and P. Zviedrans, 1984. The effect of milk protein genotypes on the cheesemaking properties of milk and on the yield of cheese. In: "Proceedings of the 4th Conference of the Australian Association of Animal Breeding and Genetics", Adelaide, p. 136-137.

Graml, R., G. Weiss, J. Buchberger and F. Pirchner, 1989. Different rates of synthesis of whey protein and casein by alleles of the β-lactoglobulin and αs1-casein locus in cattle. Genet. Sel. Evol. 21: p. 547-554.

Grisart, B., W. Coppieters, F. Farnir, L. Karim, C. Ford, P. Berzi, N. Cambisano, M. Mni, S. Reid, P. Simon, R. Spelman, M. Georges and R. Snell, 2002. Positional cloning of a QTL in dairy cattle: identification of a missense mutation in the bovine DGAT1 gene with major effect on milk yield and composition. Genome Res. 12: p. 222-231.

Grisart, B., F. Farnir, L. Karim, N. Cambisano, J.J. Kim, A. Kvasz, M. Mni, P. Simon, J.M. Frere, W. Coppieters and M. Georges, 2004. Genetic and functional confirmation of the causality of the DGAT1 K232A quantitative trait nucleotide in affecting milk yield and composition. Proc. Natl. Acad. Sci. U S A. 101: 2398-2403

Gustafsson, V. and A. Lundén, 2003. Strong linkage disequilibrium between polymorphisms in the 5'-flanking region and the coding part of the bovine β-lactoglobulin gene. J. Anim. Breed. Genet. 120: p. 68-72.

Hansen, H.O., I. Grunnet and J. Knudsen, 1984. Triacylglycerol synthesis in goat mammary gland. Factors influencing the esterification of fatty acids synthesized de novo. Biochem. J. 220: p. 521-527.

Hill, J.P., M.J. Boland and A.F. Smith, 1997a. The effect of β-lactoglobulin variants on milk powder manufacture and properties. In: "Milk protein polymorphism". International Dairy Federation. Special Issue 9702, p. 372-394.

Hill, J.P., W.C. Thresher, M.J. Boland, L.K. Creamer, S.G. Anema, G. Mandersson, D.E. Otter, G.R. Paterson, R. Lowe, R.G. Burr, R.L. Motion, A. Winkelman and B. Wickham, 1997b. The polymorphism of the milk protein β-lactoglobulin. A review. In: "Milk Composition, Production and Biotechnology", R.A.S. Welch, D.J.W. Burns, S.R. Davis, A.I. Popay and C.G. Prosser (editors), Wallingford, p. 173-202.

Hlavica, P. and M. Kehl, 1977. Studies on the mechanism of hepatic microsomal N-oxide formation, the role of cytochrome P-450 and mixed function amine oxidase in the N-oxidation of N,N-dimethylamine. Biochem. J. 164: p. 487-496.

Hobson-Frohock, A., D.G. Land, N.M. Griffiths and R.F. Curtis, 1973. Egg taints: association with trimethylamine. Nature 243: p. 304-305.

Horne, D.S., J.M. Banks and D.D. Muir, 1996. Genetic polymorphism of milk proteins: understanding the technical effects. In: "Hannah Research Institute Report", pp 70-78.

Ikonen, T., K. Ahlfors, R. Kempe, M. Ojala and O. Ruottinen, 1999a. Genetic parameters for the milk coagulation properties and prevalence of noncoaculating milk in Finnish dairy cows. J. Dairy Sci. 82: p. 205-214.

Ikonen, T., H. Bovenhuis, M. Ojala, O. Ruottinen and M. Georges, 2001. Associations between casein haplotypes and first lactation milk production traits in Finnish Ayrshire cows. J. Dairy Sci. 84: p. 507-514.

Ikonen, T., S. Morri, A.-M. Tyrisevä, O. Ruottinen and M. Ojala, 2004. Genetic and phenotypic correlations between milk coagulation properties, milk production traits, somatic cell count, casein content, and pH of milk. J. Dairy Sci. 87: p. 458-467.

Ikonen, T., M. Ojala and O. Ruottinen, 1999b. Associations between milk protein polymorphism and first lactation milk production traits in Finnish Ayrshire cows. J. Dairy Sci. 82: p. 1026-1033.

Ikonen, T., M. Ojala and E.-L. Syväoja, 1997. Effects of composite casein and β-lactoglobulin genotypes on renneting properties and composition of bovine milk by assuming an animal model. Agric. Food Sci. Finland 6: p. 283-294.

Ikonen, T., O. Ruottinen, G. Erhardt and M. Ojala, 1996. Allele frequencies of the major milk proteins in the Finnish Ayrshire and detection of a new λ-casein variant. Anim. Genet. 27: p. 179-181.

Ikonen, T., O. Ruottinen, E.-L. Syväoja, K. Saarinen, E. Pahkala and M. Ojala, 1999c. Effect of milk coagulation properties of herd bulk milks on yield and composition of Emmental cheese. Agric. Food Sci. Finland 6: p. 283-294.

Imafidon, G.I., K.F. Ng-Kwai-Hang, V.R. Harwalkar and C.-Y. Ma, 1991. Effect of genetic polymorphism on the thermal stability of ß-lactoglobulin and κ-casein mixture. J. Dairy Sci. 74: p. 1791-1802.

Jakob, E., 1994. Genetic polymorphisms of milk proteins. IDF Bulletin 298, pp. 17-27. Brussels, Belgium: International Dairy Federation.

Jakob, E. and Z. Puhan, 1992. Technological properties of milk as influenced by genetic polymorphism of milk proteins- A review. Int. Dairy J. 2: p. 157-178.

Jensen, S.K., A.K.B. Johannsen and J.E. Hermansen, 1999. Quantitative secretion and maximal secretion capacity of retinol, β-carotene and α-tocopherol into cows' milk. J. Dairy Res. 66: p. 511-522.

Johnson, P.E., L.J. Bush, G.V. Odell and E.L. Smith, 1973. The undesirable flavor in milk resulting from grazing cows on wheat pasture. Oklahoma Agricultural Experiment Station. MP-90: p. 274-277.

Kaupe, B., A. Winter, R. Fries and G. Erhardt, 2004. DGAT1 polymorphism in Bos indicus and Bos taurus cattle breeds. J. Dairy Res. 71: p. 182-187.

Karijord, Ö., N. Standal and O. Syrstad, 1982. Sources of variation in composition of milk fat. Z. Tierzüchtg. Züchtgbiol. 99: p. 81-93.

Kennedy, B.W., M. Quinton and J.A.M. van Arendonk, 1992. Estimation of effects of single genes on quantitative traits. J. Anim. Sci. 70: p. 2000-2012.

Kim, J., M. Braunschweig and Z. Puhan, 1996. Occurrence of extreme ratio of β-lactoglobulin variants A and B in Swiss Brown cattle quantified by capillary electrophoresis. Milchwissenschaft 8: p. 435-437.

Kühn, C., G. Thaller, A. Winter, O.R.P. Bininda-Edmonds, B. Kaupe, G.Erhardt, J. Bennewitz, M. Swerin and R. Fries, 2004. Evidence for multiple alleles at the *DGAT1* locus better explains a quantitative trait locus with major effect on milk fat content in cattle. Genetics 167: p. 1873-1881.

Kuss, A.W., J. Gogol and H. Geldermann, 2003. Associations of a polymorphic AP-2 binding site in the 5'-flanking region of the bovine beta-lactoglobulin gene with milk proteins. J. Dairy Sci. 86: p. 2213-2218.

Leone, P., A. Scaltriti, A. Caroli, S. Sangalli, A. Samore and G. Pagnacco, 1998. Effects of CASK E variant on milk yield indexes in Italian Holstein Friesian bulls. Anim. Genet. 29 (suppl. 1): p. 63.

Lien, S., L. Gomez-Raya, T. Steine, E. Fimland, S. Rogne and L.G. Raya, 1995. Associations between casein haplotypes and milk yield traits. J. Dairy Sci. 78: p. 2047-2056.

Lodes, A., J. Buchberger, I. Krause, J. Aumann and H. Klostermeyer, 1996. The influence of genetic variants of milk proteins on the compositional and technological properties of milk. 2. Rennet coagulation time and firmness of the rennet curd. Milchwissenschaft 51: p. 543-547.

Lum, L.S., P. Dovc and J.F. Medrano, 1997. Polymorphisms of bovine beta-lactoglobulin promoter and differences in the binding affinity of activator protein-2 transcription factor. J. Dairy Sci. 80: p. 1389-1397.

Lundén, A. and J. af Forselles, 2000. Gene frequency of λ-casein E in Swedish Red and White breeding bulls. In: "Abstract of the 27th International Conference on Animal Genetics", 22-26 July 2000, Minneapolis, Minnesota, USA, p. 63.

Lundén, A., V. Gustafsson, J.O. Bosset, R. Gauch and M. Imhof, 2002a. High trimethylamine concentration in milk from cows on standard diets is expressed as fishy off-flavour. J. Dairy Res. 69: p. 383-390.

Lundén, A., V. Gustafsson, M. Imhof, R. Gauch and J-O. Bosset, 2003. Variation in trimethylamine concentration in milk from a cow with recurrent fishy off-flavour. In: "Abstract from 3rd NIZO Dairy Conference on Dynamics of Texture, Process and Perception". 11-13 June 2003, Papendal, The Netherlands, p. 23.

Lundén, A., S. Marklund, I. Gustafsson and L. Andersson, 2002b. A nonsense mutation in the FMO3 gene underlies fishy off-flavor in cow's milk. Genome Res. 12: p. 1885-1888.

Lundén, A., M. Nilsson and L. Janson, 1997. Marked effect of β-lactoglobulin polymorphism on the ratio of casein to total protein in milk. J. Dairy Sci. 80: p. 2996-3005.

Marshall, M.O. and J. Knudsen, 1979. Specificity of diacylglycerol acyltransferase from bovine mammary gland, liver and adipose tissue towards acyl-CoA esters. Eur. J. Biochem. 94: p. 93-98.

Marshall, M.O. and J. Knudsen, 1980. Factors influencing the in vitro activity of diacylglycerol acyltransferase from bovine mammary gland and liver towards butyryl-CoA and palmitoyl-CoA. Biochim. Biophys. Acta. 617: p. 393-397.

Martin, P., M. Szymanowska, L. Zwierzchowski and C. Leroux, 2002. The impact of genetic polymorphisms on the protein composition of ruminant milks. Reprod. Nutr. Dev. 42: p. 433-459.

Marziali, A.S. and K.F. Ng-Kwai-Hang, 1986. Relationships between milk protein polymorphisms and cheese yielding capacity. J. Dairy Sci. 69: 1193-1201

Mercier, J.J. and J.L. Vilotte, 1993. Structure and function of milk protein genes. J. Dairy Sci. 76: p. 3079-3098.

Mehta, R.S., R. Bassette and G. Ward, 1974. Trimethylamine responsible for fishy flavour in milk from cows on wheat pasture. J. Dairy Sci. 57: p. 285-289.

Ng-Kwai-Hang, K.F. and F. Grosclaude, 1992. Genetic polymorphism of milk proteins. In: "Advanced Dairy Chemistry 1: Proteins", 2nd ed. P.F. Fox (editor), Elsevier, p. 405-455.

Ng-Kwai-Hang, K.F., J.F. Hayes, J.E. Moxley and H.G. Monardes, 1984. Association of genetic variants of casein and milk serum proteins with milk, fat and protein production by dairy cattle. J. Dairy Sci. 67: p. 835-840.

Ng-Kwai-Hang, K.F. and S. Kim, 1996. Different amounts of β-lactoglobulin A and B in milk from heterozygous AB cows. Int. Dairy J. 6: p. 689-695.

Ng-Kwai-Hang, K.F., H.G. Monardes and J.F. Hayes, 1990. Association between genetic polymorphism of milk proteins and production traits during three lactations. J. Dairy Sci. 73: p. 3414-3420.

Neimann-Sörensen, A., P.O. Nielsen, P.R. Poulsen and P.K. Jensen, 1973. [Investigations concerning the influence of heritability on the development of oxidized flavour in cow's milk]. Beretning 199. Statens försögsmejeri, Hilleröd, Denmark.

Näslund, J., G. Pielberg and A. Lundén, 2004. Frequency of the bovine DGAT1 (K232A) polymorphism in selection lines with high and low milk fat content. In: "Abstract from the 55th EAAP Annual meeting ", Bled, Slovenia, 5-9 September 2004, p 5.

Ojala, M., T.R. Famula and J.F. Medrano, 1997. Effects of milk protein genotypes on the variation for milk production traits of Holstein and Jersey cows in California. J. Dairy Sci. 80: p. 1776-1785.

Ojala, M., A.-M. Tyrisevä and T. Ikonen. 2005. Genetic improvement of milk quality traits for cheese production. In: Indicators of milk and beef quality, J.F. Hocquette and S. Gigli (editors), Wageningen Academic Publishers, Wageningen, The Netherlands, this book.

Okigbo, L.M., G.H. Richardson, R.J. Brown and C.A. Ernstrom, 1985. Variation in coagulation properties of milk from individual cows. J. Dairy Sci. 68: p. 822-828.

Oloffs, K., H. Schulte-Coerne, K. Pabst and H.O. Gravert, 1992. Die bedeutung der Proteinvarianten für genetische unterschiede in der Käsereitauglichkeit der Milch. Züchtungskunde 64(1): p. 20-26.
Piez, K.A., E.W. Davie, J.E. Folk and J.A. Gladner, 1961. beta-lactoglobulins A and B. I. Chromatographic separation and amino acid composition. J. Biol. Chem. 236: 2912-2916
Qin, B.Y., M.C. Bewley, L.K. Creamer, E.N. Baker and G.B. Jameson, 1999. Functional implications of structural differences between variants A and B of bovine beta-lactoglobulin. Protein Sci. 8: p. 75-83.
Ruottinen, O., T. Ikonen and M. Ojala, 2004. Associations between milk protein genotypes and fertility traits in Finnish Ayrshire heifers and first lactation cows. Livest. Prod. Sci. 85: p. 27-34.
Schaar, J., B. Hansson and H.-E. Pettersson, 1985. Effects of genetic variants of λ-casein and β-lactoglobulin on cheesemaking. J. Dairy Res. 52: p. 429-437.
Smith, S.J., S. Cases, D.R. Jensen, H.C. Chen, E. Sande, B. Tow, D.A. Sanan, J. Raber, R.H. Eckel and R.V. Farese Jr., 2000. Obesity resistance and multiple mechanisms of triglyceride synthesis in mice lacking DGAT1. Nat. Genet. 25: p. 87-90.
Spellacy, E., R.W.E. Watts and S.K. Gollamali, 1979. Trimethylaminuria. J. Inherit. Metab. Dis. 2: 85-88
Treacy, E.P., B.R. Akerman, L.M.L. Chow, R. Youil, C. Bibeau, J. Lin, A.G. Bruce, M. Knight, D.M. Danks, J.R. Cashman and S.M. Forrest, 1998. Mutations in the flavin-containing monooxygenase gene (FMO3) cause trimethylaminuria, a defect in detoxication. Hum. Mol. Genet. 5: p. 839-845.
Timmons, J.S., W.P. Weiss, D.L. Palmquist and W.J. Harper, 2001. Relationships among dietary roasted soybeans, milk components, and spontaneous oxidized flavor of milk. J. Dairy Sci. 84: p. 2440-2449.
Tyrisevä, A.-M., T. Ikonen and M. Ojala, 2003. Repeatability estimates for milk coagulation traits and non-coagulation of milk in Finnish Ayrshire cows. J. Dairy Res. 70: p. 91-98.
Tyrisevä, A.-M., T. Vahlsten, O. Ruottinen and M. Ojala, 2004. Nongoagulation of milk in Finnish Ayrshire and Holstein-Friesian cows and effect of herds on milk coagulation ability. J. Dairy Sci. 87(11):(in press)
van Eenennaam, A. and J.F. Medrano, 1991. Differences in allelic protein expression in the milk of heterozygous λ-casein cows. J. Dairy Sci. 74: p. 1491-1496.
Velmala, R., J. Vilkki, K. Elo and A. Mäki-Tanila, 1995. Casein haplotypes and their association with milk production traits in the Finnish Ayrshire cattle. Anim. Genet. 26: p. 419-425.
von Gunten, R.L., L.J. Bush, M.E. Odell, M.E. Wells and G.D. Adams, 1976. Factors related to the occurrence of trimethylamine in milk. J. Milk Food Technol. 8: p. 526-529.
Wagner, V.A., T.A. Schild and H. Geldermann, 1994. DNA variants within the 5´-flanking region of milk-protein-encoding genes. II The β-lactoglobulin-encoding gene. Theor. Appl. Genet. 89: p. 121-126.
Weller, J.I., Y. Kashi and M. Soller, 1990. Power of daughter and granddaughter designs for determining linkage between marker loci and quantitative trait loci in dairy cattle. J. Dairy Sci. 73: p. 2525-2537.
Winter, A., W. Krämer, F.A.O. Werner, S. Kollers, S. Kata, G. Durstewitz, J. Buitkamp, J.E. Womack, G. Thaller and R. Fries, 2002. Association of a lysine-232/alanine polymorphism in a bovine gene encoding acyl-CoA:diacylglycerol acyltransferase (*DGAT1*) with variation at a quantitative trait locus for milk fat content. Proc. Natl. Acad. Sci. USA 99: p. 9300-9305.
Yang, X., D. Fyodorov and E.S. Deneris, 1995. Transcriptional analysis of acetylcholine receptor α3 gene promoter motifs that bind Sp1 and AP2. J. Biol. Chem. 277: p. 50876-50884.

Genes as indicators for milk composition

H. Geldermann, A.W. Kuss and J. Gogol*
Department of Animal Breeding and Biotechnology, University of Hohenheim, Stuttgart, Germany
**Correspondence: tzunihoh@uni-hohenheim.de*

Summary

Genetic analysis of milk production on the animal level is a complex issue, which is elucidated on the basis of a few aspects. Numerous genes are involved in lactation; their gene variants were however rarely identified to be responsible for the variation of trait values between animals. Associations between carriers of genotypes or alleles and trait values in populations were extensively studied and merely seem to indicate variable effects of linked loci. QTL analyses regarding lactation traits mapped numerous effects in more or less large chromosome intervals. A number of polymorphic promoter variants were identified and some of them were associated with the quantity of allele specific mRNA in mammary gland or allele specific proteins in milk. However, even if the results look very promising, no discrimination between the analysed polymorphic sites and the context of the respective gene will be possible from studies of associations with the target traits in different individuals. Functional sites of milk protein coding genes may be separately analysed in transgenic animals or cells, but up to now very few comparisons of allelic regulatory sites were reported. New methods now provide the necessary power and flexibility to improve the causal analysis of function of numerous genes as well as of specific intragenic sites. Consequently knowledge is accumulating for lactation genetics which can be exploited for economic profit. However, also some concerns, e. g. regarding genetically modified milk or application of marker alleles in cattle breeding (e. g. loss of genetic diversity), are considered.

Keywords: candidate genes, QTL mapping, association analysis, promoter variants, transgenic models

Introduction

As reflected in many reports, the increase in milk production per cow during the last 50 years has been dramatic. Some of the improvements in lactation can probably be explained by changes in the structure and function of genes which act on the mammary gland. In principle, complex traits like lactation are under homeostatic control, where alterations in the expression of one gene are counterbalanced by the endogenous expression of other genes. This means that identifying single causative genes for lactation traits is difficult, bearing the danger of misinterpretation as milk production on the animal level depends on more than only the function of the mammary gland. Some of the genetic aspects of lactation in dairy cattle however can be focused on the involved genes, studies of association between gene variants and trait values, QTL mapping regarding lactation traits, effects of promoter variants, and analyses in transgenic cells or animals.

Genes involved in lactation and "milk traits"

Table 1 lists some of the major genes which are involved in lactation.

Milk proteins are the main source of nutrition for the neonate mammal. All genes which code for secreted proteins are expressed in epithelial tissues like mammary and salivary glands and have functions in host defence, nutrition and immunomodulation. In placental mammals caseins are the major milk proteins. Depending on the species, three to four evolutionarily related genes,

Table 1. Examples of candidate genes for lactation.

Gene	Gene product	Map position	
		Cytogenetic[1]	MARC 97[cM]
Milk protein coding			
CSN1S1	α_{s1}-Casein	6q31-q33	82.6
CSN1S2	α_{s2}-Casein		
CSN2	β-Casein		
CSN3	κ-Casein		
LALBA	α-Lactalbumin	5q21	44.5-50.5
LGB	β-Lactoglobulin	11q28	108.7
Fatty acid synthesis and secretion			
DGAT1	Diacylglycerol O-acyltransferase	14q11	0.3
BTN	Butyrophilin	23q21-q23	n.m.
LEP (OB, OBS)	Leptin (obesity)	4q32	82.8
LEPR	Leptin receptor	3q33	n.m.
FASN	Fatty acid synthetase	19q22	n.m.
ACACA	Acetyl-CoA-carboxylase α	19q13-q14	n.m.
LPL	Lipoprotein lipase	8	63.0
Mammary gland development, growth and apoptosis			
GH1 (GH)	Growth hormone	19q22prox.	65.7
GHR	Growth hormone receptor	20q17	50.0
GHRH	Growth hormone releasing hormone	13q22-q23	71.0
IGF1	Insulin-like growth factor 1	5	73.0
IGF1R	Insulin-like growth factor 1 receptor	21	13.0
IGFBP3	Insulin-like growth factor binding protein 3	4	68.0
IGFBP4	Insulin-like growth factor binding protein 4	19	60.0
PRL	Prolactin	23q23med.	43.2
PRLR	Prolactin receptor	20q17	52.5
STAT5A	Signal transducer and activator of transcription 5A	19q17	n.m.
CSNK2A2	Casein kinase 2 alpha	18	40.0
Others			
FMO3	Flavin-containing mono-oxygenase-3	n.m.	n.m.
BOLA	MHC molecules	23	35.4

[1] n.m., not mapped; med., median; prox., proximal.

[2] MG: mammary gland; AT: adipose tissue; VT: various tissues; APC: antigen presenting cells.

Major expression[2]	Function	Exemplary references for mapping information
MG	Nutrient, host defense, immunomodulation	Threadgill and Womack, 1990
MG		Gallagher *et al.*, 1994
MG		Threadgill and Womack, 1990
MG	Stabilisation of micelles	Threadgill and Womack, 1990
MG	Regulation of lactose synthesis and therefore milk volume	Threadgill and Womack, 1990
MG	Nutrient	Hayes and Petit, 1993
VT	Triglyceride synthesis	Winter *et al.*, 2002, Grisart *et al.*, 2002, Bennewitz *et al.*, 2004
MG	Fat secretion	Brunner *et al.*, 1996
AT	Feeding behaviour and energy metabolism	Pfister-Genskow *et al.*, 1996
VT		Pfister-Genskow *et al.*, 1997
VT	Fatty acid chain elongation	Roy *et al.*, 2001
VT	Fatty acid synthesis (rate limiting)	Mao *et al.*, 2001
VT	Triglyceride hydrolisation	Threadgill and Womack, 1991
VT	Regulation of energy metabolism, growth and development	Hediger *et al.*, 1990
VT		Moody *et al.*, 1995
VT		Barendse *et al.*, 1997
VT		Bishop *et al.*, 1991
VT		Moody *et al.*, 1996
VT		Maciulla *et al.*, 1997
VT		Moore *et al.*, 2003
VT	Lacto- and lipogenesis	Hallerman *et al.*, 1988
VT		Hayes *et al.*, 1996
VT	Transcription factor	Seyfert *et al.*, 2000
VT	Protein secretion	Aasland *et al.*, 2000, Lasa-Benito *et al.*, 1996, Meggio *et al.*, 1988
VT	Nitrogen-oxide-forming activity, fishy flavour of milk	Lunden *et al.*, 2002
APC	Antigen presentation	Andersson and Rask, 1988

coding for the "calcium-sensitive" caseins, and a functionally related κ-casein coding gene are located in a casein gene cluster region of between 250 and 350 kb. Comparative analysis of this cluster shows the unusual high divergence of the coding regions of the casein coding genes within species and even within breeds. This variation contrasts with the high degree of conservation observed regarding the structure of the entire casein gene cluster and for a number of non-coding regions between species. A more up to date topic of research is orientated towards the regulation of milk protein gene expression. The milk protein coding genes influence the intracellular transport, accumulation as well as secretion of their products. Their tissue specific expression is caused by gene promoter sequences containing binding sites (response elements) for ubiquitous as well as mammary gland specific transcription factors.

Lactose is synthesized in the golgi apparatus of the mammary secretory cells by the lactose synthetase complex. This complex is composed of the α-lactalbumin and the enzyme α-1,4-galactosyltransferase. The production of lactose is critical in the control of milk secretion and consequently for milk volume. Of all the bovine milk protein genes, the expression of *LALBA* is the most lactation-specific and strictly controlled.

Equivalent knowledge as for the milk protein coding genes exists for further classes of lactation associated genes. Table 1 lists some genes which are involved in milk fat synthesis and secretion. At least two enzymes catalyze the reaction in which triacylglycerol is covalently joined to long-chain fatty acyl-CoA in order to form triglycerides as major constituents of fat. *DGAT1* (diacylglycerol O-acyltransferase homolog 1) encodes one of these enzymes. Knockout mice that lack both allelic copies of *DGAT1* showed deficient lactation. In cattle, the locus was mapped in the region of a major milk fat content QTL (Grisart *et al.*, 2002, Winter *et al.*, 2002). Triglycerides in the VLDL (very low density lipoprotein) are hydrolyzed in the mammary capillaries by an enzyme called lipoprotein lipase (LPL). The resulting products are taken up by the mammary epithelial cells and used for the triglyceride synthesis. Acetyl-CoA carboxylase (ACACA) is the key enzyme for milk fat synthesis during lactogenesis. The gene for the Fatty Acid Synthetase (FASN) codes for an enzyme responsible for the chain elongation of the fatty acids.

Among the many genes which influence mammary gland development, growth and apoptosis probably those coding for hormones, receptors or DNA binding proteins belong to the most important. Members of the GH (GH1, GHRH) and IGF family (IGF-1, IGF-II, Insulin) together with their respective receptors (e. g. GHR, IGF1R) were shown to act as central regulators of energy metabolism, mammalian growth and development. Six different IGF binding proteins (IGFBP) and five different classes of membrane receptors may interact with ligands in the IGF family. STAT5A and STAT5B convert extracellular signals into gene transcription that determines the functions and differentiation of mammary gland cells (Groner and Hennighausen, 2000). The *STAT* family comprises seven genes encoding proteins of similar domain structure and modes of activation.

Thus, some knowledge is available about the structure and function of genes and how they control metabolic pathways. Many genes have significant functions for lactation traits; however information about their structure and function and how they control metabolic pathways does not necessarily indicate the importance of gene variants for the variation of trait values between individuals. So far only few DNA sequence data on genomic level are available for cattle. However the bovine genome sequencing is ongoing and will provide a basis for more data regarding DNA variability of individual gene regions, i. e. about the conserved and less conserved genomic DNA regions in cattle.

Associations between carriers of gene variants (or markers) and lactation traits

Alleles at single loci can be used to mark animals which carry distinct allelic variants and to analyse their association with trait values, either through a direct test for the desired variant or by tests for linked markers. For over 30 years, studies focused on relationships between carriers of genotypes of very diverse loci and traits of lactation. As an example Table 2 summarizes the results on associations between carriers of different milk protein coding alleles and traits of milk composition, yield and processing. However, such associations pertain not to an effect of a single gene, but to that of a chromosome interval (cluster of genes, haplotype). Therefore the results were not consistent across studies. Several reasons may be responsible for the conflicting data, e.

Table 2. Summary from literature reports on associations between carriers of different milk protein genotypes and traits of milk production for the breed Holstein (adapted from Bobe et al., *1999, FitzGerald* et al., *1999, Ng-Kwai-Hang* et al., *1990 and Ojala* et al., *1997).*

2a. Milk processing traits.

Trait	Gene locus	Superior allele or genotype
Size of micelles	*CSN3*	*BB, BC*
Curd coagulation time	*CSN1S1*	*CC*
	CSN2	*BB*
	CSN3	*BB, BC*
Curd firmness	*CSN2*	*BB*
	CSN3	*BB, BC*
	LGB	*BB*
Cheese yield	*CSN1S1*	*BB*
	CSN3	*BB*
	LGB	*BB*

2b. Composition and yield of milk protein.

Trait	Gene locus	Superior allele or genotype
Milk protein content	*CSN1S1*	*BC, CC*
	CSN2	A_2A_3
	CSN3	*BB*
	LGB	*AA, AB*
Milk protein yield	*CSN1S1*	*BC*
	CSN2	A_3B
	CSN3	*BB*
	LGB	*AA*
Casein content	*CSN1S1*	*BB, BC*
	CSN3	*BB*
	LGB	*BB*
Whey protein content (~ β-lactoglobulin content)	*CSN1S1*	*AB*
	CSN2	A_1A_3
	CSN3	*AA*
	LGB	*AA*

g. breeds, design of experimental studies, origins of individuals, statistical methods and variable haplotypes. However, new methods can be used for much improved analyses of associations, i. e. the regarding of families, phylogeny, typing of several linked genes or quantification of specific gene products (allele specific mRNAs or proteins).

QTL mapping regarding lactation traits

More advanced studies were performed within families. They included several linked loci and thus allowed a mapping of QTLs (quantitative trait loci) which affect the trait values of lactation. Much of the QTL analysis in dairy cattle was performed in order to map loci associated with milk production within breeds (mainly Holstein), rather than generating crosses between populations in order to take advantage of differences between populations, as it is done in pig QTL mapping. Usually a large number of polymorphic loci were genotyped; they covered the entire genome, a chromosome or a section within a chromosome. Almost all QTL studies in cattle used the half-sib structure available in the industry and included the offspring from a number of sires.

At loci that are heterozygous in individual sires, alleles segregate in the offspring and allow a mapping of gene effects. If trait values of the daughter performances depend on the chromosome sections which they inherited from the sire, they will be associated to the marker intervals ("daughter design"). The "granddaughter design" uses sons instead of daughters. The sons can be evaluated on the basis of records from large numbers of their daughters, and breeding values are associated with the marker loci. Confidence intervals of QTLs can be narrowed with the help of linkage disequilibrium mapping in populations.

To date more than 20 experimental studies were reported, however not all used independent material. Figure 1 assembles the results from studies regarding bovine chromosomes which

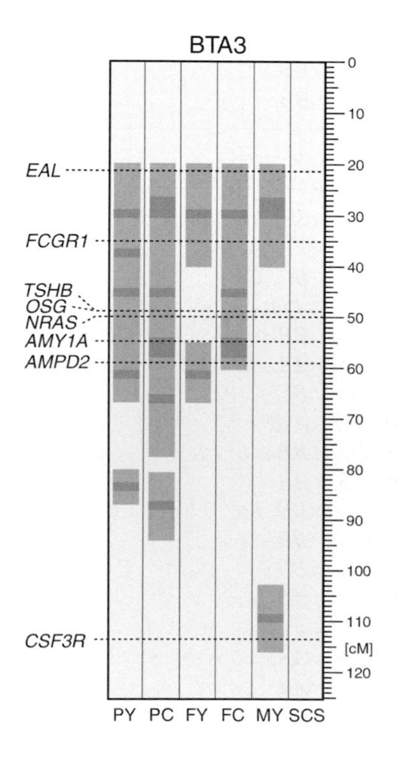

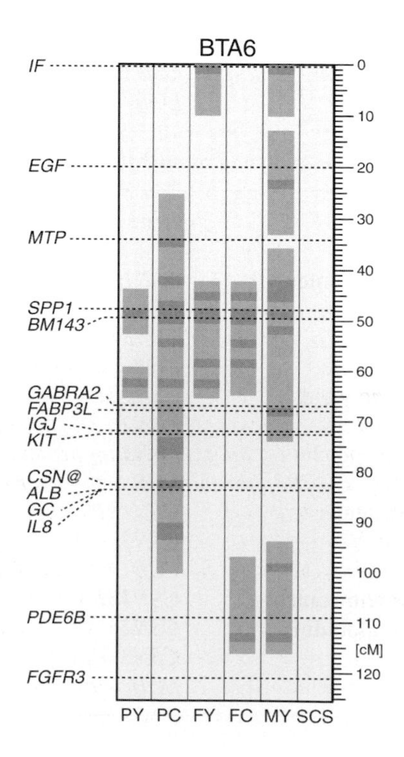

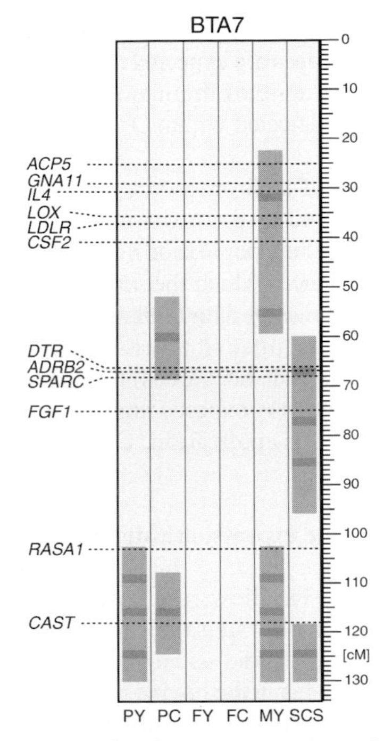

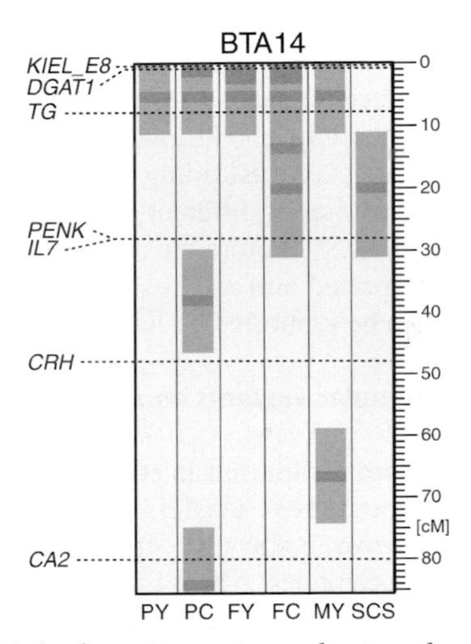

Figure 1. Intervals of Quantitative Trait Loci (QTL) for lactation traits on bovine chromosomes which were most frequently described (Ashwell et al., *2004, Ashwell and Tassell 1999, Bennewitz* et al., *2004, Farnir* et al., *2002, Freyer* et al., *2003, Geldermann* et al., *1985, Grisart* et al., *2002, Heyen* et al., *1999, Kühn* et al., *2003, Mosig* et al., *2001, Olsen* et al., *2002, Plante* et al., *2001, Riquet* et al., *1999, Rodriguez-Zas* et al., *2002b, 2002a, Ron* et al., *2001, Velmala* et al., *1999, Viitala* et al., *2003, Zhang* et al., *1998).*

PY, protein yield; PC, protein content; FY, fat yield; FC; fat content; MY, Milk yield; SCS, somatic cell score.

cM, MARC97 map units; Type I loci mapped in cattle are indicated.

BTA3: EAL, erythrocyte antigen L; FCGR1, Fc fragment of IgG, receptor for CD64; TSHB, thyroid stimulating hormone, beta polypeptide;OSG, oviduct specific glycoprotein; NRAS, neuroblastoma RAS oncogene homolog; AMY1A, amylase 1, alpha; AMPD2, adenosine monophosphate deaminase 2; CSF3R, colony stimulating factor 3, receptor.

BTA6: IF, complement component 1. EGF, epidermal growth factor. MTP, microsomal triglyceride transfer protein. SPP1, secreted phosphoprotein 1 (osteopontin); BM143, microsatellite locus; GABRA2, gamma-amino butyric acid A receptor alpha 2; FABP3L, fatty acid binding protein (heart) like; IGJ, immunglobulin J polypeptide; KIT, Hardy-Zuckerman 4 feline sarcoma viral oncogene homolog; CSN@, casein gene cluster; ALB, albumin; GC, group-specific component (vitamin D binding protein); IL8, interleukin 8; PDE6B, phosphodiesterase, cyclic GMP; FGFR3, fibroblast growth factor receptor 3.

BTA7: ACP5, tartrate-resistant acid phosphatase type 5 precursor, GNA11, guanine nucleotide binding protein, alpha 11; IL4, interleukin 4. LOX, lysyl oxidase. LDLR, low density lipoprotein receptor; CSF2,colony stimulating factor 2; DTR, diphtheria toxin receptor; ADRB2, adrenergic, beta-2, receptor; SPARC, secreted protein, acidic, cysteine-rich; FGF1, fibroblast growth factor 1, acidic; RASA1, RAS p21 protein activator (GTPase activating protein); CAST, calpastatin.

BTA14: KIEL_E8, cysteine and histidine-rich protein; DGAT1,diacylglycerol O-acyltransferase 1; TG, thyroglobulin. PENK, pro-enkephalin; IL7, interleukin 7; CRH, corticotropin releasing hormone; CA2, carbonic anhydrase II.

harboured the most frequently described QTLs. A QTL identified in such a scan has a confidence interval of several cM (mostly > 10 cM), and the causative genes are expected to be located within the marker interval, but are not directly known. Almost none of the major candidate genes (compare Table 1 with the Type 1 genes in Figure 1) were detected within QTL intervals.

A QTL describes net-effects between those haplotypes of a chromosome section which are included from the individual sire. QTL analyses are affected by epistasis (interaction between different gene loci), stratification (heterogeneous subgroups in a population), assortative mating, effects of age, heterogeneous environment and other influences which therefore led to different results between studies. QTL mapping results from population data (linkage disequilibrium mapping) can display misleading effects when complex quantitative traits are used for which several causative sites of different gene loci (and also within a distinct locus) may have influence on the trait values. Therefore the study of population parameters requires high effort, should be carefully interpreted and might still have low relevance for application. Consequently QTL mapping has to be supported by further approaches.

Effects of promoter variants on protein binding and gene expression using population data

Moving on from a situation in which the interval of a QTL allows the identification of the underlying gene or gene cluster, we may be interested in analysing the causative nucleotide variants. However, eukaryotic genes can be rather complex and usually include numerous polymorphic nucleotides. About 1 % of the nucleotide positions in the bovine milk protein coding genes were observed to be variable in 13 animals from different breeds (Geldermann *et al.*, 1996), resulting in an expectation of roughly 100 polymorphic sites within a gene of 10 kb. The question is which of the several allelic positions alters the function of the regarded gene most.

A first approach was the analysis of associations between promoter variants and protein binding as well as gene expression based on samples from populations. Polymorphisms within the protein binding sites (response elements) of milk protein coding genes were studied in different breeds. A number of polymorphic sites were identified and some of them were associated with the quantity of allele specific proteins in the milk (Table 3). Mobility shift assays and DNaseI footprinting confirmed differential binding affinity due to allele specific mutations. Only few association studies, mostly covering a small number of animals, are available concerning allele specific mRNAs, and the in vivo relevance of allele specific protein binding is not proven.

However, even if the results look very promising, no discrimination between the analysed polymorphic sites within a gene and the context of the respective gene will be possible concerning the association with target trait values in different individuals. The reason for that is that in a breed and even in a species one always has to expect disequilibria between single polymorphic nucleotides at sites within and even across genes.

Analysis of functional sites within genes in transgenic cells or animals

The transgene technology allows an isolated analysis of functional sites, their reaction with ligands, and a comparison between allelic variants of a gene. The required pre-condition however is that SNPs are known which have high probabilities of being causative. In principle the experimental approach includes a generation of animals or cell lines which carry a transgene containing a regulatory sequence ligated to a reporter gene, the quantification of reporter gene expression, and the analysis of influences on gene expression, e. g. hormones.

Table 3. Effects of promoter variants in milk protein coding genes on protein binding and association with gene expression.

Analysed position/ Response element[1]	Experimental approach[2]			Reference
	DNA tests	Protein binding	Association with / Number of animals	
CSN1S1				
-175 (a/g) / AP-1	RFLP	-	Total protein / 135	Szymanowska *et al.*, 2003
	RFLP	EMSA	Single proteins / 142	Kuss *et al.*, 2004
-728 (t/-)	?	-	mRNA / 3	Szymanowska *et al.*, 2004
	RFLP	-	Total protein / 135	Szymanowska *et al.*, 2003
	RFLP	-	Single proteins / 3	Martin *et al.*, 2002
-733 (t/c)	-	EMSA	-	Martin *et al.*, 2002 ; Szymanowska *et al.*, 2004
655 bp (5')	SSCP	-	Total protein / 678	Prinzenberg *et al.*, 2003
	RFLP	+		(Martin *et al.*, 2002)
CSN1S2				
-7 (a/c) and +7 (c/t)	-	EMSA	-	Szymanowska *et al.*, 2004
-186 (c/t)	-	EMSA	-	Szymanowska *et al.*, 2004
	RFLP	EMSA	-	Martin *et al.*, 2002
-1084 (c/t)	RFLP	-	Total protein / 135	Szymanowska *et al.*, 2003
	RFLP	-	Single proteins + mRNA/3	Martin *et al.*, 2002
-1100 (ac/ct)	?	EMSA	mRNA / 3	Szymanowska *et al.*, 2004
	-	EMSA	-	Martin *et al.*, 2002
CSN2				
-109 (g/c)	-	EMSA	-	Szymanowska *et al.*, 2004
"	RFLP	EMSA	-	Martin *et al.*, 2002
CSN3				
-385 (c/t)	RFLP	-	Total protein / 124	Kaminski, 2000
LGB				
-22 (g/a); -209 (g/c) / STAT5; -662 (g/c) / STAT5	RFLP	-	Single proteins / 71	Ehrmann *et al.*, 1997
-435 (g/c) / AP-2	RFLP	Footprint EMSA	Single proteins / ?	Lum *et al.*, 1997
"	RFLP	-	Single proteins / 142	Kuss *et al.*, 2003
208 bp (5')	SSCP	-	Total protein / 124	Kaminski and Zabolewicz, 2000

[1] Distances from the transcription starting point in bp, 5'(-) and 3'(+). Variable bases of SNPs (-: deletion) and response elements altered by the SNPs are indicated.
[2] Single proteins: allele or locus specific proteins quantified; total protein: milk protein content and yield from records of industry. -: Not investigated. ?: No information.

Expression studies in transgenic mammary gland cells or animals were performed for several of the milk protein coding genes and indicated hormone induced activation of gene promoters. RNA extracted from somatic cells in milk has been shown to be representative of gene expression in the mammary gland and thus provides a source for gene expression studies (Boutinaud and Jammes, 2002). In vitro expression systems allow the analysis of regulatory mechanisms and the study of lactoprotein gene expression under simplified and standardized conditions. However, the study of complex spatial and temporal expression patterns of genes requires the application of experimental animals. As shown in Table 4, different promoter sequences for lactoprotein coding genes were tested. As in many fields of research, transgenic mice or murine cell lines are used as test systems for investigations of regulatory DNA sites involved in lactoprotein expression.

The results show that the functions of evolutionarily conserved regulatory sites involved in bovine milk protein expression were investigated and allowed a mapping of functional sites (cis-acting regulatory elements) within genes that are relevant for lactation. However, results from very few comparisons of allelic regulatory sites were reported so far. Moreover, the test systems are time consuming and not sufficiently sensitive. There is therefore a demand for the development of homologous cell systems, their use under defined conditions and the quantification of gene expression with the help of multiplex methods like DNA- and protein-arrays.

Aspects for application and further research

Only some chromosome regions and genes were analysed for their effects on lactation (Debeljak *et al.*, 2000). However, high milk performance is more than simply the result of the mammary gland capacity and regulation. Features of e. g. body size, metabolism, udder composition as well as adaptation are involved. This means that the identification of single-gene approaches in order to enhance production characteristics is a difficult challenge and needs to regard the complex biological control of the entire organism. However, there are new methods available which seem to have the necessary power and which can be adjusted for a causal analysis of the function of numerous genes as well as of specific sites within the genes. Those methods of structural and functional genomics, like the microarrays, are well described in the literature.

Consequently more and more knowledge in lactation genetics can be applied for economic profit. On one hand, direct and accurate DNA tests identify the presence of distinct alleles and can be performed using very small samples that can be collected at any time. So e. g. an animal can be tested and selected for breeding early in life, before information of milk performance is available. On the other hand, impact on application rises from the generation of transgenic cells and animals. The examples in view of application which are discussed in the literature may be divided into the following categories:

- Improvement of yield and composition of milk components with the target to maximize milk performance, technological quality of milk, and / or offspring growth.
- Use of animals as a model for human mammary diseases and oncogenesis.
- Development of promoter sequences for production of recombinant proteins in mammary gland cells or in transgenic animals.
- Analyses of allelic variants for selection assisted by markers or causal gene variants.
- Introgression and composing of chromosome sections marked by allelic variants from donor populations in a recipient population.

As in other fields of research, there are concerns to be considered. They are being strongly discussed for genetically modified milk (public acceptance, animal welfare, safety of the product, profitability). However, further concerns for using markers or causative alleles in selection of animals may rise from insufficient knowledge of allelic compositions that are responsible for the

Table 4. Analysis of bovine regulatory DNA sequences from milk protein coding genes.

4a. Examples of studies in transgenic cells in vitro.

Gene	Analysed region[1]	Mammary epithelial cell line (species)	Major results	Reference
CSN2	-5.3/+1.6 kb	Primary cells (mouse)	Hormone and matrix dependent expression, localization in regulatory promoter elements.	Yoshimura and Oka, 1990
CSN2	-3815/+42 bp	CID9 (mouse)	Multiple regulatory promoter elements between -2605 and +42 bp.	Schmidhauser *et al.*, 1990
CSN2	-1790/+42 bp	CID9	Localization of an enhancer element (BCE1) at ca. -1.5 kb.	Schmidhauser *et al.*, 1992
CSN2	-930/+20 bp	Bovine	No expression.	Ahn *et al.*, 1995
CSN1S1	-681/+18 bp		Induction not hormone-dependent.	"
CSN2	-16 kb/+8 kb	HC11 (mouse)	Tissue-specific and developmentally regulated expression, integration-site-dependent.	Rijnkels *et al.*, 1995
CSN1S2	-8/+1.5 kb	HC11	No hormone dependent expression.	"
CSN3	-5/+19 kb	HC11	"	"
CSN3	-552/+18 bp	HC11	Localization of regulatory elements between -439 and -125 bp.	Adachi *et al.*, 1996
LGB	-759/+59 bp	HC11	Higher reporter gene expression of promoter haplotype A compared to B.	Geldermann *et al.*, 1996
CSN2	5′ (1.8 kb) + intron 1 (2 kb)	HC11	Prolactin-inducable enhancer activity of several elements in intron 1.	Kang *et al.*, 1998
LGB	-753 bp/+21 bp	HC11	Different transcriptional activity of two promoter variants.	Folch *et al.*, 1999

[1] Either distances from the transcription starting point, 5'(-) and 3'(+) or the lengths of studied fragments are given.

4b. Examples of studies in transgenic animals.

Gene	Analysed region[1]	Species of transgenic recipient	Major results	Reference
LALBA	-477/-220 bp	Mouse	Location of important cis-acting elements.	Soulier *et al.*, 1992
CSN2	-16 kb/+8 kb	Mouse	Expression tissue-specific and developmentally regulated.	Rijnkels *et al.*, 1995
CSN1S2	-8/+1.5 kb	Mouse	No proper expression.	“
CSN3	-5 to + 19 kb	Mouse	“	“
CSN1S1	5′ (14.2 kb) + coding DNA	Mouse	Tissue-specific and developmental regulated expression.	Rijnkels *et al.*, 1998
LALBA	5′ (2.0 kb) + exons (2.0 kb) + 3′ (329 bp)	Pig	Secretion of the translated protein into milk.	Bleck *et al.*, 1998
CSN2	5’ (3.8 kb)	Mouse	Accurate spatial and temporal expression in mammary gland.	Cerdan *et al.*, 1998
LGB	5′ (1.2 kb) + coding DNA (1 kb)	Mouse	Position dependent, copy-number-related expression.	Gutierrez-Adan *et al.*, 1999
CSN2	5’ (1.8 kb)	Mouse	Proper regulation in mammary gland; constitutive and sex independent expression in lung; no expression in other tissues.	Oh *et al.*, 1999
CSN2	5’(15 kb)	Mouse	Constitutive and sex independent expression in mammary gland and lung; no expression in other tissues.	“
CSN2	5′ (6.6 kb) + transcribed DNA + 3′ (2.6 kb)	Mouse	Secretion of the translated protein into milk.	Brophy *et al.*, 2003
CSN2	5′ (6.6 kb) + transcribed DNA 3′ (2.6 kb)	Mouse	Secretion of the translated hybrid protein into milk.	“

[1] Either distances from the transcription starting point, 5’(-) and 3’(+) or the lengths of studied fragments are given.

variance of a performance trait in the regarded population. For example, a strong selection on a distinct allele will lead to a relatively large homozygous chromosome interval in the population. Several closely linked loci will be affected as well and as a result the genetic diversity will be diminished without knowing the importance of the there located DNA variants for trait values. For chromosome intervals with fixed alleles further selection will produce no response, and new variants will only be generated very rarely by mutations. Therefore the genotype assisted selection can cause irreversible disadvantage and should not be used for breeding before the causal DNA variants within the influenced chromosome section are known. Moreover, simultaneous selection may be essential in order to keep the flanking gene regions variable during selection on a distinct allele; which would however require additional knowledge and efforts. Thus not only more basic research, but also advanced regulations for practical breeding are necessary.

References

Aasland, M., D.I. Vage, S. Lien and H. Klungland, 2000. Resolution of conflicting assignments for the bovine casein kinase II alpha (CSNK2A2) gene. Anim Genet. 31: p. 131-134.

Adachi, T., J.Y. Ahn, K. Yamamoto, N. Aoki, R. Nakamura and T. Matsuda, 1996. Characterization of the bovine kappa-casein gene promoter. Biosci. Biotechnol. Biochem. 60: p. 1937-1940.

Ahn, J.Y., N. Aoki, T. Adachi, Y. Mizuno, R. Nakamura and T. Matsuda, 1995. Isolation and culture of bovine mammary epithelial cells and establishment of gene transfection conditions in the cells. Biosci. Biotechnol. Biochem. 59: p. 59-64.

Andersson, L. and L. Rask, 1988. Characterization of the MHC class II region in cattle. The number of DQ genes varies between haplotypes. Immunogenetics 27: p. 110-120.

Ashwell, M.S., D.W. Heyen, T.S. Sonstegard, C.P. Van Tassell, Y. Da, P.M. VanRaden, M. Ron, J.I. Weller and H.A. Lewin, 2004. Detection of quantitative trait loci affecting milk production, health and reproductive traits in Holstein cattle. J. Dairy Sci. 87: p. 468-475.

Ashwell, M.S. and C.P.Van Tassell, 1999. Detection of putative loci affecting milk, health and type traits in a US Hostein population using 70 microsatellite markers in a genome scan. J. Dairy Sci. 82: p. 2497-2502.

Barendse, W., D. Vaiman, S.J. Kemp, Y. Sugimoto, S.M. Armitage, J.L. Williams, H.S. Sun, A. Eggen, M. Agaba, S.A. Aleyasin, M. Band, M.D. Bishop, J. Buitkamp, K. Byrne, F. Collins, L. Cooper, W. Coppettiers, B. Denys, R.D. Drinkwater, K. Easterday, C. Elduque, S. Ennis, G. Erhardt *et al.*, 1997. A medium-density genetic linkage map of the bovine genome. Mamm. Genome 8: p. 21-28.

Bennewitz, J., N. Reinsch, S. Paul, C. Looft, B. Kaupe, C. Weimann, G. Erhardt, G. Thaller, Ch. Kühn, M. Schwerin, H. Thomsen, F. Reinhardt, R. Reents and E. Kalm, 2004. The DGAT1 K232A mutation is not solely responsible for the milk production Quantitative Trait Locus on the bovine chromosome 14. J. Dairy Sci. 87: p. 431-442.

Bishop, M.D., A. Tavakkol, D.W. Threadgill, F.A. Simmen, R.C. Simmen, M.E. Davis and J.E. Womack, 1991. Somatic cell mapping and restriction fragment length polymorphism analysis of bovine insulin-like growth factor I. J. Anim Sci. 69: p. 4306-4311.

Bleck, G.T., B.R. White, D.J. Miller and M.B. Wheeler, 1998. Production of bovine alpha-lactalbumin in the milk of transgenic pigs. J. Anim Sci. 76: p. 3072-3078.

Bobe, G., D.C. Beitz, A.E. Freeman and G.L. Lindberg, 1999. Effect of milk protein genotypes on milk protein composition and its genetic parameter estimates. J. Dairy Sci. 82: p. 2797-2804.

Boutinaud, M. and H. Jammes, 2002. Potential uses of milk epithelial cells: a review. Reprod. Nutr. Dev. 42: p. 133-147.

Brophy, B., G. Smolenski, T. Wheeler, D. Wells, P. L'Huillier and G. Laible, 2003. Cloned transgenic cattle produce milk with higher levels of beta-casein and kappa-casein. Nat. Biotechnol. 21: p. 157-162.

Brunner, R.M., G. Guerin, T. Goldammer, H. Seyfert and M. Schwerin, 1996. The bovine butyrophilin encoding gene (BTN) maps to chromosome 23. Mamm. Genome 7: p. 635-636.

Cerdan, M.G., J.I. Young, E. Zino, T.L. Falzone, V. Otero, H.N. Torres and M. Rubinstein, 1998. Accurate spatial and temporal transgene expression driven by a 3.8-kilobase promoter of the bovine beta-casein gene in the lactating mouse mammary gland. Mol. Reprod. Dev. 49: p. 236-245.

Debeljak, M., S. Susnik, T. Milosevic-Berlic, J.F. Medrano and P. Dovc, 2000. Gene Technology and Milk Production. Food technol. biotechnol 38: p. 83-89.

Ehrmann, S., H. Bartenschlager and H. Geldermann, 1997. Polymorphism in the 5' flanking region of the bovine lactoglobulin encoding gene and its association with beta-lactoglobulin in the milk. J. Anim. Breed. Genet. 114: p. 49-53.

Farnir, F., B. Grisart, W. Coppieters, J. Riquet, P. Berzi, N. Cambisano, L. Karim, M. Mni, S. Moisio, P. Simon, D. Wagenaar, J. Vilkki and M. Georges, 2002. Simultaneous mining of linkage and linkage disequilibrium to fine map quantitative trait loci in outbred half-sib pedigrees: revisiting the location of a quantitative trait locus with major effect on milk production on bovine chromosome 14. Genetics 161: p. 275-287.

FitzGerald, R.J., D. Walsh, T.P. Guinee, J.J. Murphy, R. Mehra, D. Harrington and J.F. Connolly, 1999. Genetic Variants of Milk Proteins - Relevance to Milk Composition and Cheese Production. Dairy Prod. Res. Cent. p. 1-11.

Folch, J.M., P. Dovc and J.F. Medrano, 1999. Differential expression of bovine beta-lactoglobulin A and B promoter variants in transiently transfected HC11 cells. J. Dairy Res. 66: p. 537-544.

Freyer, G., P. Sorensen, C. Kuhn, R. Weikard and I. Hoeschele, 2003. Search for pleiotropic QTL on chromosome BTA6 affecting yield traits of milk production. J. Dairy Sci. 86: p. 999-1008.

Gallagher, D.S., C.P. Schelling, M.M. Groenen and J.E. Womack, 1994. Confirmation that the casein gene cluster resides on cattle chromosome 6. Mamm. Genome 5: p. 524.

Geldermann, H., J. Gogol, M. Kock and G. Tacea, 1996. DNA variants within the 5'-flanking region of bovine milk protein encoding genes. J. Anim. Breed. Genet. 113: p. 261-267.

Geldermann, H., U. Pieper and B. Roth, 1985. Effects of marked chromosome sections on milk performance in cattle. Theor. Appl. Genet. 70: p. 138-146.

Grisart, B., W. Coppieters, F. Farnir, L. Karim, C. Ford, P. Berzi, N. Cambisano, M. Mni, S. Reid, P. Simon, R. Spelman, M. Georges and R. Snell, 2002. Positional candidate cloning of a QTL in dairy cattle: identification of a missense mutation in the bovine DGAT1 gene with major effect on milk yield and composition. Genome Res. 12: p. 222-231.

Groner, B. and L. Hennighausen, 2000. Linear and cooperative signaling: roles for Stat proteins in the regulation of cell survival and apoptosis in the mammary epithelium. Breast Cancer Res. 2: p. 149-153.

Gutierrez-Adan, A., E.A. Maga, E. Behboodi, J.S. Conrad-Brink, A.G. Mackinlay, G.B. Anderson and J.D. Murray, 1999. Expression of bovine beta-lactoglobulin in the milk of transgenic mice. J. Dairy Res. 66: p. 289-294.

Hallerman, E.M., J.L. Theilmann, J.S. Beckmann, M. Soller and J.E. Womack, 1988. Mapping of bovine prolactin and rhodopsin genes in hybrid somatic cells. Anim Genet. 19: 123-131

Hayes, H., C. Le Chalony, G. Goubin, D. Mercier, E. Payen, C. Bignon and K. Kohno, 1996. Localization of ZNF164, ZNF146, GGTA1, SOX2, PRLR and EEF2 on homoeologous cattle, sheep and goat chromosomes by fluorescent in situ hybridization and comparison with the human gene map. Cytogenet. Cell Genet. 72: p. 342-346.

Hayes, H.C. and E.J. Petit, 1993. Mapping of the beta-lactoglobulin gene and of an immunoglobulin M heavy chain-like sequence to homologous cattle, sheep and goat chromosomes. Mamm. Genome 4: p. 207-210.

Hediger, R., S.E. Johnson, W. Barendse, R.D. Drinkwater, S.S. Moore and J. Hetzel, 1990. Assignment of the growth hormone gene locus to 19q26-qter in cattle and to 11q25-qter in sheep by in situ hybridization. Genomics 8: p. 171-174.

Heyen, D.W., J.I. Weller, M. Ron, M. Band, J.E. Beever, E. Feldmesser, Y. Da, G.R. Wiggans, P.M. VanRaden and H.A. Lewin, 1999. A genome scan for QTL influencing milk production and health traits in dairy cattle. Physiol. Genomics 1: p. 165-175.

Kaminski, S., 2000. Associations between polymorphism within regulatory and coding fragments of bovine kappa-casein gene and milk performance traits. J. Anim. Feed. 9: p. 73-79.

Kaminski, S. and T. Zabolewicz, 2000. Associations between bovine beta-lactoglobulin polymorphism within coding and regulatory sequences and milk performance traits. J. Appl. Genet. 41: p. 91-99.

Kang, Y.K., C.S. Lee, A.S. Chung and K.K. Lee, 1998. Prolactin-inducible enhancer activity of the first intron of the bovine beta-casein gene. Mol. Cells 8: p. 259-265.

Kühn, C., J. Bennewitz, N. Reinsch, N. Xu, H. Thomsen, C. Looft, G.A. Brockmann, M. Schwerin, C. Weimann, S. Hiendleder, G. Erhardt, I. Medjugorac, M. Forster, B. Brenig, F. Reinhardt, R. Reents, I. Russ, G. Averdunk, J. Blumel and E. Kalm, 2003. Quantitative trait loci mapping of functional traits in the German Holstein cattle population. J. Dairy Sci. 86: p. 360-368.

Kuss, A.W., J. Gogol, H. Bartenschlager and H. Geldermann, 2004. Polymorphic AP-1 binding site in bovine CASAS1 shows quantitative differences in protein binding which are associated with milk protein expression. J. Dairy Sci. submitted.

Kuss, A.W., J. Gogol and H. Geldermann, 2003. Associations of a polymorphic AP-2 binding site in the 5'-flanking region of the bovine beta-lactoglobulin gene with milk proteins. J. Dairy Sci. 86: p. 2213-2218.

Lasa-Benito, M., O. Marin, F. Meggio and L.A. Pinna, 1996. Golgi apparatus mammary gland casein kinase: monitoring by a specific peptide substrate and definition of specificity determinants. FEBS Lett. 382: p. 149-152.

Lum, L.S., P. Dovc and J.F. Medrano, 1997. Polymorphisms of bovine beta-lactoglobulin promoter and differences in the binding affinity of activator protein-2 transcription factor. J. Dairy Sci. 80: p. 1389-1397.

Lunden, A., S. Marklund, V. Gustafsson and L. Andersson, 2002. A nonsense mutation in the FMO3 gene underlies fishy off-flavor in cow's milk. Genome Res. 12: p. 1885-1888.

Maciulla, J.H., H.M. Zhang and S.K. DeNise, 1997. A novel polymorphism in the bovine insulin-like growth factor binding protein-3 (IGFBP3) gene. Anim Genet. 28: p. 375.

Mao, J., S. Marcos, S.K. Davis, J. Burzlaff and H.M. Seyfert, 2001. Genomic distribution of three promoters of the bovine gene encoding acetyl-CoA carboxylase alpha and evidence that the nutritionally regulated promoter I contains a repressive element different from that in rat. Biochem. J. 358: p. 127-135.

Martin, P., M. Szymanowska, L. Zwierzchowski and C. Leroux, 2002. The impact of genetic polymorphisms on the protein composition of ruminant milks. Reprod. Nutr. Dev. 42: p. 433-459.

Meggio, F., A.P. Boulton, F. Marchiori, G. Borin, D.P. Lennon, A. Calderan and L.A. Pinna, 1988. Substrate-specificity determinants for a membrane-bound casein kinase of lactating mammary gland. A study with synthetic peptides. Eur. J. Biochem. 177: p. 281-284.

Moody, D.E., D. Pomp and W. Barendsc, 1996. Linkage mapping of the bovine insulin-like growth factor-1 receptor gene. Mamm. Genome 7: p. 168-169.

Moody, D.E., D. Pomp, W. Barendse and J.E. Womack, 1995. Assignment of the growth hormone receptor gene to bovine chromosome 20 using linkage analysis and somatic cell mapping. Anim Genet. 26: p. 341-343.

Moore, S.S., C. Hansen, J.L. Williams, A. Fu, Y. Meng, C. Li, Y. Zhang, B.S. Urquhart, M. Marra, J. Schein, B. Benkel, P.J. de Jong, K. Osoegawa, B.W. Kirkpatrick and C.A. Gill, 2003. A comparative map of bovine chromosome 19 based on a combination of mapping on a bacterial artificial chromosome scaffold map, a whole genome radiation hybrid panel and the human draft sequence. Cytogenet. Genome Res. 102: p. 32-38.

Mosig, M.O., E. Lipkin, G. Khutoreskaya, E. Tchourzyna, M. Soller and A. Friedmann, 2001. A whole genome scan for quantitative trait loci affecting milk protein percentage in Israeli-Holstein cattle, by means of selective milk DNA pooling in a daughter design, using an adjusted false discovery rate criterion. Genetics 157: p. 1683-1698.

Ng-Kwai-Hang, K.F., H.G. Monardes and J.F. Hayes, 1990. Assciation between genetic polymorphism of milk proteins and production traits during three lactations. J. Dairy Sci. 73: p. 3414-4320.

Oh, K.B., Y.H. Choi, Y.K. Kang, W.S. Choi, M.O. Kim, K.S. Lee, K.K. Lee and C.S. Lee, 1999. A hybrid bovine beta-casein/bGH gene directs transgene expression to the lung and mammary gland of transgenic mice. Transgenic Res. 8: p. 307-311.

Ojala, M., T.R. Famula and J.F. Medrano, 1997. Effects of milk protein genotypes on the variation for milk production traits of Holstein and Jersey cows in California. J. Dairy Sci. 80: p. 1776-1785.

Olsen, H.G., L. Gomez-Raya, D.I. Vage, I. Olsaker, H. Klungland, M. Svendsen, T. Adnoy, A. Sabry, G. Klemetsdal, N. Schulman, W. Kramer, G. Thaller, K. Ronningen and S. Lien, 2002. A genome scan for quantitative trait loci affecting milk production in Norwegian dairy cattle. J. Dairy Sci. 85: p. 3124-3130.

Pfister-Genskow, M., H. Hayes, A. Eggen and M.D. Bishop, 1996. Chromosomal localization of the bovine obesity (OBS) gene. Mamm. Genome 7: p. 398-399.

Pfister-Genskow, M., H. Hayes, A. Eggen and M.D. Bishop, 1997. The leptin receptor (LEPR) gene maps to bovine chromosome 3q33. Mamm. Genome 8: p. 227.

Plante, Y., J.P. Gibson, J. Nadesalingam, H. Mehrabani-Yeganeh, S. Lefebvre, G. Vandervoort and G.B. Jansen, 2001. Detection of quantitative trait loci affecting milk production traits on 10 chromosomes in Holstein cattle. J. Dairy Sci. 84: p. 1516-1524.

Prinzenberg, E.M., C. Weimann, H. Brandt, J. Bennewitz, E. Kalm, M. Schwerin and G. Erhardt, 2003. Polymorphism of the bovine CSN1S1 promoter: linkage mapping, intragenic haplotypes and effects on milk production traits. J. Dairy Sci. 86: p. 2696-2705.

Rijnkels, M., P.M. Kooiman, P.J. Krimpenfort, H.A. de Boer and F.R. Pieper, 1995. Expression analysis of the individual bovine beta-, alpha s2- and kappa-casein genes in transgenic mice. Biochem. J. 311 (Pt 3): p. 929-937.

Rijnkels, M., P.M. Kooiman, G.J. Platenburg, M. van Dixhoorn, J.H. Nuijens, H.A. de Boer and F.R. Pieper, 1998. High-level expression of bovine alpha s1-casein in milk of transgenic mice. Transgenic Res. 7: p. 5-14.

Riquet, J., W. Coppieters, N. Cambisano, J.J. Arranz, P. Berzi, S.K. Davis, B. Grisart, F. Farnir, L. Karim, M. Mni, P. Simon, J.F. Taylor, P. Vanmanshoven, D. Wagenaar, J.E. Womack and M. Georges, 1999. Fine-mapping of quantitative trait loci by identity by descent in outbred populations: application to milk production in dairy cattle. Proc. Natl. Acad. Sci. U. S. A 96: p. 9252-9257.

Rodriguez-Zas, S.L., B.R. Southey, D.W. Heyen and H.A. Lewin, 2002a. Detection of quantitative trait loci influencing dairy traits using a model for longitudinal data. J. Dairy Sci. 85: p. 2681-2691.

Rodriguez-Zas, S.L., B.R. Southey, D.W. Heyen and H.A. Lewin, 2002b. Interval and composite interval mapping of somatic cell score, yield and components of milk in dairy cattle. J. Dairy Sci. 85: p. 3081-3091.

Ron, M., D. Kliger, E. Feldmesser, E. Seroussi, E. Ezra and J.I. Weller, 2001. Multiple quantitative trait locus analysis of bovine chromosome 6 in the Israeli Holstein population by a daughter design. Genetics 159: p. 727-735.

Roy, R., M. Gautier, H. Hayes, P. Laurent, R. Osta, P. Zaragoza, A. Eggen and C. Rodellar, 2001. Assignment of the fatty acid synthase (FASN) gene to bovine chromosome 19 (19q22) by in situ hybridization and confirmation by somatic cell hybrid mapping. Cytogenet. Cell Genet. 93: p. 141-142.

Schmidhauser, C., M.J. Bissell, C.A. Myers and G.F. Casperson, 1990. Extracellular matrix and hormones transcriptionally regulate bovine beta-casein 5' sequences in stably transfected mouse mammary cells. Proc. Natl. Acad. Sci. U.S.A 87: p. 9118-9122.

Schmidhauser, C., G.F. Casperson, C.A. Myers, K.T. Sanzo, S. Bolten and M.J. Bissell, 1992. A novel transcriptional enhancer is involved in the prolactin- and extracellular matrix-dependent regulation of beta-casein gene expression. Mol. Biol. Cell 3: p. 699-709.

Seyfert, H.M., C. Pitra, L. Meyer, R.M. Brunner, T.T. Wheeler, A. Molenaar, J.Y. McCracken, J. Herrmann, H.J. Thiesen and M. Schwerin, 2000. Molecular characterization of STAT5A- and STAT5B-encoding genes reveals extended intragenic sequence homogeneity in cattle and mouse and different degrees of divergent evolution of various domains. J. Mol. Evol. 50: p. 550-561.

Soulier, S., J.L. Vilotte, M.G. Stinnakre and J.C. Mercier, 1992. Expression analysis of ruminant alpha-lactalbumin in transgenic mice: developmental regulation and general location of important cis-regulatory elements. FEBS Lett. 297: p. 13-18.

Szymanowska, M., N. Strzalkowska, E. Siadkowska, J. Krzyzewski, Z. Ryniewicz and L. Zwierzchowski, 2003. Effects of polymorphism at 5'-noncoding regions (promoters) of alpha S1- and alpha S2-casein genes on selected milk production traits in Polish Balck-and-White cows. Animal Sciences Papers and Reports 21: p. 97-108.

Szymanowska, M., T. Malewski and L. Zwierzchowski, 2004. Transcription factor binding to variable nucleotide sequences in 5'-flanking regions of bovine casein genes. Int. Dairy J. 14: p. 103-115.

Threadgill, D.W. and J.E. Womack, 1990. Genomic analysis of the major bovine milk protein genes. Nucleic Acids Res. 18: p. 6935-6942.

Threadgill, D.W. and J.E. Womack, 1991. Synteny mapping of human chromosome 8 loci in cattle. Anim Genet. 22: p. 117-122.

Velmala, R.J., H.J. Vilkki, K.T. Elo, D.J. de Koning and A.V. Maki-Tanila, 1999. A search for quantitative trait loci for milk production traits on chromosome 6 in Finnish Ayrshire cattle. Anim Genet. 30: p. 136-143.

Viitala, S.M., N.F. Schulman, D.J. de Koning, K. Elo, R. Kinos, A. Virta, J. Virta, A. Maki-Tanila and J.H. Vilkki, 2003. Quantitative trait loci affecting milk production traits in Finnish Ayrshire dairy cattle. J. Dairy Sci. 86: p. 1828-1836.

Winter, A., W. Kramer, F.A. Werner, S. Kollers, S. Kata, G. Durstewitz, J. Buitkamp, J.E. Womack, G. Thaller and R. Fries, 2002. Association of a lysine-232/alanine polymorphism in a bovine gene encoding acyl-CoA:diacylglycerol acyltransferase (DGAT1) with variation at a quantitative trait locus for milk fat content. Proc. Natl. Acad. Sci. U.S.A 99: p. 9300-9305.

Yoshimura, M. and T. Oka, 1990. Hormonal induction of beta-casein gene expression: requirement of ongoing protein synthesis for transcription. Endocrinology 126: p. 427-433.

Zhang, Q., D. Boichard, I. Hoeschele, C. Ernst, A. Eggen, B. Murkve, M. Pfister-Genskow, L.A. Witte, F.E. Grignola, P. Uimari, G. Thaller and M.D. Bishop, 1998. Mapping quantitative trait loci for milk production and health of dairy cattle in a large outbred pedigree. Genetics 149: p. 1959-1973.

Current genomics in cattle and application to beef quality

J.F. Hocquette, I. Cassar-Malek, A. Listrat and B. Picard*
INRA, Herbivore Research Unit, Muscle Growth and Metabolism Group, Clermont-Ferrand/Theix Research Center, Theix, 63122 Saint-Genès Champanelle, France.
**Correspondence: hocquet@clermont.inra.fr*

Summary

Genetic and environmental factors profoundly alter muscle characteristics, and hence beef quality. The advent of high-throughput sequence analysis, DNA chip technology and protein analysis has revolutionised our approach to muscle physiology. For instance, global gene expression profiling at the mRNA or protein level will provide a better understanding of the mechanisms that underlie myogenesis and its control by nutrition. A powerful scientific strategy is based on genomics in combination with classic biochemical and physiological studies. This is facilitated by the use of bioinformatic tools to understand the control of a wide variety of phenotypes from genetic, environmental or nutritional origins. One of the main challenges will be to solve the problems posed by the analysis and interpretation of the large amount of data that will become available from structural (QTL, SNP) and functional (mRNA or protein levels) genomics. Major applications could be (i) the identification of new predictors of beef quality traits (for instance, tenderness and flavour), (ii) the monitoring of beef quality through the production systems (nutrition level, growth path, grass-feeding) for beef authentication, and (iii) the improvement of animal selection (markers and gene assisted selection) by including quality traits.

Keywords: beef quality, muscle physiology, genomics, indicators of quality

Introduction

Skeletal muscle is a tissue of major economic importance for meat production. Therefore, in the last century, research was conducted in some European countries to increase the growth rate of cattle and to produce lean carcasses with the ultimate objective of increasing production of meat at the expense of external fat. For instance, genetic selection in Belgium or France has been directed in favour of extreme muscle development and this has been indeed successful in increasing growth rates in beef cattle. Similarly, research was also successful with respect to adequately feeding animals so that they could fully express their higher genetic potential.

However, nowadays, consumers seek meat of high and consistent quality. The consequence is that today consistently providing the consumer with a high quality product is one of the most important issues facing beef producers and retailers. This is the reason why the beef industry is looking for indicators of quality as discussed in a separate article (Mullen and Troy, 2005).

Beef quality includes sensory quality traits (tenderness, flavour, juiciness, colour, etc), nutritional value, healthiness and technological quality (Geay *et al.*, 2001). Meat eating quality is affected both by the *post-mortem* processing treatments applied to the muscles, and by the anatomical, physical and chemical characteristics of the muscles themselves. The latter depend upon the genetic potential of the animals and the production systems. However, understanding the multi-dimensional nature of eating quality traits, such as tenderness for instance (Maltin *et al.*, 2003), remains a subject of active investigation. Classic biochemical studies of the muscle tissue demonstrated that only about 30% of the total variability in the final tenderness of beef could be explained by some muscle characteristics (Dransfield *et al.*, 2003). Therefore, more research is needed to identify new muscle properties, which may influence meat quality traits in order to control a greater part of their variability. These new putative and informative muscle characteristics would be useful in selecting and classifying animals from the best production systems to ensure good quality beef.

This paper aims to illustrate how we could take advantage of the recent development of genomics combined with muscle biochemistry to achieve this objective.

"Genomics": a new and promising scientific field

History and definitions of genomics

The advent of high-throughput analyses of gene sequence and expression has revolutionised our approach to genetics and physiology (Eggen, 2003), and in particular to muscle physiology (Eggen and Hocquette, 2004).

Firstly, improvements in the efficiency of DNA sequencing have allowed the recent sequencing of almost the complete human genome. Sequencing of genomes from other species has been achieved or is in progress. This allows for a better understanding of the genome structure. Secondly, finding single nucleotide polymorphisms is much easier than previously possible since they can be considered as a by-product of genome sequencing projects. This will help to identify genetic markers which will be useful for animal breeding. Although highly relevant, the application of genomics in genetics is being dealt with in a separate article (Kühn *et al.*, 2005). But "genomics" is not simply high-throughput genetics as indicated by Goddard (2003), it also includes what is called physiological genomics (Eggen, 2003).

Indeed, other high-throughput techniques have been developed (Figure 1). These techniques have arisen as a result of manual procedures having been replaced by robotics. Automation/miniaturization has increased the speed and the accuracy of laboratory analysis. For example, the expression of thousands of genes can be studied simultaneously by quantifying the levels of their transcripts with microarrays (Jordan, 1998). This large-scale analysis of mRNA levels is called "transcriptomics".

Furthermore, "proteomics" allows the analysis of hundreds/thousands of proteins simultaneously thanks to the development of automated mass spectrometry (Figure 1). Improvements in proteomics are of paramount importance, since the biological functions of genes operate through by proteins (Matthews, 2001). Similarly, "Metabolomics" is large-scale analysis of cellular/tissue metabolism (Figure 1). Therefore, "genomics" is nowadays a new scientific field midway between genetics and physiology.

Whereas the goal of genetics is to study the characteristics of the genome, which are transmissible to new generations, "Genomics" addresses the structure and function of genes. "Genomics" is only possible by large-scale analyses of gene characteristics at the DNA (structural genomics) and mRNA levels (expressional genomics). Initially defined as the effort to catalogue proteins expressed within tissues, the definition of "proteomics" has also evolved since it now includes the systemic study of protein cellular location, post-translational modifications and interactions between proteins (Auerbach *et al.*, 2003).

So, we can conclude that the technologies related to genomics are relatively new, close to physiology, and clearly have more of a future than a past.

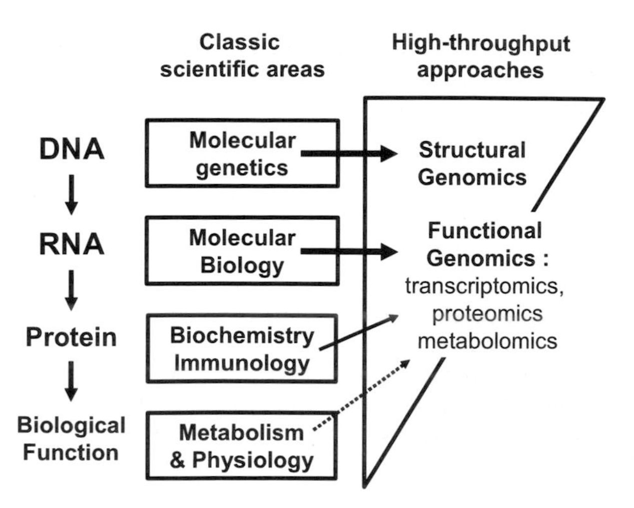

Figure 1. What is genomics? Classic scientific areas address studies at the DNA, RNA, protein and biological function levels by independent approaches (molecular genetics, molecular biology, biochemistry, metabolism), which do not interact. Structural and Functional Genomics address the same issues by high-throughput techniques. The available "large-scale" approaches (structural genomics, transcriptomics) allow integration of knowledge from genes to transcripts. In the future, other high-throughput approaches (proteomics and metabolomics) will be more developed which will be useful to integrate knowledge from genes to proteins and biological functions.

Practical aspects, difficulties and limits of functional genomics

Techniques of functional genomics include a great number of methods for gene expression profiling (Moody, 2001) at the transcript or protein levels. DNA arrays are nowadays the most popular approaches to transcriptomic studies. As a representative example, we will here discuss advantages and limits of this approach since many of them can be extrapolated to other large-scale techniques including those of proteomics. However, proteomics, which is another emerging technology of interest, is being discussed in a separate article (Bendixen *et al.*, 2005).

Array technology is based on the principle that probes (cDNA from EST libraries or oligonucleotides) can be arrayed on membranes or glass slides for hybridisation with labelled targets, which originate from the biological samples of interest. The flatness of the glass support makes it possible to miniaturise the procedure. The resulting images of hybridisation are captured and analysed using software that quantifies the signal of each spot. The intensity of each spot is proportional to the amount of mRNA of the corresponding gene expressed in the studied sample. Thus, it is possible to survey thousands of genes in parallel. The two major applications are (i) identification of differentially expressed genes between different biological samples and (ii) expression monitoring in different physiological conditions. Whereas the first application is well known, we must underline that understanding patterns of expressed genes is expected to improve our knowledge of the highly complex networks in various physiological and pathological situations, especially in muscles. Even if this procedure appears theoretically simple, many problems have to be solved, and many scientists working in Animal Science are not aware of them.

A first set of problems is the availability of biological molecular material. Indeed, unlike analysis of the human genome, the development of genome analysis of the livestock species is not well advanced (Kutzer *et al.*, 2003). Therefore, some groups have performed cross-species hybridisation studies. For instance, skeletal muscle (Sudre *et al.*, 2003a) and oocyte (Dalbiès-Tran and Mermillod, 2003) RNAs from cattle have been hybridised on human cDNA microarrays. But, cross-species hybridisation may generate spurious results. Nevertheless, in spite of this loss of information, these approaches were successful provided that researchers paid careful attention to data analysis. The recent availability of bovine expressed sequence tags from liver and intestine (Dorroch *et al.*, 2001; Herath *et al.*, 2004), embryos (Potts *et al.*, 2003; Renard *et al.*, 2004), uterus and ovaries (Anderson *et al.*, 2004) and several pooled tissues (Smith *et al.*, 2001) as well as the availability of different bovine cDNA arrays (Moody *et al.*, 2003) do not justify this type of strategy any more. But these molecular tools are not well adapted for monitoring beef quality due to the lack of muscle-specific genes of physiological importance. This is the reason why arrays have been constructed from cDNA libraries derived from bovine muscles (Sudre *et al.*, 2005) or muscle and adipose tissue (Reverter *et al.*, 2003; Cho *et al.*, 2002). However, the number of array elements was much lower than that derived from multiple tissues (Suchyta *et al.*, 2003). Nowadays, different groups promote oligoarrays. Once the bovine genome is sequenced, this strategy will be the best to produce generic arrays.

A second set of problems covers the specific technical difficulties. For instance, a great proportion of the cDNAs printed on arrays cannot be hybridised with labelled cDNA targets (reverse-transcribed from tissue RNA). This may be due to low expression of the corresponding genes in the examined tissues, to low quality of some cDNAs or to technical problems related to the hybridisation. About 20-40% of positive signals have been indeed reported in skeletal muscles with various arrays and a small proportion of them generally correspond to differentially expressed genes (Hampson and Hughes, 2001). In addition, several technical problems (Murphy, 2002) and artefacts (Drobyshev *et al.*, 2003) have been reported, thus the technology needs to be improved. Part of these difficulties originate from the image analyse procedure (Wang *et al.*, 2001) and data analysis including normalisation and statistics (Törönen *et al.*, 2003). Consequently, results from oligonucleotide and cDNA-based microarray systems may differ at least in part due to spurious signals, to technical difficulties and to the principles of the different methodologies (Mah *et al.*, 2004).

It is now accepted that the quality (Fischer, 2002) and the comparison of results from different laboratories require standardisation procedures for both transcriptomic (Brazma *et al.*, 2001) and proteomic (Taylor *et al.*, 2003) analysis. However, this is obviously difficult to define. Secondly, any specific gene declared as differentially expressed can be regarded as provisional until it can be confirmed by independent methods. Lastly, important biological results have been obtained recently by using genomic approaches despite the technical difficulties listed above. Some examples will be described later on.

Increasing usefulness of bio-informatics

Molecular genetics, genomics and proteomics all take advantage of the huge amount of data that arises from sequencing of genomes, DNA or proteins. In addition, the high-throughput techniques of genomics generate tremendous amounts of results in their own right. Consequently, powerful data management tools and computational techniques are required now more than ever to store, share, analyse, and compare this enormous amount of biological information. Bioinformatics combines the tools of mathematics, computer science and biology with the aim of associations and interpretation within and between sets of data of various origins (genomics, physiology, etc).

For the past 20 years, DNA sequence information has been entered into public databases. The number of expressed sequenced tags in GeneBank has now surpassed 500,000 for cattle. Other specific bioinformatic databases have been developed. One problem is the lack of fidelity due to the huge amounts of information, which makes it difficult to check and collate on a large-scale (Halgren *et al.*, 2001). Therefore, the use of sequence verified cDNA clones for the production of arrays is strongly warranted. A major application of existing databases is the comparison of DNA sequences from various species, which is powerful for inferring genome function and deciphering genome evolution. Computer programs have been developed to achieve this objective (Thomas *et al.*, 2002; Harhay and Keele, 2003). Some of the existing databases and associated bioinformatic tools are devoted specifically to applications within livestock species (Hu *et al.*, 2001; http://www.ark-genomics.org/)

A second group of bioinformatic tools are software programs developed for the analysis of data, which originate from arrays and proteomics. Three major sources of variability have been identified: technical, physiological, and sampling variability (Novak *et al.*, 2002). Unfortunately, replications of analysis or of individuals is often hindered by financial or technical constraints. Therefore, a strategy of pooling samples may be chosen (Peng *et al.*, 2003). It is now clear that technical variability is so high that at least three identical arrays are needed to reduce the rate of misleading conclusions (Lee *et al.*, 2000; Wernisch *et al.*, 2003). Statistics in transcriptomics is also a great matter of research (Lee and Whitmore, 2002; Cui and Churchill, 2003; Reverter *et al.*, 2003) and specific software packages for statistical analysis of arrays have been developed (Tusher *et al.*, 2001; Wu *et al.*, 2002). Scatterplot, principal component analysis, and clustering methods are also very useful to interpret patterns of gene expression (Kalocsai and Shams, 2001).

Bioinformatic tools are also of paramount importance for sets of data from different origins (genetics, genomics, biology). This is the most difficult, but the most interesting part of bioinformatics. A well-known example was modelling metabolic reprogramming of yeast accompanying the metabolic shift from fermentation to respiration (DeRisi *et al.*, 1997). The combination of proteomic and transcriptomic approaches may also be useful to understand metabolic networks as for bacteria (Eymann *et al.*, 2002). With bioinformatic-based methods only, we can characterize a novel protein by using sequence data or by searching in the genome for possible protein matches (Matthews, 2001). However, the combination of genomic and physiological data is obviously much easier for model species than for poorly studied mammals such as cattle. Secondly, phenotypes of evolved and therefore complex animals are much more difficult to describe and to understand. Thirdly, the manipulation of individuals with a putative interesting gene detected by genomics will be easier with small animals than with larger ones. However, despite these difficulties, models of physiological characteristics have been developed by using microarray data and previous physiological knowledge (Winslow and Boguski, 2003). Therefore, genomics and integration of knowledge do not have to be limited to a few select organisms (Crawford, 2001).

To summarize, thanks to the development of Genomics, high-throughput techniques are contributing to the current shift from research on individual genes with major biological effects to the investigation of the "expression signature" and, therefore, the elucidation of comprehensive functional networks. Research is also shifting from the genes themselves to their products (the proteins). But, so far, only a limited number of studies have been performed in cattle (Table 1) including the identification of indicators of beef quality.

"Genomics" applied to bovine muscle tissue

While DNA-based techniques are best suited for the detection of bacterial contaminations (for beef safety), species identification (for racial traceability), and genetic selection (for quality improvement across generations), functional genomics will have major applications with respect to the characterization of species, the tissue (especially the muscle type and hence the beef cut),

Table 1. Examples of some major experimental results in functional genomics in cattle which are currently available in the literature.

Technical approach	Identification of genes differentially expressed:	Literature
• Array technology	Following infection by pathological organisms.	Reviewed by Moody, Rosa and Reecy (2003).
• Array technology	Between animals resistant or susceptible to diseases or parasite infections.	Reviewed by Moody, Rosa and Reecy (2003).
• Array technology	In placenta and uterine tissues between non-pregnant and pregnant cows.	Ishiwata *et al.*, (2003).
• Array technology	In mammary gland between non-lactating and lactating cows.	Suchyta *et al.*, (2004).
• Array technology	In blood neutrophils of parturient cows.	Madsen *et al.*, (2004).
• Array technology	In bovine oocytes during maturation.	Dalbiès-Trand and Mermillod (2003).
• Transcriptomics, Proteomics	Between different tissues and organs.	Suchyta *et al.*, (2003); Cho *et al.*, (2002); Talamo *et al.*, (2003).
• Differential RNA display	Between different genetic types.	Dorroch *et al.*, (2001).
• Array technology	During muscle ontogenesis.	Sudre *et al.*, (2003b).
• SSH and cDNA array, proteomics	Between normal double-muscled bovines.	Potts *et al.*, (2003); Bouley *et al.*, (2005).
• Differential RNA display-PCR	During development of bovine intramuscular adipocytes.	Childs *et al.*, (2002).
• Array technology	Between different feeding regimes.	Reverter *et al.*, (2003).

the health, physiological and nutritional status of the animals according the place where they are bred (Table 1). Thus, genomics has many applications in food authentication (Martinez and Friis, 2004) in addition to those in genetics (Goddard, 2003). But the first applications of genomics in muscle physiology were historically the discovery of novel genes specifically expressed in muscles and not in other tissues (Piétu *et al.*, 1996). We must emphasize that, unlike transcriptomics, proteomics allows the study of *post-mortem* processes that may be relevant to meat quality as described in pigs (Lametsch *et al.*, 2002).

The main function of skeletal muscle is postural support and movement. In farm animals, it is also converted into meat for human consumption. Skeletal muscle is composed of muscle fibres with different functional properties optimised for different tasks. The diversity of skeletal muscle fibres arises from (i) the myofibrillar apparatus, for instance the myosin heavy chain isoforms (ii) the sarcolemma and the sarcoplasmic reticulum in which many calcium and ion channels are present (thereby inducing differences in electrical membrane properties), and (iii) the metabolic systems devoted to the production of free energy (i.e. ATP) (Bottinelli and Reggiani, 2000). The three major muscle fibre types are called type I (slow oxidative), IIA (fast oxydo-glycolytic) and IIX and IIB (fast glycolytic). The existence of different fibre types is one of the main determinants of muscle performance *in vivo*. Slow fibres are indeed used for posture control and for long-duration contractions, whereas fast fibres are used for powerful and short contractions.

We will here illustrate firstly one general application of genomics: the identification of genes specifically expressed in red or white muscles. Furthermore, we will detail applications of genomics (i) in genetics and (ii) in animal husbandry and nutrition of bovines.

Comparison of muscle types

The first aim when looking at gene expression in various muscle types is clearly a new attempt to understand, on a genomic scale, the expression profiles that underline the broad skeletal muscle classes or allotypes. This is a basic research objective. The second aim is to get a better knowledge of the biological mechanisms, which control nutrient partitioning and hence body composition (Cortright *et al.*, 1997). This may be of practical use in Medical Science to control body composition of human beings, and in Animal Science to control body composition of farm animals. Another aim in Animal Science is to get a better understanding of muscle characteristics, which determine meat quality traits since the major factor of variability of tenderness is the muscle type rather than cattle breed or animal type (heifer, young bulls or cull cows) (Dransfied *et al.*, 2003). Furthermore, some beef quality indicators, such as shear force, which is an indication of tenderness, are poorly correlated between different muscles types (Shackelford *et al.*, 1995). Therefore, understanding differences between muscle types may also be of practical use in Meat Science.

In mice, microarray analysis identified 49 mRNA sequences that were differentially expressed between white and red skeletal muscles, including newly identified genes. These genes can be classified into different biological functions: energy metabolism (14 genes), transcription factors and regulators (10 genes), contractile structure (7 genes), Ca^{2+} homeostasis and signalling (4 genes) (Campbell *et al.*, 2001). Not surprisingly, the majority of the differentially expressed genes were classified as metabolic genes, which confirms that differences in the ATP-generating process are important between muscle types. Similarly, in pigs, 70 genes were declared as over-expressed in red muscles and 47 of them were of mitochondrial origin. Among the others, some genes were involved in various cell functions (gluconeogenesis, transcription, translation, signal transduction) or were related to unknown biological function. On the other hand, 45 clones were over-expressed in white muscles, such as the fast isoforms of sarcomeric proteins (myosin heavy chain 2a, 2x, 2b, myosin regulatory light chain, fast-troponins C and T3, etc) and genes involved in glycolysis (such as glyceraldehyde 3-phosphate dehydrogenase) (Bai *et al.*, 2003).

Differences between bovine muscle types were also compared by using human macroarrays (Sudre *et al.*, 2003a and b). Although little differences were detected between muscles as these arrays were initially prepared to explore muscular dystrophy in humans, some genes of interest for muscle biology were revealed. Some of them were indeed linked to the contractile (nebulin, actinin-associated LIM protein) or metabolic (ICDH β-subunit precursor) properties of muscles. Two interesting differentially expressed genes were LEU5 and Trip 15. LEU5 is a tumour suppressor gene associated with B-cell chronic lymphocytic leukemia. Trip 15 is a thyroid receptor interacting protein. They are probably involved in the regulation of muscle development. The differences between muscle types appeared to depend on the stage of development and the genetic types of animals (Sudre *et al.*, 2003a and 2003b). This confirms biochemical studies (Sudre *et al.*, 2003b; Picard *et al.*, 2002).

This expanding knowledge of diversity in gene expression between muscle types underlines the complex determinism of muscle phenotypes. Similar studies should also be performed to detect the specificity of intramuscular adipose tissue (which is important for beef flavour) in comparison to other fat tissues of the carcass undesirable for beef producers.

Improving beef quality by selecting animals

The genetic practical issues of genomics have been described elsewhere (Goddard, 2003; Renand *et al.*, 2003 and Smith *et al.*, 2003) and in a separate article within this publication (Kühn *et al.*, 2005). Only basic principles will thus be recalled here. Genetic improvement is a permanent and cumulative procedure, therefore investing in selection can be highly profitable for livestock producers (Liinamo and Neeteson-van Nieuwenhoven, 2003). The objective of breeders is to exploit the variability among animals by selecting those with a superior genetic potential for the trait of interest. However, the selected traits must be easy to measure on a great number of animals by low cost techniques and they must be sufficiently heritable (i.e. transmissible to the next generation).

One of the first problems with this approach is the high number of criteria, which have to be considered in order to improve beef quality. As indicated before, they may include tenderness, intramuscular fat content, colour, etc. Unfortunately, no routinely used biochemical methods are available to breeders to assess those meat quality traits. Another big problem may be the low phenotypic or genetic correlations of tenderness estimates between muscles, which suggests that genetic improvement of overall tenderness at the animal level may be difficult. Review of the scientific literature also indicates that genetic variation of many meat quality traits, especially tenderness, is of similar magnitude between breeds and within breeds (Burrow *et al.*, 2001). Furthermore, estimates of tenderness by taste panel or shear force measurement are only moderately heritable ($h^2 = 0.20$ to 0.25; Burrow *et al.*, 2001) whereas that of marbling may be higher (Bertrand *et al.*, 2001).

Taking all these difficulties into account, research is currently being conducted in order to identify genes that influence quality traits, and therefore that can be used to select animals. This is a promising future because genotyping is now easy to perform, and allelic variation is the genetic mechanism which induces variability in phenotypes. Clearly, one of the benefits of genomics will be a better knowledge of the genes of interest in an organism of interest by using genomic information from different species (Harhay and Keele, 2003).

The traditional approach to genetics is to map quantitative trait loci (QTL) to a chromosomal region using linkage and linkage disequilibrium. The region of interest will then be fine mapped by using new genetic markers in order to reduce the confidence interval that contains the genes of interest. The final step is the identification of the gene itself, which determines variability in the studied phenotype and of the causal mutation (Goddard, 2003). This approach takes advantage of the knowledge of potential functional candidate genes present in the short chromosome region studied. This strategy also takes advantage of the structural knowledge of the genome based on cytogenetics, genetics, radiation hybrids, physical and comparative maps of various genomes from different species (Harhay and Keele, 2003 ; Eggen and Hocquette, 2004).

One important benefit of high-throughput techniques of genomics will be the identification and the practical use of polymorphisms within DNA sequences of those genes (Garnier *et al.*, 2003). So far, polymorphisms in some genes have been identified as important for tenderness or beef marbling (i.e. the amount of visual intramuscular fat) (Barendse *et al.*, 2002a). They include the genes encoding m-calpaïn (Page *et al.*, 2002) and calpastatin (Barendse *et al.*, 2002b) for tenderness and also the genes encoding DGAT1 (Thaller *et al.*, 2003) for marbling. Some mitochondrial DNA substitutions were also found to be associated with *longissimus* muscle area and marbling score in Japanese Black cattle (Mannen *et al.*, 2003). However, applications, such as DNA-based tests of generic merit for beef quality traits, have been slow to develop (Smith *et al.*, 2003). But, it is speculated that further large gains are possible in the future, thanks to the identification of many

QTL and SNPs (single nucleotide polymorphisms), the decrease in the cost of genotyping and the increased knowledge of functional candidate genes, which will be discovered from functional genomics approaches (array data, proteomic studies) (Goddard, 2003). Indeed, a private company recently characterized a dense map of novel genetic markers based on SNPs in beef cattle. Novel genetic tests will be developed and commercialized to identify cattle bearing traits that satisfy consumer demands for consistency and quality in beef. These tools are expected to result in superior beef for consumers (http://www.cargill.com/today/releases/2004/04_06_07meta.htm). One of the other long-term aims of genome analysis in domestic animals is to obtain genetic screening tests that will improve the health and welfare by selective breeding (Kutzer *et al.*, 2003). In the future, SNP chips would probably be a tool of choice for genotyping animals.

Monitoring beef quality by using gene expression profiling in muscles

Functional genomics studies gene characteristics, which are either mutation-dependent or mutation-independent. In other words, because gene expression depends on both genetic and environmental factors, applications of genomics are for both breeders and farmers. But, clearly, functional genomics has a long way to go in cattle. Therefore, to address specific objective of Meat Science, scientists should take advantage of the technical and scientific progress, which has been made so far in yeast, bacteria, animal models and, more recently, in mammals, while designing specific experiments in the species of interest.

From a technical point of view, many improvements in array technologies or proteomics techniques have been made so far either for sample preparation, molecular hybridisation, protein separation or data analysis. These changes should be progressively incorporated into experimental design and genomic techniques applied to farm animals. Similarly, the construction of a bovine muscle protein map (Bouley *et al.*, 2004) is facilitated by previous knowledge of muscle maps from other species. Some more examples will be detailed here.

Firstly, genomics applied to beef should also exploit molecular resources from other species including muscle-specific arrays (Sudre *et al.*, 2003a and b). This allowed us to identify up to 110 genes differentially expressed from 110 days post-conception to 15 months of age after birth. Among these genes, 33% have unknown functions so far. Most of the differentially expressed genes are either up-regulated or down-regulated at 260 days of foetal development. This confirms the importance of the last three months of gestation in bovine muscle myogenesis (Picard *et al.*, 2002). Indeed, the number of muscle fibres is roughly fixed at two thirds of foetal development and muscle fibres differentiate before birth making cattle a relatively mature species at birth compared to other mammals (Picard *et al.*, 2002). Not only did the macroarray experiment confirm this general view, but it also helped in the discovery of putative interesting genes regulated throughout development and involved in cell homeostasis (e.g. α2,8-Sialyltransferase) or cell regulation (e.g. activin A, the thyroid receptor interacting protein 15) and in metabolic (e.g. oxoglutarate dehydrogenase) or contractile (e.g. nebulin) muscle properties (Sudre *et al.*, 2003a). Further studies are needed to explore the biological functions of the novel identified genes.

Secondly, some technical efforts are specifically needed for beef, in order to better address the scientific questions specific to meat quality. Two major reasons justify this strategy: (i) except for one of them (Sudre *et al.*, 2005), the available bovine libraries (Moody *et al.*, 2003) are not muscle specific and therefore are not suited to the objectives of monitoring beef quality, (ii) the quality and reliability of results arising from array experiments should be improved by using bovine probes. The muscle specific library was prepared from selected various muscle types (oxidative or glycolytic) sampled from different bovine animals of various age (foetal and post-natal) and different breeds. Among the 1440 cDNAs characterised so far, 1019 from several gene families could be exploited: mitochondrial genes (133), genes encoding ribosomal proteins (100), contractile proteins, metabolic enzymes, etc. Finally, 353 cDNAs from this library and 75 cDNAs individually prepared by RT-PCR were considered as a first bovine muscle cDNA repertoire. (Sudre *et al.*, 2005). This repertoire is currently used for monitoring beef quality.

From a scientific point of view, scientists involved in genomics should be able on one hand to exploit the results obtained in species, which are not of interest to them, and on the other hand design their own experimental designs to address specific questions.

Some relevant models of interest for growth and meat quality have been studied so far in species other than cattle. For instance, gene expression was studied during senescence in muscles from mice (Lee *et al.*, 1999) and later on from humans: 17 genes were shown to be regulated in a similar way in both species, but for 32 other genes, the effect of age is not the same (Welle *et al.*, 2001). In the field of beef, we should examine all the genes differentially expressed between muscle types whichever the species as described above. They are potentially of paramount importance to understand muscle biology. In addition, it is clear that the scientific objectives addressed by the Meat Industry are common whichever the species: all meat researchers aim to improve tenderness, flavour, healthiness even if the hierarchy of quality traits may differ (tenderness is more important for beef than for pork). As an example, a strict selection for increased growth rate and reduced fat content of the carcass had similar consequences on meat quality (for instance, a reduced intramuscular fat content) in pigs as in cattle (Garnier *et al.*, 2003).

One point of great importance is the choice of good animal experimental models for genomic studies. Indeed, variability within and between breeds must be exploited to detect, with confidence, important genes for quality traits. As an example, the comparison of divergent phenotypes for feeding conditions (Sudre *et al.*, 2003b; Reverter *et al.*, 2003) is highly relevant for farmers and nutritionists.

To address the consequences of selection for increased growth rate and reduced fat content of the carcass, transcriptomic analyses of two muscles (*rectus abdominis* (RA) oxidative and *semitendinosus* (ST) glycolytic) from young bulls divergently selected on muscle growth potential were performed. The selection process in favour of muscle growth decreased oxidative muscle activity especially in *rectus abdominis*, which is the most oxidative muscle of both. Some genes were differentially expressed in RA and/or ST between the two groups of bulls with high or low muscle growth potential. Many of them are involved in muscle structure (e.g. sarcosin, titin) or in cellular regulation (e.g. thyroid hormone receptor interacting protein 10, LEU5, heat shock protein 90a) and were more expressed in muscles from low muscle growth potential bulls (Sudre *et al.*, 2003b).

Similarly, proteomic analysis of bovine *semitendinosus* muscle was performed with the ultimate objective of identifying markers of muscle hypertrophy. To achieve this goal, muscles were compared between normal and double-muscled Belgian Blue bulls. The first results indicated that quantitative variations concern less than 5% of the analysed proteins. Out of the 26 identified proteins, 12 were under-expressed and 14-over-expressed in muscles from double-muscled animals. Among them, the troponin T slow skeletal muscle isoform was under-expressed and the phosphoglucomutase over-expressed (Bouley *et al.*, 2003). These results confirmed the shift of muscle characteristics towards the fast-glycolytic type in hypertrophied muscles. These new markers should be incorporated in the list of putative markers of interest for further studies at the DNA, RNA and protein levels.

In addition, some sectors of the beef industry are looking for gene markers that would identify animals that have a high propensity to accumulate intramuscular fat in order to produce a tasty meat. Some research aimed to obtain various degrees of intramuscular fat development by different finishing periods of high grain feeding (Childs *et al.* 2002). Differential-display polymerase chain reaction has allowed the identification of a known gene (NAT1 which is a translational suppressor) in this system, which was previously unsuspected to play a role in fat accumulation. Putative functional genes were found to be differentially expressed (*e.g.* ATP citrate lyase) or, surprisingly, not differentially expressed (*e.g.* PPARγ) between extreme animals (Childs *et al.* 2002). Other transcriptomic studies identified some genes (*e.g.* 12-lipoxygenase, prostaglandin D synthase) as key candidates involved in the control of intramuscular fat accumulation (Cho *et al.*, 2002). These examples underline again the gain in knowledge, which may be drawn from good experimental models and finally integrative biology from genes to phenotypes.

Conclusion

Expression profiling at the mRNA or protein levels has already provided valuable data sets in the field of muscle biology in various species with different ultimate objectives including characterisation of muscle types and meat quality traits. All these genes or proteins, that are important for muscle biology whatever the species, are potentially interesting candidates for improving our understanding of bovine muscle biology.

However, muscle tissue is a complex tissue with different cell types (multi-nucleated fibres, adipocytes, etc) in various proportions between muscles or even within the same muscle. This may induce a high sample-to-sample variability. In addition, further research is needed to investigate which cell population is responsible for any change in gene expression and to check functional relevance of the genes of interest, which will be discovered.

Nevertheless, the development of genomics, proteomics, etc, brings a huge potential for improving beef quality or producing differentiated products. Long-term applications in cattle breeding, husbandry and nutrition have to be considered. As for Medical Science, it can be anticipated that, in the long-term, individual prediction of beef quality could be achieved based on muscle-specific genomic data (e.g. the gene or protein “expression signature”) according to the production systems. Again, this first needs the most informative molecular markers to be identified. Moreover, we must also meet ethical, legal, environmental, consumer and any societal concerns. For instance, the general public perception should not confuse genomics with GMO (genetic modified organisms). So, we are at the beginning of a long road on both scientific and societal aspects. The success of this new research will also depend on our ability to integrate knowledge from different disciplines thanks to the development of bioinformatic tools.

References

Anderson, S.I., H.A. Finlayson and A.L. Archibald, 2004. Development of cDNA and EST resources for studying reproduction and embryo development in pigs and cattle. Accession Numbers AJ669550 to 697608. http://www.ncbi.nlm.nih.gov/entrez/query.fcgi?CMD=PagerandDB=nucleotide.

Auerbach, D., M. Fetchko and I. Stagljar, 2003. Proteomic approaches for generating comprehensive protein interaction maps. Targets 2: p. 85-92.

Bai, Q., C. McGillivray, N. da Costa, S. Dornan, G. Evans, M.J. Stear and K.C. Chang, 2003. Development of a porcine skeletal muscle cDNA microarray: analysis of differential transcript expression in phenotypically distinct muscles. BMC Genomics 4: p. 8 http://www.biomedcentral.com/1471-2164/4/8.

Barendse, W., 2002a. DNA markers for meat tenderness. Patent WO02064820, http://ep.espacenet.com (Septembre 2002)

Barendse, W., 2002b. Assessing lipid metabolism. Patent WO9923248, http://ep.espacenet.com (Septembre 2002).

Bendixen, E., R. Taylor, K. Hollung, K.I. Hildrum, B. Picard and J. Bouley, 2005. Proteomics, an approach towards understanding the biology of meat quality. In: Indicators of milk and beef quality, J.F. Hocquette and S. Gigli (editors), Wageningen Academic Publishers, Wageningen, The Netherlands, this book.

Bertrand, J.K., R.D. Green, W.O. Herring and D.W. Moser, 2001. Genetic evaluation for beef carcass traits. J. Anim. Sci. 79 E. Suppl.: p. E190-E200.

Bottinelli, R. and C. Reggiani, 2000. Human skeletal muscle fibers: molecular and functional diversity. Progress Biophysics Mol. Biol. 73: p. 195-262.

Bouley, J., C. Chambon and B. Picard, 2003. Proteome analysis applied to the study of muscle developement and sensorial qualities of bovine meat. Sci. Alim. 23: p. 75-79.

Bouley, J., C. Chambon and B. Picard, 2004. Mapping of proteins in bovine skeletal muscle using two-dimensional gel electrophoresis and mass spectrometry. Proteomics 4: p. 1811-1824.

Bouley, J., B. Meunier, C. Chambon, S. De Smet, J.F. Hocquette and B. Picard, 2005. Proteomic analysis of bovine skeletal muscle hypertrophy. Proteomics, in press.

Brazma, A., P. Hingamp, J. Quackenbush, G. Sherlock, P. Spellman, C. Stoeckert, J. Aach, W. Ansorge, C.A. Ball, H.C. Causton, T. Gaasterland, P. Glenisson, F.C.P. Holstege, I.F. Kim, V. Markowitz, J.C. Matese, H. Parkinson, A. Robinson, U. Sarkans, S. Schulze-Kremer, J. Stewart, R. Taylor, J. Vilo and M. Vingron, 2001. Minimum information about a microarray experiment (MIAME)-toward standards for microarray data. Nature Genetics 29: p. 365-371.

Burrow, H.M., S.S. Moore, D.J. Johnston, W. Barendse and B.M. Bindon, 2001. Quantitative and molecular genetic influences on properties of beef: a review. Aust. J. Exp. Agr. 41: p. 893-919.

Campbell, W.G., S.E. Gordon, C.J. Carlson, J.S. Pattison, M.T. Hamilton and F.W. Booth, 2001. Differential global gene expression in red and white skeletal muscle. Am. J. Physiol-Cell Ph. 280: p. C763-C768.

Childs, K.D., D.W. Goad, M.F. Allan, D. Pomp, C. Krehbiel, R.D. Geisert, J.B. Morgan and J.R. Malayer, 2002. Differential expression of NAT1 translational repressor during development of bovine intramuscular adipocytes, Physiol. Genomics 10: p. 49-56.

Cho, K.K., K.H. Han, S.K. Kang, S.H. Lee and Y.J. Choi, 2002. Applications of cDNA microarray in ruminants. The 4th Korea-Japan Joint Symposium on Rumen Metabolism and Physiology, May 21-24, 2002, Jeju, Korea.

Cortright, R.N., D.M. Muoio and G.L. Dohm, 1997. Skeletal muscle lipid metabolism: a frontier for new insights into fuel homoeostasis (review). J. Nutr. Biochem. 8: p. 228-245.

Crawford, D.L., 2001. Functional genomics does not have to be limited to a few select organisms. Genome Biology 2: 1001.1-1001.2. http://genomebiology.com/2001/2/1/interactions/1001.

Cui, X. and G.A. Churchill, 2003. Statistical tests for differential expression in cDNA microarray experiments. Genome Biology 4: 210. http://genomebiology.com/2003/4/4/210.

Dalbies-Tran, R., and P. Mermillod, 2003. Use of heterologous complementary DNA array screening to analyse bovine oocyte transcriptome and its evolution during in vitro matura. Biol. Reprod. 68: p. 252-261.

DeRisi, J.L., V.R. Iyer and P.O. Brown, 1997. Exploring the metabolic and genetic control of gene expression on a genomic scale. Science 278: p. 680-686.

Dorroch, U., T. Goldammer, R.M. Brunner, S.R. Kata, C. Kühn, J.E. Womack and M. Schwerin, 2001. Isolation and characterization of hepatic and intestinal expressed sequence tags potentially involved in trait differentiation between cows of different metabolic type. Mamm. Genome 12: p. 528-537.

Dransfield, E., J.F. Martin, D. Bauchart, S. Abouelkaram, J. Lepetit, J. Culioli, C. Jurie and B. Picard, 2003. Meat quality and composition of three muscles from French cull cows and young bulls. Anim. Sci. 76: p. 387-399.

Drobyshev, A.L., M. Hrabé de Anegelis and J. Beckers, 2003. Artefacts and reliability of DNA microarrays expression profiling data. Current Genomics 4: p. 615-621.

Eggen, A., 2003. Basics and tools of genomics. Outlook Agric. 32: p. 215-217.

Eggen, A. and J.F. Hocquette, 2004. Genomic approaches to economic trait loci and tissue expression profiling: application to muscle biochemistry and beef quality. Meat Sci. 66: p. 1-9.

Eymann, C., G. Homuth, C. Scharf and M. Hecker, 2002. Bacillus subtilis functional genomics: global characterization of the stringent response by proteome and transcriptome analysis. J. Bacteriol. 184: p. 2500-2520.

Fischer, H.P., 2002. Turning quantity into quality: novel quality assurance strategies for data produced by high-throughput genomics technologies. Targets 1: p. 139-146.

Garnier, J.P., R. Klont and G. Plastow, 2003. The potential impact of current animal research on the meat industry and consumer attitudes towards meat. Meat Sci. 63: p. 79-88.

Geay, Y., D. Bauchart, J.F. Hocquette and J. Culioli, 2001. Effect of nutritional factors on biochemical, structural and metabolic characteristics of muscles in ruminants; consequences on dietetic value and sensorial qualities of meat (review). Reprod. Nutr. Dev. 41: 1-26. Erratum, 41: p. 377.

Goddard, M.E., 2003. Animal breeding in the (post-) genomic era. Anim. Sci. 76: p. 353-365.

Halgren, R.G., M.R. Fielden, C.J. Fong and T.R. Zacharewski, 2001. Assessment of clone identity and sequence fidelity for 1189 IMAGE cDNA clones. Nucleic Acids Res. 29: p. 582-588.

Hampson, R. and S.M. Hughes, 2001. Muscular expressions: profiling genes in complex tissues. Genome Biology 2: p. 1033.1-1033.3. http://genomebiology.com/2001/2/12/reviews/1033.

Harhay, G.P. and J.W. Keele, 2003. Positional candidate gene selection from livestock EST databases using gene ontology. Bioinformatics 19: p. 249-255.

Herath, C.B., S. Shiojima, H. Ishiwata, S. Katsuma, T. Kadowaki, K. Ushizawa, K. Imai, T. Takahashi, A. Hirasawa, G. Tsujimoto and K. Hashizume, 2004. Pregnancy-associated changes in genome-wide gene expression profiles in the liver of cow throughout pregnancy. Biochem. Biophys. Res. Commun. 16: p. 666-80.

Hu, J., C. Mungall, A. Law, R. Papworth, J.P. Nelson, A. Brown, I. Simpson, S. Leckie, D.W. Burt, A.L. Hillyard and A.L. Archibald, 2001. The ARKdb: genome databases for farmed and other animals. Nucleic Acids Res. 29: p. 106-110.

Ishiwata, H., S. Katsuma, K. Kizaki, O.V. Patel, H. Nakano, T. Takahashi, K. Imai, A. Hirasawa, S. Shiojima, H. Ikawa, Y. Suzuki, G. Tsujimoto, Y. Izaike, J. Todoroki and K. Hashizume, 2003. Characterization of gene expression profiles in early bovine pregnancy using a custom cDNA microarray. Mol. Reprod. Dev. 65: p. 9-18.

Jordan, B.R., 1998. Large-scale expression measurement by hybridization methods: from high-density membranes to "DNA chips". J. Biochem. 124: p. 251-258.

Kalocsai, P. and S. Shams, 2001. Use of bioinformatics in arrays. In: Methods in Molecular Biology, vol. 170: DNA Arrays: Methods and Protocols, J.B. Rampal (editor), Humana Press Inc., Totowa, New Jersey, USA, p. 223-236.

Kühn, Ch., H. Leveziel, G. Renand, T. Goldammer, M. Schwerin and J. Williams, 2005. Genetic markers for beef quality. In: Indicators of milk and beef quality, J.F. Hocquette and S. Gigli (editors), Wageningen Academic Publishers, Wageningen, The Netherlands, this book.

Kutzer, T., T. Leeb and B. Brenig, 2003. Current State of Development of Genome Analysis in Livestock. Current Genomics 4: p. 487-525.

Lametsch, R., P. Roepstorff and E. Bendixen, 2002. Identification of protein degradation during *post-mortem* storage of pig meat. J. Agric. Food. Chem. 50: p. 5508-5512.

Lee, C.K., R.G. Klopp, R. Weindruch and T.A. Prolla, 1999. Gene expression profile of aging and its retardation by caloric restriction. Science 285: p. 1390-1393.

Lee, M-L.T., F.C. Kuo, G.A. Whitmore and J. Sklar, 2000. Importance of replication in microarray gene expression studies: statistical methods and evidence from repetitive cDNA hybridizations. Proc. Natl. Acad. Sci. USA 97: p. 9834-9839.

Lee, M-L.T. and G.A. Whitmore, 2002. Power and sample size for DNA microarray studies. Stat. Med. 21: p. 3543-3570.

Liinamo, A.E. and A.M. Neeteson-van Nieuwenhoven, 2003. The economic value of livestock production in the European Union. Farm Animal Industrial Platform (FAIP).

Madsen, S.A., L.C. Chang, M.C. Hickey, G.J.M. Rosa, P.M. Coussens and J.L. Burton, 2004. Microarray analysis of gene expression in blood neutrophils of parturient cows. Physiol. Genomics 16: p. 212-221.

Mah N., A. Thelin, T. Lu, S. Nikolaus, T. Kuhbacher, Y. Gurbuz, H. Eickhoff, G. Kloppel, H. Lehrach, B. Mellgard, C.M. Costello and S. Schreiber S. 2004. A comparison of oligonucleotide and cDNA-based microarray systems. Physiol. Genomics 16: 361-70

Maltin, C., D. Balcerzak, R. Tilley and M. Delday, 2003. Determinants of meat quality: tenderness. Proc. Nutr. Soc. 62: p. 337-347.

Mannen, H., M. Morimoto, K. Oyama, F. Muka and S. Tsuji, 2003. Identification of mitochondrial DNA substitutions related to meat quality in Japanese Black cattle. J. Anim. Sci. 81: p. 68-73.

Martinez, I. and T.J. Friis, 2004. Application of proteome analysis to seafood authentication. Proteomics 4: p. 347-354.

Matthews, D.E., 2001. What is more exciting than "genomics"? "Proteomics". Curr. Opin. Clin. Nutr. Metab. Care 4: p. 339.

Moody, D.E. 2001. Genomic techniques: an overview of methods for the study of gene expression. J. Anim. Sci. 79(E. Suppl.): p. 128-135.

Moody, D.E., A.J. Rosa and J.M. Reecy, 2003. Current status of livestock DNA microarrays. AgBiotechNet 5: p. 1-8.

Mullen, A. M. and D.J. Troy, 2005. Current and emerging technologies for the prediction of meat quality. In: Indicators of milk and beef quality, J.F. Hocquette and S. Gigli (editors), Wageningen Academic Publishers, Wageningen, The Netherlands, this book.

Murphy, D., 2002. Gene expression studies using microarrays: principles, problems, and prospects. Adv. Physiol. Educ. 26: p. 256-270.

Novak, J.P., R. Sladek and T.J. Hudson, 2002. Characterization of variability in large-scale gene expression data: implications for study design. Genomics 79: p. 104-113.

Page, B.T., E. Casas, M.P. Heaton, N.G. Cullen, D.L. Hyndman, C.A. Morris, A.M. Crawford, T.L. Wheeler, M. Koohmaraie, J.W. Keele and T.P.L. Smith, 2002. Evaluation of single-nucleotide polymorphisms in CAPN1 for association with meat tenderness in cattle. J. Anim. Sci. 80: p. 3077-3085.

Peng, X., C.L. Wood, E.M. Blalock, K.C. Chen, P.W. Landfield and A.J. Stromberg, 2003, Statistical implications of pooling RNA samples for microarray experiments. BMC Bioinformatics 4: 26. http://www.biomedcentral.cm/1471-2105/4/26.

Picard, B., L. Lefaucheur, C. Berri, and M.J. Duclos, 2002. Muscle fibre ontogenesis in farm animal species. Reprod. Nutr. Dev. 42: p. 415-431.

Piétu, G., O. Alibert, V. Guichard, B. Lamy, F. Bois, E. Leroy, R. Mariage-Samson, R. Houlgatte, P. Soularue and C. Auffray, 1996. Novel gene transcripts preferentially expressed in human muscles revealed by quantitative hybridization of a high density cDNA array. Genome Res. 6, p. 492-503.

Potts, J.K., S.E. Echternkamp, T.P.L. Smith and J.M. Reecy, 2003. Characterization of gene expression in double-muscled and normal-muscled bovine embryos. Anim. Genet. 34: p. 438-444.

Renand, G., C. Larzul, E. Le Bihan-Duval and P. Le Roy, 2003. L'amélioration génétique de la qualité de la viande dans les différentes espèces: situation actuelle et perspectives à court et moyen terme, INRA Prod. Anim. 16, p. 159-173.

Renard, J.P., H.A. Lewin, J. Yang, A. Hernandez, O. Sandra, R.E. Everts and I. Hue. 2004. Endometrium ESTs (bcbp). Accession Numbers CR848861 to 853261.
http://www.ncbi.nlm.nih.gov/entrez/query.fcgi?CMD=PagerandDB=nucleotide

Reverter A, K.A. Byrne, H.L. Bruce, Y.H., Y.H. Wang, B.P. Dalrymple, S.A. Lehnert, 2003. Mixture model-based cluster analysis of DNA microarray gene expression data on Brahman and Brahman composite steers fed high-, medium-, and low-quality diets. J. Anim. Sci. 81: p. 1900-1910.

Shackelford, S.D., T.L. Wheeler and M. Koohmaraie, 1995. Relationship between shear force and trained sensory panel tenderness ratings of 10 major muscles from Bos indicus and Bos taurus cattle. J. Anim. Sci. 73: p. 3333-3340.

Smith, T. P.L., W.M. Grosse, B.A. Freking, A.J. Roberts, R.T. Stone, E. Casas, J.E. Wray, J. White, J. Cho, S.C. Fahrenkrug, G.L. Bennett, M.P. Heaton, W.W. Laegreid, G.A. Rohrer, C.G. Chitko-McKown, G. Pertea, I. Holt, S. Karamycheva, F. Liang, J. Quackenbush and J.W. Keele, 2001. Sequence evaluation of four pooled-tissue normalized bovine cDNA libraries and construction of a gene index for cattle. Genome Res. 11: p. 626-630.

Smith, T.P.L., R.M. Thallman, E. Casas, S.D. Shackelford, T.L. Wheeler and M. Koohmaraie, 2003. Therory and application of genome-based approaches to improve the quality and value of beef. Outlook Agric. 32: p. 253-265.

Suchyta, S.P., S. Sipkovsky, R. Kruska, A. Jeffers, A. McNulty, M.J. Coussens, R.J. Tempelman, R.G. Halgren, P.M. Saama, D.E. Bauman, Y.R. Boisclair, J.L. Burton, R.J. Collier, E.J. DePeters, T.A. Ferris, M.C. Lucy, M.A. McGuire, J.F. Medrano, T.R. Overton, T.P. Smith, G.W. Smith,T. S. Sonstegard, J.N. Spain, D.E. Spiers, J. Yao and P.M. Coussens, 2003. Development and testing of a high-density cDNA microarray resource for cattle. Physiol. Genomics 15: p. 158-164.

Suchyta, S.P., S. Sipkovsky, R.G. Halgren, R. Kruska, M. Elftman, M. Weber-Nielsen, M.J. Vandehaar, L. Xiao, R.J. Tempelman and P.M. Coussens, 2004. Bovine mammary gland expression profiling using a cDNA microarray enhanced for mammary-specific transcripts. Physiol. Genomics 16: p. 8-18.

Sudre, K., C. Leroux, G. Piétu, I. Cassar-Malek, E. Petit, A. Listrat, C. Auffray, B. Picard, P. P. Martin and J.F. Hocquette, 2003a. Transcriptome analysis of two bovine muscles during ontogenesis. J. Biochem. 133: p. 745-756.

Sudre, K., I. Cassar-Malek, C. Leroux, A. Listrat, Y. Ueda, C. Jurie, G. Renand, P. Martin and J.F. Hocquette, 2003b. Transcriptome analysis of muscle in order to identify genes which determine muscle characteristics and sensory quality traits of beef. Sci. Alim. 23: p. 65-69.

Sudre, K., C. Leroux, I. Cassar-Malek, J.F. Hocquette and P. Martin, 2005. A collection of bovine cDNA probes for gene expression profiling in muscle. Mol. Cell. Probes. In press.

Talamo, F., C. D'Ambrosio and S. Arena, 2003. Proteins from bovine tissues and biological fluids: defining a reference electrophoresis map for liver, kidney, muscle, plasma and red blood cells. Proteomics 3: p. 440-460.

Taylor, C.F., N.W. Paton, K.L. Garwood, P.D. Kirby, D.A. Stead, Z. Yin, E.W. Deutsch, L. Selway, J. Walker, I. Riba-Garcia, S. Mohammed, M.J. Deery, J.A. Howard, T. Dunkley, R. Aebersold, D.B. Kell, K.S. Lilley, P. Roepstorff, J.F. Yates, A. Brass, A.J. Brown, P. Cash, S.J. Gaskell, S.J. Hubbard and S.G. Oliver, 2003. A systematic approach to modeling, capturing, and disseminating proteomics experimental data. Nat. Biotechnol. 21: p. 247-254.

Thaller, G., C. Kühn, A. Winter, A. Ewald, O. Bellmann, J. Wegner, H. Zühlke and R. Fries, 2003. DGAT1, a new positional and functional candidate gene for intramuscular fat deposition in cattle. Anim. Genet. 34: p. 354-357.

Thomas, J.W., A.B. Prasad, T.J. Summers, S.Q. Lee-Lin, V.V.B. Maduro, J.R. Idol, J.F. Ryan, P.J. Thomas, J.C. McDowell and E.D. Green, 2002. Parallel construction of orthologous sequence-ready clone contig maps in multiple species. Genome Res. 12: p. 1277-1285.

Törönen, P., G. Wong, and E. Castren, 2003. Methods for quantification and clustering of gene expression data. Current Genomics 4: p. 445-463.

Tusher, V.G., R. Tibshirani and G. Chu, 2001. Significance analysis of microarrays applied to the ionizing radiation response, Proc. Natl. Acad. Sci. USA 98: p. 5116-5121.

Wang, X., S. Ghosh and S.W. Guo, 2001. Quantitative quality control in microarray image processing and data acquisition. Nucleic Acids Res. 29: 15e75.

Welle, S., A. Brooks and C.A. Thornton, 2001. Senescence-related changes in gene expression in muscle: similarities and differences between mice and men. Physiol. Genomics 5: p. 67-73.

Wernisch, L., S.L. Kendall, S. Soneji, A. Wietzorrek, T. Parish, J. Hinds, P.D. Butcher and N.G. Stoker, 2003. Analysis of whole-genome microarray replicates using mixed models. Bioinformatics 19: p. 53-61.

Winslow, R.L. and M.S. Boguski, 2003. Genome informatics: current status and future prospects. Circ. Res. 92: p. 953-961.

Wu, H., M.K. Kerr, X. Cui and G.A. Churchill, 2002. MAANOVA: A software package for the analysis of spotted cDNA microarray experiments. http://www.jax.org/staff/churchill/labsite/pubs/Wu_maanova.pdf.

Proteomics, an approach towards understanding the biology of meat quality

E. Bendixen[1], R. Taylor[2], K. Hollung[3], K.I. Hildrum[3], B. Picard[2] and J. Bouley[2]*
[1]Danish Institute of Agricultural Sciences, P.O. Box 50, 8830 Tjele, Denmark, [2]INRA, Theix, 63122 Saint-Genès-Champanelle, France, [3] Matforsk, 1430 Ås, Norway.
**Correspondence: Emoke.Bendixen@agrsci.dk*

Summary

Sequencing the complete genome of many species within the last few years has allowed proteomics, like all other approaches to functional genomics to rapidly expand. The technologies within proteomics are presently developing with unexpected speed, and these technologies promise a quantum leap in many applications of life sciences. Within the field of animal production the current focus is to use these technologies to describe the function of individual genes, and how heredity and environment interact to control cellular functions and consequently the physiological traits that are relevant for production of farm animals. Understanding the biology behind the complex traits of meat quality still remains a major challenge in cattle breeding and production. A number of presently running projects aim at finding molecular markers for meat quality by combining the proteome and transcriptome technologies, with the ultimate objective to link the genotype and phenotype behind meat quality. In this review we introduce some of the most frequently used technologies of comparative proteomics, and how they have been applied to studies of muscle physiology and meat quality traits in farm animals. Finally, we will introduce some of the ongoing initiatives of muscle growth and meat science proteomics at DIAS (Denmark), INRA (France), and MATFORSK (Norway).

Keywords: proteomics, meat, muscle

Introduction

Proteomics is a scientific field in rapid development, hence the term proteomics is currently expanding from its classical status as "biochemistry at an unprecedented high throughput scale" to the much broader definition of "panoramic protein characterisation". For a recent review, see (Tyers and Mann, 2003).

In this review we will limit the term proteomics to that of large-scale and high throughput biochemistry, and we will firstly introduce the most frequently used technologies for characterising cellular protein expression patterns. Secondly, we will discuss examples of how proteomics already has been used in the research areas of animal production and meat quality. The agricultural industry, as well as research institutions have already started to invest in proteome technologies and projects, and in this review we will discuss some of the ongoing initiatives of meat science proteomics at DIAS (Denmark), INRA (France), and MATFORSK (Norway).

Proteomics in the postgenomic era

According to the classic definition of Wilkins *et al.*, **proteome** is the **prote**in complement of a gen**ome** (Wilkins *et al.*, 1996). Global characterisation of cellular protein expression was already the aim of the first 2DE (two dimensional electrophoresis) technologies (O'Farrell, 1975), but the current development of proteomics was not unleashed until the late 1990s, by the wealth of information that came from the human genome sequencing project. Since then, the genomes of other species were quickly to follow, and presently the genomes of more than 220 organisms, have been completed and more than 950 are in progress (http://genomesonline.org). Not only did the genome data make clear that DNA sequencing is only an essential step towards understanding the function of genes, but also the large genome databases allowed protein identification at an unprecedented high speed and low cost, which is the core of proteomics.

Characterising the function of genes is a major challenge of the post-genomic era, and within this research area, termed functional genomics, a wide range of tools are currently being developed.

Proteomics is a toolbox that is complementary to other functional genomic tools, such as transcriptomics and metabolomics, which all share the aim to translate genome information into useful biological insight.

Currently there are two classic approaches to proteome characterisation, namely comparative proteomics and mapping proteomics. **Mapping proteomics** is in many ways comparable to genome sequencing projects, where the goal is to characterise and make comprehensive databases of "cellular proteomes". However invaluable as information-mining resources, the mapping of complete proteomes is presently a complicated task, partly due to the complex variety of modification forms most proteins possess (Mann and Jensen, 2003), and also because (in contrast to the genome) the proteome is a dynamic entity that constantly changes with time and physiological state. In this sense, every single cell and organism has an infinite number of proteomes. The recently published first drafts of the yeast (Ghaemmaghami *et al.*, 2003; Huh *et al.*, 2003) and mitochondria proteomes (Lescuyer *et al.*, 2003) indicate the complexity of this task.

The aim of **comparative proteomics** is to characterise the biological mechanisms that form the link between observable phenotypes and the genotypes, hence comparative proteome studies can be regarded as making moment-by-moment snapshots of cellular responses at the protein level (Hunter *et al.*, 2002). This review is focused on introducing the field of comparative proteomics, and how it can be applied to animal production sciences, with a special attention on production and quality of meat.

Technologies in comparative proteomics

In this chapter we will give a short introduction to some of the most frequently used methods and tools for extraction, separation, quantification and characterisation of proteomes.

Technology: 2DE

2DE (two dimensional electrophoresis) is a separation method that allows the display and parallel analysis of thousands of proteins extracted from complex samples such as biopsies, tissues and cell cultures. In a 2DE analysis, complex spot patterns are formed, where every single spot represents an individual protein that migrates to its specific coordinates, due to its specific molecular weight and charge. The intensity of an individual spot indicates how much the cell has produced of that actual protein. Figure 1 shows a typical 2DE pattern of bovine muscle proteins. Detailed 2DE maps of porcine (Lametsch *et al.*, 2002; Morzel *et al.*, 2004), cattle (Bouley *et al.*, 2004) and chicken (Doherty *et al.*, 2004) muscles have already been published.

Attempting a panoramic characterisation of cellular proteomes was already the aim of the earliest 2DE analyses (O'Farrell, 1975), and since then, 2DE methods has been developed into more reproducible and stable IPG (immobilised pH gradients) based systems (Gorg *et al.*, 2000). Methods for visualisation of the protein patterns includes coomassie-, silver-, and fluorescence-based methods, as reviewed by (Rabilloud, 2000). In particular the DIGE system, based on

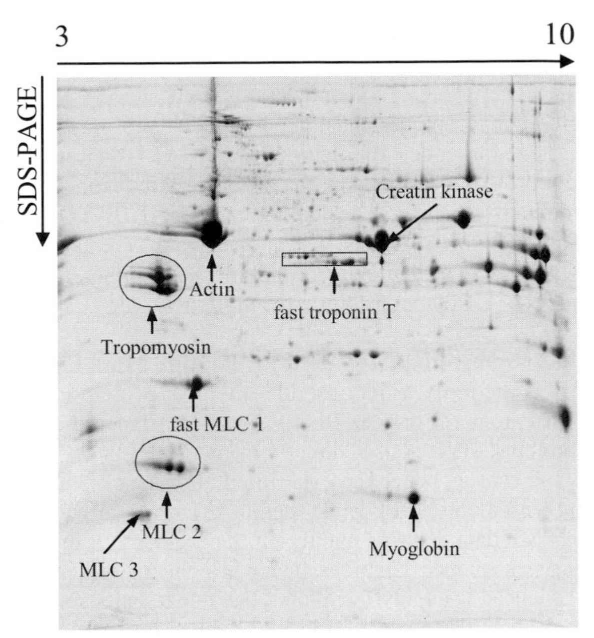

Figure 1. Two dimensional electrophoresis (2-DE) of bovine Semitendinosus skeletal muscle. 2-DE was performed using an IPG 3-10 strip in the first dimension and SDS-PAGE (11 % T) in the second (Bouley et al., 2004).

differential fluorescence-tagging of comparable samples, prior to their multiplexed separation and analysis, has improved the quantitative properties of 2DE based comparative proteomics (Unlu *et al.*, 1997). 2DE-based proteomics is a multi-step procedure unfit for automation, but in spite of this, and many other shortcomings, 2DE technology has remained an important cornerstone of proteomics (Klein *et al.*, 2004). This is partly due to the fact that 2DE is an excellent preparative medium for further protein characterisation, but also because better alternatives for a visual protein-display methods have not yet been presented.

A serious shortcoming of 2DE is that analysis is constrained to only a limited subset of the cellular protein complement. This is partly due to chemical properties of IPG-based 2DE systems that discriminate mainly against basic (Gorg, 1999), and hydrophobic proteins (Santoni *et al.*, 2000) and therefore excludes analysis of most receptors and transmembrane proteins. Moreover, the limited dynamic range of 2DE, only spans 2-3 orders of magnitude (pico-nanomolar range) while the ranges of cellular protein expression in most tissues spans over 10^6- fold. Hence 2DE analysis of unfractionated tissue samples is constrained to the analysis of the most abundant proteins. This is clearly demonstrated by the characterisation of the Yeast proteome, as 50% of the proteins have a PI above 8, and 30% of the predicted ORFs encode a transmembrane protein (Pedersen *et al.*, 2003). A wide range of methods for pre-fractionation of complex samples prior to 2DE analysis has been developed to bypass this problem (Gorg *et al.*, 2002; Spandidos and Rabbitts, 2002), but the current rapid development of comparative mass spectrometry methods has already to some extent replaced some of the classic 2DE-based applications, as will be further discussed in the followings.

Technology: Mass spectrometry

The wide range of mass spectrometry-based methods for protein characterisation that has been developed within the past 20 years has been a major driving force in the field of proteomics, and developing techniques for comparative mass spectrometry, currently dominates the technological frontiers of proteomics.

The discoveries of soft ionisation techniques, most importantly electrospray ionisation (**ESI**) (Fenn *et al.*, 1989), and matrix-assisted laser desorption ionisation (**MALDI**) (Karas, 1996), were major steps towards using MS for protein identification, for which a shared Nobel prize was awarded in 2002. These ionisation methods are the core technologies of **MALDI-TOF** (time of flight) and **ESI-MSMS** (tandem mass spectrometry)-based instruments that has remained the most powerful for proteomics, partly because these technologies complement each other, and their combined use has become a classic approach in proteomics. For a recent review, see (Yates, 2004).

MALDI-TOF instruments are most frequently used for **Peptide Mass Fingerprinting (PMF)** analyses, which currently is the most widely used method for protein identification, and the workhorse of any 2DE-based proteome project. Briefly, PMF-based identification is an analysis of proteolytically digested proteins, often excised directly from a 2DE matrix. Tryptic digests yields peptides of optimal size and charge, hence the experimental masses obtained by MALDI-TOF analysis can be compared with theoretical peptide masses of *in silico*-digested protein sequence databases. The quality of PMF data depends mainly on the accuracy of the MS data, and of the availability of comprehensive sequence databases, hence protein analysis in organisms where the genome has been fully sequenced is greatly facilitated. But due to extensive sequence homology between species, PMF analysis is also a well established method for characterising proteins from organisms with less well characterised genomes, which at the moment includes all domestic animals. Proteins from pig, (Morzel *et al.*, 2004), cattle (Bouley *et al.*, 2004) and chicken (Doherty *et al.*, 2004) have been successfully characterised in spite of lacking fully sequenced genomes. For porcine proteins, typically 90% of 2DE separated proteins could be identified with a typical sequence coverage of 20% (Lametsch *et al.*, 2002). For cattle, 70% of the 2DE proteins could be identified by PMF analysis, with a typical sequence coverage around 30%. Only 40% of these were based on alignment to bovine sequences, hence extensive cross-species identification by PMF is widely applicable. A similar conclusion was drawn from a thorough analysis of bovine proteins extracted from a wide range of tissues (Talamo *et al.*, 2003). A shortcoming of MALDI technology is that the instruments are not easily automated or interfaced with liquid separation systems.

ESI-MS/MS instruments are most frequently used to obtain peptide sequence information, and contrary to MALDI instruments, ESI-instruments are easily interfaced to a wide range of liquid online separation platforms, and are therefore a key technology in MS-based comparative proteomics, as will be discussed in the followings.

Technology: Comparative mass spectrometry

Considering the shortcomings of 2DE technology, there is a great need to develop alternatives to 2DE-based comparative proteomics.

However, mass spectrometry of proteins and peptides is a qualitative rather than quantitative method, mainly due to unpredictable ionisation capabilities of different peptides. A wide range of approaches for bypassing the non-quantitative nature of MS analyses are currently being developed, for recent reviews, see (Aebersold and Mann, 2003; Ong *et al.*, 2003).

One of the most promising is the Isotope Coded Affinity Tag (**ICAT**) approach that allows differential chemical tagging of proteins. By using heavy versus light isotope-coded tags in two comparable samples, the relative abundance of the individual proteins can be analysed as peak-doublets in the same MS-spectra, hence reducing the analytic noise and increasing the quantitative nature of compared MS data. Similarly, Stable Isotope Labeling by Amino acids in Cell culture (**SILAC**) based methods allow differential metabolic labelling followed by MS analysis of comparable cell culture samples.

Reducing the complexity of proteomes is widely used in comparative proteomics, and a frequently used approach is to introduce multi-dimensional protein separation steps by interfacing liquid chromatography methods to ESI-MS/MS instruments. Classically, serial combinations of nano-HPLC columns containing cation exchange and reverse phase resins were used (Link, 2002), but a wide range of separation methods can be used to selectively enrich specific proteome subsets, like phosphopeptides or glycosylated peptides.

Technology: Protein arrays

Although slow to develop, there is a great need to find platforms for high throughput protein expression analysis that are easily automated. Currently, a number of approaches are based on immobilising a variety of "baits", (like antibodies, lectins and many others) in ordered arrays at specially treated surfaces. For a recent review see (Cutler, 2003). Further advances are needed before this technology will find wide appeal, but the current focus on nanotechnology and microfluidics may allow this to happen in the near future.

Bioinformatics and data analysis in comparative proteomics

Storing and making sense of enormous amounts of data is a major challenge in proteomics as well as in other post-genomic sciences. As the throughput and capacity of analytical technologies rapidly increase, data overload is easily created. Computational biology, commonly termed bioinformatics, has become increasingly important, and high speed computers, computational techniques and sophisticated algorithms (for analysing, storing, sorting, searching and integrating data) is currently being developed with amazing speed. For a recent review see (Ouzounis and Valencia, 2003). Moreover, the complexity of statistical analysis of comparative proteome data also requires innovation of statistical tools. Bioinformatics and statistical analyses of expression data are scientific areas in their own rights, and covering the recent development of post-genomic data handling is beyond the scope of this review. However, a few issues of importance for comparative proteomics will be briefly introduced in order to highlight the necessity and complexity of this field.

Data analysis

Digital Image analysis: Increasingly sophisticated software for analysing 2DE images has greatly improved the quantification of comparative 2DE analysis. A wide range of software is commercially available and information from 2DE databases is accessible at publicWeb sites.
MS-Data: Thousands of MS spectra are generated during a comparative proteome study, and extracting information from MS data includes multiple analytical steps, like noise extraction, mass calibration, and deconvolution of complex peak patterns, for which improved algorithms and software is continuously being created. For a recent review, see (Blueggel *et al.*, 2004). In addition, MS-based protein identification requires alignment of the experimental data to *in silico*-digested sequence databases. Although probability-based scoring methods are continuously improved in order to facilitate automatic data interpretation, the very time consuming step of manual review of scoring data is still a critical step in avoiding false identification results.

Statistical analyses of comparative proteome data, where the number of variables (typically the expression level of hundreds or thousands of proteins) greatly outnumbers the repeated observations (typically 10-20 animal samples). This implies that data normalization and statistical approaches must be improved for accurate detection of differential protein expression (Karp *et al.*, 2004). In addition, automatic processing and software development is needed for handling large data sets. In this respect, statistical analyses of proteome data have very much in common with data analysis of microarray-based mRNA analyses, hence many of the methods developed for microarray data may also be applicable to analysis of proteome data. In order to extract biological knowledge from microarray data, a wide variety of pattern recognition methods has been developed, of which hierachical clustering, self organising maps, and K-means clustering are the most widely used. For a review, see (Valafar, 2002). Some of these statistical tools are already available through commercially available 2DE-softwares, but their use is not yet widely used in proteomics.

Data integration

Understanding the complex data that emerge from comparative proteomics often requires integration of many additional sources of data, including phenotype descriptions, as well as structural and functional genome data. **Systems biology** is the emerging field of science that aims to integrate data from biological sciences at many different levels, and across a wide range of disciplines, in order to understand the interrelations and pathways that create biological systems and organisms. This has been summarised in Figure 2. Data integration and data mining is a major

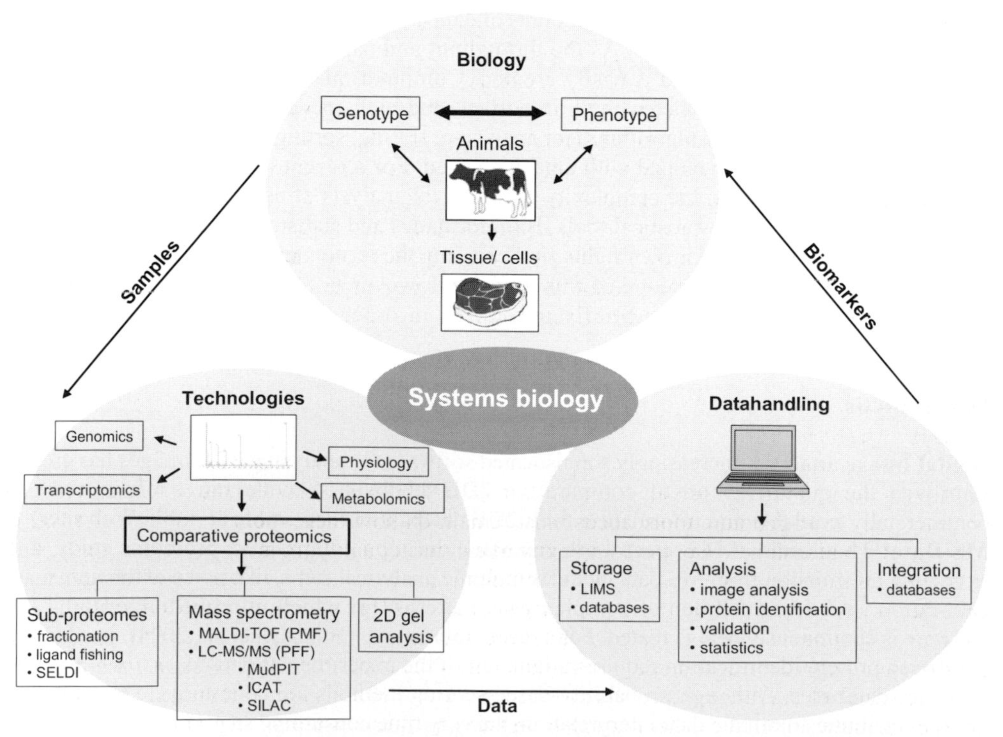

Figure 2. Elements and workflow in systems biology. The aim is to characterise the function of living organisms through computational methods that allow data integration from observational biology as well as a wide range of post genome technologies.

challenge in all post-genomic sciences, and the development of increasingly sophisticated algorithms and computational methods that aid the discovery process is rapidly developing. For a recent review, see (Hood, 2003).

Improved database technologies has greatly facilitated integration and mining of different data types, like linking differentially expressed proteins and genes into assemblies and pathways. For a review see (Blueggel *et al.*, 2004). Very few studies have as yet demonstrated how complex, integrated data sets can be distilled; however, an early example is the study of the yeast proteome (Hazbun *et al.*, 2003).

Currently, the open source database mainly cover studies of human as well as classic model organisms like mouse, rat, and yeast. Although not much information is to be found about cattle and pig proteomics, these repositories offer tools and information that are useful for making sense of protemics data from all organisms.

Applying proteomics for production and quality of meat

Proteome markers for muscle growth and meat quality traits will allow scientists to build and test better hypotheses, and thereby find better solutions to challenges in production and quality of meat. Presently, this area of research is in a start-up phase, and not much information can yet be extracted from the literature. However, proteome studies of metabolism, growth and development of muscle tissues from man, and of classic model organisms like mouse, are well characterised at the proteome level, and of great interest for implementing proteome technologies to understanding muscle biology of live-stock species.

A major challenge within the area of meat quality is to understand the variability of tenderness. Although many of the biochemical factors are well known (Koohmaraie, 1996, Maltin *et al.*, 2003), and a number of QTLs for tenderness have been determined (Burrow *et al.*, 2001), the complex mechanisms of *post-mortem* processes, including the interactions of soluble muscle proteins, pH and ion transport during the very early *post-mortem* phase, and their effect on proteolysis, remains a major challenge. A second challenge is to characterise factors that influence the water holding capacity, which still remains a major quality problem of the pork industry. Also the relationship of growth rate to both carcass yield and meat quality traits remains poorly characterised. In the following we will discuss, in a little more detail, how comparative proteomics has been related to these issues.

How can some of the challenges in meat production and quality be addressed by a proteomic approach?

In order to achieve better product quality and a more sustainable animal production through proteome and post-genomic tools, focus must remain on how these tools contribute to building useful biological insight. Also, the complexity of e.g. meat quality traits must also be considered. It is well established that meat quality traits like tenderness and water holding capacity are complex traits influenced by genetic (Burrow *et al.*, 2001) as well as environmental factors including feeding, welfare, pre-slaughter stress and processing technologies like curing and chilling, hanging methods, and electric stimulation. Meat quality is often classified according to the appearance of the raw material, like texture and color (Hildrum *et al.*, 1999), or according to consumers preferences like taste, juicyness, and texture (McKeith *et al.*, 1985).

However useful for producers, slaughterhouses, retailers and consumers, these traits may be insufficiently precise for classifying comparative proteome groups directly, as complex traits will give rise to complex variation in proteome profiles. One of the ways to minimise this problem could be through comparative proteome studies of genetically well defined groups of animals, and optimally this would imply analysis of effects controlled by single gene effects. Only a limited list of single gene effects on meat quality has yet been described, however, these variant animals are valuable for finding new molecular markers for meat quality traits through proteome and expression analyses.

Currently, a wealth of QTL information related to meat quality traits is being established, which will be of great value for finding the biological mechanisms that link phenotypes to genotype, and help to spot the biological mechanisms and pathways that are important for production and quality traits of domestic animals. Mutants are informative resources for studying biological mechanisms, and the current focus on large scale characterisation of genetic variants in domestic animals will allow more well defined comparative proteome studies in the near future.

When established, proteome markers for growth-, development-, and meat quality traits may be further developed into technological markers that can facilitate authentification, more precise labelling and sorting of meat quality, and for optimising meat processing technologies. Some markers will also be suited for further development of molecular markers for use in selective breeding.

Growth and development: Skeletal muscle tissues are heterogeneous population of muscle fibers, generally classified according to their metabolic traits as well as their contractile properties, as reviewed by (Picard *et al.*, 2002). Expression of fiber types is species-specific, and in all species, muscle fiber properties are modified during post-natal life, relative to age, sex, and breed.

In general there is a relationship of fiber type with sensory measures of juiciness and flavor, but the relationship to tenderness is not clear, as shown for lamb (Valin and Touraille, 1982), beef (Maltin *et al.*, 1997), and pork (Henckel *et al.*, 1997). A negative correlation between tenderness and muscle fiber area has been reported by several authors, but no significant relation or a positive one was found by others. For reviews about determinants of tenderness, see Renand *et al.*, (2001) and Maltin *et al.* (2003). Animal models important for growth characteristics have been characterised, including double muscling in cattle (Fiems *et al.*, 1995), and the callipyge trait in sheep (Taylor and Koohmaraie, 1998). As these animals are unique in their tenderness traits, they have contributed greatly to our understanding of how muscle growth and meat quality are interrelated. Bovine muscle hypertrophy is characterised by a higher proliferation of secondary generation of myoblasts, resulting in smaller, but more fast glycolytic fibers in the adult animal (Deveaux *et al.*, 2001).

A recent proteome study was aimed at finding markers for muscle hypertrophy in semitendinosus muscles from young Belgian Blue bulls (Bouley *et al.*, 2005). This study showed that thirteen proteins were significantly altered in response to the myostatin deletion, suggesting that myostatin negatively controls mainly fast-twitch glycolytic muscle fibre number.

The perimysium organisation (Boccard, 1981) as well as calpain/calpastatin activity (Uytterhaegen *et al.*, 1994), and fat metabolism (McPherron and Lee, 2002) seem to be influenced in the muscle tissues of hypertrophic animals, hence further studies should be informative about how muscle physiology and meat quality traits are interrelated.

Callipyge mutation in sheep causes muscle hypertrophy due to a mutation in chromosome 18 (Freking *et al.*, 2002) that results in a two- to three-fold increase in calpastatin (Koohmaraie *et al.*, 1995). The *post-mortem* structural changes in muscle from callipyge sheep seem to be similar to that of wild type animals, but occur at a slower rate , emphasizing the important role of the calpain/calpastatin system in tenderisation of meat. Proteome studies of callipyge sheep are not yet done, but this will certainly yield new information about *post-mortem* processes in meat.

Understanding *post-mortem* metabolism, is of great importance, as it is well established that the biochemical events that take place during the early *post-mortem* period greatly affect meat quality. Changes related to physio-chemical factors (Goll *et al.*, 1964), histo-chemical properties (Brocks *et al.*, 1998), temperature (Monin *et al.*, 1995), genotype (Mariani *et al.*, 1996), and many other factors, influence *post-mortem* metabolism, but the relationship to meat quality has remained poorly understood. In general, any factor which alters the onset of rigor can have a significant effect on tenderness, including electrical stimulation (Hildrum *et al.*, 1999), stress (Scanga *et al.*, 1998), chilling rate, and nutritional status of the living animal (Locker, 1985). A series of papers (Lametsch *et al.*, 2002; Lametsch *et al.*, 2003; Lametsch and Bendixen, 2001; Morzel *et al.*, 2004) reported proteome changes of *post-mortem* processes in pork. *Post mortem* markers detected during the first 48 hours of post-slaughter storage, included structural proteins (actin, myosin and troponin T) as well as metabolic enzymes, like myokinase, pyruvate kinase and glycogen phosphorylase (Lametsch *et al.*, 2002), and accumulation of specific metabolic enzymes, as well as specific actin and myosin fragments were observed to correlate to meat tenderness (Lametsch *et al.*, 2003). These studies must be regarded as a starting point, and further studies are needed to understand some of the complex mechanisms that rule *post-mortem* metabolism of muscles and meat.

Calpains role in tenderness: It is well established that calpains play a key role in tenderisation of meat (Taylor *et al.*, 1995), and that the rate-limiting factor is actually calpastatin-mediated inhibition of *post-mortem* μ-calpain activity (Koohmaraie, 1996). Callipyge sheep have very tough meat and also a low *post-mortem* calpain activity but are normal in other *post-mortem* changes (Koohmaraie *et al.*, 1995). A very important recent finding is that transgenic mice that over-express calpastatin have very limited *post-mortem* calpain activity (Koohmaraie, USDA, personal communication). Recent proteome studies have identified calpain-mediated degradation of specific myofibril proteins *in vitro*, including actin, desmin, troponin, and several tropomyosin isoforms (Lametsch *et al.*, 2004).

Water holding capacity is an important meat quality trait, closely related to variations in the *post-mortem* metabolism, however, little progress has been made within the past decades. Several single-gene models of porcine drip loss phenotypes exist, including variant of the ryanodine receptor (Ervasti *et al.*, 1991), and variants of the PRKAG3 gene, commonly termed the RN gene (Le Roy *et al.*, 2000). Recently, a comparative proteome study of the RN-gene effect has shown that the expression profiles of several enzymes of the glycogen storage pathways are differentially regulated in a pattern that indicates that regulation of glucose transport is severely affected in mutant animals (Hedegaard *et al.*, 2004). Further studies of these mutations are of great interest in order to find and explain molecular mechanisms that influence drip loss in porcine meat.

Technological processing: Understanding the protein changes that are induced during technological processing, and why the outcome of processing is so variable, will make these technologies increasingly useful. Numerous studies have shown that variation in *post-mortem* glycolytic rates in muscles from different carcasses yields aged meat of varied tenderness (O'Halloran *et al.*, 1997). As slowly glycolysing muscles yield tough meat, electrical stimulation is widely used to accelerate *post-mortem* glycolysis, however, the underlying mechanisms of the stimulation are not well understood. As a consequence, this process is rather erratic and deleterious to fast glycolysing muscles (Hwang *et al.*, 2003), due to lack of knowledge on how to optimize the technologies. Moreover, a wide range of protein modifications are also important for meat processing techniques like fermentation, marination, drying, heating, freezing and high-pressure processing. As with electrical stimulation, the principles and mechanisms behind the processes are often insufficiently understood. Proteomics could be a powerful tool in shedding light over these and allow optimisation of meat processing technologies.

Ongoing initiatives

DIAS

Cattle: In order to characterise mechanisms in growth and development, a current proteome initiative is aimed at comparing proteome profiles of muscle tissues in young bulls that are differentially fed during early growth. Some of the characterised markers will be related to *post-mortem* characteristics of these animals.

Also the effect of protease inhibitors, like calpastatin and proteasome inhibitors, will be studied by *in vitro* incubations, and analysed by proteome profiling.

Pig: Post-natal muscle growth is being studied by manipulations of established primary cell cultures from porcine muscles. Growth and proliferation in these cell cultures will be studied by comparative proteome profiling of tissue cultures.

Chicken: A pilot-scale study is aimed at comparing proteome signatures of growth and differentiation in chicken muscles. Samples will be taken from breeds that possess extremely different growth rates, and proteome studies will be used to compare muscle metabolism patterns in the breast muscle of these animals. Proteome patterns will be related to meat quality characters.

INRA

In order to understand mechanisms that control meat quality, a current study is aimed at characterising *semitendinosus* muscle development in young bulls of different French breeds, in order to identify molecular markers for beef tenderness. Muscle samples for this study have been classified by sensory panel data. Studies are also underway to characterise protein expression in control *vs.* Texel sheep, a breed which shows muscle hypertrophy. Pig muscle *post-mortem* metabolism and the relationship of stress to meat quality are parts of two ongoing studies.

Another project aims to identify the changes of proteins at the principal stages of foetal life (proliferation of the different generation of myoblasts and differentiation) from muscle samples and from primary cultures of myoblasts.

MATFORSK

A previous study identified a heritability of 0.40 on beef tenderness. A current project is aimed at characterising the molecular basis of this heritability. The study includes 600 bulls, (descendants of 40 breeding bulls). Tenderness, as well as calpain/calpastatin activities and pH decline during the first 10 h *post-mortem* will be analysed. Comparative proteomics, as well as Microarray analyses and SNP-typing will be performed on selected samples from animals representing extreme tender/tough meat quality. Proteome patterns will be related to tenderness and pH phenotypes. With proteomics, also the *post-mortem* changes will be characterised, and compared to those previously found in porcine muscle (Lametsch *et al.*, 2003). Moreover, proteome changes related to stunning/electrical stimulation will be studied at a pilot scale.

Future directions

From proteome to systems biology

Genes and proteins do not function independently, but participate in complex networks that give rise to cellular functions, tissues, organs and organisms. The challenge of defining these systems is far beyond the capabilities of any single functional genomic tool, hence systems biology is the science of integrating data from many levels of biology and many technologies, including those of post-genomic tools, in order to understand complex biological systems. In this respect, production animals are genetically well characterised and are a tremendous resource for establishing model organisms with well characterised diversities.

Conclusions

Although proteomics is currently experiencing unpredicted technological developments, it is unlikely that a single ultimate best-way for doing comparative proteomics will appear. Proteomics should rather be regarded as an ever expanding toolbox, hence we must take advantage of the developing technologies and, in every single application, combine the technological opportunities that best suit the actual study. This truly requires interdiciplinary collaborations between a broad range of sciences, including those of physiology, genetics, cell biology, computer sciences, as well as from the animal production and meat processing industry.

References

Aebersold, R. and M. Mann, 2003. Mass spectrometry-based proteomics. Nature 422: p. 198-207.

Blueggel, M., D. Chamrad and H.E. Meyer, 2004. Bioinformatics in proteomics. Curr. Pharm. Biotechnol. 5: p. 79-88.

Boccard, R., 1981. Facts and reflexions on muscular hypertrophy in cattle: Double muscling or Culard. Meat Sci. 2: p. 1-28.

Bouley, J., C. Chambon and B. Picard, 2004. Mapping of bovine skeletal muscle proteins using two-dimensional gel electrophoresis and mass spectrometry. Proteomics. 4: p. 1811-1824.

Bouley, J., B. Meunier, C. Chambon, S. De Smet, J.H. Hocquette and B. Picard, 2005. Proteomic analysis of bovine skeletal muscle hypertrophy. Proteomics in press.

Brocks, L., B. Hulsegge and G. Merkus, 1998. Histochemical characteristics in relation to meat quality properties in the *Longissimus lumborum* of fast and lean growing lines of large white pigs. Meat Sci. 50: p. 411-420.

Burrow, H.M., S.S. Moore, D.J. Johnston, W. Barendse and B.M. Bindon, 2001. Quantitative and molecular genetic influences on properties of beef: a review. Austr. J. Exp. Agr. 41: p. 893-919.

Cutler, P. 2003. Protein arrays: the current state-of-the-art. Proteomics 3: p. 3-18.

Deveaux V., I. Cassar-Malek and B. Picard, 2001. Comparison of contractile characteristics of muscle from Holstein and double-muscled Belgian-Blue foetuses. Comp. Biochem. Physiol. Part A, 131: p. 21-29.

Doherty, M.K., L. McLean, J.R. Hayter and J.M. Pratt, 2004. The proteome of chicken skeletal muscle: Changes in soluble protein expression during growth in a layer strain. Proteomics (in press).

Ervasti, J.M., M.A. Strand, T.P. Hanson, J.R. Mickelson and C.F. Louis, 1991. Ryanodine receptor in different malignant hyperthermia-susceptible porcine muscles. Am. J. Physiol. 260: p. C58-C66.

Fenn, J.B., M. Mann, C.K. Meng, S.F. Wong and C.M. Whitehouse, 1989. Electrospray ionization for mass spectrometry of large biomolecules. Science 246: p. 64-71.

Fiems, L.O., J.V. Hoof, L. Uytterhaegen, C.V. Boucque and D. Demeyer, 1995. Comparative quality of meat from double-muscled and normal beef cattle. In: "Expression of Tissue Proteinases and Regulation of Protein Degradation as Related to Meat Quality". eds. A Ouali, D Demeyer and F Smulders ECCEAMST series. p. 381-391.

Freking, B.A., S.K. Murphy, A.A. Wylie, S.J. Rhodes, J.W. Keele, K.A. Leymaster, R.L. Jirtle, and T.P. Smith, 2002. Identification of the single base change causing the callipyge muscle hypertrophy phenotype, the only known example of polar overdominance in mammals. Genome Res. 12: p. 1496-1506.

Ghaemmaghami, S., W.K. Huh, K. Bower, R.W. Howson, A. Belle, N. Dephoure, E.K. O'Shea, and J.S. Weissman, 2003. Global analysis of protein expression in yeast. Nature 425: p. 737-741.

Goll, D.E., D.W. Henderson and E.A. Kline. 1964. *Post-mortem* changes in physical and chemical properties of bovine muscle. J. Food. Sci. 29: p. 590-596.

Gorg, A,. 1999. IPG-Dalt of very alkaline proteins. Methods Mol. Biol. 112: p. 197-209.

Gorg, A., G. Boguth, A. Kopf, G. Reil, H. Parlar and W. Weiss, 2002. Sample prefractionation with Sephadex isoelectric focusing prior to narrow pH range two-dimensional gels. Proteomics 2: p. 1652-1657.

Gorg, A., C. Obermaier, G. Boguth, A. Harder, B. Scheibe, R. Wildgruber and W. Weiss, 2000. The current state of two-dimensional electrophoresis with immobilized pH gradients. Electrophoresis 21: p. 1037-1053.

Hazbun, T.R., L. Malmstrom, S. Anderson, B.J. Graczyk, B. Fox, M. Riffle, B.A. Sundin, J.D. Aranda, W.H. McDonald, C.H. Chiu, B.E. Snydsman, P. Bradley, E.G. Muller, S. Fields, D. Baker, J.R. Yates, III and T.N. Davis,. 2003. Assigning function to yeast proteins by integration of technologies. Mol. Cell 12: p. 1353-1365.

Hedegaard J., P. Horn, R. Lametsch, H.S. Møller, P. Roepstorff, C. Bendixen and E. Bendixen, 2004. UDP-Glucose pyrophosphorylase is upregulated in carriers of the porcine RN^- mutation in the AMP-activated protein kinase. Proteomics 4: p. 2448-2454.

Henckel, P., O.N. Erlandsen, P. Barton-Gade and C. Bejerholm, 1997. Histo- and biochemical characteristics of the *Longissimus dorsi* muscle in pigs and their relationship to performance and meat quality. Meat Sci. 47: p. 311-321.

Hildrum, K.I., M. Solvang, B.N. Nilsen, T. Frøystein and J. Berg, 1999. Combined effects of chilling rate, low voltage electrical stimulation and freezing on sensory properties of bovine M. longissimus dorsi. Meat Sci. 52: p. 1-7.

Hood, L., 2003. Systems biology: integrating technology, biology, and computation. Mech. Ageing Dev. 124: p. 9-16.

Huh, W.K., J.V. Falvo, L.C. Gerke, A.S. Carroll, R.W. Howson, J.S. Weissman and E.K. O'Shea, 2003. Global analysis of protein localization in budding yeast. Nature 425: 686-691

Hunter, T.C., N.L. Andon, A. Koller, J.R. Yates and P.A. Haynes, 2002. The functional proteomics toolbox: methods and applications. J. Chromatogr. B Analyt. Technol. Biomed. Life Sci. 782: p. 165-181.

Hwang, I.H., C.E. Devine and D.L. Hopkins, 2003.The biochemical and physical effects of electrical stimulation on beef and sheep meat tenderness. Meat Sci. 65: p. 677-691.

Karp, N.A., D.P. Kreil and K.S. Lilley, 2004. Determining a significant change in protein expression with DeCyder during a pair-wise comparison using two-dimensional difference gel electrophoresis. Proteomics. 4: p. 1421-1432.

Klein, E., J.B. Klein and V. Thongboonkerd, 2004. Two-dimensional gel electrophoresis: a fundamental tool for expression proteomics studies. Contrib. Nephrol. 141: p. 25-39.

Koohmaraie, M., S.D. Shackelford, T.L. Wheeler, S.M. Lonergan and M.E. Doumit, 1995. A muscle hypertrophy condition in lamb (callipyge): characterization of effects on muscle growth and meat quality traits. J. Anim. Sci. 73: p. 3596-3607.

Koohmaraie, M., 1996. Biochemical factors regulating the toughening and tenderization processes of meat. Meat Sci. 43: p. 193-201.

Lametsch, R. and E. Bendixen, 2001. Proteome analysis applied to meat science: characterizing *postmortem* changes in porcine muscle. J. Agric. Food Chem. 49: p. 4531-4537.

Lametsch, R., A. Karlsson, K. Rosenvold, H.J. Andersen, P. Roepstorff and E. Bendixen, 2003. Postmortem proteome changes of porcine muscle related to tenderness. J. Agric. Food Chem. 51: p. 6992-6997.

Lametsch, R., Roepstorff, P., Møller, H.S. and E. Bendixen, 2004. Identification of Myofibrillar Substrates for u-Calpain. Meat Sci. 68: p. 515-521.

Lametsch, R., P. Roepstorff and E. Bendixen, 2002. Identification of protein degradation during *post-mortem* storage of pig meat. J. Agric. Food Chem. 50: p. 5508-5512.

Le Roy, P., J.M. Elsen, J.C. Caritez, A. Talmant, H. Juin, P. Sellier and G. Monin, 2000. Comparison between the three porcine RN genotypes for growth, carcass composition and meat quality traits. Genet. Sel. Evol. 32: p. 165-186.

Lescuyer, P., J.M. Strub, S. Luche, H. Diemer, P. Martinez, A. Van Dorsselaer, J. Lunardi and T. Rabilloud, 2003. Progress in the definition of a reference human mitochondrial proteome. Proteomics. 3: p. 157-167.

Link, A.J., 2002. Multidimensional peptide separations in proteomics. Trends Biotechnol. 20: p. S8-13.

Locker R.H., 1985. Cold-induced toughness of meat. In: "Advances in Meat Research Vol. 1 - Electric Stimulation" eds. AM Pearson and TR Dutson, AVI Publishing Co. Inc. Westport, CT, p. 1-44.

Maltin, C.A., C.C. Warkup, K.R. Matthews, C.M. Grant, A.D. Porter and M.I. Delday, 1997. Pig muscle fibre characteristics as a source of variation in eating quality. Meat Sci. 47: p. 237-248.

Maltin, C., D. Balcerzak, R. Tilley and M. Delday. 2003. Determinants of meat quality: tenderness. Proc. Nutr. Soc. 62: p. 337-347.

Mann, M. and O.N. Jensen, 2003. Proteomic analysis of post-translational modifications. Nat. Biotechnol. 21: p. 255-261.

Mariani, P., K. Lundström, U. Gustafsson, A.-C. Enfält, R. K. Juneja and L. Andersson, 1996. A major locus (RN) affecting muscle glycogen content is located on pig Chromosome 15. Mamm. Genome 7: p. 52-54.

McKeith, F.K., D.L. DeVol, R.S. Miles, P.J. Bechtel and T.R. Carr, 1985. Chemical and sensory properties of thirteen major beef muscles. J. Food Sci. 50: p. 869-872.

McPherron, A.C. and S.J. Lee, 2002. Suppression of body fat accumulation in myostatin-deficient mice. J. Clin. Invest. 109: p. 595-601.

Monin, G., E. Lambooy and R. Klont,1995. Influence of temperature variation on the metabolism of pig muscle *in situ* and after exercise. Meat Sci. 40: p. 149-158.

Morzel, M., C. Chambon, M. Hamelin, T. Santé-Lhoutellier, T. Sayd and G. Monin, 2004. Proteome changes during pork meat ageing following use of two different pre-slaughter handling procedures. Meat Sci. 67: p. 698-696.

O'Farrell, P.H., 1975. High resolution two-dimensional electrophoresis of proteins. J. Biol. Chem. 250: p. 4007-4021.

O'Halloran, G.R., D.J. Troy and D.J. Buckley, 1997. The relationship between early *post mortem* pH and the tenderisation of beef muscles. Meat Sci. 45: p. 239-251.

Ong, S.E., L.J. Foster and M. Mann, 2003. Mass spectrometric-based approaches in quantitative proteomics. Methods 29: p. 124-130.

Ouzounis, C.A. and A. Valencia, 2003. Early bioinformatics: the birth of a discipline - a personal view. Bioinformatics. 19: p. 2176-2190.

Pedersen, S.K., J.L. Harry, L. Sebastian, J. Baker, M.D. Traini, J.T. McCarthy, A. Manoharan, M.R. Wilkins, A.A. Gooley, P.G. Righetti, N.H. Packer, K.L. Williams and B.R. Herbert, 2003. Unseen proteome: mining below the tip of the iceberg to find low abundance and membrane proteins. J. Proteome. Res. 2: p. 303-311.

Picard, B., L. Lefaucheur, C. Berri and M.J. Duclos, 2002. Muscle fibre ontogenesis in farm animal species. Reprod. Nutr. Dev. 42: p. 415-431.

Rabilloud, T. 2000. Detecting proteins separated by 2-D gel electrophoresis. Anal. Chem. 72: p. 48A-55A.

Renand G., B. Picard, C. Touraille, P. Berge and J. Lepetit, 2001. Relationships between muscle characteristics and meat quality traits of young Chatolais bulls. Meat Sci. 59: 49-60

Santoni, V., M. Molloy and T. Rabilloud, 2000. Membrane proteins and proteomics: un amour impossible? Electrophoresis 21: p. 1054-1070.

Scanga, J.A., K.E. Belk, J.D. Tatum, T. Grandin and G.C. Smith, 1998. Factors contributing to the incidence of dark cutting beef. J. Anim. Sci. 76: p. 2040-2047.

Spandidos, A. and T.H. Rabbitts, 2002. Sub-proteome differential display: single gel comparison by 2D electrophoresis and mass spectrometry. J. Mol. Biol. 318: p. 21-31.

Talamo, F., C. D'Ambrosio, S. Arena, P. Del Vecchio, L. Ledda, G. Zehender, L. Ferrara and A. Scaloni, 2003. Proteins from bovine tissues and biological fluids: defining a reference electrophoresis map for liver, kidney, muscle, plasma and red blood cells. Proteomics 3: p. 440-460.

Taylor, R.G., G.H. Geesink, V.F. Thompson, M. Koohmaraie and D.E. Goll, 1995. Is Z-Disk Degradation Responsible for Postmortem Tenderization? J. Anim. Sci. 73: p. 1351-1367.

Taylor, R.G. and M. Koohmaraie, 1998. Effects of *postmortem* storage on the ultrastructure of the endomysium and myofibrils in normal and callipyge longissimus. J. Anim. Sci. 76: p. 2811-2817.

Tyers, M. and M. Mann, 2003. From genomics to proteomics. Nature 422: p. 193-197.

Unlu, M., M.E. Morgan and J.S. Minden, 1997. Difference gel electrophoresis: a single gel method for detecting changes in protein extracts. Electrophoresis 18: p. 2071-2077.

Uytterhaegen, L., E. Claeys and D. Demeyer, 1994. Effects of exogenous protease effectors on beef tenderness development and myofibrillar degradation and solubility. J. Anim. Sci. 72: p. 1209-1223.

Valafar, F., 2002. Pattern recognition techniques in microarray data analysis: a survey. Ann. N. Y. Acad. Sci. 980: p. 41-64.

Valin, C. and C. Touraille, 1982. Prediction of lamb meat quality traits based on muscle biopsy fibre typing. Meat Sci. 6: p. 257-263.

Wilkins, M.R., C. Pasquali, R.D. Appel, K. Ou, O. Golaz, J.C. Sanchez, J.X. Yan, A.A. Gooley, G. Hughes, I. Humphery-Smith, K.L. Williams and D.F. Hochstrasser, 1996. From proteins to proteomes: large scale protein identification by two-dimensional electrophoresis and amino acid analysis. Biotechnology (N.Y.) 14: p. 61-65.

Yates, J.R., 2004. Mass spectral analysis in proteomics. Annu. Rev. Biophys. Biomol. Struct. 33: p. 297-316.

Muscle metabolism in relation to genotypic and environmental influences on consumer defined quality of red meat

D.W. Pethick[1], D.M. Fergusson[2], G.E. Gardner[3], J.F. Hocquette[4], J.M. Thompson[3] and R. Warner[5]*
Cooperative Research Centre for Cattle and Beef Quality, [1]Murdoch University, Perth, 6150, Western Australia, [2]CSIRO Livestock Industries, Armidale, New South Wales, Australia, 2350, [3]University of New England, Armidale, 2351, New South Wales, Australia, [4]INRA, Herbivore Research Unit, Theix, France, [5] Department of Primary Industries, 600 Sneydes Road, Werribee Victoria 3030 Australia.
**Correspondence: d.pethick@murdoch.edu.au*

Summary

This paper discusses the management of consumer defined beef palatability using a carcass grading scheme which utilizes the concept of total quality management. The scheme called Meat Standards Australia (MSA) has identified the Critical Control Points (CCPs) from the production, pre-slaughter, processing and value adding sectors of the beef supply chain and quantified their relative importance using large-scale consumer testing. These CCPs have been used to manage beef palatability in two ways. Firstly, CCPs from the pre-slaughter and processing sectors have been used as mandatory criteria for carcasses to be graded. Secondly, other CCPs from the production and processing sectors have been incorporated into a model to predict palatability for individual muscles. The CCPs from the production (breed, ossification and implants of hormonal growth promotants), pre-slaughter and processing (pH/temperature window, alternative carcass suspension, marbling and ageing) sectors are reviewed. The paper then discusses the interacting roles of nutrition and genotype as determinants of muscle energy pattern with respect to glycogen and fat metabolism. In particular the roles of fibre type and/or pattern of muscle energy metabolism is discussed in relation to the high ultimate pH syndrome (dark cutting beef), the rate of *post-mortem* glycolysis and the response to electrical stimulation. Finally the development of intramuscular fat is discussed in terms of growth and development, biochemical regulation and nutritional modification.

Keywords: beef, tenderness, palatability, Meat Standards Australia, genotype, breed, glycogen, ultimate pH, pH decline, intramuscular fat, muscle fibre type, metabolism

Introduction

Meat palatability is a function of production, processing, value adding and cooking method used to prepare the meat for consumption by the consumer. Failure of one or more links in the beef supply chain increases the risk of a poor eating experience for the consumer. A guarantee for eating quality can only be given if the links that most affect eating quality are controlled along the meat production chain.

An example of a 'paddock to plate' quality assurance system which manages meat quality along the entire length of the meat production chain is the new grading scheme called Meat Standards Australia (MSA), which is presently being implemented for the Australian domestic beef market by Meat and Livestock Australia (MLA).

This paper will initially overview the main factors used in the MSA system to describe the development and implementation of a quality assurance system which manages and describes the palatability of meat for the consumer. A more detailed analysis of MSA can be found in Thompson

(2002). Next the paper describes the impact of genetic and nutritional manipulation on the metabolism of muscle in relation to glycogen and lipid metabolism, which are important factors which may affect beef quality traits (Hocquette *et al.*, 1998).

A total quality management approach to meat quality

The MSA grading scheme uses a total quality management approach to identify critical control points (CCPs) and to predict the quality of the final product. Much of the research undertaken by MSA was not new. New components included the use of a large-scale consumer testing system that allowed the effects of the CCPs to be quantified using a standard evaluation procedure. A second new feature was the introduction of a cuts-based-grading system to improve the accuracy of predicting palatability in beef and the need to grade all muscles in the carcass. Analysis of the MSA database showed that the variation in palatability explained by muscles was approximately 60 times greater than that explained by the variation between animals for the same muscle.

The consumer testing system

At the commencement of MSA the decision was made to use sensory results derived from untrained consumers as the means to describe palatability of beef. Although objective measurements (such as shear force) have the advantage of being relatively cheap, they are rather simplistic one dimensional measures of a complex set of interactions which occur when cooked meat is chewed and masticated in the mouth. Furthermore, studies in France showed that shear force may explain only up to 48% of total variability in tenderness and this proportion depends on the breed and the production system (Brouard *et al.*, 2001).

The consumer sensory testing protocol used by MSA (Polkinghorne *et al.*, 1999) was based on existing protocols in use by the American Meat Science Association protocols (AMSA 1995). Briefly, untrained consumers were asked to score tenderness, juiciness, flavour and overall acceptability on a scale of 0 to 100. They also graded the sample on the following word associations; unsatisfactory, good everyday (3 star), better than everyday (4 star), or premium quality (5 star). To combine the 4 sensory dimensions into a single palatability or meat quality score (MQ4), weightings were formulated from a discriminant analysis (0.4, 0.1, 0.2 and 0.3 for tenderness, juiciness, flavour and overall acceptability, respectively). The palatability scores were then used to calculate the optimum boundaries for the grades assigned by the consumers with 45.5 separating ungraded (fail) and 3 star categories, 63.5 for 3 and 4 star, and 76.5 and above for 5 star. The current iteration of the model in June 2004 is based on a data base which contains responses from 60,100 consumers.

Components of the MSA model

The specifications for producers and processors to supply carcasses which are eligible for grading by MSA include compliance with a set of conditions aimed at reducing pre-slaughter stress and optimizing processing conditions.

Producers need to be registered and must adhere to MSA Cattle Handling Guidelines to minimize stress. They must declare the *Bos Indicus* % content of their cattle, and whether the cattle can be classed as milk fed calves. The time of loading must be supplied, the cattle trucked direct to slaughter, not mixed in lairage, and killed the day after dispatch.

Abattoir procedures are audited within a QA system to ensure pH and temperature relationships are within the prescribed window to achieve optimal palatability. To minimise variation in cooling rates carcasses must have an even distribution of fat with at least 3 mm of fat at the rib site. All carcasses must have an ultimate pH below 5.7 and a USDA ossification score (Romans *et al.*, 1994) below 300.

A summary of the input factors which drive the palatability prediction model are:

- *Bos indicus* %: This is specified on the producer declaration and/or estimated as the hump height which is measured on the carcass and related to carcass weight. The magnitude of the *Bos indicus* effect varies with muscle. There were no other breed effects that could not be explained by the other commercial production parameters measured (marbling score, rib fat, ossification, pHu).
- Sex: A sex adjustment (steers versus heifers) is made which varies with muscle and is relatively small, being of the order of 2 palatability units. As yet beef from entire males is not included within the model.
- USDA Ossification score: This is used as an estimate of animal physiological age and proved more revealing than using traditional dentition classes which have previously been used by the Australian beef industry. As ossification score increases from 100 to 200 the consumer score declines 5, 10 and 12 points for the *m. longissimus thoracis (LT)*, *m. gluteus medius* (GM) and *m. semitendinosus* (ST) respectively indicating a greater effect of animal age in the leg muscles.
- Milk Fed Veal (MFV): Muscles from calves weaned immediately prior to slaughter (at approximately 8-10 months of age) receive a higher score than from weaned cattle of equivalent ossification score. The magnitude of the MFV effect is typically 5 to 6 palatability units.
- Carcass Hanging Method: This effect is applied on an individual muscle basis, with different values for each muscle and hang combination. Hanging methods are AT (Achilles tendon) or TS (Tenderstretch from the ligament or related procedures). Differences in palatability between AT and TS carcasses are in the order of 5-6 points for muscles which are under tension due to the TS process (GM and LT muscles).
- Intramuscular fat or marbling: As marbling score and rib fat were positively correlated, both parameters are used to assess the impact of marbling on palatability of individual cuts. An increase in USDA marble score from 250 to 550 (equivalent to an increase from 0 to 3 marble score on the AUS-MEAT system) results in an increase of 8 palatability units for the LT muscle. The adjustment made for marbling depends on the muscle as different muscles express different levels of intramuscular fat.
- Ultimate pH: A small improvement in eating quality occurs as pH declines from the threshold of 5.7 (ca. 1 palatability unit).
- Ageing: The rate of ageing is estimated differently for each muscle within each hanging option. MSA product cannot be sold to consumers before 5 days post slaughter and aging to 21 days increases the consumer score by up to 4 units.
- Hormonal growth promotants: Hormonal growth promotants as combinations of oestrogenic or androgenic steroids are sometimes used in Australia to increase the rate of lean tissue deposition. Recent MSA research has shown that use of these hormones reduce the palatability of beef particularly in the LT muscle. The effects are lower in other muscles and reduced further by aging from 5 out to 21 days.
- Cooking method and muscle: Palatability for individual muscles is predicted for a specific cooking method. Larger muscles generally have several cooking options. Grilling (25 mm thickness) low connective tissue cuts resulted in the highest palatability scores. Roasting low connective cuts gave similar scores to grilling, whereas for the high connective cuts roasting gave higher palatability scores than did grilling. Stir frying (10 mm) and thin slicing (4 mm) gave similar results to grilling for low connective muscles, but relatively higher scores in the high connective tissue muscles. The magnitude of the muscle effect is large and in the order of 30-40 palatability units regardless of cooking method.

Muscle metabolism and the critical control points

The critical control points underpinning palatability are directly dependant on key biochemical parameters including on one hand, muscle characteristics of alive animals and, on the other hand, *post-mortem* muscle biochemistry. The former include glycogen metabolism, intramuscular adipocyte number and size (marbling), connective tissue chemistry, and pigment content which contributes to colour. The latter includes rates of proteolysis during ageing. Among the muscle characteristics linked to metabolism of alive animals, the pH of meat (which depends on glycogen metabolism before slaughtering) and intramuscular fat content are the major factors associated with beef quality traits. This is true in both Australia (as shown before) and Europe. Indeed, studies with young bulls in France have shown that pH at 3 hours *post-mortem* and intramuscular fat content explain 52% and 56% of the variation in tenderness and flavour respectively (Dransfield *et al.*, 2003). Furthermore, any change in growth rate (Cassar-Malek *et al.*, 2004) and feeding conditions (Listrat *et al.*, 2001a) were shown to affect primarily muscle metabolic activity. This section will thus focus on glycogen and fat metabolism of muscle.

Glycogen metabolism and ultimate pH

The MSA model has a requirement that the ultimate pH of the LT muscle (measured at the $12/13^{th}$ rib) must be 5.7 or less. As ultimate pH increases beef becomes less juicy, has altered cooking properties, lacks visual appeal and has reduced shelf life (Shorthose, 1989). In the pH range of 5.8 - 6.2, beef is also tougher (Purchas and Aungsupakorn, 1993). To achieve an ultimate pH of 5.5, muscle needs to contain at least 50-60 μmoles/g of glycogen immediately pre-slaughter to form sufficient lactic acid to lower pH (Tarrant, 1989).

Glycogen reserves at slaughter are a function of the initial levels of glycogen (i.e. on farm level) and the losses due to stresses placed on the animal during the immediate pre-slaughter period. It is proposed that both genotype and nutrition play an interacting role in determining this balance between the initial level and loss of glycogen from muscle.

One mechanism is based on differential metabolism of glycogen in the contrasting muscle fibre types. Muscles consist of distinct fibre types that can be differentiated on the basis of their contractile and metabolic properties. They are: (i) slow-twitch oxidative fibres (type I); (ii) fast-twitch oxidative-glycolytic fibres (type IIa); and (iii) fast-twitch glycolytic fibres (type IIb) (Peter *et al.*, 1972). The metabolism of glycogen is different in the various fibres such that type I fibres have low levels of glycogen. Type IIa fibres have higher levels of glycogen, a high rate of glycogen resynthesis and are least affected by stress. Type IIb fibres have lower glycogen levels, slow rates of glycogen synthesis and are therefore the most susceptible to stress induced glycogen depletion (Monin, 1981b, Pethick *et al.*, 1999). These differences can be largely explained by the different enzyme compliment of each fibre type of mammals (Saltin and Gollnick, 1983) including cattle (Talmant *et al.*, 1986). Thus the very high activity of glycogen phosphorylase in combination with low activities of glycogen synthase and hexokinase mean that type IIb muscle fibres rapidly deplete and slowly replete glycogen levels when compared to type IIa. Clear evidence of this is shown in Figure 1 where the glycogen content of muscle was compared between normal and sheep suffering a congenital deficiency of glycogen phosphorylase within muscle known as McArdles disease (Kumar, 1998). The data shows that basal levels of glycogen increase dramatically in the absence of glycogen phosphorylase and are constant with glycogen turnover being a regulator of concentration. Further work is needed to access the rates of glycogen turnover in cattle. Some authors have speculated that the expression and/or activity of glucose transporters (which control the entry of blood circulating glucose in muscle cells) might regulate muscle glycogen content, and hence the final quality of beef (Hocquette and Abe, 2000).

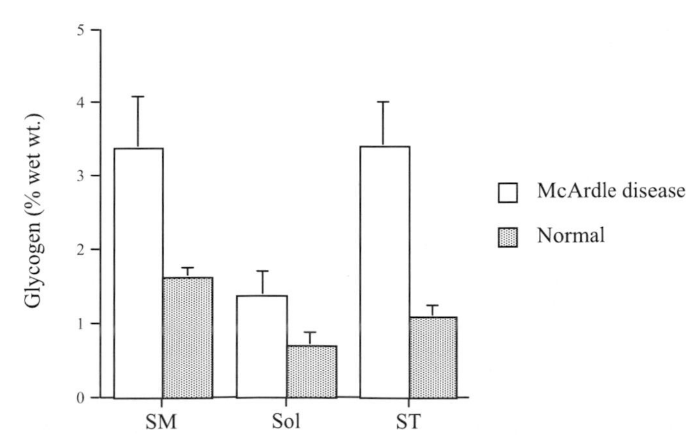

Figure 1. The level of glycogen in the m. semimembranosus *(SM),* m. soleus *(Sol)* m. semitendinosus *(ST) of normal and McArdle disease affect sheep. Data is mean ± sem for 6 sheep per treatment.*

The effects of fibre type and or pattern of energy metabolism on the response of muscle to nutrition are also dramatic such that the more aerobic *m. semimembranosus* (SM) shows a strong linear relationship between the extent of glycogen repletion during a 72h period after exercise depletion whereas the more anaerobic ST muscle showed no significant repletion during the same time period (Gardner *et al.*, 2001). There are also important chronic effects of nutrition on the level of muscle glycogen (Pethick *et al.*, 1999).

The occurrence of dark-cutting in beef carcasses in Australia has been reported to have a seasonal effect although the peak months of dark-cutting vary between years and region. Our studies have investigated the effect of season (Pethick *et al.*, 1999). There was a strong seasonal influence on the concentration of glycogen in muscle with consistently low levels in winter and summer and high levels in spring. This drop in muscle glycogen concentration is partly explained by declining animal growth rate, which is driven by changes in pasture availability and quality. There was a positive correlation between live weight change and muscle glycogen concentration (r^2=0.69, P=0.04) and it was concluded that a growth rate of above 1kg/day is needed to assure a level of muscle glycogen that is sufficient to help reduce the incidence dark cutting for cattle grazing in southern Australia.

Recently we have completed an experiment employing the exercise depletion/repletion methodology of Gardner *et al.* (2001) to investigate rates of glycogen repletion in different breeds of cattle. Cross-bred cattle (9 months of age) from either Peidmontese, Angus or Wagyu sires were maintained on either roughage or concentrate rations. Marked differences were evident between the breeds in both levels of glycogen depletion through exercise (trotting at 9 km/h for 5 x 15min intervals with a 15 minute rest period between each interval resulting in a blood lactate of 2-3 mM at the end of exercise), with the Angus sired cattle depleting almost 60% more muscle glycogen than Piedmontese ($P<0.05$), and rates of glycogen repletion 72 hours following exercise, with the Wagyu sired cattle repleting about 50% more muscle glycogen ($P<0.05$, Figure 2). These responses are apparent after adjustment for live weight, fat depth (P8), metabolisable energy intake, and starting muscle glycogen concentration, and therefore cannot be attributed to these factors. Given that Wagyu are noted for a propensity to accrete fat at the intramuscular adipocyte, the repletion response appears to be consistent with a general trend for increased substrate deposition (either fat or glycogen) by this breed at the level of the muscle. We are currently profiling the muscle tissue for fibre type and aerobic/anaerobic enzyme activity to test the

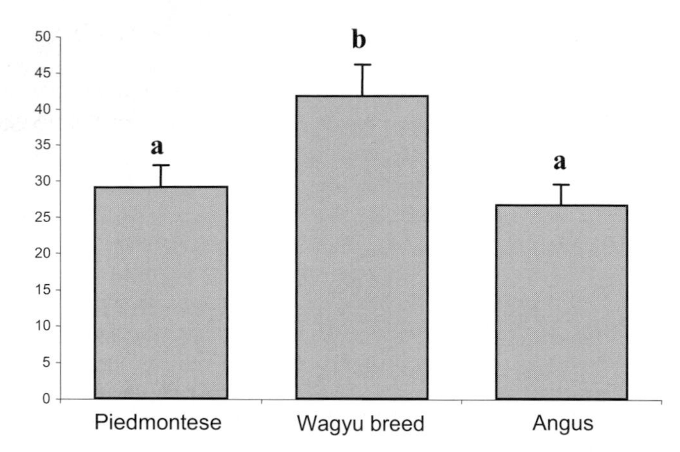

Figure 2. The effect of sire breed on muscle glycogen repletion during 72 h following exercise in cattle. Values are least square means ± sem (adjusted).

hypothesis that these differences are related to fibre type and/or pattern of energy metabolism within muscle.

In a recent study we have investigated the effect of sire on the level of glycogen both at pasture and subsequently at slaughter. Ten month old cross bred cattle sired by one of 3 Simmental bulls (22 progenies per bull) were used in the study. Muscle glycogen content was measured on biopsy samples taken at pasture and then on muscle collected immediately *post-mortem*. Sire 3 had a significant effect on *m. semitendinosus* (ST) glycogen concentration at slaughter but not at pasture (Table 1, $P<0.05$). Progeny from sire 3 had about 15% more glycogen within the ST when compared to the other two sires. The fact that this response was found only for the ST suggests that the differences were associated with stress, that is sire 3 progeny were most probably less stress sensitive. The basis of this response could clearly be complex but points to the possibility of selecting sires on the basis that their progeny loose less glycogen from muscle when subjected to commercial stressors.

Table 1. Effect of sire on glycogen content (%) of the m. semimembranosus *(SM) and* m. semitendinosus *(ST) at different times pre-slaughter. Values are Means ± sem.*

	Sire			Significance of Effect
	1	2	3	
SM (Slaughter)	1.50 ± .037	1.54 ± .055	1.61 ± .034	ns
ST (Slaughter)	1.15 ± .042[a]	1.14 ± .062[a]	1.28 ± .039[b]	.044
SM (pre-slaughter)	1.61 ± .051	1.55 ± .081	1.70 ± .054	ns
ST (pre-slaughter)	1.33 ± .074	1.25 ± .118	1.33 ± .078	ns

Values within rows followed by different letters are different ($P<0.05$).

Glycogen metabolism and the pH/temperature window

The pH/temperature window was one of the initial specifications for the MSA 'carcass pathways' grading scheme. The concept of the window originated from the results of Locker and Hagyard (1963) who showed that myofibrillar shortening occurred when pre-rigor muscle was held at either low or high temperatures. At low muscle temperatures extensive shortening occurred and the subsequent increased toughness was termed 'cold shortening'. Pearson and Young (1989) considered that for cold shortening to occur the muscle pH had to be greater than 6.0 with ATP still available for muscle contraction and the muscle temperature to be less than 10°C. At high muscle temperatures some shortening also occurred, in some cases (but not all) leading to increased toughness (Uruh *et al.*, 1986; Simmons *et al.*, 1997). This effect was termed rigor or heat shortening and was considered to be due to the combination of high temperature and low pH in the muscle causing early exhaustion of proteolytic activity and so reduced tenderisation during aging (Dransfield, 1993; Simmons *et al.*, 1996; Hwang and Thompson, 2001) and increased drip loss (Denhertogmeischke *et al.*, 1997). These studies lead to the development of the MSA pH/temperature window, whereby controlled use of chilling and/or electrical inputs during processing were managed to achieve a pH/temperature relationship of greater than pH 6 for muscle temperatures greater than 35°C, and a pH of less than 6 for muscle temperatures less than 12°C (Figure 3). The pH/temperature window is currently measured for the LL muscle at the level of the 12/13th rib. Currently further work is underway to optimize the pH/temperature window of the deeper muscles (such as the GM muscle) which are more susceptible to heat shortening.

Electrical stimulation can elicit several benefits in tenderness because (i) it can be used to prevent of cold shortening, (ii) it causes increased fracturing and disruption of the myofibrillar structure of muscle, and (iii) it accelerates *post-mortem* proteolysis (Fergusson *et al.*, 2001). Whereas electrical stimulation is widely used in Australia, its use in Europe is minimal , especially in the South of Europe. This contributes among other factors (breeds, fattening systems) to the differences in the final quality of beef between these two parts of the world. When the MSA pH/temperature window was implemented as part of the abattoir audit it was found that many Australian abattoirs had an over use of electrical stimulation, with carcasses entering the heat shortening region. This was due in part to other electrical inputs being installed in the slaughter chain (eg immobilisers and rigidity probes), which along with electrical stimulation, accelerate glycolytic rate (Petch and Gilbert, 1997). It is clear that differences between abattoirs in the positioning of the stimulator,

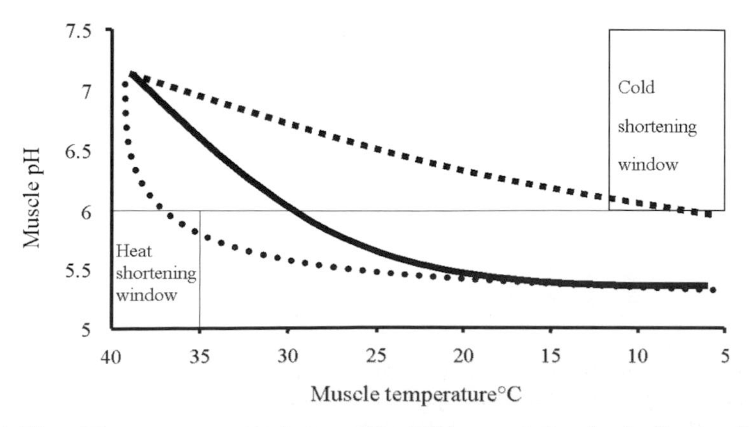

Figure 3. The pH/temperature window used by MSA to optimise the decline in pH relative to the temperature of the muscle. The solid line represents an optimal rate of decline, the dashed line a cold shortening, and the dotted line, a heat shortening scenario.

effectiveness of contact electrodes, and speed of the chain make it impossible to recommend a uniform protocol for stimulation and so MSA chose to regularly audit individual abattoirs.

The extent to which genotype and nutrition can influence the extent of pH drop due to electrical stimulation and then the subsequent rate of pH decline is not clear. Traditionally the rate of pH decline is thought to be driven by the rate of ATP hydrolysis via various muscle ATP'ase systems. In a recent study sheep were used to test the influence of muscle glycogen level pre slaughter on the pH decline post slaughter in 3 different muscles. Delta pH was calculated as the difference between the electrically stimulated and non-stimulated sides immediately following stimulation. Rate of pH decline was modeled using an exponential function from which pH at 3 hours post slaughter (pH3) was calculated. Temperature and pH readings were corrected to account for temperature differences and the impact of temperature on rate of pH decline. Muscle glycogen concentration was found to affect both delta pH and pH3. The dependence of delta pH on muscle glycogen concentration differed between muscles, with no relationship in the *m. semimembranosus* (SM) and *m. longissmus thoracis* (LT), but a strong positive relationship in the more anaerobic ST ($P<0.01$, Figure 4).

Generally delta pH values were higher in the ST ($P<0.01$) indicating that the ST is capable of very fast rate of ATP hydrolysis and so glycolytic rates but only when adequate substrate (muscle glycogen) is available. Alternatively, the more aerobic SM and LT muscles do not have the same glycolytic potential, and therefore the availability of muscle glycogen is not the rate limiting factor, over the ranges measured in this study.

pH3 was markedly reduced as muscle glycogen concentration increased (Figure 5), indicating a faster rate of pH decline. This response was curvilinear ($P<0.001$), plateauing at a muscle glycogen concentration of about 60 μmoles/g (after adjusted for the starting pH immediately following stimulation in both stimulated and non-stimulated sides). pH3 also differed between muscles, with the ST having the fastest rates of pH decline ($P<0.001$), with pH3 values approx 0.1 pH units lower than the SM and LT at all glycogen concentrations (Figure 5). Electrical stimulation generally resulted in lower pH3 values and subsequently faster rates of pH fall ($P<0.01$, Figure 5), however it also made the relationship between pH3 and muscle glycogen concentration more pronounced ($P<0.05$). Generally these results support a role for muscle glycogen concentration to impact on pH change in the more anaerobic muscle types. The impact of muscle glycogen

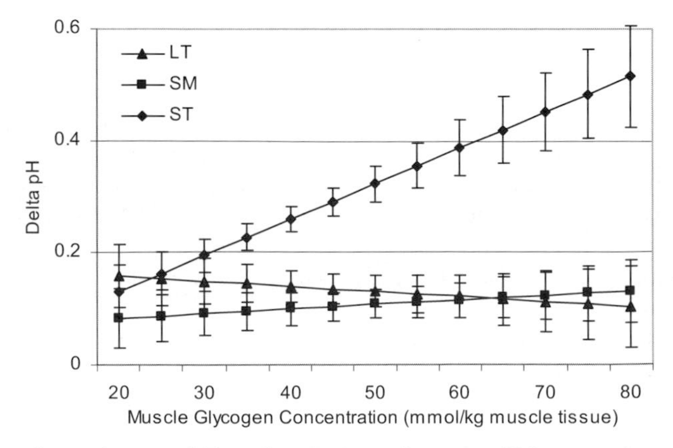

Figure 4. Effect of muscle type (LT, m. longissimus thoracis*; SM,* m. semimembranosus*; ST,* m. semitendinosus*) on the response to electrical stimulation (delta pH) at different muscle glycogen concentrations.*

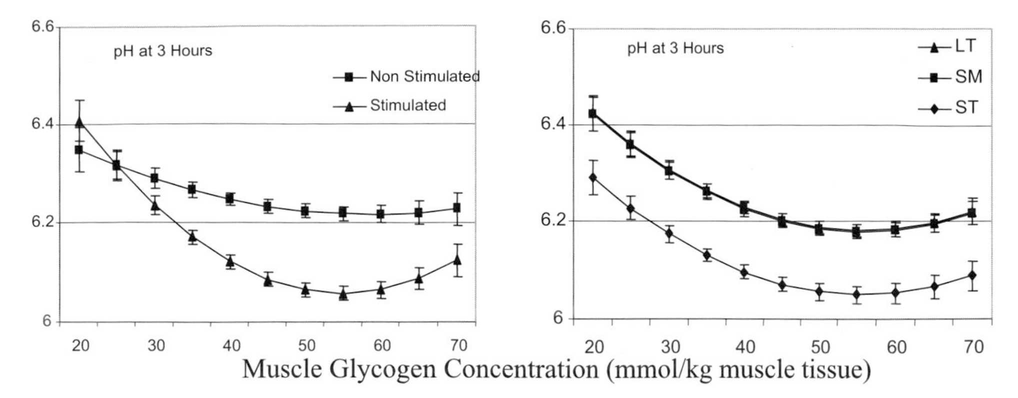

Figure 5. Effect of muscle glycogen concentration on pH at 3 hours post-mortem (rate of pH decline) in stimulated and non-stimulated carcasses (main effect across all muscles), and in the m. semimembranosus *(SM),* m. semitendinosus *(ST), and* m. longissimus thoracis *(LT) (main effect across stimulated and non-stimulated carcasses).*

concentration is further supported by data collected in cattle over a 6 month period at a commercial abattoir by Daly *et al.* (2002). In both stimulated and non-stimulated carcasses increasing muscle glycogen concentration resulted in faster rates of pH decline and this response was curvilinear in stimulated groups, plateauing at about 90 umoles/g muscle glycogen.

There is some further evidence of genotypic effects on muscle glycogen metabolism. For example, double-muscled cattle which have more glycolytic muscle tissue than normal cattle (see below) and are also characterized by an accelerated *post-mortem* glycolysis resulting in a rapid fall of the *post-mortem* pH and a rapid rise in lactic acid (Clinquart *et al.*, 1994; Fiems *et al.*, 1995). Additionally, double-muscled cattle are more sensitive to stress, which may result prior to slaughter in reduced carbohydrate stores (Monin, 1981a), especially in hypertrophic muscles (Fernandez *et al.*, 1997).

Given this information we conclude that nutrition will effect pH decline, especially in response to electrical stimulation, via making a contribution to elevating the initial glycogen content of muscle. Furthermore we also predict that genotype will influence pH decline, especially the response to electrical stimulation (delta pH). Any genetic effects to increase the type IIb fibre proportion would increase the response to electrical stimulation. Future studies need to quantify such effects so as the processing sector can better optimize pH decline via chilling and electrical inputs especially in countries where there is a broad base of animal genotypes and production systems such as in Australia. In the longer term, better control of electrical inputs, in conjunction with a prediction model to allow the stimulation requirements to be specified for different classes of cattle being processed at specific abattoirs, may be the answer.

Genotype as a determinant of muscle fibre type and pattern of energy metabolism

Given that the metabolic properties of muscle fibres will strongly effect the metabolism of muscle glycogen in the living and *post-mortem* animal what evidence is there that genotypic differences can effect muscle fibre type proportions and the final pattern of energy metabolism ? Double-muscled cattle display strong muscle hypertrophy (about 20%) and lower fat deposition in the carcass. Their muscles are more glycolytic since they contain a higher proportion of fast glycolytic fibers than bovines with 'normal' muscle mass. Double muscle cattle are also characterized by an enhanced muscle sensitivity to insulin (Hocquette *et al.*, 1998), and by lower levels of triiodothyronine, insulin and glucose plasma concentrations (Hocquette *et al.*, 1999), underlining the importance of the metabolic and hormonal status in the control of carcass composition and muscle characteristics. As a general rule, whichever the species and by whatever mechanism, this type of selection induces an increase in glycolytic muscle energy metabolism and an enhanced sensitivity to insulin (Hocquette *et al.,* 1998). Indeed, the comparison of two lines of young Charolais bulls obtained by divergent selection on growth rate and feed efficiency demonstrated that an increased lean to fat ratio was associated with lower intramuscular fat content (Renand *et al.*, 1994), lower oxidative metabolism especially in oxidative muscles (Cassar-Malek *et al.*, 2003), a greater number of fibers, a higher proportion of fast glycolytic fibers and a lower proportion of slow fibers (Duris *et al.*, 1999). Furthermore, the Blonde d'Aquitaine breed, in which neither deletion nor mutation in the myostatin gene have been yet identified (Grobet *et al.*, 1998), shows similar muscle characteristics to those of double muscle cattle (Listrat *et al.*, 2001b).

In conclusion, whatever the genetic basis (breeds with or without mutations in the myostatin gene, genetic selection on growth or muscle parameters within a breed), muscle hypertrophy is associated with a higher glycolytic muscle metabolism and a lower intramuscular fat content. The extent to which these genetic differences will impinge on glycogen metabolism is not known but would seem highly probable based on the above discussion. Thus beef cattle with a more glycolytic muscle metabolism would need more careful management with respect to pre slaughter nutrition and stress management so as to prevent the elevated ultimate pH syndrome. In addition such cattle will show a different pH/temperature decline post slaughter with the response to electrical stimulation being greater and modulated by the pre slaughter initial glycogen concentration. There would almost certainly be effects on retail colour stability in the form of metmyoglobin formation which is effected by myoglobin content and *post-mortem* oxygen consumption rate (Trout, 2002), both of which will be influenced by the nature of the energy metabolism within muscle.

Intramuscular fat or marbling

Role in palatability

Although marbling is generally an integral part of any beef grading scheme the literature suggests that it has only a minor association with palatability. Dikeman (1987) concluded that marbling accounted for only 10 to 15% of the variance in palatability. The MSA research would agree and showed that the contribution of marbling to palatability was significant but importantly just one of several factors. However, Thompson (2001) concluded that as variations in tenderness are controlled by schemes such as MSA, marbling will become a more important determinant of palatability due to its specific contribution to juiciness and flavour of grilled steaks for Australian consumers.

There is also a concern that at very low levels of intramuscular fat found in young highly muscled lean cattle (double-muscled genotypes, young bulls from Belgian Blue or Blonde d'Aquitaine for instance), that the meat is perceived as dry and less tasty. This is especially true for French and Belgian beef breeds which have been selected in favour of high muscle growth potential and low fat deposition. Savell and Cross (1986) concluded that the minimum requirement for ether extractable fat in order to achieve acceptable consumer satisfaction for grilling cuts was 3% on a fresh uncooked basis.

Development of intramuscular fat

A common conclusion from animal developmental studies is that intramuscular fat is late developing (Vernon 1981). Indeed the usually quoted developmental order is abdominal, then intermuscular, then subcutaneous, then finally intramuscular. However, because fat is deposited at a greater rate than lean tissues later in life, the concentration of fat in muscle will inevitably increase later in an animal's life. Therefore the commercial trait, marbling, visible intramuscular fat or actual percentage intramuscular fat is late maturing. This does not mean that the rate of fat accretion in intramuscular adipocytes is also late maturing. The study of Johnson *et al.* (1972), showed that the proportional distribution of fat between carcass pools is found to be constant over a wide range of carcass fat contents (in the range from 5 to over 150 kg total fat) indicating that the major fat depots grow in the same proportion as animals fatten. The results of Pugh *et al.* (2005) are also consistent with this observation.

The development of intramuscular fat is shown in Figure 6. The data suggests a period of minimal change of intramuscular fat content in young ages followed by a linear increase between a carcass weight of about 200 - 400 kg at least for American Angus x Hereford (Duckett *et al.*, 1993), Australian Angus (Pugh *et al.*, 2005) or Japanese Black x Holstein (Aoki *et al.*, 2001) type cattle undergoing prolonged grain feeding. Based on this data we hypothesise two key genetic drivers of intramuscular fat development (i) the initial or 'starting' intramuscular fat content at ≤ 200kg carcass weight, and (ii) the potential for muscle growth.

The initial or 'starting' intramuscular fat content at ≤ 200kg carcass weight is likely driven by the genetic predisposition for development of adipocytes at the intramuscular site relative to other depots. Importantly there is a proportional developmental difference that is maintained when the American or Australian cattle are compared to the Japanese Black cross cattle such that the starting (2 vs 4%) and final (13 vs 27%) intramuscular fat contents are proportionally different at about 2 fold (Figure 6). Cellular studies in rabbits (Gondret *et al.*, 1998) have shown that intramuscular fat develops due to both an increased number and size of clustered adipocytes with associated increases in lipogenic enzyme activity (Gondret *et al.*, 1997). The cattle data would suggest that the potential for cellular development of adipocytes is fixed relatively early in life and there after changes in either size and or number of cells occurs in proportion to the initial cell number and/or lipogenic proteins and so we propose the theoretical response shown is Figure 7. This would clearly indicate that a variety of 'fat' measurements taken on muscle tissue in early life would hold great potential for predicting subsequent intramuscular fat development. Examples would include intramuscular fat content (perhaps by non invasive methods such as ultrasound), marker

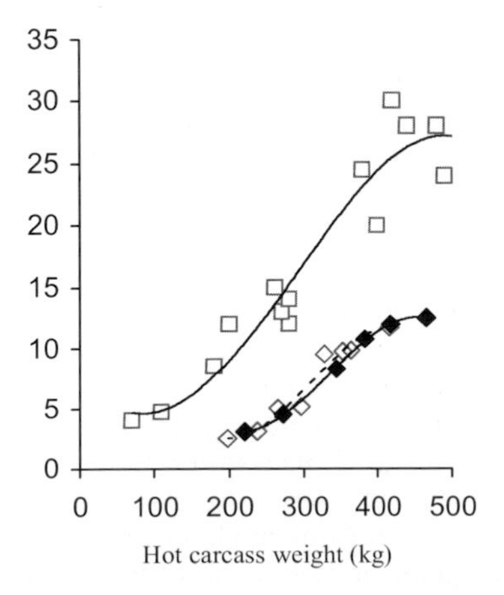

Figure 6. The relationship between carcass weight and intramuscular fat content of the m. longissimus lumborum of *American Angus x Hereford (◊, Duckett et al. 1993), Australian Angus (♦, Pugh et al. 2005) and Japanese Black x Holstein cross cattle (□, Aoki et al. 2001).*

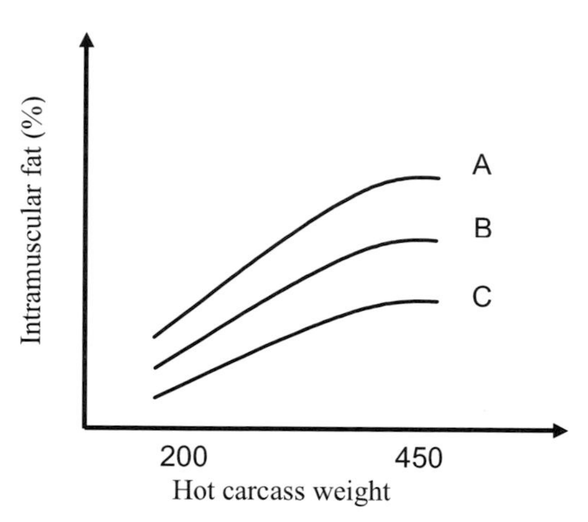

Figure 7. The effect of initial intramuscular fat (%) content (A >B >C) on final value in cattle of similar muscle growth potential.

of adipocytes such fatty acid carrier proteins and/or functional lipogenic enzymes involved in fatty acid biosynthesis. Certainly cattle breeds with a higher propensity to deposit fat have higher expression of key lipogenic enzyme either within adipose tissue (expressed per mg protein) or muscle (Bonnet *et al.*, 2003).

Another feature associated with the development of intramuscular fat is the fibre type or metabolic pattern of energy metabolism expressed by the muscle tissue. Within the one animal genotype the more glycolytic muscle types (e.g. ST) have lower levels of intramuscular fat (Gondret *et al.*, 1998; Hocquette *et al.*, 2003). Across genotypes a similar response can be found. Thus in the study by Hocquette *et al.*, (2003) where 2 muscle types were contrasted across 3 breeds of cattle with disparate propensity to accumulate intramuscular fat, there was a strong correlation between intramuscular fat and the aerobic markers cytochrome-*c* oxidase and isocitrate dehydrogenase as well as the adipose specific fatty acid binding protein. Of course studies across genotypes which correlate the extent of anaerobic muscle metabolism to the level of intramuscular fat accumulation are confounded by the observation that highly muscled cattle are more glycolytic and also have less carcass fat. However, these studies as well as others in rabbits (Gondret *et al.*, 2004) suggested that intramuscular fat content results from a balance between catabolic and anabolic pathways rather than from the regulation of a specific biochemical pathway. It has thus been speculated that a high fat turnover (which is a characteristics of oxidative muscles) would favour fat deposition (Hocquette *et al.*, 2003).

The propensity to lay down muscle tissue will also be an important component which will determine the expression of intramuscular fat. As animals develop or fatten the rate of muscle gain declines while the rate of fat gain is maintained (Owens *et al.*, 1995). Logically therefore increased intramuscular fat (%), relies on continued fat synthesis within muscle combined with a decreasing rate of muscle growth as animals mature. Therefore we propose that the developmental curve for intramuscular fat will be shifted to the right in animals with a high propensity to grow muscle (and in this case with the same propensity to marble at maturity, Figure 8). This 'right shift' would also occur in response to metabolic modifiers such as hormonal growth promotants, ß agonists and organic chromium supplementation, all of which can increase muscle growth. In very highly muscled animals it might be that intramuscular fat does not reach the 'linear accumulation' phase discussed in Figure 6 within normal commercial slaughter weights (that is the right shift described in Figure 8 is profound). This would appear to be the case for the modern pig

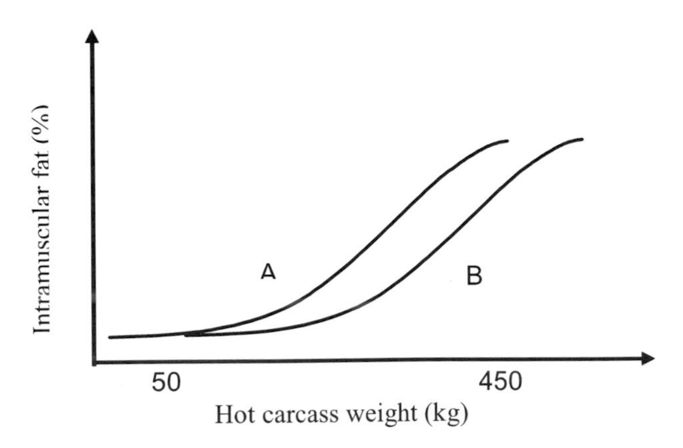

Figure 8. Hypothetical graph showing the development of intramuscular fat in cattle of different mature body weight (B >A).

genotypes where intramuscular fat % does not increase over a wide range of commercial carcass weights (Dunshea and D'Souza, 2003).

The role of nutrition in the regulation of intramuscular fat deposition has been reviewed by Pethick *et al.* (2004). The key factors centre on the availability of net energy for fat deposition and possibility that glucose availability might accelerate lipogenesis within intramuscular adipocytes.

Conclusion

Nowadays, farming and agri-food sectors are faced with a general saturation of meat markets in developed countries including Europe and with an increasing demand from consumers for high-quality products. The major current questions are thus how to define and increase the quality of animal products to satisfy the requirements of modern consumers while maintaining efficient production systems.

This is the reason why research must now aim to construct decision support tools for use throughout the beef production chain to provide consistent quality beef to consumers. The Meat Standards Australia is a successful system which is the culmination of many years of research. It has been, however, constructed based on the characteristics of the beef production chain in Australia. The beef production systems will differ in other countries and factors such as the *Bos indicus* effect and the influence of hormonal growth promotants would not need to be addressed by European systems. However many of the production factors within the MSA system would be directly applicable to beef supplychains within Europe such as ossification, pH and temperature decline, ultimate pH, marbling, days of meat aging, muscle and cooking method. Basic research will still need to be conducted to better understand how to manipulate the major animal characteristics which may have an important impact on beef quality. The pattern of glycogen and fat metabolism within muscle are among them. It is clear that glycogen metabolism is differently regulated depending on breed, selection for muscle yield, nutrition and the use of electric stimulation. It is also obvious that beef contains less fat in many European countries than in Australia or America. However we believe that the basic biological mechanisms are the same for any animal type. Therefore, understanding the development of intramuscular fat in the Australian context is of general interest for the rest of the world.

References

AMSA, 1995. Research Guidelines for cookery, sensory evaluation, and instrumental tenderness evaluation of fresh meat. Am. Meat Sci. Assoc. Chicago, Illinois.

Aoki, Y., N. Nakanishi, T. Yamada, N. Harashima and T. Yamazaki, 2001. Subsequent performance and carcass characteristics of Japanese Black x Holstein crossbred steers reared on pasture from weaning at three months of age. Bull. Nat. Grass. Res. Inst. 60: p. 40-55.

Bonnet, M., Y. Faulconnier, J.F. Hocquette, C. Leroux, P. Boulesteix, Y. Chilliard and D.W. Pethick, 2003. Lipogenesis in subcutaneaous adipose tissue and in oxidative or glycolytic muscles from Angus, Black Japanese x Angus and Limousin steers. In: Progess in Research on Energy and Protein Metabolism, W.B.Souffrant, C.C. Metges (editors). EAAP Publication N° 109, Wageningen Academic Publishers., Wageningen, The Netherlands, p. 469-472.

Brouard, S., G. Renand and F. Turin, 2001. Relationships between *Longissimus lumbarum* muscle characteristics and tenderness of three local beef cattle breeds. Renc. Rech. Ruminants 8: p. 41-52.

Cassar-Malek, I., K. Sudre, J. Bouley, A. Listrat, Y. Ueda, C. Jurie, Y. Briand, M. Briand, B. Meunier, C. Leroux, V. Amarger, D. Delourme, G. Renand, B. Picard, P. Martin, H. Levéziel and J.F. Hocquette, 2003. Integrated approach combining genetics, genomics and muscle biology to manage beef quality. Annual Meeting of The Bristish Society of Animal Science, York (England), 24-26th March 2003. Proceedings of the British Society of Animal Science, p. 52.

Cassar-Malek I., J.F. Hocquette, C. Jurie, A. Listrat, R. Jailler, D. Bauchart., Y. Briand and B. Picard 2004. Muscle-specific metabolic, histochemical and biochemical responses to a nutritionally induced discontinuous growth path. Anim. Sci. 204: p. 79-59.

Clinquart, A., C. Van Eenaeme, T. Van Vooren, J. Van Hoof, J.L. Hornick and L. Istasse, 1994. Meat quality in relation to breed (Belgian blue *vs* Holstein) and conformation (double-muscled *vs* dual purpose type). Sci. Alim. 14: p. 401-407.

Daly, B.L., I. Richards, P.G. Gibson, G.E. Gardner and J.M. Thompson, 2002. Rate of pH decline in bovine muscle *post-mortem* - a benchmarking study. Proc. 48th ICoMST, Rome, Italy. Vol. 2, p. 560-561.

Denhertogmeischke, M.J.A., F.J.M. Smulders, J.G. Vanlogtestijn and F. Vanknapen, 1997. Effect of electrical stimulation on the water-holding capacity and protein denaturation of two bovine muscles. J. Anim. Sci. 75: p. 118-124.

Dikeman, M.E., 1987. Fat reduction in animals and the effects on palatability and consumer acceptance of meat products. Rec. Meat Conf. 40: p. 93-103.

Dransfield, E., 1993. Modelling *post-mortem* tenderisation-IV: Role of calpains and calpastatin in conditioning. Meat Sci. 34: p. 217-234.

Dransfield, E., J.F. Martin, D. Bauchart, S. Abouelkaram, J. Lepetit, J. Culioli, C. Jurie and B. Picard, 2003. Meat quality and composition of three muscles from French cull cows and young bulls. Anim. Sci. 76: p. 387-399.

Duckett, S.K., D.G. Wagner, L.D. Yates, H.G. Dolezal and S.G. May SG, 1993. Effects of time on feed on beef nutrient composition. J. Anim. Sci. 71: p. 2079-88.

Dunshea, F.D. and D.N. D'Souza, 2003. Fat deposition and metabolism in the pig. In: Manipulating Pig Production IX, J.E. Paterson (editor), Publ. Australian Pig Science Assoc. (Inc.). 127-150 p.

Duris, M.P., G. Renand and B. Picard, 1999. Genetic variability of fetal bovine myoblasts in primary culture. Histochem. J. 31: p. 753-760

Fergusson, D.M., H.L. Bruce, J.M. Thompson, A.F. Egan, D. Perry and W.R. Shorthose, 2001. Factors affecting beef palatability - farm gate to chilled carcass. Aust. J. Exp. Agric. 41: p. 879-891

Fernandez, X., J.F. Hocquette and Y. Geay, 1997. Glycolytic metabolism in different muscles of normal and double-muscled calves. In: Book of Abstracts of the 48th EAAP Annual Meeting, J.A.M. van Arendonk (editor), Wageningen Pers, Wageningen, The Netherlands, p. 70.

Fiems, L.O., J. Van Hoof, L. Uytterhaegen, C.V. Boucque and D. Demeyer, 1995. Comparative quality of meat from double-muscled and normal beef cattle. In: Expression of tissue proteinases and regulation of protein degradation as related to meat quality. A. Ouali, D. Demeyer and F.J.M. Smulders (editors). ECCEAMST, Utrecht, The Netherlands, p. 381-393.

Gardner, G.E., B.L. McIntyre, G. Tudor and D.W. Pethick, 2001. The impact of nutrition on bovine muscle glycogen metabolism following exercise. Aust. J. Agric. Res. 52: p. 461-470.

Gondret, F., J. Mourot and M. Bonneau, 1997. Developmental changes in lipogenic enzymes in muscle compared to liver and extramuscular adipose tissues in the rabbit (Oryctolagus cuniculus). Comp. Biochem. Physiol. 117B: p. 259-265.

Gondret, F., J. Mourot and M. Bonneau, 1998. Comparison of intramuscular adipose tissue cellularity in muscles differing in their lipid content and fibre type composition during rabbit growth. Livest. Prod. Sci. 54: p. 1-10.

Gondret F., J.F. Hocquette and P. Herpin P., 2004. Age-related relationships between muscle fat content and metabolic traits in growing rabbits. Reprod. Nutr. Dev. 44, p. 1-16.

Grobet, L., D. Poncelet, L.J. Royo, B. Brouwers, D. Pirottin, C. Michaux, F. Ménissier, M. Zanotti, S. Dunner and M. Georges, 1998. Molecular definition of an allelic series of mutations disrupting the myostatin function and causing double-muscling in cattle. Mamm. Genome 9: p. 210-213.

Hocquette, J.F. and H. Abe, 2000. Facilitative glucose transporters in livestock species. Reprod. Nutr. Dev. 40: p. 517-53.

Hocquette, J.F., P. Bas, D. Bauchart, M. Vermorel and Y. Geay, 1999. Fat partitioning and biochemical characteristics of fatty tissues in relation to plasma metabolites and hormones in normal and double-muscled young growing bulls. Comp. Biochem. Phys. A 122: p. 127-138. Erratum, 1999. Comp. Biochem. Phys. A 123: p. 311-312.

Hocquette, J.F., C. Jurie, Y. Ueda, P. Boulesteix, D. Bauchart and D.W. Pethick, 2003. The relationship between muscle metabolic pathways and marbling of beef. In: Progess in Research on Energy and Protein Metabolism, W.B.Souffrant, C.C. Metges (editors). EAAP Publication N° 109, Wageningen Academic Publishers, Wageningen, The Netherlands, p. 513-516.

Hocquette, J.F., I. Ortigues-Marty, D.W. Pethick, P. Herpin and X. Fernandez, 1998. Nutritional and hormonal regulation of energy metabolism in skeletal muscles of meat-producing animals. Livest. Prod. Sci. 56: p. 115-143.

Hwang, I.H. and J.M. Thompson, 2001. The effect of time and type of electrical stimulation on the calpain system and meat tenderness in beef *longissimus dorsi* muscle. Meat Sci. 58: p. 135-144.

Johnson, ER, R.M. Butterfield and W.J. Pryor, 1972. Studies of fat distribution in the bovine carcass. I. The partition of fatty tissues between depots. Aust. J. Agric. Res. 23: p. 381.

Kumar, R., 1998. Expression of glycogenin in ovine muscle under differing metabolic conditions. Honours thesis, Murdoch University, Western Australia.

Listrat, A., C. Jurie, I. Cassar-Malek, L. Bouhraoua, B. Picard, D. Micol and J.F. Hocquette, 2001a. Effect of grass feeding on muscle characteristics of finishing Charolais steers. Reprod. Nutr. Dev. 42, p. 508.

Listrat, A., B. Picard, R. Jailler, H. Collignon, J.R. Pecatte, D. Micol and D. Dozias, 2001b. Grass valorisation and muscular characteristics of blond d'Aquitaine steers. Anim. Res. 50: p. 1-14.

Locker, R.H. and C.J. Hagyard, 1963. A cold shortening effect in beef muscle. J. Sci. Food 14: p. 787-793.

Monin, G., 1981a. Double muscling and sensitivity to stress. In: The problem of dark-cutting in beef. Current topics in Veterinary Medicine and Animal Science, Vol. 10, D.E. Hood and P.V. Tarrant (editors), Martinus Nijhoff Publishers, The Hague, The Netherlands, p. 199-208.

Monin, G., 1981b. Muscle metabolic type and the DFD condition. In: The problem of dark-cutting in beef. Current topics in Veterinary Medicine and Animal Science, Vol. 10, D.E. Hood and P.V. Tarrant (editors), Martinus Nijhoff Publishers, The Hague, The Netherlands, p. 63-85.

Owens, F.N., D.R. Gill, D.S. Secrist and S.W. Coleman, 1995. Review of some aspects of growth and development of feedlot cattle. J. Anim. Sci. 73: p. 3152-3172

Pearson, A.M. and R.B. Young, 1989. Muscle and meat biochemistry. Academic Press, San Diego.

Peter, J.B., R.J. Barnard, V.R. Edgerton, C.A. Gillespie and K.E. Stempel, 1972. Metabolic profiles of three fibre types of skeletal muscle in guinea pigs and rabbits. Biochemistry 11: 2627-2633

Petch, P.E. and K.V. Gilbert, 1997. Interaction of electrical processes applied during slaughter and dressing with stimulation requirements. In: Proc. 43th ICoMST, Auckland, New Zealand 43: p. 684.

Pethick, D.W., L. Cummins, G.E. Gardner, B.W. Knee, M. McDowell, B.L. McIntyre, G. Tudor, P.J. Walker and R.D. Warner, 1999. The regulation by nutrition of glycogen in the muscle of ruminants. Recent Adv. Anim. Nutr. Austral.12: p. 145-152.

Pethick, D.W., G.S. Harper and V.H. Oddy, 2004. Growth development and nutritional manipulation of marbling in cattle: a review. Austr. J. Exp. Agric. 44: p.705-715.

Polkinghorne, R., R. Watson, M. Porter, A. Gee, J. Scott and J. Thompson, 1999. Meat Standards Australia, A 'PACCP' based beef grading scheme for consumers. 1) The use of consumer scores to set grade standards. In: Proc. 45th ICoMST, Yokohama, Japan 45: 14-15

Pugh, A.K., B.L. McIntyre, G. Tudor and D.W. Pethick, 2005. Understanding the effect of gender and age on the pattern of fat deposition in cattle. In: Indicators of milk and beef quality, J.F. Hocquette and S. Gigli (editors), Wageningen Academic Publishers, Wageningen, The Netherlands, this book.

Purchas, R.W. and R. Aungsupakorn, 1993. Further investigations into the relationship between ultimate pH and tenderness in beef samples from bulls and steers. Meat Sci. 34: 163-178

Renand, G., P. Berge, B. Picard, J. Robelin, Y. Geay, D. Krauss and F. Ménissier, 1994. Genetic parameters of beef production and meat quality traits of young Charolais bulls progeny of divergently selected sires. In: Proc. 5th World Congress in Genetics Applied to Livestock production 19: p. 446-449.

Romans, J.R., W.J. Costello, C.W. Carlson, M.L. Greaser and K.W. Jones (editors), 1994. The meat we eat. Interstate Publishers, Inc: Danville, Ill, p. 369-375.

Saltin, B. and P.D. Gollnick, 1983. Skeletal muscle adaptability: significance for metabolism and performance. In: Handbook of Physiology, Section 10, Skeletal Muscle, LD. Peachey, R.H. Adrian and S.R. Geiger (editors) Publ. American Physiological Society, Maryland, p. 555-631.

Savell, J.W. and H.R. Cross, 1986. The role of fat in the palatability of beef, pork and lamb. Meat Res. Update 1 (4): p. 1-10. Publ. Dept. of Animal Science, The Texas AandM University System, Texas, USA.

Shorthose, W.R., 1989. Dark cutting in sheep and beef carcasses under the different environments in Australia. In: Dark cutting in cattle and sheep, S.U. Fabiansson, W.R. Shorthose and R.D. Warner (editors) Australian Meat and Livestock Research and Development Corporation: Sydney, p. 68-73.

Simmons, N.J., J.M. Cairney and C.C. Daly, 1997. Effect of pre-rigor temperature and muscle prestraint on the biophysical properties of meat tenderness. In: Proc. 43th ICoMST, Auckland, New Zealand 43: p. 608-609.

Simmons, N.J., K. Singh, P.M. Dobbie and C.E. Devine, 1996. The effect of pre-rigor holding temperature on calpain and calpastatin activity and meat tenderness. In: Proc. 42nd ICoMST, Lillehammer, Norway, 42: p. 414-415.

Talmant, A., G. Monin, M. Briand, L. Dadet and Y. Briand, 1986. Activities of metabolic and contractile enzymes in 18 bovine muscles. Meat Sci. 18: p. 23-40.

Tarrant, P.V., 1989. Animal behaviour and environment in the dark cutting condition. In: Dark cutting in cattle and sheep, S.U. Fabiansson, W.R. Shorthose and R.D. Warner (editors) Australian Meat and Livestock Research and Development Corporation: Sydney, p. 8-18.

Thompson, J.M., 2001. The relationship between marbling and sensory traits. In 'Marbling symposium 2001. H. Burrow *et al.*, (editors), Published by the Cooperative Research Centre for Cattle and Beef Quality, Armidale, NSW, Aust., p. 30-35.

Thompson, J.M., 2002. Managing meat Tenderness. Meat Sci. 60: p. 365-369.

Trout, G.R., 2002. Biochemistry of lipid and myoglobin oxidation in *postmortem* muscle and processed meat products - effects on rancidity. In: Proc. 48th ICoMST, Rome, Italy. 1: p. 50-55.

Uruh, J.A., C.L. Kastner, D.H. Kropf, M.E. Dikeman and M.C. Hunt, 1986. Effects of low voltage electrical stimulation during exsanguination on meat quality and display colour stability. Meat Sci. 18: p. 281-293.

Vernon, R.G., 1981. Lipid metabolism in the adipose tissue of ruminants. In: Lipid metabolism in Ruminant Animals, W.W. Christie (editor), Pergamon Press, Oxford, p. 279-362.

Beef tenderness: significance of the calpain proteolytic system

E. Veiseth[1] *and M. Koohmaraie*[2*]
[1]*Department of Mathematical Sciences and Technology, Agricultural University of Norway, P.O. Box 5003, N-1432 Aas, Norway,* [2]*Roman L. Hruska U.S. Meat Animal Research Center, ARS, USDA, Clay Center, NE 68933-0166, USA*
[*]*Correspondence: koohmaraie@email.marc.usda.gov*

Summary

The purpose of this paper is to review the role of the calpain proteolytic system in beef tenderization during *post-mortem* cooler storage. Ultimate beef tenderness is dependent on three factors: the background toughness, the toughening phase, and the tenderization phase. However, the large variation in beef tenderness at the consumer level is mainly a result of variation in the tenderization phase. Proteolysis of key myofibrillar and myofibril-associated proteins leads to a weakening of the rigid structure of the myofibrils, and meat becomes more tender. Although several proteolytic systems have been suggested to play a role in *post-mortem* proteolysis in skeletal muscle, ample evidence shows that the calpain proteolytic system is responsible for this process. Moreover, μ-calpain activity regulated by the calpain-specific inhibitor calpastatin starts shortly after slaughter, and proceeds throughout the storage period. Towards the end of this paper, different *in vitro* and *in vivo* models used to study the calpain proteolytic system in *post-mortem* skeletal muscle are discussed.

Keywords: calpain, calpastatin, muscle, *post-mortem*, proteolysis

Introduction

Overall beef meat quality is determined by multiple factors, including water-holding capacity, colour, palatability, nutritional value and safety. The importance of these traits varies depending on both the end product as well as the consumer profile. The eating quality, or palatability, of beef is influenced by flavour, juiciness and tenderness. Consumers in the USA ranked tenderness to be the most important attribute of beef eating quality (Miller *et al.*, 1995). Further, it has been shown that consumers can distinguish between tough and tender meat (Huffman *et al.*, 1996), and that they are willing to pay a premium for tender meat (Boleman *et al.*, 1997; Lusk *et al.*, 2001). One particular cut of meat that demonstrates the importance of beef tenderness is the tenderloin (i.e. *psoas major*). Although the tenderloin is found to be one of the least flavourful and juicy cuts of meat, it is the most highly valued retail cut due to its supreme tenderness (Savell and Shackelford, 1992).

The objective of this review paper is to discuss the role of the calpain proteolytic system in *post-mortem* tenderization of beef meat. A brief overview of other factors affecting ultimate beef tenderness will also be given.

Factors influencing ultimate beef tenderness

In this paper, ultimate beef tenderness is defined as the tenderness level found after cooler storage of meat, at a time when further cooler storage would not lead to additional tenderization. Significant variations in ultimate beef tenderness can be found when comparing different muscles within an animal (Ramsbottom *et al.*, 1945; Strandine *et al.*, 1949). Further, variation in ultimate beef tenderness is also seen when comparing the same muscle from different breeds (Shackelford *et al.*, 1991). Variation in ultimate beef tenderness can exist at the time of slaughter, can be created during the *post-mortem* storage period, or can be a combination of both. The three factors that determine ultimate beef tenderness is the background toughness, the toughening phase, and the tenderization phase. While the toughening and tenderization phases take place during the *post-mortem* storage period, the background toughness exists at the time of slaughter and does not change during the storage period. The effect of the opposing toughening and tenderization phase on meat tenderness is illustrated in Figure 1.

Background toughness

Background toughness of meat is defined as "the resistance to shearing of the unshortened muscle" (Marsh and Leet, 1966), and variation in the background toughness results from the connective tissue component of muscle. In particular, it seems as if the organization of the perimysium affects this toughness, since a general correlation between perimysium organization and the tenderness of muscles has been found in both chicken and beef (Strandine *et al.*, 1949). Moreover, double muscled cattle contain less perimysium and have larger fasciculi than normal cattle (Boccard, 1981), and these factors seem to be related to the improved tenderness of these animals. Nevertheless, extensive evidence shows that the variability in toughness found between similar muscles from individual animals is independent of connective tissue properties (Whipple *et al.*, 1990b; Silva *et al.*, 1999; Ngapo *et al.*, 2002).

Toughening phase

The toughening phase is caused by sarcomere shortening during *rigor* development. This process usually occurs within the first 24 h *post-mortem*, after which sarcomere lengths do not change (Wheeler and Koohmaraie, 1994; 1999). The relationship between sarcomere shortening and meat toughness was first reported in 1960 (Locker, 1960). Later it was shown that there is a strong negative relationship between sarcomere length and meat toughness when sarcomeres are shorter than 2 μm, and that the relationship is poorer at longer sarcomere lengths (Herring *et al.*, 1967; Bouton *et al.*, 1973). A theory relating the tightly bound actin-myosin complex to the increased meat toughness during the first 24 h *post-mortem* has also been suggested (Goll *et al.*, 1995). However, this theory has been weakened by a report showing that muscles prevented from contraction during *rigor* development does not go through the normal toughening phase even

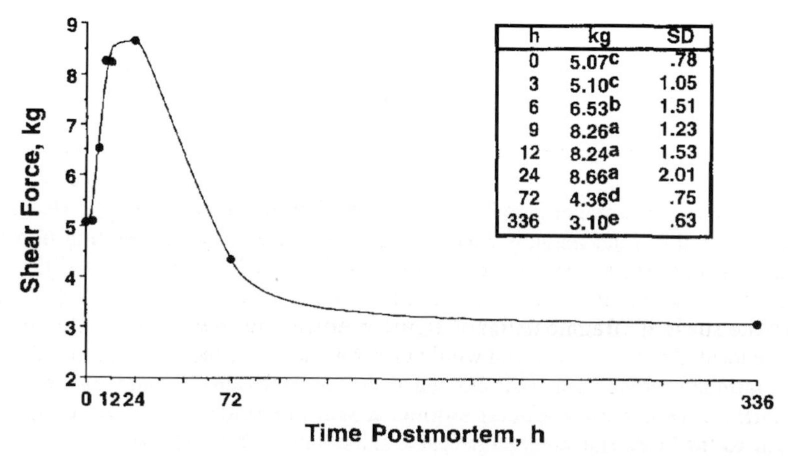

h	kg	SD
0	5.07c	.78
3	5.10c	1.05
6	6.53b	1.51
9	8.26a	1.23
12	8.24a	1.53
24	8.66a	2.01
72	4.36d	.75
336	3.10e	.63

Figure 1. Tenderness of ovine longissimus, measured as Warner-Bratzler shear force, at various times post-mortem. *Means without a common superscript differ ($P < 0.05$; from Wheeler and Koohmaraie, 1994).*

though these muscles also form the tightly bound actin-myosin complex at the onset of *rigor mortis* (Koohmaraie *et al.*, 1996). Thus, it is not the formation of the tightly bound actin-myosin complex at *rigor mortis* that results in the increased toughness within the first 24 h *post-mortem*, but rather the contraction of sarcomeres immediately prior to the onset of *rigor mortis*.

Tenderization phase

Is has been known for almost a century that meat tenderness improves during cooler storage, and it was suggested that the changes taking place in meat during storage were largely due to enzymatic activity (Hoagland *et al.*, 1917). It is now well established that *post-mortem* proteolysis of myofibrillar and myofibril-associated proteins is responsible for this process. While the toughening phase is similar in all carcasses, the tenderization phase is highly variable. There is a large variation in both the rate and extent of *post-mortem* tenderization of meat, and this results in the inconsistency in meat tenderness found at the consumer level.

Post-mortem proteolysis and meat tenderization

Multiple studies have been performed to identify which proteins are degraded during *post-mortem* storage of meat and what specific changes are responsible for tenderization. Initially researchers suggested that degradation of the Z-discs was related to meat tenderization (Davey and Gilbert, 1969), however a major component of Z-discs, α-actinin, showed no sign of degradation until after 2 weeks of *post-mortem* storage at 4°C (Hwan and Bandman, 1989). Degradation of costameres (e.g. desmin, dystrophin, vinculin), intermediate filaments (e.g. desmin, vimentin), titin, and nebulin are now thought to be responsible for *post-mortem* tenderization of meat (Taylor *et al.*, 1995). The rigid structure of the myofibrils is weakened once some of these proteins are degraded, and meat becomes more tender.

The strong association between *post-mortem* proteolysis and tenderization of meat during cooler storage has been demonstrated in several studies. Differences in meat tenderness found between *Bos taurus* (tender) and *Bos indicus* (tough) cattle are related to the reduced rate of *post-mortem* protein degradation in the tougher breed (Whipple *et al.*, 1990b; Shackelford *et al.*, 1991). Treatment of sheep and cattle with a β-adrenergic agonist (BAA) gave similar results, where meat from treated animals showed no *post-mortem* proteolysis and no tenderization (Fiems *et al.*, 1990; Kretchmar *et al.*, 1990). Ultimately, species differences in *post-mortem* tenderization rate in pigs, sheep, and cattle were directly related to the rate of myofibrillar protein degradation in the three species (Ouali and Talmant, 1990; Koohmaraie *et al.*, 1991b).

In skeletal muscle there are three endogenous proteolytic systems that can be responsible for *post-mortem* proteolysis during cooler storage of meat: the calpain system, the lysosomal cathepsins, and the multicatalytic proteinase complex (MCP). In addition to being endogenous in skeletal muscle, these proteolytic systems would have to fulfil two other requirements before they could be considered responsible for *post-mortem* proteolysis in meat (Koohmaraie, 1994; Goll *et al.*, 1983). Firstly, the proteases must have access to the substrates, and secondly, they must be able to reproduce the substrate fragmentation pattern observed after *post-mortem* storage of meat. Cathepsins are located in lysosomes, and would therefore not have access to the myofibrils unless they are released from the lysosomes. However, evidence for the release of cathepsins from lysosomes during *post-mortem* storage of meat is conflicting. A shift in cathepsin activity from the lysosomal to the soluble fraction of bovine muscle indicates release of cathepsins from lysosomes (O'Halloran *et al.*, 1997). In contrast, it has been shown that cathepsins still were located in lysosomes after electrical stimulation and 28 days of storage of meat at 4°C (LaCourt *et al.*, 1986). Regardless of their release or not, the major reason why a role of cathepsins in meat tenderization is doubtful is their proteolytic abilities. Actin and myosin are very good substrates for cathepsin, and these two proteins are not degraded during *post-mortem* storage of meat (Koohmaraie, 1994). Additionally, studies where different protease inhibitors were injected into lamb *longissimus* have shown that inhibition of cathepsins does not affect *post-mortem* proteolysis and meat tenderization (Hopkins and Thompson, 2001ab). A significant role of MCP could also be excluded, since myofibrils are very poor substrates for this protease (Koohmaraie, 1992a). Furthermore, several publications have shown that increased Ca^{2+} concentrations in *post-mortem* muscle greatly improves meat tenderness through increased proteolysis (Koohmaraie *et al.*, 1988; Polidori *et al.*, 2000), however neither cathepsins nor MCP are activated by Ca^{2+} (Koohmaraie, 1992b). Based on this, the calpain proteolytic system is the only proteolytic system to fulfil all the requirements and is thought to be the system responsible for *post-mortem* proteolysis and tenderization of meat.

Although most reports investigating the mechanism behind meat tenderization support the importance of *post-mortem* proteolysis of myofibrillar and myofibril-associated proteins, a calcium theory of tenderization has also been suggested (Takahashi, 1992; 1996). According to this theory, meat tenderization during *post-mortem* storage is a result of fragmentation of structural proteins, weakening of the Z-discs, and lengthening of *rigor*-shortened muscles caused directly by calcium through non-enzymatic processes. However, several groups have now refuted this theory. *Post rigor* lengthening of sarcomeres was not detected in lamb *longissimus* even though a considerable variation was seen in tenderization of the muscles (Geesink *et al.*, 2001). Moreover, it was shown that degradation of troponin-T correlated well with meat tenderization, although this particular protein, according to Hattori and Takahashi (1982), was not directly affected by calcium. Additionally, incubations of single bovine muscle fibres in 1 or 10 mM Ca^{2+} decreased the strength of the fibres. However this weakening was prevented by addition of protease inhibitors, suggesting that the weakening was a result of proteolysis and not a direct effect of calcium (Christensen *et al.*, 2004).

The calpain proteolytic system

The first report of the calpain proteolytic system came 40 years ago, when a Ca^{2+}-activated proteolytic enzyme was purified from rat brain (Guroff, 1964). A similar Ca^{2+}-activated proteolytic enzyme was later purified from pork skeletal muscle (Dayton *et al.*, 1976ab). Some years later, the existence of a second Ca^{2+}-activated protease with a lower Ca^{2+}-requirement was purified from canine cardiac muscle (Mellgren, 1980). These proteases were named calpains (Murachi, 1989), and are now referred to as μ-calpain and m-calpain according to their Ca^{2+}-requirement (Cong *et al.*, 1989). An inhibitor of calpain, calpastatin, was later isolated from human erythrocytes and bovine cardiac muscle (Takano and Murachi, 1982; Otsuka and Goll, 1987). Over the last decade multiple calpain-like genes have been identified in different tissues and species, however little or nothing is known about the proteins encoded by these genetic variants (Goll *et al.*, 2003).

Both μ-calpain and m-calpain are composed of two subunits with molecular weights of 28 and 80 kDa (Dayton *et al.*, 1976ab; Dayton *et al.*, 1981; Emori *et al.*, 1986). While μ-calpain and m-calpain have identical small subunits encoded by a single gene, they possess genetically different large subunits (Ohno *et al.*, 1990). An important characteristic of μ-calpain and m-calpain is their ability to undergo autolysis in the presence of calcium. Autolysis reduces the Ca^{2+}-requirements for half maximal activity of μ-calpain and m-calpain (Suzuki *et al.*, 1981ab; Dayton, 1982; Nagainis *et al.*, 1983), however the specific activity of the enzymes does not change (Edmunds *et al.*, 1991). Autolysis of the large subunit of μ-calpain produces a 78-kDa fragment followed by a 76-kDa fragment (Inomata *et al.*, 1988). However, autolysis of the large subunit of m-calpain produces a 78-kDa fragment only (Brown and Crawford, 1993). The small subunit of calpain produces a fragment of 18 kDa upon autolysis (McClelland *et al.*, 1989). It seems as if autolysis of the large subunit gives rise to the reduced Ca^{2+}-requirements of autolyzed calpains (Brown and Crawford, 1993; Elce *et al.*, 1997).

Calpastatin is the endogenous specific inhibitor of μ-calpain and m-calpain (Maki *et al.*, 1988). Several isoforms exist of this protein, however the predominant form found in skeletal muscle consists of four repetitive calpain-inhibiting domains in series with an N-terminal leader region (Lee *et al.*, 1992). A 125-kDa isoform of calpastatin is found in ovine, porcine, and bovine skeletal muscle (Geesink *et al.*, 1998). Calpastatin requires calcium to bind and inhibit calpains (Cottin *et al.*, 1981; Imajoh and Suzuki, 1985). However since calpastatin does not bind calcium itself, the Ca^{2+}-requirement of calpain-calpastatin interaction must originate from the protease. Calpastatin is also a substrate for calpains and can be degraded in the presence of calcium (Mellgren *et al.*, 1986; Doumit and Koohmaraie, 1999). Degradation of calpastatin by calpains does not lead to loss of inhibitory activity, and even after extensive proteolysis it retains inhibitory function (DeMartino *et al.*, 1988; Nakamura *et al.*, 1989).

For more detailed information regarding the calpain proteolytic system, see review by Goll *et al.* (2003).

Calpains and meat tenderization

Much of the evidence for the involvement of calpains in meat tenderization has come from studies showing an involvement of calcium in *post-mortem* tenderization of meat. As early as 1969 it was demonstrated that Ca^{2+}-treatment resulted in the disappearance of Z-discs within the myofibrils (Davey and Gilbert, 1969). Moreover, it was shown that Ca^{2+} ions provoked myofibrillar fragmentation, and that the chelating agent ethylenediaminetetraacetic acid (EDTA) inhibited this fragmentation (Busch *et al.*, 1972). However, more direct evidence for the importance of calpains in meat tenderization was demonstrated when whole carcasses or cuts of meat were infused with calcium chloride. Infusion of whole lamb carcasses with calcium chloride immediately after death greatly accelerated the *post-mortem* tenderization process (Koohmaraie *et al.*, 1988; Polidori *et al.*, 2000), as did calcium chloride injections into *longissimus* from *Bos indicus* cattle (Koohmaraie *et al.*, 1990). Moreover, shear force values of beef round muscles at 1, 8, and 14 days *post-mortem* were greatly reduced as a result of calcium chloride injection shortly after slaughter, and it has been shown that injections of calcium chloride 24 h *post-mortem* are just as effective as infusions performed at 0 h *post-mortem* (Wheeler *et al.*, 1991; 1992). Another report, however, found that calcium injection at 0 h *post-mortem* was more efficient at reducing shear force values measured at 10 days *post-mortem* than injections performed at 12 or 24 h *post-mortem* (Boleman *et al.*, 1995). By comparing carcasses infused with solutions of calcium chloride and sodium chloride of equal ionic strengths, it was proven that the tenderization was caused by calcium and not by an increase in ionic strength (Koohmaraie *et al.*, 1989). Since calpains are Ca^{2+}-dependent, all of the above is compelling evidence for the calpains involvement in *post-mortem* meat tenderization. Stronger evidence of their role in meat tenderization was established when infusions of zinc chloride (an inhibitor of calpains) into lamb carcasses prevented the *post-mortem* tenderization process (Koohmaraie, 1990). Injections of zinc chloride into beef strip loins have also been proven to drastically inhibit meat tenderization (Lawrence *et al.*, 2003).

Some of the changes seen in meat during *post-mortem* cooler storage include the disappearance of desmin and troponin-T and the appearance of 28-32-kDa fragments, while both actin and myosin are not affected (Koohmaraie, 1994). Incubations of purified myofibrils with calpain *in vitro* have produced the exact proteolytic pattern found in *post-mortem* meat (Koohmaraie *et al.*, 1986; Huff-Lonergan *et al.*, 1996; Geesink and Koohmaraie, 1999a). Moreover, incubations of bovine single muscle fibres with μ-calpain resulted in thinner Z-lines and reduced fibre strength, demonstrating that μ-calpain is capable of reducing the mechanical strength of muscle fibres (Christensen *et al.*, 2003).

Recently a role of calpain 3 (i.e. p94) has been suggested in *post-mortem* proteolysis leading to meat tenderization (Ilian *et al.*, 2001; 2004ab). A good correlation was reported between calpain 3 mRNA level and ultimate tenderness of bovine and ovine muscles (Ilian *et al.*, 2001). However, no relationship was found between calpain 3 levels, determined by western blot using a specific calpain 3 antibody, and meat tenderness of porcine *longissimus* (Parr *et al.*, 1999). Mutations in the calpain 3 gene in humans are linked with limb-girdle muscular dystrophy type 2a (LGMD 2a), a disease associated with excessive protein catabolism (Richard *et al.*, 1995). These mutations result in loss of catalytic activity of calpain 3 (Ono *et al.*, 1998). The suggestion that elevated levels of calpain 3 are associated with increased tenderization seems contradictory to the enzymes behaviour in LGMD 2a. Thus, based on these data alone, calpain 3 is not likely to play a role in meat tenderization. Furthermore, calpastatin which has been clearly documented to play a key regulatory role in *post-mortem* proteolysis and meat tenderness does not inhibit calpain 3 (Sorimachi *et al.*, 1993). Geesink *et al.*, (2004) using calpain 3 knockout mice, demonstrated that absence of calpain 3 had not effect on *post-mortem* proteolysis. Therefore, it is clear that calpain 3 does not play a relevant role in *post-mortem* proteolysis. The hypothesis of Ilian and co-workers with respect for the role of calpain 3 in *post-mortem* proteolysis and meat tenderization (Ilian *et al.*, 2001; 2004a,b) can essentially be summarized as (1) unlike with the ubiquitous form of the calpain there are no methods for purification and quantification of calpain 3, (2) native calpain 3 has never been isolated from skeletal muscle, hence neither the proteolytic capacity nor substrates are known, (3) the only method available are mRNA quantification and autolysis of calpain 3 and their correlation to *post-mortem* proteolysis and meat tenderization, and (4) because mRNA level and pattern of calpain 3 is correlated with *post-mortem* proteolysis, calpain 3 must be involved in this process. Correlations and cause and effect are very different phenomena. The data presented by Geesink *et al.* (2004) indicated that while calpain 3 autolysis occurred as reported repeatedly by Ilian and co-workers, such autolysis is independent of degradation of proteins that are involved in *post-mortem* meat tenderization. Therefore because: (1) absence of calpain 3 does not affect *post-mortem* proteolysis, (2) overexpression of calpastatin which shuts down *post-mortem* proteolysis has no effect of the pattern of autolysis of calpain 3 (Kent *et al.*, 2004), and (3) autolysis of calpain 3 in *post-mortem* muscle is the sole basis for suggesting a role for calpain 3 in *post-mortem* muscle proteolysis (Ilian *et al.*, 2001a, b, 2004a, b), it appears that calpain 3 plays a minor, if any, role in *post-mortem* proteolysis of the proteins analyzed.

Calpain activity in *post-mortem* muscle

A body of evidence has been gathered which indicates that the calpain proteolytic system is responsible for meat tenderization during cooler storage, and numerous studies have been performed to determine the activity of μ-calpain, m-calpain, and calpastatin in *post-mortem* muscle. In general, reports show that μ-calpain and calpastatin activities rapidly decline during *post-mortem* storage of meat, while m-calpain activity seems to be stable (Ducastaing *et al.*, 1985; Geesink and Koohmaraie, 1999b; Kretchmar *et al.*, 1990). However, some reports have shown that m-calpain activity also declines *post-mortem* (Sensky *et al.*, 1996; Beltrán *et al.*, 1997). These declines in m-calpain activity have later been shown to be artefacts, caused by the use of inappropriate extraction buffers, rather than an actual decline in activity (Veiseth and Koohmaraie, 2001). Since exposure of calcium leads to inactivation of both μ-calpain and m-alpain through autolysis, the fact that m-calpain activity is stable in *post-mortem* muscle while μ-calpain activity is not has been used to suggest that μ-calpain rather than m-calpain is responsible for *post-mortem* proteolysis in meat (Koohmaraie *et al.*, 1987). This assertion was more recently confirmed when analysis using casein zymography showed that m-calpain did not undergo autolysis during a 15-day *post-mortem* storage period, and that only μ-calpain is active during the ageing period (Veiseth *et al.*, 2001). Typical changes in μ-calpain, m-calpain, and calpastatin activities in *post-mortem* muscle are shown in Figure 2.

Calpain activity in *post-mortem* muscle is affected by factors such as temperature and pH. Both temperature and pH decline in muscle during the first 24 h *post-mortem*, and this has raised questions whether μ-calpain is active at the typical pH and temperature of *post-mortem* muscle. However, it has been shown that μ-calpain retained 24 to 28% of its maximum activity, found at pH 7.5 and 25°C, under conditions simulating *post-mortem* muscle (Koohmaraie *et al.*, 1986). Additionally, incubation of bovine myofibrils with μ-calpain at pH 5.6 and 4°C produced the same pattern of proteolysis as observed in bovine *longissimus* during *post-mortem* storage (Koohmaraie *et al.*, 1986; Huff-Lonergan *et al.*, 1996).

As mentioned earlier, the large variation in meat tenderness found at the consumer level is caused by a large variation in both the rate and the extent of *post-mortem* tenderization. Although pH and temperature affect μ-calpain activity in *post-mortem* muscle, this would not introduce variation in meat tenderness, since their declines in *post-mortem* muscle do not vary significantly between carcasses subjected to the same pre- and post-slaughter conditions. However, it is possible to manipulate the pH and temperature declines in carcasses by using different *post-mortem* cooling regimes or electrical stimulation. Post-mortem cooling rate of carcasses influences meat tenderness through its effect on sarcomere shortening during *rigor* mortis development (Redmond *et al.*, 2001; Van Moeseke *et al.*, 2001; Devine *et al.*, 2002), and also through its effect on proteolysis (Devine *et al.*, 1999; Geesink *et al.*, 2000). Electrical stimulation of carcasses accelerates *post-mortem* pH decline, which again affects *rigor mortis* development due to its effect on the intracellular Ca^{2+} concentration in muscle (Whiting, 1980). Although μ-calpain activity is affected by pH, the effect of electrical stimulation on *post-mortem* proteolysis is less clear, and conflicting results have been reported (Hwang *et al.*, 2003).

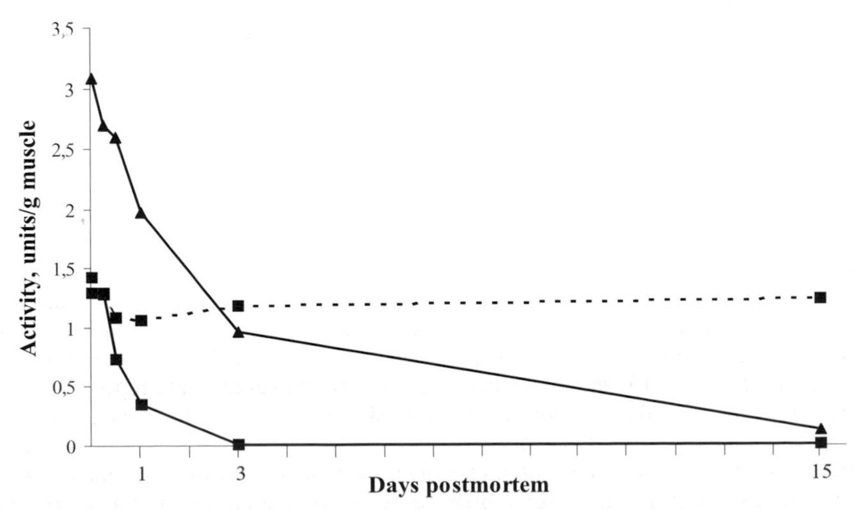

Figure 2. Changes in μ-calpain (–■–), m-calpain (- -■- -), and calpastatin (–▲–) activities during post-mortem *cooler storage of ovine longissimus (graphs based on data reported by Veiseth et al., 2004b).*

The principal regulator of calpain activity in *post-mortem* muscle is calpastatin, and a negative relationship exists between calpastatin activity and *post-mortem* proteolysis in muscle. A concern that has been raised regarding μ-calpain activity in *post-mortem* muscle is the excess of calpastatin found in muscle. Myofibril incubations with μ-calpain and calpastatin have shown that calpastatin limits the rate and the extent of proteolysis and autolysis of μ-calpain (Geesink and Koohmaraie, 1999a). Importantly however, it was also shown that a 1:4 ratio of μ-calpain to calpastatin was not sufficient to completely inhibit μ-calpain activity. In another study, co-eluted μ-calpain and calpastatin extracted from ovine *longissimus* muscle at different times *post-mortem* showed a net proteolytic activity at all times *post-mortem* (Veiseth *et al.*, 2004b). This not only indicated that μ-calpain could be active in the presence of excess calpastatin, but also that μ-calpain was still active at 15 days *post-mortem*.

Autolysis of μ-calpain has been detected as early as 3 h *post-mortem* in ovine *longissimus*, indicating that μ-calpain has sufficient calcium levels for proteolytic activity shortly after slaughter (Veiseth *et al.*, 2001). Activity of μ-calpain has also been demonstrated in ovine *longissimus* after 56 days of cooler storage (Geesink and Koohmaraie, 1999b). These results indicate that μ-calpain can be active within a short time after slaughter and throughout an extended storage period. Significant amount of *post-mortem* proteolysis has been observed as early as 9 h in ovine *longissimus*, resulting in a significant improvement in myofibril fragmentation index (MFI) at 12 h *post-mortem* (Veiseth *et al.*, 2004b). A significant increase in MFI of bovine *longissimus* has also been detected at 12 h *post-mortem* (Koohmaraie *et al.*, 1987). Although MFI is an indirect measure of meat tenderness, it has been shown to correlate well with both Warner-Bratzler shear force (WBSF) and trained sensory panel tenderness ratings in bovine *longissimus* (Whipple *et al.*, 1990a; Crouse *et al.*, 1991). Improvements in MFI at 12 h *post-mortem* in both ovine and bovine *longissimus* therefore indicate that the tenderization process starts early *post-mortem*.

Model systems for the involvement of calpain activity in meat tenderization

Multiple animal models have been used to investigate the role of calpains and calpastatin in meat tenderization. A muscle hypertrophy condition in lambs, referred to as callipyge, leads to increased weights of all major leg and loin muscles (Jackson *et al.*, 1993). Meat from callipyge lamb was found to remain tough throughout the *post-mortem* storage period, while WBSF of meat from normal lamb decreased during the same storage period (Koohmaraie *et al.*, 1995). This lack of tenderization in meat from callipyge lambs is a result of a greatly reduced rate and extent of *post-mortem* proteolysis due to elevated levels of calpastatin in these animals (Geesink and Koohmaraie, 1999b; Duckett *et al.*, 2000). Muscles from animals fed BAA also do not undergo *post-mortem* proteolysis, and results in tough meat. BAA-administration has given comparable results in both lamb and cattle, and the reduced *post-mortem* proteolysis has been attributed to the reduced activity of calpains caused by increased calpastatin levels (Fiems *et al.*, 1990; Koohmaraie *et al.*, 1991a). Animal age also affects the calpain proteolytic system in skeletal muscles. In general, these studies have revealed that calpain and calpastatin activities decline with increasing animal age (Ou and Forsberg, 1991; Shackelford *et al.*, 1995). Recently, it was reported that calpastatin activity in *longissimus* from lambs ranging 2 to 10 months of age decreased, and that this resulted in increased *post-mortem* proteolysis and meat tenderization in the older animals (Veiseth *et al.*, 2004a).

Comparison of species has also been used as a model to study the calpain system in relation to meat tenderization. Meat tenderization occurred faster in pigs than in lambs, while beef was the slowest, and this difference in meat tenderization was a reflection of the rate of *post-mortem* proteolysis in the different species. Analysis of the calpain proteolytic system in muscles from these three species revealed that calpastatin levels were low in pork, intermediate in lamb, and

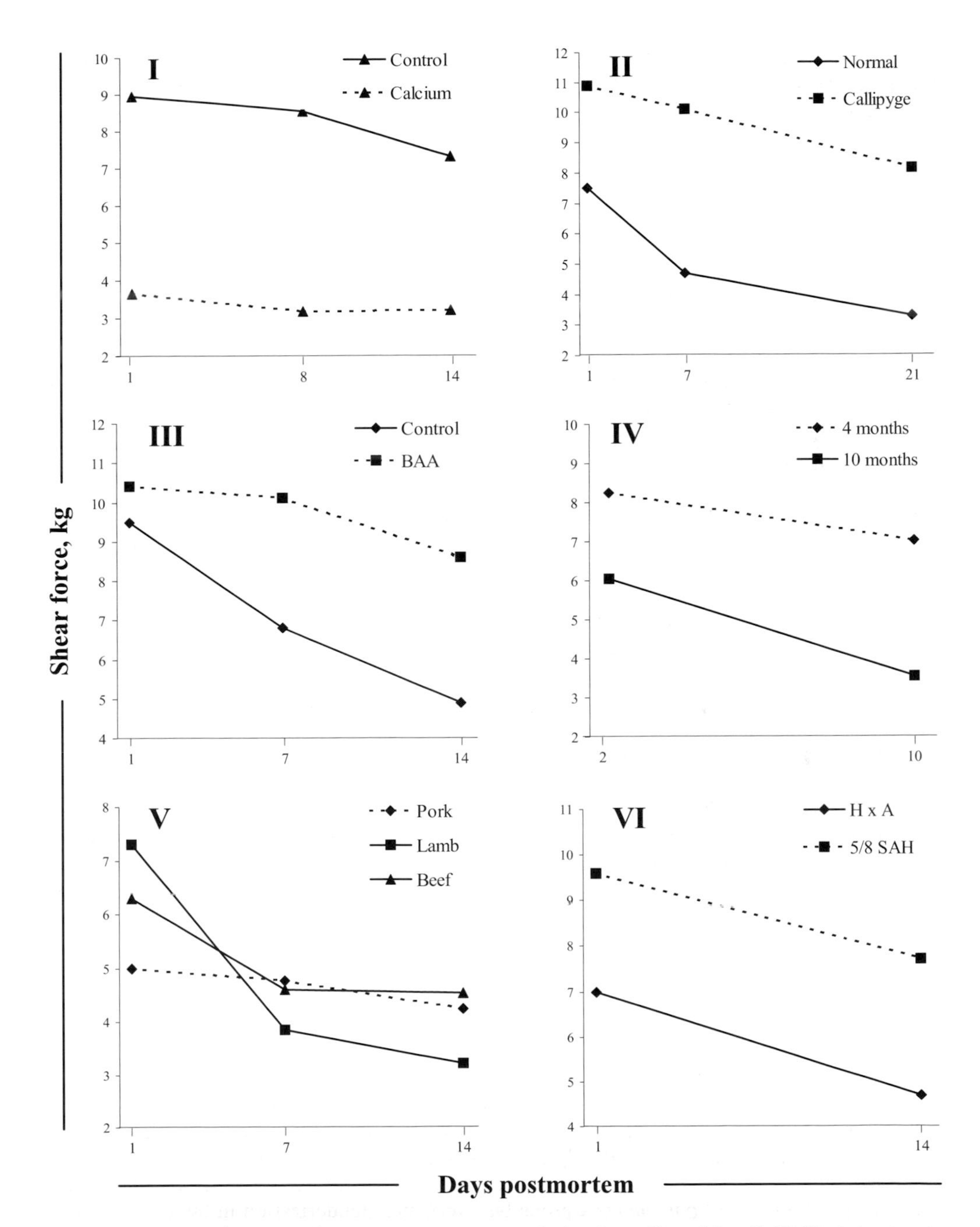

Figure 3. Rate and extent of post-mortem *meat tenderization affected by: I) $CaCl_2$ injection in cattle (data from Wheeler et al., 1991), II) callipyge phenotype in lambs (data from Koohmaraie et al., 1995), III) β-adrenergic agonist (BAA) treatment in lambs (data from Koohmaraie et al., 1991a), IV) animal age in lambs (data from Veiseth et al., 2004a), V) species (data from Koohmaraie et al., 1991b), and VI) breeds in cattle (data from Whipple et al., 1990b).*

high in beef. Thus, the difference in the rate of *post-mortem* meat tenderization between these three species is caused by a variation in calpastatin activity in their muscles at the time of slaughter (Ouali and Talmant, 1990; Koohmaraie *et al.*, 1991b). Meat obtained from *Bos indicus* carcasses has been demonstrated to be tougher meat than that obtained from *Bos taurus* carcasses. *Bos indicus* cattle have reduced *post-mortem* proteolysis compared to the more tender *Bos taurus* cattle, and this reduction is associated with a higher content of calpastatin in those animals (Whipple *et al.*, 1990b; Ferguson *et al.*, 2000). In fact, calpastatin activity at 24 h *post-mortem* in *longissimus* from crossbred cattle explained 44% of the variation in WBSF and trained sensory panel tenderness ratings (Whipple *et al.*, 1990a). The effects of some *in vitro* and *in vivo* models on meat tenderization are illustrated in Figure 3.

The key finding, independent of model, has been that muscles with elevated levels of calpastatin compared to calpain have reduced *post-mortem* proteolysis and produce tough meat. However, the most direct evidence thus far for calpains involvement in *post-mortem* proteolysis in skeletal muscle, comes from a study of transgenic mice overexpressing calpastatin (Kent *et al.*, 2004). In these mice, calpastatin activity in hind limb muscles was 370-fold greater than in control mice, while μ-calpain and m-calpain was unaffected by the transgene. Calpastatin overexpression resulted in a slower autolysis of μ-calpain, and *post-mortem* proteolysis measured by degradation of desmin and troponin-T was inhibited. Based on their results, the authors concluded that calpain activity alone is responsible for *post-mortem* proteolysis in skeletal muscle.

Conclusion

Tenderness is a very important quality trait for consumer acceptance of whole meat products, and the estimated cost of inadequate beef tenderness in the USA is 200-300 million $US annually. It is therefore crucial to reduce the variation in beef tenderness at the consumer level. Substantial evidence exists that μ-calpain activity regulated by calpastatin is responsible for *post-mortem* proteolysis of key myofibrillar and myofibril-associated proteins in skeletal muscle leading to meat tenderization. Further, *post-mortem* proteolysis mediated by μ-calpain starts shortly after slaughter and proceeds throughout the storage period. Information on the activity and regulation of this proteolytic system can be used to improve the rate and extent of *post-mortem* proteolysis, and thereby reduce the amount of tough beef reaching the consumers. Additionally, approaches to predict ultimate tenderness levels should be further explored.

References

Beltrán, J.A., I. Jaime, P. Santolaria, C. Sañudo, P. Albertí and P. Roncalés, 1997. Effect of stress-induced high *post-mortem* pH on protease activity and tenderness of beef. Meat Sci. 45: p. 201-207.

Boccard, R, 1981. Facts and reflections on muscular hypertrophy in cattle: double muscling or culard. In: Developments in Meat Science - 2, R. Lawrie (editor), Applied Science Publishers Ltd., London, p. 1-28.

Boleman, S.J., S.L. Boleman, T.D. Bidner, K.W. McMillin and C.J. Monlezun, 1995. Effects of *post-mortem* time of calcium chloride injection on beef tenderness and drip, cooking, and total loss. Meat Sci. 39: p. 35-41.

Boleman, S.J., S.L. Boleman, R.K. Miller, J.F. Taylor, H.R. Cross, T.L. Wheeler, M. Koohmaraie, S.D. Shackelford, M.F. Miller, R.L. West, D.D. Johnson and J.W. Savell, 1997. Consumer evaluation of beef of known categories of tenderness. J. Anim. Sci. 75: p. 1521-1524.

Bouton, P.E., P.V. Harris, W.R. Shorthose and R.I. Baxter, 1973. A comparison of the effects of aging, conditioning and skeletal restraint on the tenderness of mutton. J. Food Sci. 38: p. 932-937.

Brown, N. and C. Crawford, 1993. Structural modifications associated with the change in Ca^{2+} sensitivity on activation of m-calpain. FEBS. Lett. 322: p. 65-68.

Busch, W.A., M.H. Stromer, D.E. Goll and A. Suzuki, 1972. Ca^{2+}-specific removal of Z lines from rabbit skeletal muscle. J. Cell. Biol. 52: 367-381.

Christensen, M., L.M. Larsen, P. Ertbjerg and P.P. Purslow, 2004. Effect of proteolytic enzyme activity and heating on the mechanical properties of bovine single muscle fibres. Meat Sci. 66: p. 361-369.

Christensen, M., R.D. Young, M.A. Lawson, L.M. Larsen and P.P. Purslow, 2003. Effect of added μ-calpain and *post-mortem* storage on the mechanical properties of bovine single muscle fibres extended to fracture. Meat Sci. 66: p. 105-112.

Cong, J., D.E. Goll, A.M. Peterson and H.-P. Kapprell, 1989. The role of autolysis in activity of the Ca^{2+}-dependent proteinases (μ-calpain and m-calpain). J. Biol. Chem. 264: p. 10096-10103.

Cottin, P., P.L. Vidalenc and A. Ducastaing, 1981. Ca^{2+}-dependent association between a Ca^{2+}-activated neutral proteinase (CaANP) and its specific inhibitor. FEBS Lett. 136: 221-224

Crouse, J.D., M. Koohmaraie and S.D. Seideman, 1991. The relationship of muscle fibre size to tenderness of beef. Meat Sci. 30: p. 295-302.

Davey, C.L. and K.V. Gilbert, 1969. Studies in meat tenderness. 7. Changes in the fine structure of meat during aging. J. Food Sci. 34: p. 69-74.

Dayton, W.R., 1982. Comparison of low- and high-calcium-requiring forms of the calcium-activated protease with their autocatalytic breakdown products. Biochim. Biophys. Acta. 709: p. 166-172.

Dayton, W.R., D.E. Goll, M.G. Zeece, R.M. Robson and W.J. Reville, 1976a. A Ca^{2+}-activated protease possibly involved in myofibrillar protein turnover. Purification from porcine muscle. Biochemistry. 15(10): p. 2150-2158.

Dayton, W.R., W.J. Reville, D.E. Goll and M.H. Stromer, 1976b. A Ca^{2+}-activated protease possibly involved in myofibrillar protein turnover. Partial characterization of the purified enzyme. Biochem. 15: p. 2159-2167.

Dayton, W.R., J.V. Schollmeyer, R.A. Lepley and L.R. Cortés, 1981. A calcium-activated protease possibly involved in myofibrillar protein turnover. Isolation of a low-calcium-requiring form of the protease. Biochim. Biophys. Acta. 659: p. 48-61.

DeMartino, G.N., R. Wachendorfer, M.J. McGuire and D.E. Croall, 1988. Proteolysis of the protein inhibitor of calcium-dependent proteases produces lower molecular weight fragments that retain inhibitory activity. Arch. Biochem. Biophys. 262: p. 189-198.

Devine, C.E., S.R. Payne, B.M. Peachey, T.E. Lowe, J.R. Ingram and C.J. Cook, 2002. High and low rigor temperature effects on sheep meat tenderness and ageing. Meat Sci. 60: p. 41-146.

Devine, C.E., N.M. Wahlgren and E. Tornberg, 1999. Effect of rigor temperature on muscle shortening and tenderisation of restrained and unrestrained beef m. *longissimus thoracis et lumborum*. Meat Sci. 51: p. 1-72.

Doumit, M.E. and M. Koohmaraie, 1999. Immunoblot analysis of calpastatin degradation: evidence for cleavage by calpain in *post-mortem* muscle. J. Anim. Sci. 77: p. 1467-1473.

Ducastaing, A., C. Valin, J. Schollmeyer and R. Cross, 1985. Effects of electrical stimulation on *post-mortem* changes in the activities of two Ca dependent neutral proteinases and their inhibitor in beef muscle. Meat Sci. 15: p. 193-202.

Duckett, S.K., G.D. Snowder and N.E. Cockett, 2000. Effect of the callipyge gene on muscle growth, calpastatin activity, and tenderness of three muscles across the growth curve. J. Anim. Sci. 78: 2836-2841

Edmunds, T., P.A. Nagainis, S.K. Sathe, V.F. Thompson and D.E. Goll, 1991. Comparison of the autolyzed and unautolyzed forms of μ- and m-calpain from bovine skeletal muscle. Biochim. Biophys. Acta. 1077: p. 197-208.

Elce, J.S., C. Hegadorn and J.S.C. Arthur, 1997. Autolysis, Ca^{2+} requirement, and heterodimer stability in m-calpain. J. Biol. Chem. 272: p. 11268-11275.

Emori, Y., H. Kawasaki, S. Imajoh, S. Kawashima and K. Suzuki. 1986. Isolation and sequence analysis of cDNA clones for the small subunit of rabbit calcium-dependent protease. J. Biol. Chem. 261: p. 9472-9476.

Ferguson, D.M., S.-T. Jiang, H. Hearnshaw, S.R. Rymill and J.M. Thompson, 2000. Effect of electrical stimulation on protease activity and tenderness of M. *longissimus* from cattle with different proportions of *Bos indicus* content. Meat Sci. 55: p. 265-272.

Fiems, L.O., B. Buts, C.V. Boucqué, D.I. Demeyer and B.G. Cottyn, 1990. Effect of a β-agonist on meat quality and myofibrillar protein fragmentation in bulls. Meat Sci. 27: p. 29-39.

Geesink, G.H., A.-D. Bekhit and R. Bickerstaffe, 2000. Rigor temperature and meat quality characteristics of lamb longissimus muscle. J. Anim. Sci. 78: p. 2842-2848.

Geesink, G.H. and M. Koohmaraie, 1999a. Effect of calpastatin on degradation of myofibrillar proteins by μ-calpain under *post-mortem* conditions. J. Anim. Sci. 77: p. 2685-2692.

Geesink, G.H. and M. Koohmaraie, 1999b. Post-mortem proteolysis and calpain/calpastatin activity in callipyge and normal lamb biceps femoris during extended *post-mortem* storage. J. Anim. Sci. 77: p. 1490-1501.

Geesink, G.H., D. Nonneman and M. Koohmaraie, 1998. An improved purification protocol for heart and skeletal muscle calpastatin reveals two isoforms resulting from alternative splicing. Arch. Biochem. Biophys. 356: p. 19-24.

Geesink, G.H., R.G. Taylor, A.E.D. Bekhit and R. Bickerstaffe, 2001. Evidence against the non-enzymatic calcium theory of tenderization. Meat Sci. 59: p. 417-422.

Geesink, G. H., R. G. Taylor and M. Koohmaraie. 2004. p94/Calpain 3 is not involved in *post-mortem* proteolysis. Meat Science (submitted).

Goll, D.E., G.H. Geesink, R.G. Taylor and V.F. Thompson, 1995. Does proteolysis cause all *post-mortem* tenderization, or are changes in the actin/myosin interaction involved? In: Proc. Int. Cong. Meat Sci. Technol., San Antonio, TX, 41: p. 537-544.

Goll, D.E., Y. Otsuka, P.A. Nagainis, J.D. Shannon, S.K. Sathe and M. Muguruma, 1983. Role of muscle proteinases in maintenance of muscle integrity and mass. J. Food Biochem. 7: p. 137.

Goll, D.E., V.F. Thompson, H. Li, W. Wei and J. Cong, 2003. The calpain system. Physiol. Rev. 83: p. 731-801.

Guroff, G., 1964. A neutral, calcium-activated proteinase from the soluble fraction of rat brain. J. Biol. Chem. 239: p. 149-155

Hattori, A. and K. Takahashi, 1982. Calcium-induced weakening of skeletal muscle z-disks. J. Biochem. 92: 381-390.

Herring, H.K., R.G. Cassens, G.G. Suess, V.H. Brungardt and E.J. Briskey, 1967. Tenderness and associated characteristics of stretched and contracted bovine muscles. J. Food Sci. 32: p. 317-323.

Hoagland, R., C.N. McBryde and W.C. Powick, 1917. Changes in fresh beef during cold storage above freezing. USDA Bull. No. 433: p. 1-100.

Hopkins, D.L. and J.M. Thompson, 2001a. Inhibition of protease activity 2. Degradation of myofibrillar proteins, myofibril examination and determination of free calcium levels. Meat Sci. 59: p. 199-209.

Hopkins, D.L. and J.M. Thompson, 2001b. Inhibition of protease activity. Part 1. The effect on tenderness and indicators of proteolysis in ovine muscle. Meat Sci. 59: p. 175-185.

Huff-Lonergan, E., T. Mitsuhashi, D.D. Beekman, F.C. Parrish Jr., D.G. Olson and R.M. Robson, 1996. Proteolysis of specific muscle structural proteins by μ-calpain at low pH and temperature is similar to degradation in *post-mortem* bovine muscle. J. Anim. Sci. 74: p. 993-1008.

Huffman, K.L., M.F. Miller, L.C. Hoover, C.K. Wu, H.C. Brittin and C.B. Ramsey, 1996. Effect of beef tenderness on consumer satisfaction with steaks consumed in the home and restaurant. J. Anim. Sci. 74: p. 91-97.

Hwan, S.-F. and E. Bandman, 1989. Studies of desmin and α-actinin degradation in bovine semitendinosus muscle. J. Food Sci. 54: p. 1426-1430.

Hwang, I.H., C.E. Devine and D.L. Hopkins, 2003. The biochemical and physical effects of electrical stimulation on beef and sheep meat tenderness. Meat Sci. 65: p. 677-691.

Ilian, M.A., A.E. Bekhit and R. Bickerstaffe, 2004a. The relationship between meat tenderization, myofibril fragmentation and autolysis of calpain 3 during *post-mortem* aging. Meat Sci. 66: p. 387-397.

Ilian, M.A., A.E. Bekhit, B. Stevenson, J.D. Morton, P. Isherwood and R. Bickerstaffe. 2004b. Up- and down-regulation of longissimus tenderness parallels changes in the myofibril-bound calpain 3 protein. Meat Sci. 67: p. 433-445.

Ilian, M.A., J.D. Morton, M.P. Kent, C.E. Le Couteur, J. Hickford, R. Cowley and R. Bickerstaffe, 2001. Intermuscular variation in tenderness: association with the ubiquitous and muscle-specific calpains. J. Anim. Sci. 79: p. 122-132.

Imajoh, S. and K. Suzuki, 1985. Reversible interaction between Ca^{2+}-activated neutral protease (CANP) and its endogenous inhibitor. FEBS Lett. 187: p. 47-50.

Inomata, M., Y. Kasai, M. Nakamura and S. Kawashima, 1988. Activation mechanism of calcium-activated neutral protease. J. Biol. Chem. 263: p. 19783-19787.

Jackson, S.P., M.F. Miller and R.D. Green, 1993. The effect of a muscle hypertrophy gene on muscle weights of ram lambs. J. Anim. Sci. 71(Suppl. 1): 146(Abstr.).

Kent, M.P., M.J. Spencer and M. Koohmaraie, 2004. Post-mortem proteolysis is reduced in transgenic mice overexpressing calpastatin. J. Anim. Sci. 82: p. 794-801.

Koohmaraie, M., 1990. Inhibition of *post-mortem* tenderization in ovine carcasses through infusion of zinc. J. Anim. Sci. 68: p. 1476-1483.

Koohmaraie, M., 1992a. Ovine skeletal muscle multicatalytic proteinase complex (Proteasome): purification, characterization, and comparison of its effects on myofibrils with μ-calpains. J. Anim. Sci. 70: p. 3697-3708.

Koohmaraie, M., 1992b. The role of Ca2+-dependent proteases (calpains) in *post mortem* proteolysis and meat tenderness. Biochimie 74: p. 239-245.

Koohmaraie, M., 1994. Muscle proteinases and meat aging. Meat Sci. 36: 3-104

Koohmaraie, M., A.S. Babiker, A.L. Schroeder, R.A. Merkel and T.R. Dutson, 1988. Acceleration of *post-mortem* tenderization in ovine carcasses through activation of Ca^{2+}-dependent proteases. J. Food Sci. 53: p. 1638-1641.

Koohmaraie, M, J.D. Crouse and H.J. Mersmann, 1989. Acceleration of *post-mortem* tenderization in ovine carcasses through infusion of calcium chloride: effect of concentration and ionic strength. J. Anim. Sci. 67: p. 934-942.

Koohmaraie, M., M.E. Doumit and T.L. Wheeler, 1996. Meat toughening does not occur when rigor shortening is prevented. J. Anim. Sci. 74: p. 2935-2942.

Koohmaraie, M., J.E. Schollmeyer and T.R. Dutson, 1986. Effect of low-calcium-requiring calcium activated factor on myofibrils under varying pH and temperature conditions. J. Food Sci. 51: p. 28-32.

Koohmaraie, M., S.C. Seideman, J.E. Schollmeyer, T.R. Dutson and J.D. Crouse, 1987. Effect of *post-mortem* storage on Ca^{++}-dependent proteases, their inhibitor and myofibril fragmentation. Meat Sci. 19: p. 187-196.

Koohmaraie, M., S.D. Shackelford, N.E. Muggli-Cockett and R.T. Stone, 1991a. Effect of the β-adrenergic agonist $L_{644,969}$ on muscle growth, endogenous proteinase activities, and *post-mortem* proteolysis in wether lambs. J. Anim. Sci. 69: p. 4823-4835.

Koohmaraie, M., S.D. Shackelford, T.L. Wheeler, S.M. Lonergan and M.E. Doumit, 1995. A muscle hypertrophy condition in lamb (callipyge): characterization of effects on muscle growth and meat quality traits. J. Anim. Sci. 73: p. 3596-3607.

Koohmaraie, M., G. Whipple and J.D. Crouse, 1990. Acceleration of *post-mortem* tenderization in lamb and Brahman-cross beef carcasses through infusion of calcium chloride. J. Anim. Sci. 68: p. 1278-1283.

Koohmaraie, M., G. Whipple, D.H. Kretchmar, J.D. Crouse and H.J. Mersmann, 1991b. Post-mortem proteolysis in longissimus muscle from beef, lamb and pork carcasses. J. Anim. Sci. 69: p. 617-624.

Kretchmar, D.H., M.R. Hathaway, R.J. Epley and W.R. Dayton, 1990. Alterations in *post-mortem* degradation of myofibrillar proteins in muscle of lambs fed a β-adrenergic agonist. J. Anim. Sci. 68: p. 1760-1772.

LaCourt, A.A., A. Obled, C. Deval, A. Ouali and C. Valin, 1986. Post-mortem localization of lysosomal peptide hydrolase. Cathepsin B. In: Cysteine proteinases and their inhibitors, V. Turk (editor), Walter de Gruyter and Co., New York, p. 239-248.

Lawrence, T.E., M.E. Dikeman, J.W. Stephens, E. Obuz and J.R. Davis, 2003. In situ investigation of the calcium-induced proteolytic and salting-in mechanisms causing tenderization in calcium-enhanced muscle. Meat Sci. 66: p. 9-75.

Lee, W.J., H. Ma, E. Takano, H.Q. Yang, M. Hatanaka and M. Maki, 1992. Molecular diversity in amino-terminal domains of human calpastatin by exon skipping. J. Biol. Chem. 267: p. 8437-8442.

Locker, R.H., 1960. Degree of muscular contraction as a factor in tenderness of beef. Food Res. 25: p. 304-307.

Lusk, J.L., J.A. Fox, T.C. Schroeder, J. Mintert and M. Koohmaraie, 2001. In-store valuation of steak tenderness. Amer. J. Agr. Econ. 83: p. 539-550.

Maki, M., E. Takano, T. Osawa, T. Ooi, T. Murachi and M. Hatanaka, 1988. Analysis of structure-function relationship of pig calpastatin by expression of mutated cDNAs in *Escherichia coli*. J. Biol. Chem. 263: p. 10254-10261.

Marsh, B.B. and N.G. Leet, 1966. Studies in meat tenderness. III. The effects of cold shortening on tenderness. J. Food Sci. 31: **p.** 450-459.

McClelland, P., J.A. Lash and D.R. Hathaway, 1989. Identification of major autolytic cleavage sites in the regulatory subunit of vascular calpain II. J. Biol. Chem. 264: p. 17428-17431.

Mellgren, R.L., 1980. Canine cardiac calcium-dependent proteases: resolution of two forms with different requirements for calcium. FEBS Lett. 109: p. 129-133.

Mellgren, R.L., M.T. Mericle and R.D. Lane, 1986. Proteolysis of the calcium-dependent protease inhibitor by myocardial calcium-dependent protease. Arch. Biochem. Biophys. 246: p. 233-239.

Miller, M.F., K.L. Huffman, S.Y. Gilbert, L.L. Hamman and C.B. Ramsey, 1995. Retail consumer acceptance of beef tenderized with calcium chloride. J. Anim. Sci. 73: p. 2308-2314.

Murachi, T., 1989. Intracellular regulatory system involving calpain and calpastatin. Biochem. Int. 18: p. 263-294.

Nagainis, P.A., S.K. Sathe, D.E. Goll and T. Edmunds, 1983. Autolysis of high-Ca^{2+} and low Ca^{2+}-forms of the Ca^{2+}-dependent proteinase from bovine skeletal muscle. Fed. Proc. 42: p. 1780.

Nakamura, M., M. Inomata, S. Imajoh, K. Suzuki and S. Kawashima, 1989. Fragmentation of an endogenous inhibitor upon complex formation with high- and low-Ca^{2+}-requiring forms of calcium-activated neutral proteases. Biochemistry 28: p. 449-455.

Ngapo, T.M., P. Berge, J. Culioli, E. Dransfield, S. De Smet and E. Claeys, 2002. Perimysial collagen crosslinking and meat tenderness in Belgian Blue double-muscled cattle. Meat Sci. 61: p. 91-102.

O'Halloran, G.R., D.J. Troy, D.J. Buckley and W.J. Reville, 1997. The role of endogenous proteases in the tenderization of fast glycolysing muscle. Meat Sci. 47: p. 187-210.

Ohno, S., S. Minoshima, J. Kudoh, R. Fukuyama, Y. Shimizu, S. Ohmi-Imajoh, N. Shimizu and K. Suzuki, 1990. Four genes for the calpain family locate on four distinct human chromosomes. Cytogenet. Cell Genet. 53: p. 225-229.

Ono, Y, H. Sorimachi and K. Suzuki, 1998. Structure and physiology of calpain, an enigmatic protease. Biochem. Biophys. Res. Commun. 245: p. 289-294.

Otsuka, Y. and D.E. Goll, 1987. Purification of the Ca^{2+}-dependent proteinase inhibitor from bovine cardiac muscle and its interaction with the millimolar Ca^{2+}-dependent proteinase. J. Biol. Chem. 262: p. 5839-5851.

Ou, B.R. and N.E. Forsberg, 1991. Determination of skeletal muscle calpain and calpastatin activities during maturation. Am. J. Physiol. 261: p. E677-E683.

Ouali, A. and A. Talmant, 1990. Calpains and calpastatin distribution in bovine, porcine and ovine skeletal muscles. Meat Sci. 28: p. 331-348.

Parr, T., P.L. Sensky, G.P. Scothern, R.G. Bardsley, P.J. Buttery, J.D. Wood and C. Warkup, 1999. Relationship between skeletal muscle-specific calpain and tenderness of conditioned porcine longissimus muscle. J. Anim. Sci. 77: p. 661-668.

Polidori, P., M.T. Marinucci, F. Fantuz, C. Renieri and F. Polidori, 2000. Tenderization of wether lambs meat through pre-rigor infusion of calcium ions. Meat Sci. 55: p. 197-200.

Ramsbottom, J.M., E.J. Strandine and C.H. Koonz, 1945. Comparative tenderness of representative beef muscles. Food Res. 10: p. 497-509.

Redmond, G.A., B. McGeehin, J.J. Sheridan and F. Butler, 2001. The effect of ultra-rapid chilling and subsequent ageing on the calpain/calpastatin system and myofibrillar degradation in lamb M. *longissimus thoracis et lumborum*. Meat Sci. 59: p. 293-301.

Richard, I., O. Broux, V. Allamand, F. Fougerousse, N. Chiannilkulchai, N. Broug, L. Brenguier, C. Devaud, P. Pasturaud, C. Roudaut, D. Hillaire, M.-R. Passos-Bueno, M. Zatz, J.A. Tischfield, M. Fardeau, C.E. Jackson, D. Cohen and J.S. Beckmann, 1995. Mutations in the proteolytic enzyme calpain 3 cause limb-girdle muscular dystrophy type 2A. Cell 81: p. 27-40.

Savell, J.W. and S.D. Shackelford, 1992. Significance of tenderness to the meat industry. In: Proc. Recip. Meat Conf., Fort Collins, CO, 45: p. 43-46.

Sensky, P.L., T. Parr, R.G. Bardsley and P.J. Buttery, 1996. The relationship between plasma epinephrine concentration and the activity of the calpain enzyme system in porcine longissimus muscle. J. Anim. Sci. 74: p. 380-387.

Shackelford, S.D., M. Koohmaraie, M.F. Miller, J.D. Crouse and J.O. Reagan, 1991. An evaluation of tenderness of the longissimus muscle of Angus by Hereford versus Brahman crossbred heifers. J. Anim. Sci. 69: p. 171-177.

Shackelford, S.D., Wheeler, T.L. and M. Koohmaraie, 1995. The effects of *in utero* exposure of lambs to a β-adrenergic agonist on prenatal and postnatal muscle growth, carcass cutability, and meat tenderness. J. Anim. Sci. 73: p. 2986-2993.

Silva, J.A., L. Patarata and C. Martins, 1999. Influence of ultimate pH on bovine meat tenderness during ageing. Meat Sci. 52: p. 453-459.

Sorimachi, H., N. Toyama-Sorimachi, T.C. Saido, H. Kawasaki, H. Sugita, M. Miyasaka, K. Arahata, S. Ishiura, and K. Suzuki. 1993. Muscle-specific calpain, p94, is degraded by autolysis immediately after translation, resulting in disappearance from muscle. J. Biol. Chem. 268: p. 10593-10605.

Strandine, E.J., C.H. Koonz and J.M. Ramsbottom, 1949. A study of variations in muscles of beef and chicken. J. Anim. Sci. 8: p. 483-494.

Suzuki, K., S. Tsuji, S. Ishiura, Y. Kimura, S. Kubota and K. Imahori, 1981a. Autolysis of calcium-activated neutral protease of chicken skeletal muscle. J. Biochem. 90: p. 1787-1793.

Suzuki, K., S. Tsuji, S. Kubota, Y. Kimura and K. Imahori, 1981b. Limited autolysis of Ca^{2+}-activated neutral protease (CANP) changes its sensitivity to Ca^{2+} ions. J. Biochem. 90: p. 275-278.

Takahashi, K., 1992. Non-enzymatic weakening of myofibrillar structures during conditioning of meat: calcium ions at 0.1 mM and their effect on meat tenderization. Biochimie 74: 247-250

Takahashi, K., 1996. Structural weakening of skeletal muscle tissue during *post-mortem* ageing of meat: the non-enzymatic mechanism of meat tenderization. Meat Sci. 43: p. S67-S80.

Takano, E. and T. Murachi, 1982. Purification and some properties of human erythrocyte calpastatin. J. Biochem. 92: p. 2021-2028.

Taylor, R.G., G.H. Geesink, V.F. Thompson, M. Koohmaraie and D.E. Goll, 1995. Is Z-disk degradation responsible for *post-mortem* tenderization? J. Anim. Sci. 73: p. 1351-1367.

Van Moeseke, W., S. De Smet, E. Claeys and D. Demeyer, 2001. Very fast chilling of beef: effects on meat quality. Meat Sci. 59: p. 31-37.

Veiseth, E. and M. Koohmaraie, 2001. Effect of extraction buffer on estimating calpain and calpastatin activity in *post-mortem* ovine muscle. Meat Sci. 57: p. 325-329.

Veiseth, E., S.D. Shackelford, T.L. Wheeler and M. Koohmaraie, 2001. Effect of *post-mortem* storage on μ-calpain and m-calpain in ovine skeletal muscle. J. Anim. Sci. 79: p. 1502-1508.

Veiseth, E., S.D. Shackelford, T.L. Wheeler and M. Koohmaraie, 2004a. Factors regulating lamb *longissimus* tenderness are affected by age at slaughter. Meat Sci. 68: p. 635-640.

Veiseth, E., S.D. Shackelford, T.L. Wheeler and M. Koohmaraie, 2004b. Indicators of tenderization are detectable by 12 h *post-mortem* in ovine longissimus. J. Anim. Sci. 82: p. 1428-1436.

Wheeler, T.L., J.D. Crouse and M. Koohmaraie, 1992. The effect of *post-mortem* time of injection and freezing on the effectiveness of calcium chloride for improving beef tenderness. J. Anim. Sci. 70: p. 3451-3457.

Wheeler, T.L. and M. Koohmaraie, 1994. Prerigor and postrigor changes in tenderness of ovine longissimus muscle. J. Anim. Sci. 72: p. 1232-1238.

Wheeler, T.L. and M. Koohmaraie, 1999. The extent of proteolysis is independent of sarcomere length in lamb longissimus and psoas major. J. Anim. Sci. 77: p. 2444-2451.

Wheeler, T.L., M. Koohmaraie and J.D. Crouse, 1991. Effects of calcium chloride injection and hot boning on the tenderness of round muscles. J. Anim. Sci. 69: p. 4871-4875.

Whipple, G., M. Koohmaraie, M.E. Dikeman and J.D. Crouse, 1990a. Predicting beef-longissimus tenderness from various biochemical and histological muscle traits. J. Anim. Sci. 68: p. 4193-4199.

Whipple, G., M. Koohmaraie, M.E. Dikeman, J.D. Crouse, M.C. Hunt and R.D. Klemm, 1990b. Evaluation of attributes that affect longissimus muscle tenderness in *Bos taurus* and *Bos indicus* cattle. J. Anim. Sci. 68: p. 2716-2728.

Whiting, R.C., 1980. Calcium uptake by bovine muscle mitochondria and sarcoplasmic reticulum. J. Food Sci. 45: p. 288-292.

Milk indicators for recognizing the types of forages eaten by dairy cows

B. Martin[1], A. Cornu[1,2], N. Kondjoyan[2], A. Ferlay[1], I. Verdier-Metz[3], P. Pradel[4], E. Rock[5], Y. Chilliard[1], J.B. Coulon[1] and J.L. Berdagué[2]*
[1]INRA-Centre de Clermont-Fd/Theix, Unité de Recherches sur les Herbivores, Theix, 63 122 Saint-Genès-Champanelle, France, [2]INRA-Centre de Clermont-Fd/Theix, Unité de Recherches sur la Viande, Theix, Laboratoire Flaveur, 63 122 Saint-Genès-Champanelle, France, [3]INRA-Centre de Clermont-Fd/Theix, Laboratoire de Recherches Fromagères, 36 route de Salers, 15 000 Aurillac, France, [4]INRA-Centre de Clermont-Fd/Theix, Domaine de la Borie, 15 190 Marcenat, France, [5]INRA-Centre de Clermont-Fd/Theix, Unité Maladies Métaboliques et Micronutriments, Theix, 63 122 Saint-Genès Champanelle, France.
**Correspondence: bmartin@clermont.inra.fr*

Summary

Over the recent years, various teams of the INRA centre of Clermont Ferrand/Theix have tried to develop analytical techniques to recognise the dairy cows' feeding. The techniques consist in identifying and measuring components in milk or cheese, which could directly or indirectly originate from dairy cows' feed and which could be used to trace the type of forage eaten by cows. Those components are terpenes, carotenes and milk fatty acids. The purpose of this paper is to synthesise the data obtained in recent trials and to discuss the efficiency and the pertinence of each compound family according to the context. Terpene profiles from milk or cheese alike were particularly effective in recognising the presence of diversified grassland forage in the feed. Analysis of β-carotene and/or colour measurement in milk or cheese permits distinguishing "maize silage" or "hay and concentrate" milk from "grass silage" or "grazed grass" milk. Milk fatty acid composition also carries out valuable information in particular to single out the concentrate rich diets from the others. Nevertheless, those compounds taken individually only offer partial solutions to recognise one diet among the others. In a trial where 6 diets were compared in controlled conditions (concentrate-rich diets, diets based on maize silage, rye-grass hay, rye-grass silage, mountain natural grassland hay and grazed grass), the best results were obtained by combining analyses of terpene compounds and fatty acids, which permitted successful categorization of 100% of the milk. These early results indicate that properly selected, analytical techniques applied to the milk lipid fraction could be, after optimisation and validation of the method, very efficient tools to characterize milks according to the main forages types consumed by cows.

Keywords: dairy products, traceability, forages, terpenes, fatty acids

Introduction

Within the European context where the dairy product market becomes increasingly segmented, specifications and requirements for dairy cows' feeding have multiplied over the recent years. After involving almost exclusively dairy products with protected denomination of origin (PDO), protected geographic identification (PGI) or issued from organic farming, these requirements now have a much broader scope. Distributors and/or producers of more generic products are also establishing them to meet consumers' demand. Following recent food crises (BSE, dioxin...), consumers are increasingly involved in the way cattle are fed and some of them are interested in products whose production methods and conditions are more friendly to the environment, regional tradition and/or that possess special intrinsic qualities (taste, nutritional value). The traceability of these products requires adequate labelling, to warrant consumers' trust. Compliance with specifications is ensured by certification bodies or by professionals themselves. Nevertheless, insofar as producers' milk is sometimes paid more in those chains, fraud becomes an issue. Fraud can be evidenced either by direct inspections in farms or by performing analytical tests on the products themselves.

Over the recent years, various teams of the INRA centre of Clermont Ferrand/Theix have tried to develop analytical techniques to recognise the geographical origin of dairy products and/or cattle feeding. The purpose of this paper is to describe the techniques currently being developed and which could be used to trace the type of forage eaten by dairy cows. The techniques consist in identifying and measuring components in milk or dairy products, which could directly or indirectly originate from dairy cows' feed. Those components usually are terpenes, carotenes and milk fatty acids. Other techniques based on the H, O or C isotopes ratios (Renou *et al.*, 2004) or other milk compounds varying according to the cows diets, like lactones or phytenes (Urbach, 1990) or flavonoids will not be described.

Discriminating between forage botanical types

A technique based on terpene analysis in dairy products was implemented to recognise the botanical type of forage eaten by dairy cows. The terpenes considered are made up by 10 or 15 carbon atoms (monoterpenes or sesquiterpenes, respectively). Terpenes almost exclusively originate from plants and are the major components of essential oils. They are also commonly found among the volatile compounds desorbing from plants at room temperature. These molecules are involved in plant pollination and in plant resistance to predation (repellents) and infection (antimicrobial agents). Their identification in dairy products dates back a long time. Dumont and Adda, (1978) and Dumont *et al.*, (1981) showed that Comté and Beaufort cheeses produced in alpine highlands were richer in terpenes than their valley counterparts. These authors ascribed those differences to the plants species consumed by cattle in highlands. With new analytical development, those components regained interest, both for their possible impact on cheese sensory characteristics (Viallon *et al.*, 1999; Bugaud *et al.*, 2001; Carpino *et al.*, 2004) and as potential tracing agents of cattle diets and moreover mountain cheeses (Bosset *et al.*, 1994).

Terpene variability according to forage and transfer into dairy products

Terpene concentration in forage is mainly governed by its botanical composition: graminaceae-based forages are terpene-poor whereas mountain diversified pastures forages including a large number of dicotyledons contain much more terpenes (Mariaca *et al.*, 1997; Bugaud *et al.*, 2000). In forages cropped in mountain pastures, the ratio between terpene-rich and terpene-poor plants can be as high as 1800:1 (Cornu *et al.*, 2001). Terpene concentration also varies seasonally although seasonal variations are clearly not as wide in the same plant as between different plants (ratio 1:2) (Cornu *et al.*, 2001).

Terpene transfer from forages to milk fat was explored by Viallon *et al.*, (2000) who showed that transfer was very quick. When a cocksfoot hay-based diet complemented with 3 kg/d of a dried aromatic plant (*Achillea millefollium*) is fed to dairy cows, the transfer of terpenes into the milk can be observed as early as at the first milking following the first meal, and the maximum concentration is reached three days following the initiation of the diet. Terpene analysis in milk fat therefore provides instant data on the consumption of terpene-rich forages by cattle.

Comparing between terpene fractions in various forages and corresponding cheeses revealed that forage terpenes are transferred into cheese with some minor alterations (Viallon *et al.*, 1999; Bugaud *et al.*, 2001; Schlichtherle-Cerny *et al.*, 2004). Hay that are characterised by a wide diversity of plant species, dicotyledons in particular, contain higher concentration and wider diversity of terpenes. Such composition variability can be also found in corresponding cheeses (Figure 1) (Viallon *et al.*, 1999).

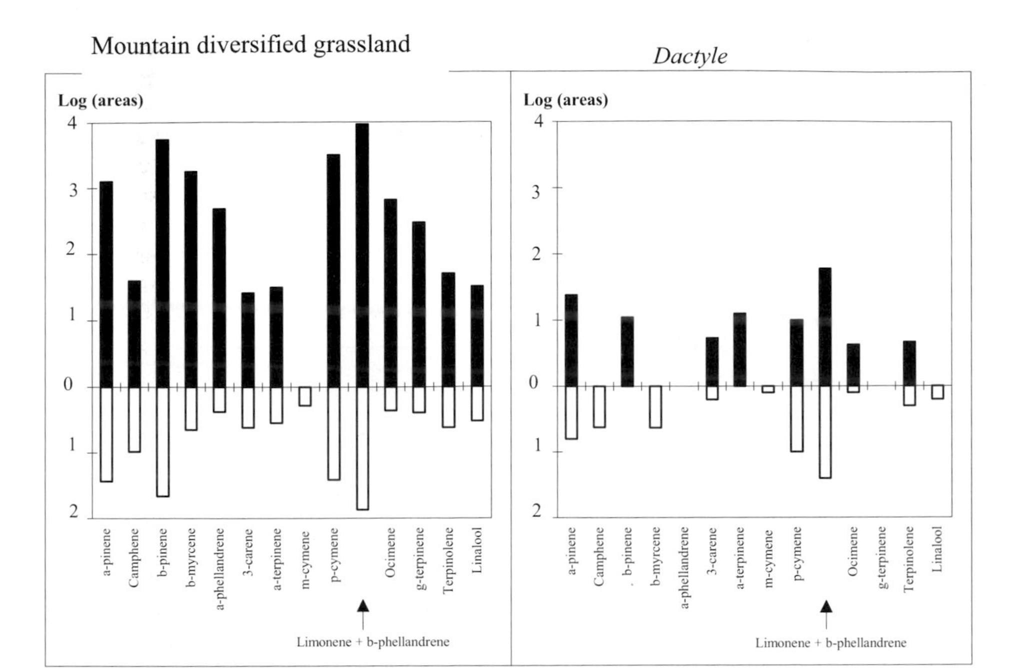

Figure 1. Comparison of terpene profiles between forages (■) and cheeses (□) according to the type of hay (from Viallon et al., *1999).*

Examples of application

The use of terpenes as a means to recognize dairy cow's diet was proposed by Bosset *et al.*, (1994) to recognize Etivaz cheeses produced when cows were fed native highland pasture and by Moio *et al.*, (2001) who suggested that sesquiterpenes could be considered as chemical markers of the milk produced at pasture during the summer period. Similar results were obtained by Bendall (2001) with milk from pasture containing plant compounds (mono- and sesquiterpenes, phytenes, phytol) while milk from cows fed on preserved forages contained high amounts of lactones. More recently, Carpino *et al.*, (2004) also reported that some terpenes were only detected in Ragusano cheese made with milk from cows consuming native Sicilian pasture plants. The effectiveness of terpenes to recognize dairy cow's diet was also tested recently in two different trials; one from milk fat terpenes and the other one from terpenes found in cheeses.

The first trial (Fernandez *et al.*, 2003) was performed with milk sampled from herds in plain (Brittany-France) or highland (Auvergne-France) farms to verify whether the region of origin could be discriminated by analysing the terpene fraction in milk fat. Milk samples were collected during the summer season, when the cows were in pasture, and in winter, when the cows were indoors. In Brittany, winter diets were based on maize silage and in the summer, the cows were grazing graminaceae and legumes from seasonal meadows. In Auvergne, winter diets were based on permanent and diversified meadows conserved in the form of hay or silage in winter and directly grazed in the summer. Terpene analysis (sesquiterpenes in particular) made it possible to discriminate between the geographic origins of milk regardless of seasonal factors (Figure 2). In that case, geographical distinction was based on the botanical type of forage fed to cattle.

These analyses, applied to Salers cheeses from 30 farmhouse producers scattered over the Cantal department (France), revealed that the diversity and total amount of terpenes contained in cheeses

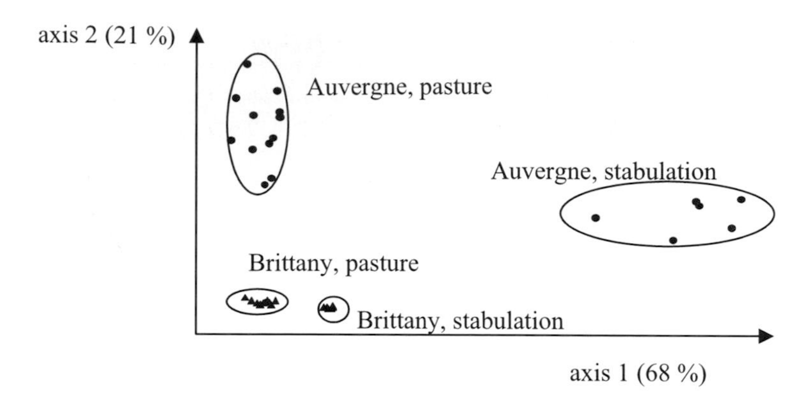

Figure 2. Milk discrimination according to the production area (Auvergne = ●; Brittany = ▲) and seasonal feeding management. Discriminating factor analysis was performed with 6 sesquiterpenes : δ-elemene, β-bourbonene, β-caryophyllene, β-chamigrene, γ-cadinene and β-sesquiphellandrene (from Fernandez et al.*, 2003).*

were closely linked to such geographical factors as pasture elevation and orientation. This link is due to the fact those geographical factors are also associated to the proportion of permanent meadows in the forage systems and to the botanical composition of permanent meadows (Murat, unpublished, Cornu *et al.*, 2002).

These results are proofs of the particular effectiveness of that method in recognising the presence of diversified grassland forage in the feed, from milk or cheeses alike.

Discriminating between the main diets

To verify whether the most common forages used for dairy cattle could be distinguished through milk analysis, a trial was organised under experimental conditions. Six diets corresponding to a wide array of nutritional situations were compared (concentrate-rich diets (barley and soybean meal and hay), diets based on maize silage, rye-grass hay, rye-grass silage, permanent mountain grassland hay and grass) using equivalent groups of eight dairy cows. In the first diet, concentrates represented 65 % of the dry matter ingested by the cows whereas in the other diets their proportion varied from 0 to 15 %. Terpenes, fatty acids, β-carotene, lutein and vitamins A and E were analysed in the cows' individual milk samples. The analyses of terpenes and fatty acids were made as described by Fernandez *et al.*, (2003) and Ferlay *et al.*, (2002) respectively and carotenoids and vitamins A and E were analysed simultaneously using the method described by Lyan *et al.*, (2001). All results underwent common treatment. Through successive analyses of variance to compare diets with all others one-to-one, the compounds most influenced by the type of diet ($p<0.0005$) could be identified. Those compounds are listed in Table 1. So, 10 terpenes, β-carotene and 5 fatty acids provided relevant information on the type of diets on a case-by-case basis. In contrast, vitamins and lutein did not provide any additional information.

Terpenes

The milk samples collected from the cows grazing the diversified meadow contained between 6 and 23 times more terpenes (depending on terpenes), than the milk samples obtained from the cows eating other forages, including hay from diversified meadow. The first attempts to discriminate samples according to the terpene composition of milk made it possible to successfully classify them (Cornu *et al.*, 2002) (Figure 3). However, the detailed analysis of the clustering revealed some

Table 1. Most discriminating variables of each of the six diet types (p<0.0005).

Diets	Variables best discriminating one diet among all others (in decreasing order of pertinence)	Characteristic of the milk[1]
• Concentrate (65%) + Cocksfoot hay (35 %)	1/ Linoleic acid (w/o conjugates)	rich
	2/ *cis*9 *trans*13 C18:2	poor
	3/ Rumenic acid	poor
• Maize silage	1/ Σ C18 :1 other than oleic and vaccenic acids	rich
	2/ Linolenic acid	poor
	3/ terpene 31	rich
• Rye-grass silage	1/ terpenes 39	rich
	2/ terpene 40	rich
	3/ terpene 34	rich
	4/ β-carotene	rich
• Rye-grass hay	1/ terpene 21	rich
	2/ Σ C18 :1 other than oleic and vaccenic acids	rich
	3/ Rumenic acid	rich
• Mountain-diversified grassland hay	1/ β-carotene	poor
	2/ terpene 17	rich
• Mountain-diversified grassland pasture	1/ terpene 1	rich
	2/ terpene 37	rich
	3/ terpene 28	rich
	4/ terpene 11	rich

[1] "Poor" or "rich" qualifies the milk corresponding to the considered diet and compound in comparison to the milks corresponding to all the other diets.

similarities between the preserved diets. The likeness of the terpene profiles found in milk samples obtained from the same grass preserved as hay or silage confirmed a previous finding that the preservation method modifies terpene proportions but only moderately affects their nature (Verdier-Metz *et al.*, 1998). The profiles obtained remain close, even under experimental conditions where a single type of forage composed the diet. Therefore, in farms where grass silage is rarely given alone, using terpenes as a means to detect grass silage presence in the diets has little chance of giving fully satisfying results, even if it is a strong point in a number of specifications.

Moreover, in that experiment, milk samples obtained from the same grassland hay stored for variable periods time (8 or 11 months) displayed similar profiles although average terpene amounts tended to decrease by about 30% in the milk issued from cows fed the hay that was stored for longer. The possible terpene disappearance during forage storage could explain the confusion between initially terpene-rich forages stored for a long time and initially terpene-poor forages stored for a short period.

These first results showed that measuring the terpene fraction of milk is not sufficient to perfectly discriminate between all diets.

β-carotene and colour

β-carotene is the main pigment responsible for the yellow coloration of bovine dairy products. It exclusively originates from forage and its concentration mainly depends on the plant species and development stage and on the mode of forage preservation. Grass, especially when leafy, contains

rumenic acid made it possible to distinguish in particular the concentrate rich diet or the rye-grass silage diet from the rye-grass hay diet (Figure 4). This result confirms the interest of CLA isomers as potential indicators of cow feeding, as it was suggested in Alpine conditions by Collomb *et al.*, (2004) who observed in milk a correlation between 3 CLA isomers (9*cis*, 11*trans*; 11*trans*,13*cis* and 8*trans*,10*cis*) and pasture elevation. Other C18 fatty acids like vaccenic acid or other C18:1 isomers are likely to be good tracers insofar as they widely vary according to the type of forage given to dairy cows.

Simultaneous use of several analytical methods

Compound taken individually only offer partial solutions to recognise a diet among others. Considering the complexity of the problem, using several of the above-described techniques simultaneously appears necessary to discriminate milk samples according to the main types of forage fed to the cows. Indeed, no single tracer has been identified to confront the various facets of the problem. The best results have been obtained by the combined use of terpene compounds and milk fatty acids. Using the results of those analyses, performed by chromatography of the lipid fraction of milk, data analysis elicited a restricted selection of compounds (8 terpenes and 4 fatty acids) whose combination provided efficient discriminations. The most effective models obtained under learning, validation and testing conditions, once implemented, permitted successful discrimination of 100% of the milk samples analysed (Figure 5). Those models were of the non-linear, multi-layer neuronal network type whose input variables and architecture were respectively selected and optimised by powerful computerized procedures.

Such positive results indicate that milk discrimination according to the main types of animal diets has been possible with this database. In our knowledge, such results obtained with a large diversity of diets had never been described in literature before. These results are probably partly ascribable to the milk samples, produced under standardised and controlled experimental conditions, from which reference databases should be elaborated. Analytically, the characterizations thus obtained

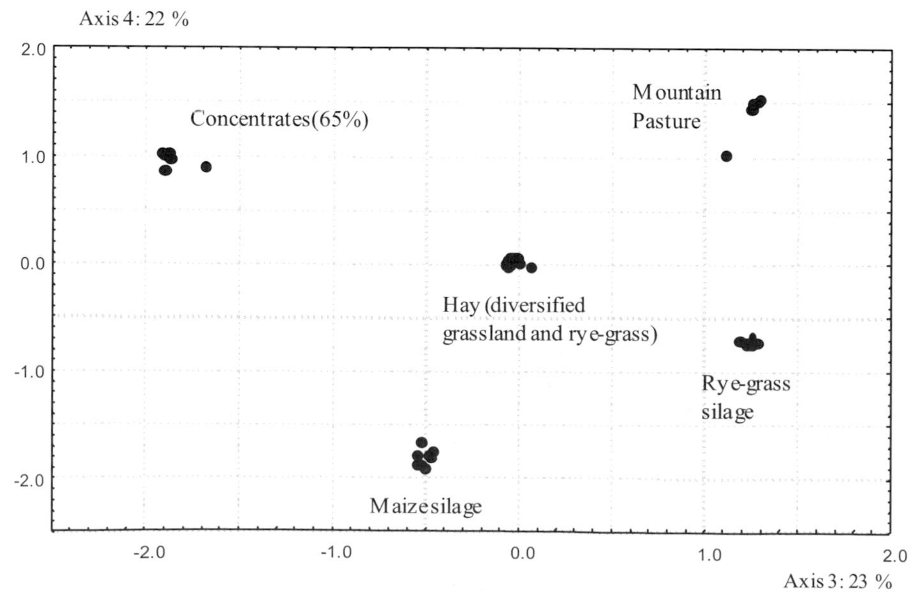

Figure 5. Milk discrimination according to the type of diet from 8 terpenes and 4 fatty acids. Array of points obtained by Principal Component Analysis on the multiplayer neuronal network outputs.

are rather cumbersome and relatively expensive to implement, as all chromatography analyses. Therefore the technique described here cannot be applied to systematic control procedures but could, after optimisation, help identifying the type of feed provided to dairy cows. Nevertheless, before any possible use, this method has to be validated with a large number of samples collected in bulk milks and its robustness has to be tested over several seasons. In addition, bringing this diagnostic method to general use will require analytical efforts to identify tracers (plant or metabolic tracers) obtained with a single analytical technique, so as to reduce costs.

Conclusion

Studies conducted on the traceability of dairy cows' feeding are fully consistent with the current context of consumers' inquisitive attitude towards the way animals are fed. The methods in consideration however will not detect any specific banned feed in a diet (GMO,..). Early results indicate that properly selected analytical techniques applied to the milk lipid fraction have already yielded in experimental conditions efficient tools to identify the main forage types. Further work is now needed to validate the method proposed and to test its robustness. Furthermore, these analyses are also cumbersome and costly and therefore can only be considered for use in a restricted number of high-stake cases, but their mere existence may in itself have a dissuasive effect on frauds.

References

Bendall, J.G., 2001. Aroma compounds of fresh milk from New Zealand Cows fed different diets. J. Agric. Food Chem. 49: p. 4825-4832.

Bosset J.O., U.Bütikofer, R. Gauch and Sieber, R., 1994. Caractérisation de fromages d'alpages subalpins suisses: mise en évidence par GC-MS de terpènes et d'hydrocarbures aliphatiques lors de l'analyse par «Purge and Trap» des arômes volatils de ces fromages. Schweiz. Milchw. Forschung 23: p. 37-41.

Bugaud, C., A. Bornard, A. Hauwuy, B. Martin, J.C.Salmon, L. Tessier, and S. Buchin, 2000. Relations entre la composition botanique des vegetations de montagne et leur composition en composes volatils. Fourrages 162: p. 141-155.

Bugaud, C., S. Buchin, A. Hauwuy, and J.B. Coulon, 2001. Relationships between flavour and chemical composition of Abondance cheese derived from different types of pasture. Lait 81: p. 757-774.

Bugaud, C., S. Buchin, Coulon, J.B., Hauwuy, A. and D. Dupont, 2001.Influence of the nature of alpine pastures on plasmin activity, fatty acid and volatile compound composition of milk. Lait 81: p. 401-414.

Carpino, S., Mallia, S., Terra S., Melilli, C., Licitra, G., Acree, T.E., Barbano, D.M. and P.J. van Soest, 2004. Composition and aroma compounds of Ragusano cheese: native pasture and total mixed rations. J. Dairy Sci. 87: p. 816-830.

Chilliard, Y., A. Ferlay, and M. Doreau, 2001. Effect of different types of forages, animal fat or marine oils in cow's diet on milk fat secretion and composition, especially conjugated linoleic acid (CLA) and polyinsaturated fatty acids. Livest. prod. Sci. 70: p. 31-48.

Chilliard, Y. and A. Ferlay, 2004. Dietary lipids and forage interactions on cow and goat milk fatty acid composition and sensory properties. Reprod. Nutr. Dev. 44: p.467-492.

Collomb, M., Bütikofer, U., Sieber, R., Bosset, J.O. and B. Jeangros, 2001. Conjugated linoleic acid and trans fatty acid composition of cows' milk fat produced in lowlands and highlands. J. Dairy Res. 68: p. 519-523.

Collomb, M., Bütikofer, U., Sieber, R., Jeangros, B., Bosset, J.O., 2002. Correlations between fatty acids in cows' milk fat produced in the lowlands, mountains and highlands of Switzerland and botanical composition of the fodder. Int. Dairy J. 12: p. 661-666.

Collomb, M., Sieber, R. and U. Bütikofer, 2004. CLA isomers in milk fat from cows fed diets with high levels of unsaturated fatty acids. Lipids 39: p. 355-364.

Cornu, A., A.P. Carnat, B. Martin, J.B.Coulon, J.L. Lamaison, and J.L. Berdagué, 2001. Solid phase microextraction of volatile components from natural grassland plants. J. Agric. Food Chem. 49: p. 203-209.

Cornu, A., B. Martin, I. Verdier-Metz, P. Pradel, J.B. Coulon, and J.L. Berdagué, 2002.Use of terpene profiles in dairy produce to trace the diet of dairy cows. Grassland Science in Europe 9: p. 550-551.

Dumont, J.P. and J. Adda, 1978. Occurrence of sesquiterpenes in mountain cheese volatiles. J. Agric. Food Chem. 26: p. 364-367.

Dumont, J.P., J. Adda, and P. Rousseau, 1981. Exemple de variation de l'arôme à l'intérieur d'un même fromage : le Comté. Lebensm. Wiss. u. Technol. 14: p. 198-202.

Ferlay, A., B. Martin, P. Pradel, P. Capitan, J.B. Coulon, and Y. Chilliard, 2002.Effect of the nature of forages on cow milk fatty acids having a positive role on human health. Grassland Science in Europe 9: p. 556-557.

Fernandez, C., C. Astier, E. Rock, J.B Coulon,. and J.L. Berdagué, 2003 Characterisation of milk by analysis of its terpene fractions. Int. J. Food Sci. Technol. 38: p. 445-451.

Jiang, J., Bjoerck, L. Fondén, R. and M. Emanuelson, 1996. Occurrence of conjugated cis-9,trans-11-octadecadienoic acid in bovine milk: effect of feed and dietary regimen. J. Dairy Sci. 79: p. 438-445.

Lyan, B., Azaïs-Braesco, V., Cardinault, N., Borel, P., Alexandre-Gouabau, M.C. and P. Grolier, 2001. Simple method for clinical determination of 13 carotenoids in human plasma using an isocratic high-performance liquid chromatographic method, J. of Chromatogr. B 751: p. 297-303.

Loor, J.J., Ueda, K., Ferlay, A., Chilliard Y. and M. Doreau, 2002. Patterns of biohydrogenation and duodenal flow of trans fatty acids and conjugated linoleic acids (CLA) are altered by dietary fiber level and linseed oil in dairy cows. J. Dairy Sci. 85 (Suppl. 1): p. 314.

Mariaca, R.G., T.F.H. Berger, R. Gauch, M.I. Imhof, B. Jeangros, and J.O. Bosset, 1997. Occurrence of volatile mono- and sesquiterpenoids in highland and lowland plant species as possible precursors for flavor compounds in milk and dairy products, J. Agric. Food Chem. 45: p. 4423-4434.

Moio, L., Rillo, L., Ledda, A. and F. Addeo, 1996. Odorous constituents of ovine milk in relationship to diet. J. Dairy Sci. 79: p. 1322-1331.

Pillonel, L., Collomb, M., Tabacchi, R. and Bosset J.O. 2002. Analytical methods for the discrimination of the geographic origin of Emmental cheese. Free fatty acids, triglycerides and fatty acid composition of cheese fat. Mit. Lebens. und Hygiene 93: p. 217-231.

Prache, S. and M. Theriez, 1999. Traceability of lamb production systems: carotenoids in plasma and adipose tissue. Anim. Sci. 69: p. 29-36.

Prache, S., A. Priolo, H. Tournadre, R. Jailler, H. Dubroeucq, D. Micol, and B. Martin, 2002. Traceability of grass-feeding by quantifying the signature of carotenoid pigments in herbivores meat, milk and cheese. Grassland Science in Europe 9: p. 592-593.

Renou, J.P., Deponge, C., Gachon, P., Bonnefoy, J.C., Coulon, J.B., Garel, J.P., Vérité, R. and P. Ritz, 2004. Characterization of animal products according to geographic origin and feeding diet using nuclear magnetic resonance and isotope ratio mass spectrometry: cow milk. Food Chem. 85: p. 63-66.

Schlichtherle-Cerny H., M.I. Imhof, E. Fernandez-Garcia and J.O. Bosset, 2004. Changes in terpene composition from pasture to cheese. Cheese Art 2004, 6th International Meeting on Mountain Cheese, Ragusa, Donnafugata Castle, June 1st-2nd, 2004. CD-ROM, Corfilac Ed.

Urbach, G., 1990. Effect of feed on flavor in dairy foods. J. Dairy Sci. 73: 3639-3650

Verdier-Metz I., J.B. Coulon, P. Pradel, C. Viallonand J.L. Berdagué, 1998. Effect of forage conservation (hay or silage) and cow breed on the coagulation properties of milks and on the characteristics of ripened cheeses. J. Dairy Res. 65: p. 9-21.

Viallon C., B. Martin, I. Verdier-Metz, P. Pradel, J.P. Garel, J.B. Coulon, and J.L. Berdagué, 2000. Transfer of monoterpenes and sesquiterpenes from forages into milk fat. Lait 80: p. 635-641.

Viallon C., I. Verdier-Metz, C. Denoyer, P. Pradel, J.B. Coulon and J.L. Berdagué, 1999. Desorbed terpenes and sesquiterpenes from forages and cheeses. J. Dairy Res. 66: p. 319-326.

Zeppa, G., Giordano, M., Gerbi, V. and M. Arlorio, 2003. Fatty acid composition of Piedmont "Ossolano" cheese. Lait 83: p. 167-173.

High-fat rations and lipid peroxidation in ruminants: consequences on the health of animals and quality of their products

D. Durand, V. Scislowski, D. Gruffat, Y. Chilliard and D. Bauchart*
INRA-Centre de Recherches de Clermont-Ferrand-Theix, Unité de Recherches sur les Herbivores, 63122 Saint-Genès-Champanelle, France
**Correspondence: durand@clermont.inra.fr*

Summary

The health value of ruminant products (milk, meat) can be improved by different dietary strategies in order to lower atherogenic fatty acids (FA) such as some saturated FA (14:0 and 16:0) and transvaccenic acid (18:1Δ11t) and favour beneficial polyunsaturated FA (PUFA) with a particular emphasis on n-3 PUFA. Hence, fresh grass which is rich in linolenic acid (18:3n-3) and vegetable oils (supplied as seeds or free oil) which are rich in linoleic (18:2n-6, from soybean or sunflower seed) or linolenic (from linseed) acids are the main dietary sources of PUFA which can be readily incorporated into lipids of milk or muscle tissues. Nevertheless, the high sensitivity of PUFA to lipoperoxidation in plasma and tissues of ruminants can increase the risk of alteration of the animal health in addition to a lower nutritional quality of their products. Thus, further investigations are needed to define the most efficient combination of antioxidant sources in the context of dietary PUFA supplementations in lactating and meat producing ruminants.

Keywords: lipid peroxidation, ruminant, high-fat rations, beef quality, milk quality

Introduction

Evidence of causal links between the quality of diets and the occurrence of some diseases such as cancer, atherosclerosis and diabetes is often sufficiently strong that recommendations are often made by government health authorities to help to reduce their incidence. High levels of fat consumption are considered to favour several of so-called "Diseases of Western Civilisation" in human, especially in the case of coronary heart disease (Department of Health, 1994; Williams, 2000). In this context, ruminant products have always had a bad press due to their relatively high saturated pro-atherogenic fatty acids content and low PUFA/SFA ratio (Wood *et al.*, 1999). Hence, many studies in ruminant animals have been conducted in order to improve the nutritional quality of bovine products.

The aim of this review is to 1) compare lipid contents of ruminant products with dietary needs by humans 2) summarise the known effects of different diets on dairy and meat products 3) evaluate the consequences of lipid peroxidation on animal health and quality of products.

Fatty acid content of ruminant products and dietary needs in humans

Human nutritional requirements for fatty acids

From birth to old age, vegetable and animal (milk and meat) fats provide some essential components of the human diet since they contain saturated fatty acids (FA) as an energy source for cells but also metabolically important essential polyunsaturated fatty acids (PUFA) such as linoleic (18:2n-6) and linolenic (18:3n-3) acids as well as the longer chain PUFA arachidonic (20:4n-6, AA), eicosapentaenoic (20:5n-3, EPA) and docosahexaenoic (22:6n-3, DHA) acids. PUFA are fundamental components of membrane phospholipids, specific precursors of cellular function molecules (prostaglandins, thromboxanes, leucotrienes) especially involved in reproduction, cardiac physiology, blood coagulation, inflammation, physiology of endocrine and exocrine glands, and functions of blood platelets or of the retina.

Recent investigations suggest that there is a need to consider the quality as well as the quantity of fat supplied in the diets of the Western population. Recommended nutritional supplies for saturated, monounsaturated and polyunsaturated FA (18:2n-6, 18:3n-3) (Legrand, 2002) are given in Table 1. Total fatty acid intake was limited to about 30 to 33% of the energy intake and amounted to 66 and 80 g FA/d for women and men, respectively.

Concerning saturated FA (SFA), it is suggested that their intake should be limited to about 8% of energy intake (ie 25% total FA ingested). Qualitatively, the different SFA present in ruminant products are not similar in term of biological effects on the health of humans. Short chain fatty acids (from 4 to 10 carbons) are now considered as neutral in the context of hypercholesterolemia as well as stearic acid (18:0), whereas lauric (12:0) and especially myristic (14:0) and palmitic (16:0) acids are clearly hypercholesterolemic agents (Grundy and Denke, 1990).

Cis oleate (18:1n-9 cis) is the main monounsaturated fatty acid present in plant and animal fats. It plays a significant role in the physiology of membranes as a constituent of their phospholipids (modulation of activity of enzymes, transporters or receptors) and, at the difference of saturated FA, it reduces cholesterolemia in human (Legrand, 2002).

Trans fatty acids mainly correspond to monounsaturated FA in ruminant fats (resulting from biohydrogenation of dietary PUFA by rumen bacteria) or in margarines, shortenings and salad oils (result of the industrial partial hydrogenation of vegetable oils). They provoke an increased risk of coronary heart disease since they induce both higher levels of plasma atherogenic lipoproteins (LDL and lipoprotein (a), Oomen *et al.*, 2001) and lower levels of plasma HDL (non atherogenic) (Lichtenstein, 2000). However, recent studies indicate that they do not increase susceptibility of LDL to peroxidation (Lichtenstein, 2000) which is an additionnal factor of risk for atherosclerosis. The most abundant *trans* fatty acids in both hydrogenated oils and ruminant fats are *trans*-C18:1. A study undertaken in 14 European countries has shown that the intake of *trans* FA varied between countries and was generally the lowest in Southern Europe (below 1% of energy) and the highest (around 2%) in Northern Europe (especially in Iceland). The relative risk of coronary heart disease for ruminant *trans* FA seemed to be similar to that of industrial sources of *trans* FA (Mensink, 2002).

Table 1. Dietary recommendations (expressed as g/d or as % of total energetic supply, TES) in saturated FA, monounsaturated FA, linoleic acid (18:2n-6), linolenic acid (18:3n-3) and total FA for men, women and elderly subjects (Legrand, 2002).

		Saturated fatty acids	Monounsaturated fatty acids	18:2n-6	18:3n-3	Total
Men	g / d	19.5	49	10	2	81
	% TES	8	20	4.0	0.8	33
Women	g / d	16	40	8	1.6	66
	% TES	8	20	4.0	0.8	33
elderly subjects	g / d	15	38	7.5	1.5	62.5
	% TES	8	20	4.4	0.9	33.7

The potential for reducing saturated FA through substitution with monounsaturated FA (MUFA) has increased in recent years. Nevertheless, there has continued to be debate over whether MUFA or PUFA should replace saturated FA in human diets. MUFA lowered plasma LDL levels with no effect on HDL level, whereas PUFA lowered both LDL and HDL levels. Actual MUFA supply in nutritional needs in human amounted to near 50 g/d (60% total FA) (Legrand, 2001).

A considerable amount of evidence has accumulated to support the view that the long chain n-3 PUFA (EPA and DHA) have beneficial cardiovascular and anti-inflammatory properties but their levels in diets are insufficient in the most Western countries. The recommended dietary PUFA supply is estimated to 12.5 g/d (ie 15% total FA) with the value of the 18:2n-6/18:3n-3 ratio not higher than 5 in order to preserve adequate supply of these two precursors for these essential PUFA families (Legrand, 2001).

Fatty acid content of ruminant products

Ruminant fat represents an important source of lipids in Western countries, particularly milk, fat which provides 15-25% of total dietary fats for humans, thus contributing to 25-35% of total saturated FA supply (O'Donnell, 1993). On the other hand, beef is estimated to contribute only about 5% of total fat intake, but is regarded as a rich source of saturated fat.

Milk fat is synthesised by mammary gland either from FA taken up from the blood (60%) or by de novo FA synthesis (40%). FA composition of milk is reported in Figure 1. Briefly, saturated FA averaged 70% of total FA composed of 12% 14:0, 31% 16:0 and 9 % 18:0. Monounsaturated FA averaged 25% of total FA mainly as C18:1. Polyunsaturated FA averaged only 5% of total FA

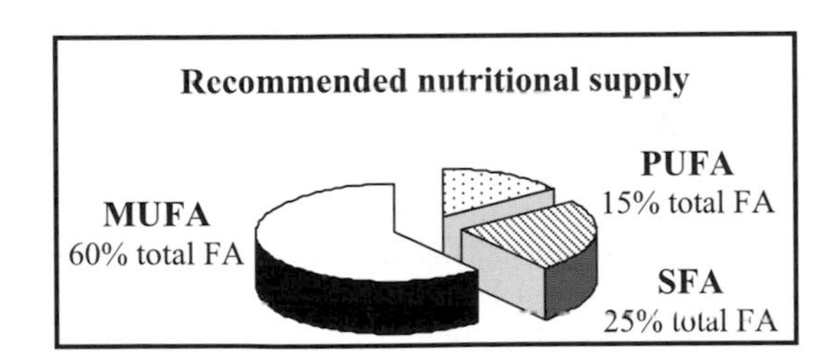

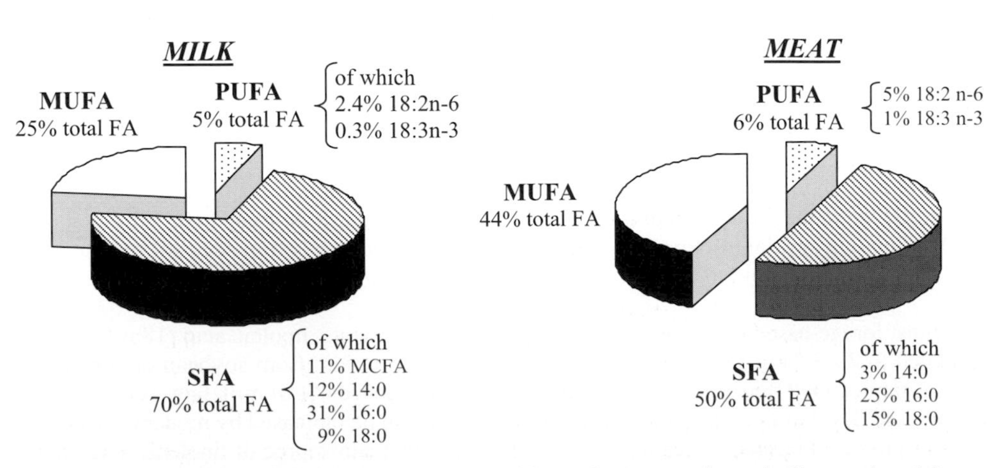

Figure 1. Comparison of fatty acid composition of fats in bovine products (milk, meat) and dietary fatty acid recommendations for human.

with 2.4% 18:2n-6 and 0.3% 18:3n-3. This composition differ strongly from human recommendations which needs more MUFA (60% vs 25%) and PUFA (15% vs 5%) and less SFA (25% vs 65%).

FA composition of meat fat is nearer to the composition of FA recommended for human than FA composition of milk fat. SFA represented 50 % of total FA with only 3% of lauric acid (the most undesirable SFA for human) and 15 % of stearic acid considered as a neutral FA. MUFA reach 40-45% of total FA which is still below recommendations. As in milk fat, PUFA are very poorly represented in bovine meat fat (6%).

Consequently, the very low amounts of PUFA and high amounts of saturated fatty acids in ruminant products explain why recently a range of nutritional studies have been conducted to improve the health value of milk and beef fat, particularly by (1) increasing PUFA with particular emphasis on n-3 PUFA and decreasing the n-6:n-3 ratio, (2) decreasing atherogenic SFA and (3) maintaining or reducing trans FA.

Dietary strategies and nutritional quality of ruminant products

In ruminant, fatty acid composition of fats in meat and milk reflects rumen metabolism of dietary lipids. Rumen bacteria hydrogenate double bonds of a large amount of unsaturated dietary fatty acids so that muscle FA in cattle and sheep are more saturated and less unsaturated than those in monogastric animals. In particular, 18:2n-6, which is (with 18:3n-3) a major FA in plants, is far less abundant in ruminant products.

Manipulation of milk fatty acid composition

Different strategies have been developed in the last 30 years to reduce saturated FA contents of milk fat. This has been reviewed by Chilliard *et al.*, (2000) which summarised the known effects of dietary treatments applied to lactating cows on the fatty acid composition of milk fat (Table 2). Most of the FA arising from de novo synthesis are saturated (4:0 to 16:0), because delta-9 desaturase has a very low activity towards fatty acids shorter than 18 carbon chain length, although small proportions of 14:0 and 16:0 can be desaturated into 14:1 and 16:1. Long-chain FA, originated from dietary lipids or from body mobilisation, are potent inhibitors of mammary FA synthesis leading to a decrease in the percentage of medium-chain FA in milk lipids. This is due to both an higher secretion of long-chain FA from the blood and a lower de novo synthesis of FA.

Polyunsaturated FA 18:2n-6 and 18:3n-3 are not synthesised by ruminant tissues and hence their concentration in milk is mainly dependent on dietary intake. Thus, numerous studies have analysed different dietary sources of lipids rich in PUFA and tested different factors able to decrease their biohydrogenation in the rumen. Large amounts of encapsulated safflower oil added to rations for dairy cattle led to the production of milk fat rich in linoleic acid (18:2n-6) up to 35% of total milk FA. Feeding encapsulated rapeseed, sunflower seed, soybean or cotton seed oil resulted in a proportion of linoleic acid to total fat equals to 15 to 20%.

With most forage-based diets given without any lipid supplements, linoleic acid (18:n-6) in milk amounted to 2 - 3 % of total FA. With diets enriched with lipids (from soybean or sunflower) given as unprotected free oils or as oil seeds linoleic acid reached 4% of milk fatty acids. As with linoleic acid, level of linolenic acid (18:3 n-3) in milk fat can be increased by moderate supply of lipids as protected linseed oil leading to 6.4%. Grass is the main source of linolenic acid. When animals are fed fresh grass, the level of linolenic acid in milk can be four times higher than with conventional diets (2.4 vs 0.7%), but its amount in grass highly varied extensively with season and grass varieties which explains large variations of 18:3n-3 in milk. Among non-forage foodstuffs, linseed represents the main source of 18:3n-3 for ruminants (>50% total FA). With such supplement, the level of this PUFA increases in milk fat to the same extend than with free oil (Table 2) (review of Chilliard *et al.*, 2000).

Table 2. Compared effects of the addition of oil seed and free oil in rations of dairy cattle on milk fatty acid composition (expressed as the difference in percentage between values obtained with control and lipid-supplemented diets). From Ferlay and Chilliard, personal communication Kennelly, 1996 ; Casper et al., *1988 ; Steele* et al., *1971 ; Tesfa* et al., *1991 ; Murphy et al., 1990*

Origin	Linseed		Sunflower		Soybean		Rapeseed	
Type	Oil	Seed	Oil	Seed	Oil	Seed	Oil	Seed
Lipids (% diet DM)	*3.0*	*4.0*	*3.0*	*4.2*	*4.2*	*3.2*	*4.3*	*5.3*
6:0 to 8:0	-0.7	-0.3	-0.9	-2.1	-1.6	-0.4	-0.7	-0.4
10:0 to 14:0	-4.2	-4.6	-5.9	-8.6	-13.1	-8.4	-4.7	-5.2
16:0	-8.9	-5.2	-9.7	-8.7	-14.7	-11.5	-4.7	-8.9
18:0	+2.1	+4.9	+2.9	+4.1	+3.9	+7.7	+3.1	+5.5
18:1n-9	+8.5	+5.4	+11.8	+15.8	+25.9	+13.9	+8.5	+10.4
18:2n-6	-0.3	+0.2	+0.3	+1.3	+0.9	+1.8	+0.6	-0.6
18:3n-3	+0.3	+0.3	-0.1	+0.3	nd	+0.1	0.0	0.0

Manipulation of beef fatty acid composition

As observed in milk, PUFA: SFA value (an important nutritional index) in meat (0.11) is below the recommended value for human (0.45). On the other hand, the n-6:n-3 PUFA ratio found in cattle and sheep meat is closer to the recommended value (near 5) than that found in pig meat. This difference can be explain by the relatively high 18:3n-3 level in ruminants, especially in meat lipids of fattening steers (Bauchart *et al.*, 2001) given fresh grass in which this fatty acid is prevailing (Bauchart *et al.*, 1984). Nevertheless, as for milk composition, the proportion of lipid and 18:3n-3 in grass varies across the season (Bauchart *et al.*, 1984) and this is reflected in meat lipids (Wood *et al.*, 1999). A further difference between pig and ruminants is that the long-chain n-3 PUFA are not incorporated into triacylglycerols to the same extent in ruminants. They are restricted to the phospholipids (mainly included in membranes) and, therefore, are found mostly in muscle lipids but not fat tissue, excepted at very low levels in the total lipid fraction (Enser *et al.*, 1996).

Many studies have found important effects of dietary fatty acids on PUFA content of muscle lipids in growing or finishing cattle (Scollan *et al.* 2005). Unsaturated oils provoke comparable effects to those observed in milk fat, ie a decrease in short- and medium-chain FA contents, and an increase in long-chain unsaturated FA, mainly 18:1n-9. With free oil, 18:2n-6 is preferentially incorporated in intramuscular phospholipids, reflecting the metabolic importance of this lipid fraction in muscle (Demeyer and Doreau, 1999). Therefore, it is possible to modify the content of 18:2n-6 and 18:3n-3 of muscle lipids by dietary lipid supplements (Bas and Sauvant, 2001) as illustrated in Figure 2.

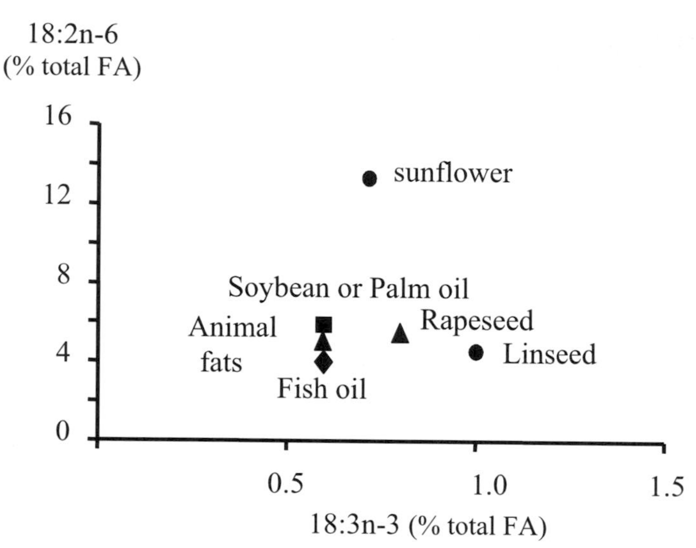

Figure 2. Polyunsaturated fatty acid content (linoleic acid : 18:2n-6; linolenic acid: 18:3n-3) of muscle lipids in bovines fed different sources of dietary lipids (Bas and Sauvant, 2001).

Nevertheless, PUFA are sensitive to peroxidation processes both in plasma and in tissues negative which can have deleterious consequences on the health of animals and on sensory and nutritional qualities of their products. Hence, mechanisms of oxidative stress and the role of antioxidants in the prevention of lipid peroxidation processes are described in the following section.

Fatty acid peroxidation and animal health

Oxidative stress mechanism

The oxygen can be either toxic or essential for all aerobic organisms. This phenomena has been termed the "oxygen paradox". The term "reactive oxygen metabolites" (ROM) has been applied to free radicals and their metabolites. ROM are produced 1) in physiological conditions as a result of electron escape from the electron-transport chain in mitochondria, 2) during activation of immune cells, 3) as the result of xenobiotic metabolism. Eventhough, a complete prevention of ROM formation in biological systems is impossible and, even, undesirable, the regulation of this process is the main aim of the integrated antioxidant system of the body.

Some ROM are produced endogenously by normal metabolic processes, but emission of ROM can be increased markedly by exogenous factors. Any imbalance between production of ROM and their safe disposal can initiate lipid peroxidation involving radical chain reactions. The different steps of lipid peroxidation, described by Miller *et al.* (1993), are presented in Table 3.

Different causes can activate peroxidation processes such as critical physiological periods (peri-partum, gestation or neo-natal periods, ...), feeding (high lipid intake during finishing period, micronutrient deficiency, ...) and breeding conditions (climatic shocks, infectious state, ...) (see review of Aurousseau, 2002).

Similarly, fat supplementation in ruminant diets led to higher level of PUFA in blood compartment which are preferential targets of free radicals (Scislowski *et al.*, 2004) and may activate

Table 3. Initiation and propagation of reactive oxygen metabolites (ROM)(from Miller et al.*, 1993).*

Reactions	Products
O_2 + 1 electron $\rightarrow O_2^{\circ -}$	Anion superoxide
$2O_2^{\circ -} + 2 H^+ \rightarrow O_2 + H_2O_2$	Hydrogen peroxide
$O_2^- + Fe^+ \rightarrow O_2 + Fe^{2+}$	Reduced iron
$H_2O_2 + Fe^{2+} \rightarrow Fe^{3+} + OH^- + {}^{\circ}OH$	Hydroxyl radical
${}^{\circ}OH$ + RH or LH $\rightarrow H_2O + R^{\circ}$ or L°	Fatty acid or other organic oxidised molecule
$R^{\circ} + LH \rightarrow RH + L^{\circ}$	oxidised fatty acid
L° or $R^{\circ} + O_2 \rightarrow LOO^{\circ}$ or ROO°	Peroxyl radical
$LOO^{\circ} + LH \rightarrow L^{\circ} + LOOH$	Lipid hydroperoxide

LH = fatty acid, L° = dienyl radical, RH = organic component, R° = radical of organic component, OH^- = hydroxyl anion.

peroxidation processes. The mechanism of fatty acid peroxidation consisted in several free radical chain reactions involving initiation, propagation and termination phases (Frankel, 1984) (Figure 3).

The initiation consists in the loss of a hydrogen radical to form an dienyl radical (L°).

The propagation step comes on with a reaction with the L° radical and oxygen to form a peroxyl radical (LOO°) which then extracts an hydrogen from another fatty acid to form lipid hydroperoxides (LOOH) and regenerate a new L° radical, leading to an explosive propagation of chain reaction. The terminal step involves reactions between two radicals or one L° radical and a

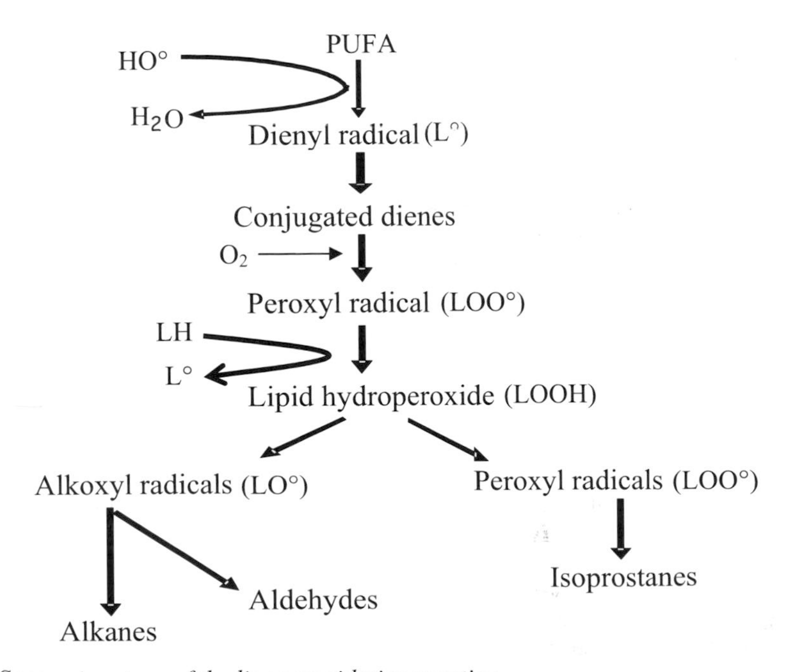

Figure 3. Successive steps of the lipoperoxidation reaction.

Fresh grass feeding led to an enrichment of blood in PUFA but also in vitamin E which helped to minimise peroxidation of plasma lipids (Mouty *et al.*, 2001).

On the opposite, addition of "protected oils" in diets strongly decreased the time of resistance against lipoperoxidation (Lag Phase, LP) due to insufficient amount of antioxidants facing high amounts of dietary PUFA absorbed by intestine and subsequently transferred in plasma lipids.

PUFA peroxidation and quality of ruminant products

Lipid oxidation increasing the rancidity of feed components has long been recognised as a problem generated during storage of animal fats and vegetable oils. These changes in the meat characteristics associated with the oxidative deterioration of animal fats include the development of unpleasant tastes and odours as well as changes in colour, rheological properties, and potential formation of toxic compounds (Geay *et al.*, 2001). Numerous studies have analysed hydroperoxide decomposition which involves a very complicated set of reaction pathways in the meat (see review of Frankel, 1991).

These reactions led to formation of volatile products which are known to have an impact on aroma. The nature and the relative proportion of compounds in the volatile fraction depend on numerous factors. Among these factors, fatty acid structure is the most important because it affects the number and the proportion of hydroperoxide isomers. Other important factors involve conditions in which peroxides are formed and decomposed, including the mechanisms of oxidation (autoxidation, thermo-oxidation, photo-oxidation, ...) and physicochemical conditions of the medium (temperature, pH, level of iron, ...) (Frankel, 1991). Among the numerous volatile products produced by the oxidation of unsaturated fatty acids, the principal aroma compounds are aldehydes, several unsaturated ketones and furan derivatives (Grosch, 1987). Volatile compounds arising from n-3 PUFA tend to have low odour threshold values. These PUFA which are highly sensitive to oxidation may have a larger impact on flavour than n-6 PUFA, even if they are present in lower proportions in lipids of muscle than n-6 PUFA (see review of Gandemer, 1999).

As for meat, flavour induced by milk oxidation reduces the acceptability of dairy products by the consumer. Spontaneous oxidised flavour (SOF) develops without the addition of exogenous compounds or without exposure to light. SOF resulted from the production of volatile aldehydes and other carbonyl compounds following formation of hydroperoxides (Frankel, 1991) from the oxidation of pentadienyl groups in PUFA. The oxidative process is autocatalytic, requiring only one initiating radical to start production of lipid hydroperoxides. Thus, SOF increased with time of storage. Milk having high level of PUFA is more susceptible to autoxidation than the conventional milk. Production of oxidised flavour at 8 day post sampling was positively correlated with levels of 18:2n-6 ($r=0.49$), 18:3n-3 ($r= 0.55$) and total PUFA ($r= 0.50$) in milk fat (Timmons *et al.*, 2001).

Antioxidative factors in ruminant products.

As demonstrated in blood, different enzymes can prevent formation of radicals and, scavenge radicals or peroxides (see review by Lindmark-Mansson and Akesson, 2000). The antioxidative enzymes described in milk and meat are the same as the ones described in plasma, ie superoxide dismutase (SOD), catalase and glutathione peroxidase (GSHPx).

Non-enzymatic antioxidants of milk are principally lactoferrin (iron-binding), vitamins C and E and carotenoids. Vitamin C content in milk can be markedly influenced both by storage and heat whereas vitamin E (fat-soluble) is more stable (Landmark-Mansson and Akesson, 2000).

Non-enzymatic antioxidants in muscle include essentially vitamin E and histidine-containing dipeptides (ie carnosine and anserine). Carnosine and anserine inhibit lipid oxidation by a combination of free radical scavenging and metal chelation. They are less affected by diet than vitamin E but varied widely with the muscle type (Chan and Decker, 1994). The combined supply of carnosine and vitamin E resulted in the greatest lipid (and cholesterol) stability in muscles of chickens and resultant meat products (O'Neill *et al.*, 1999).

Fatty acid composition, antioxidant content and peroxidation in milk

Numerous studies have been conducted in dairy cows in order to improve the PUFA content in milk (see review of Chilliard *et al.*, 2000) and some have evaluated the interactions betweeen PUFA content in milk and milk fat oxidation. Focant *et al.*(1998) have shown that supplementation of diets with extruded linseed (550g/day/animal) increased the proportion of 18:2n-6 (+ 35%) and 18:3n-3 (+ 60%) in milk fat and this was associated with a lower resistance of lipids to peroxidation (- 40%). A daily oral supplement of vitamin E (9616 IU) increased its concentration in milk by about 45%, sufficient to prevent milk fat oxidation as shown by a resistance to peroxidation similar to that of control animals (Figure 4).

In our laboratory, the uptake of vitamin E by the mammary gland has been measured in different dietary conditions (Durand *et al.*, 2004, unpublished data). Linseed oil supplementation (3% diet DM) in diets of dairy cattle increased the efficiency of vitamin E uptake by the mammary gland (6% *vs* 1.5%). These results tend to indicate that milk vitamin E is not directly correlated to the level of plasma vitamin E, but depend on the fatty acid characteristics of milk lipids. These results are in agreement with those observed in the lactating rat of which mammary gland exhibited a greater efficiency of vitamin E uptake than adipose tissue (Martinez *et al.*, 2002).

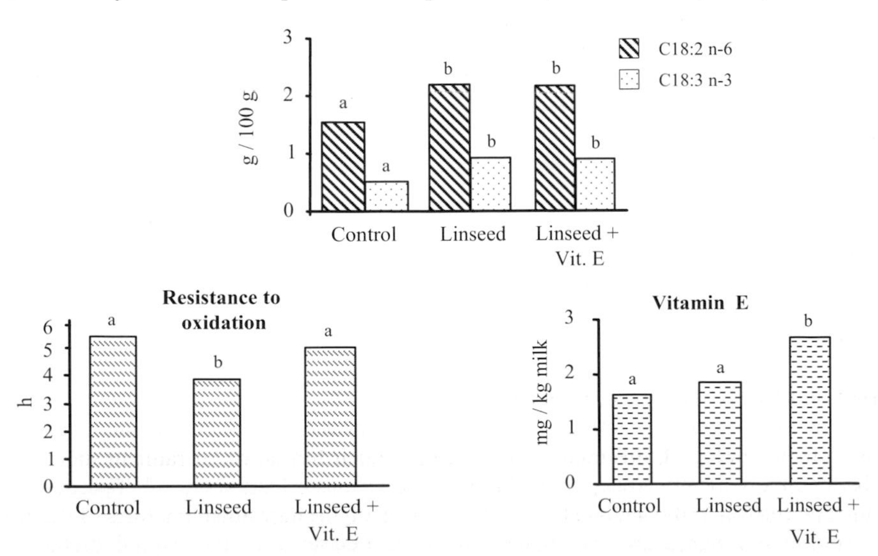

Figure 4. Effect of dietary linseed and vitamin E supplements on milk PUFA (18:2n-6 and 18:3n-3) composition and resistance to peroxidation (Focant et al., *1998).*

Fatty acid composition, antioxidant content and peroxidation in meat

Similarly to milk production, different breeding strategies have been tested in order to improve nutritional value of meat from bovine. In this context, grass feeding or fat supplementation during the finishing period are the most efficient strategies. The consequences on colour stability, in relation with lipid peroxidation, are not clear.

Indeed, some studies reported no negative effect or a significant reduction of colour stability. We have examined the fatty acid composition and several antioxidant factors in *rectus abdominis* muscle from steers given grass compared to linseed or protected linseed oil-enriched diets (Table 6).

Large incorporation of n-6 (from sunflower) and n-3 (from linseed) PUFA [given as protected oils (compared to oil seeds) or as fresh grass rich in n-3 PUFA] into lipids of RA muscle did not stimulate activities of muscle antioxidant enzymes (Durand *et al.*, 2001, Mouty *et al.*, 2001, Mouty *et al.*, personal communication). Therefore, protection of large amounts of dietary PUFA incorporated into muscle lipids (especially with protected oils) were strictly dependent on the amount of antioxidant molecules stored by muscle cells.

Indeed, oil seeds are relatively poor in antioxidants (especially linseed) in comparaison to grass which contains high amount of vitamine E. This results in a higher sensitivity of the muscles to lipoperoxidation. When PUFA in meat is enhanced by feeding oil seeds or using ruminally protected oils, it is necessary to feed the animals with additional antioxidant in order to control the increased oxidative potential of the meat. This area requires further investigation.

Table 6. Effect of the source of dietary polyunsaturated fatty acids (PUFA) compared to control basal diets on rectus abdominis muscle concentration of PUFA, on characteristics of lipoperoxidation (TBAR's index) and of antioxidant efficiency (vitamin E; antioxidant enzyme activities) in finishing steers (Durand et al., 2001, Mouty et al., 2001, Mouty et al., unpublished data).

	Source of dietary PUFA		
	Grass	extruded linseed	protected linseed oil
PUFA	+ 300%	+10%	+300%
antioxidant enzyme activities	≅	≅	≅
vitamin E	+ 160%	≅	≅
TBARS	≅	+ 45%	+92%

Conclusion

Different nutritional strategies (grass-based diet, oil seed or free oil-supplemented diets) can improve the health value of fatty acids in milk and meat lipids of ruminants, especially by decreasing their atherogenic FA content and favouring their beneficial FA content such as PUFA. Nevertheless, these dietary strategies also induce deleterious effects such as lipoperoxidation on the animal performance and welfare and have an impact on the nutritional qualities of the products of the animals. Hence, new investigations are needed in animals offered dietary PUFA supplements to determine 1) the mechanisms involved in the relationships between oxidative stress and lipoperoxidation processes taking place in the blood as well as in the tissues (muscles, mammary gland, liver), 2) new strategies in terms of antioxidant supply by the identification of efficient molecules (nature, source) and the study of their synergy for the protection against PUFA peroxidation.

References

Aurousseau B. ,2002. Les radicaux libres dans l'organisme des animaux d'élevage : conséquences sur la reproduction, la physiologie et la qualité de leurs produits. *(Oxygen radicals in farm animals: physiological effects and consequences on animal products)*. INRA Prod. Anim. 15 : p. 67-82.

Bas, P. and D. Sauvant; 2001. Variations de la composition des dépôts lipidiques chez les bovins *(Variation of lipid composition of adipose tissues and muscles in cattle)*. INRA Prod. Anim. 14: p. 311-322.

Bauchart D., Vérité R., and B. Rémond, 1984. Long-chain fatty acid digestion in lactating cows fed fresh grass from spring to autumn. Can. J. Anim. Sci. 64 (suppl.): p. 330-331.

Bauchart D, Durand D., Mouty D., Dozias D., Ortigues-Marty I. and D. Micol, 2001. Concentration and fatty acid composition of lipids in musclesand liver in fattening steers fed a fresh grass based diet. Renc. Rech. Ruminants 8: p. 108.

Casper, D.P., D.J. Schingoethe, R.P. Middaugh and R.J. Baer, 1988. Lactational responses of dairy cows to diets containing regular and high oleic acid sunflower seeds J. Dairy Sci. 71: p. 1267-1274.

Chan K.M. and E.A. Decker, 1994. Endogenous skeletal muscle antioxydants. Crit. Rev. Food Sci. Nut. 34: p. 403-426.

Chilliard, Y., A. Ferlay, R.M. Mansbridge and M. Doreau, 2000. Ruminant milk fat plasticity : nutritional control of saturated, polyunsaturated, *trans* and conjugated fatty acids. Ann. Zootech. 49: p. 181-205.

Departement of Health, 1994. Report on health and social subjects No. 46. Nutritional aspects of cardiovascular disease. London: HMSO

Demeyer, D. and M. Doreau, 1999. Targets and procedures for altering ruminant met and milk. Proc. Nutr. Soc. 58: p. 593-607.

Durand D., D. Gruffat-Mouty, V. Scislowski and D. Bauchart, 2001. Influence de la supplémentation de la ration en huiles végétales riches en acides gras polyinsaturés sur la lipoperoxydation plasmatique et musculaire chez le bouvillon en finition. (*Influence of oil enriched diets on plasma and muscle peroxidation in finishing steers*). Viandes Prod. Carnés (Hors Série), 75 (abstract)

Enser, M., K.G. Hallett, B. Hewett, G.A.J. Fursey and J.D. Wood, 1996. Fatty acid content and composition of English beef, lamb and pork at retail. Meat Sci. 44: p. 329-341.

Frankel, E.N., 1984. Review: Recent advances in lipid oxidation . J. Sci. Food Agric. 54: p. 495-511.

Focant M., E. Mignolet, M. Marique, F. Clabots, T. Breyne, D. Dalmans and Y. Larondelle, 1998. The effect of vitamine E supplementation of cow diets containing rapeseed and linseed on prevenation of milk fat oxidation. J. Dairy Sci. 81: p. 1095-1101.

Gandemer G., 1999. Lipids and meat quality: lipolysis, oxidation, Maillard reaction and flavour. Sci. Aliments 19: p. 439-458.

Geay Y., D. Bauchart, J.F. Hocquette and J. Culioli, 2001. Effect of nutritional factors on biochemical, structural, and metabolic characteristics of muscles in ruminants; consequences on dietetic value and sensorial qualities of meat. Reprod. Nut. Dev. 41: 1-26; Erratum 41, p. 377.

Grosch W., 1987. Reactions of hydroperoxides products of low molecular weight. In : Chan H.W.S. (editor), Autoxidation of unsaturated lipids, 95-139, Academic Press, London.

Grundy S.M. and M.A Denke, 1990. Dietary influence on serum lipids and lipoproteins. J. Lipid Res. 31: p. 1149-1172.

Kennelly, J.J., 1996. The fatty acid composition of milk fat as influenced by feeding oilseeds, Feed Sci. Tech. 60: p. 137-152.

Lindmark-Mansson H. and B. Akesson, 2000. Antioxidative factors in milk. Brit. J. Nutr. 84 (suppl. 1), p. S103-S110.

Legrand, P., 2002. Recent knowledge on saturated fatty acids: a better understanding. Sci. Aliments 22: p. 351-354.

Legrand, P. 2001. ANCs for fat. Sci. Aliments. 21: p. 348-360.

Lichtenstein A. H., 2000. Trans fatty acids and cardiovascular disease risk. Curr. Opi. Lipidol. 11: p. 37-42.
Martinez, S., C. Barbas and E. Herrera, 2002. Uptake of α-tocopherol by the mammary gland but not by white adipose tissue is dependent on lipoprotein lipase activity around parturition and during lactation in the rat. Metabolism. 51: p. 1444-1451.
Mensink, R.P.,2002. *Trans* fatty acids: state of the art. Sci. Aliments. 22: p. 365-369.
Miller, J.K., E. Brzezinska-Slebodzinska, and F.C. Madsen. 1993. Oxidative stress, antioxidants, and animal function. J. Dairy Sci. 76:2812-2823.
Mouty D., D. Durand, D. Dozias, D. Micol, I. Ortigues-Marty and D. Bauchart, 2001. Lipoperoxidation and antioxidant status in plasma, liver and muscles of fattening steers given a fresh grass-based diet. Renc. Rech. Ruminants. 8: p. 106.
Murphy, J.J., G.P. Mc Neill, J.F. Connolly and P.A. Gleeson, 1990. Effect on cow performance and milk fat composition of including full fat soybeans and rapeseeds in the concentrate mixture for lactating dairy cows. J. Dairy Sci. 57: p. 295-306.
O'Donnell, J.A., 1993. Future of milk fat modification by production or processing: integration of nutrition food science, and animal science. J. Dairy Sci. 76: p. 1797-1801.
O'Neill L.M., K. Galvin, P. A. Morrissey and D.J. Buckley, 1999. Effect of carnosine, salt and dietary vitamin E on the oxidative stability of chicken meat. Meat Sci. 52: p. 89-94.
Oomen, C.M., M.C. Ocke, E.J.M. Feskens, M-A.J. Erp-Baart, F.J. Kok and D. Kromhout, 2001. Association between trans fatty acid intake and 10-years risk of coronary heart disease in the Zutphen elderly study: a prospective population-based study. Lancet 357: p. 746-751.
Rusting R., 1993. Les causes du vieillisssement (*Causes of ageing*). Pour la Science 184: p. 54-62.
Scislowski V., D. Durand, D. Gruffat-Mouty, C. Motta and D. Bauchart, 2004. Linoleate supplementation in steers modify lipid composition of plasma lipoproteins but does not alter their fluidity. Brit. J. Nut. 91: p. 575-584.
Scollan, N.D., R. I. Richardson, S. De Smet, A.P. Moloney, M. Doreau, D. Bauchart and K. Nuernberg, 2005. Enhancing the content of beneficial fatty acids in beef and consequences for meat quality. In: Indicators of milk and beef quality, J.F. Hocquette and S. Gigli (editors) this book.
Steele, W., R.C. Noble and J.H. Moore, 1971. The effects of dietary soybean oil on milk-fat composition in the cow. J. Dairy Sci. 38: p. 49-56.
Surai, P.F., 2002. Antioxidant protection in the intestine: a good beginning is half the battle. In: nutritional biotechnology in the feed and food industries, T. P. Lyons and K. A. Jacques (editors), Nottingham University Press. p. 301-321.
Tefsta, A.T., M. Tuori and L. Syrjälä-Qvist, 1991. High rapeseed oil feeding to lactating cows and its effects on milk yield and composition. J. Dairy Sci. 49: p. 65-81.
Timmons J.S., W.P. Weiss, D.L. Palmquist and W. J. Harper, 2001. Relationships among dietary soybeans, milk components, and spontaneous oxidized flavor of milk. J. Dairy Sci. 84: p. 2440-2449.
Williams, C.M., 2000. Dietary fatty acids and human health. Ann. Zootech. 49: p. 165-180.
Wood, J.D., M. Enser, A.V. Fisher, G.R. Nute, R.I. Richardson and P.R. Sheard, 1999. Manipulating meat quality and composition. Proc. Nutr. Soc. 58: p. 363-370.

Enhancing the content of beneficial fatty acids in beef and consequences for meat quality

N.D. Scollan[1], I. Richardson[2], S. De Smet[3], A.P. Moloney[4], M. Doreau[5], D. Bauchart[5] and K. Nuernberg[6]*
[1]Institute of Grassland and Environmental Research, Plas Gogerddan, Aberystwyth, Wales, SY23 3EB; [2]Division of Farm Animal Science, University of Bristol, Langford, England BS40 5DU; [3]Department of Animal Production, Ghent University, Proefhoevestraat 10, 9090 Melle, Belgium; [4]Teagasc, Grange Research Centre, Dunsany, Co. Meath, Ireland, [5] INRA-Unité de Recherches sur les Herbivores, Theix, 63122 Saint-Genès-Champanelle, France and [6]Institute for Biology of Farm Animals, Wilhelm-Stahl-Allee 2, D-18196, Dummerstorf, Germany.
**Correspondence: nigel.scollan@bbsrc.ac.uk*

Summary

There is a demand for food which offers human nutritional benefits. This has increased interest in producing healthier beef characterised by a (1) higher ratio of polyunsaturated (PUFA) to saturated fatty acids (P:S ratio), (2) more favourable balance between *n*-6 and *n*-3 PUFA (lower *n*-6:*n*-3 ratio) and (3) higher content of conjugated linoleic acid (CLA). Beef can have a low intramuscular fat content ranging between 0.6 - 5%. Incorporating linseed (rich in α-linolenic acid, 18:3*n*-3) in the diet increases the content of 18:3*n*-3 and eicosapentaenoic acid (EPA, 20:5n-3) but not docosahexaenoic acid (DHA, 22:6*n*-3) in beef muscle and adipose tissue, resulting in a lower *n*-6:*n*-3 ratio. Protecting PUFA from ruminal biohydrogenation results in further enhancement of PUFA in the meat resulting in higher P:S ratios and lower *n*-6:*n*-3 ratios. Grass in comparison to concentrate-fed beef has naturally high levels of 18:3*n*-3 and its longer chain (C20-C22) derivatives contributing towards higher P:S ratios and lower *n*-6:*n*-3 ratios. The main CLA isomer in beef is CLA *cis*-9, *trans*-11 and it is mainly associated with the neutral lipid fraction and hence is positively correlated with degree of fatness. The majority of CLA *cis*-9, *trans*-11 in beef is synthesised in the tissue by delta-9 desaturase from ruminally produced *trans*-vaccenic acid. Feeding PUFA-rich diets increases the content of CLA *cis*-9, *trans*-11 in beef. Increasing the content of *n*-3 PUFA can influence colour shelf life and sensory attributes of the meat. As the content of *n*-3 PUFA increases then sensory attributes such as "greasy" and "fishy" score higher and colour shelf life may be reduced. Under these situations, high levels of dietary antioxidant are necessary to help stabilise the effects of incorporating high levels of long chain PUFA into meat. Grass provides natural antioxidants (vitamin E) which help to maintain high *n*-3 PUFA levels in the meat while decreasing deterioration in quality during processing and retail display.

Keywords: health, beef, fatty acids, meat quality

Introduction

Consumer interest in the nutritional aspects of health has increased interest in developing methods to manipulate the fatty acid composition of ruminant products. This is because these products are considered to be a major source of fat and especially saturated fatty acids in the human diet. The relationships between dietary fat and incidence of lifestyle diseases, particularly coronary heart disease are well established and this has contributed towards the development of specific health messages by governments for some food components, including fats (Simopoulos, 2001). It is suggested that the contribution of fat and saturated fatty acids to dietary energy intake should not exceed 0.35 and 0.10 of total intake, respectively, the ratio of polyunsaturated to saturated fatty acids (P:S ratio) should be around 0.4 and the ratio of *n*-6 to *n*-3 polyunsaturated fatty acids

(PUFA) should be less than 4 (Department of Health, 1994; Simopoulos , 2001; Leaf *et al.*, 2003). Although it is the fat content and fatty acid composition of the whole diet which is important, research has focused on altering individual foods to change them in line with these guidelines.

Ruminant meats such as beef and lamb are often criticised by nutritionists for having high amounts of saturated fatty acids and low PUFA. The P:S ratio in beef is approximately 0.1, the ideal being about 0.4. This causes beef to receive critical comments on the grounds of human health. Set against this, the ratio of *n*-6:*n*-3 fatty acids is beneficially low, approximately 2.0. This reflects the considerable amounts of *n*-3 PUFA in beef, particularly α-linolenic (18:3*n-3*) and the long chain PUFA, eicosapentaenoic acid (20:5*n-3*) and, docosahexaenoic acid (22:6*n-3*). Meat, fish, fish oils and eggs are the only significant sources of these *n*-3 long chain PUFA (C20-22) for man. Although meat has lower concentrations of these fatty acids compared to oily fish, it is a very significant source for many people, since fish consumption is low (British Nutrition Foundation, 1999). Ruminants also produce conjugated linoleic acids (CLA) which offer a range of nutritional benefits (Enser, 2001).

Dietary PUFA are exposed to microbial biohydrogenation during passage through the gastro-intestinal tract of the ruminant. This process results in the production of saturated fatty acids which is one of the key reasons why ruminant fats tend to be highly saturated in nature. However, it also results in the formation of CLA and trans monoene intermediates (Demeyer and Doreau, 1999). Understanding the events surrounding fatty acid metabolism in the rumen is central to enabling effective strategies to manipulate the fatty acid composition of beef to be designed.

Increasing PUFA in ruminant tissues increases the susceptibility to oxidative breakdown of muscle lipid during conditioning and retail display. High oxidation changes flavour and promotes muscle pigment oxidation, which reduces shelf life. However, some oxidation is required for optimum flavour development in beef. The extent of lipid oxidation is limited by antioxidants in tissues which include vitamin E (either added to the diet or present naturally) and other phenolic compounds from the diet with antioxidant activity.

The purpose of this paper is to summarise the main activities and results of a European Commission project, Healthy Beef (Enhancing the content of beneficial fatty acids in beef and improving meat quality for the consumer: QLK1-CT-2000-01423). The project examined strategies (nutritional, genetic) to enhance the nutritional value of beef by increasing its content of *n*-3 PUFA and CLA and the implications for meat quality in particular colour shelf life and sensory attributes of such enhancements. The role of the rumen in manipulating the composition of dietary fatty acids was also studied.

Factors influencing the fatty acid composition of beef

Effect of nutrition

The main method of manipulating the fatty acid composition of beef is by changing dietary ingredients which are known sources of long chain PUFA, such as 18:3*n*-3, 20:5*n*-3 and 22:6*n*-3. There are four main sources of fatty acids in ruminant diets: (1) fresh and ensiled forages, (2) oils and oilseeds, (3) fish oil and marine algae and (4) fat supplements. Across Europe the contribution of forage to the total diet is usually large and even though the oil content of the forage is low, it represents the main supply of dietary lipids. Green plants are the primary source of *n*-3 fatty acids and forages such as grass and clover contain a high proportion (50-75%) of total fatty acids as α-linolenic acid (18:3*n*-3). Diets containing either whole oilseeds or seed oils, such as rapeseed (rich in 18:1*n*-9) , soybean and sunflower (rich in 18:2*n*-6) and linseed (rich in 18:3*n*-3) have been widely used to manipulate the fatty acid composition of ruminant products. Fish oils are rich in the long chain PUFA, 20:5*n*-3 and 22:6*n*-3. Plants have the unique ability to synthesise *de novo* 18:3*n*-3 which is the building block of the *n*-3 series of essential fatty acids. Elongation and desaturation of this fatty acid results in the synthesis of 20:5 *n*-3 and 22:6*n*-3. The formation of these long chain *n*-3 PUFA's by marine algae and their transfer through the food chain to fish, accounts for the high amounts of these important fatty acids in fish oils. The use of fat supplements such as *Megalac®* (rich in 16:0, palmitic acid; Volac Ltd., Royston, UK) in beef rations is less common compared to their use in the diets of dairy cows.

An important focus in the project was the potential to modify beef fatty acids by feeding grass (and other forages) and/or concentrates rich in 18:3*n*-3 from linseed or linseed oil. Feeding grass (followed by a finishing period feeding a concentrate containing linseed) compared to intensive concentrate (grain) caused significant changes in the fatty acid composition of intramuscular fat of *longissimus* muscle in German Holstein (GH) and German Simmental (GS) bulls (Table 1; Nuernberg *et al.*, 2004; Dannenberger *et al.*, 2004). The percentage and the absolute concentration of total *n-3* fatty acids was increased on pasture resulting in a low *n*-6*:n*-3 ratio. The absolute concentrations of all saturated fatty acids, all monounsaturated fatty acids, PUFA and all *n-6* fatty acids were significantly reduced by feeding grass compared to concentrate. Grass relative to concentrate feeding not only increased 18:3*n*-3 in muscle phospholipid but also 20:5*n*-3, 22:5*n*-3 and 22:6*n*-3 (Dannenberger *et al.*, 2004). Concentrates rich in 18:2*n*-6 lead to higher concentrations of 18:2*n*-6 and associated longer chain derivatives (20:4*n*-6). These beneficial responses with grass have been found to be related to time on diet (Table 2; Noci *et al.*, 2003).

Table 1. Nutritionally important fatty acids in total lipid of longissimus thoracis muscle in German Holstein and German Simmental bulls fed on concentrate or pasture (Nuernberg et al., *2004).*

	German Holstein				German Simmental				P[1]
	Concentrate		Pasture		Concentrate		Pasture		
	mean	s.e.m.	mean	s.e.m.	mean	s.e.m.	mean	s.e.m.	
%									
Sum SFA	43.6	0.66	45.6	0.68	44.5	0.68	43.9	0.71	
Sum 18:1trans[2]	2.83	0.23	4.37	0.24	3.19	0.24	4.28	0.24	F
CLA *cis*-9, *trans*-11	0.75	0.04	0.84	0.04	0.72	0.04	0.87	0.04	F
Sum *n*-3 PUFA	0.96	0.29	3.25	0.30	0.90	0.30	4.70	0.31	B,F
Sum *n*-6 PUFA	6.14	0.82	6.30	0.85	7.73	0.85	9.80	0.88	B
mg/100 g muscle									
Sum SFA	1367	112.0	1046	112.1	1126	112.1	747	124.4	B,F
Sum 18:1trans	94.4	11.60	102.0	11.61	80.4	11.60	78.7	12.87	
CLA *cis*-9, *trans*-11	17.1	1.7	17.3	1.77	13.3	1.77	11.5	1.83	B
Sum *n*-3 PUFA	27.8	2.3	65.1	2.3	20.5	2.27	57.4	2.52	B,F
Sum *n*-6 PUFA	179	7.5	125	7.5	168	7.5	113	8.3	F
n-6/*n*-3 Ratio	6.47	0.1	1.92	0.1	8.32	0.15	2.00	0.17	B,F,B*F
P:S Ratio	0.11	0.02	0.16	0.020	0.14	0.02	0.19	0.02	

[1]B = Breed; F = Feed effects significant at $P < 0.05$

[2]Sum of C18:1*trans*= sum of the isomers C18:1*trans*-6-11

Table 2. Nutritionally important fatty acids of longissimus thoracis *muscle in Friesian steers fed on grass for differing times (0, 40, 99 or 158 days; Noci* et al.*, 2003).*

	Days at Grass				s.e.d.	P[1]
	0	40	99	158		
%						
Sum SFA	45.4	45.8	45.5	43.2	0.77	**L,Q
18:1*trans-11*	1.35	1.93	2.27	3.01	0.18	L
CLA *cis-9, trans-11*	0.50	0.50	0.57	0.71	0.06	***L
Sum *n-6* PUFA	3.25	3.20	2.97	3.31	0.23	NS
Sum *n-3* PUFA	1.79	2.06	1.91	2.43	0.17	**L
mg/100 g muscle						
Sum SFA	1117	1060	1262	1090	80.8	*
18:1*trans-11*	32.5	44.9	60.2	76.6	4.54	***L
CLA *cis 9, trans 11*	12.3	12.1	15.2	18.4	1.79	***L
Sum *n-6* PUFA	77.3	79.3	76.8	78.6	3.87	NS
Sum *n-3* PUFA	39.1	44.3	51.7	59.7	3.07	***L
n-6:n-3 ratio	2.00	1.79	1.56	1.32	0.10	***L
P:S ratio	0.12	0.14	0.12	0.15	0.009	*

[1]L and Q are significant linear and quadratic effects of days at grass, respectively; SFA = saturated fatty acids.
* $P < 0.05$; ** $P < 0.01$; *** $P < 0.001$.

Feeding linseed or linseed oil in a concentrate results in increases in 18:3*n*-3 in the tissues and 20:5*n*-3 but not 22:6*n*-3 (Scollan *et al.*, 2001). Studies investigating the effects of including grazed and conserved grass compared to maize silage and linseed oil in the diet of double-muscled Belgian Blue bulls found that grass and linseed were beneficial in improving the content of *n*-3 PUFA resulting in significant reductions in the *n*-6:*n*-3 ratio. Longer-term feeding on grass with a finishing period that included linseed in the diet was beneficial in increasing the content of 20:5*n*-3 and 22:6*n*-3 (Raes *et al.*, 2004).

In general, nutritional manipulation did not increase the P:S ratio in the meat above the 0.1-0.15 normally observed. This relates to the high degree of biohydrogenation of dietary PUFA in the rumen. Providing 18:3*n*-3 as linseed oil directly into the small intestine, avoiding the efficient hydrogenating powers of the rumen micro-organisms demonstrated the potential to deposit high levels of PUFA in beef muscle (Durand *et al.*, 2005). The study confirmed other results from this project and in the literature that relative to a control diet (C) feeding linseed (L) in the diet increases 18:3*n*-3 in muscle by approximately 1.5-2.0 fold. However, infusing an equivalent amount of 18:3*n*-3 (as linseed oil, LO) into the small intestine increased the percentage and amount (mg/100 g muscle) of 18:3*n*-3 in total lipid to 1.0, 1.5 and 8.7 and 15.8, 26.3 and 176.5 for diets C, L and LO, respectively. The LO treatment resulted in a high P:S ratio (0.495 relative to the recommended target of > 0.4) and low n-6:n-3 ratio (1.04 relative to the recommended target of < 2-3). Similarly, using ruminally protected lipids (rich in both 18:2*n*-6 and 18:3*n*-3) resulted in meat characterised by higher P:S ratio and lower *n*-6:*n*-3 ratio relative to a control (Scollan *et al.*, 2003; 2004). Feeding a protected lipid supplement with a *n*-6:*n*-3 ratio of 1.1:1 (Scollan *et al.*, 2003) compared to 2.4:1 (Scollan *et al.*, 2004) resulted in a much lower *n*-6:*n*-3 ratio in the meat (1.88 *v* 3.59, respectively). It was possible to increase the 20:5*n*-3 and 22:6*n*-3 fatty acids by feeding them as ruminally protected products, but this had little effect on the P:S ratio and a small improvement in the *n*-6:*n*-3 ratio only at the highest level fed (Richardson *et al.*, 2004).

Conjugated linoleic acid

There is currently intense scientific interest in conjugated linoleic acid. For man, the major dietary sources of CLA are foods derived from ruminants with typically 75 and 25% coming from dairy products and red meat, respectively (Ritzenthaler *et al.*, 2001). The CLA *cis*-9, *trans*-11 typically accounts for 75-90% of total CLA's in meat and this is particularly important as studies have established that this isomer is anticarcinogenic (Ip *et al.*, 1999). Effects of other isomers have been demonstrated, for example CLA *trans*-10, *cis*-12 is a potent inhibitor of fat synthesis in both lactating and growing animals (Bauman *et al.*, 2003).

Our studies confirmed that the main CLA isomer in beef is CLA *cis*-9, *trans*-11 and is mainly associated with the neutral lipid fraction (tyically 92% of total CLA in muscle lipid) and hence is positively correlated with degree of fatness (Figure 1a). As with CLA in milk (Bauman *et al.*, 2003), results have confirmed that the majority of CLA found in muscle is synthesised from 18:1*trans*-11 (*trans*-vaccenic acid, TVA) via delta-9 desaturase in the tissue rather than directly from rumen (see discussion on rumen below). This is reflected in the strong relationship between tissue content of 18:1*trans*-11 and CLA (Figure 1b), reflecting a strong precursor-product relationship. Hence, the CLA in tissue is influenced by (1) CLA and 18:1*trans* produced in the rumen and (2) conversion of 18:1*trans*-11 to CLA in the tissue. Of particular note in this project was the effect of grass feeding resulting in higher CLA and a positive association between CLA content and duration at pasture before slaughter (see Table 2). Other studies have reported that supplementing diets with sunflower oil or seeds, linseed or soya bean oil increases CLA and 18:1*trans*-11 in beef (see Mir *et al.*, 2003).

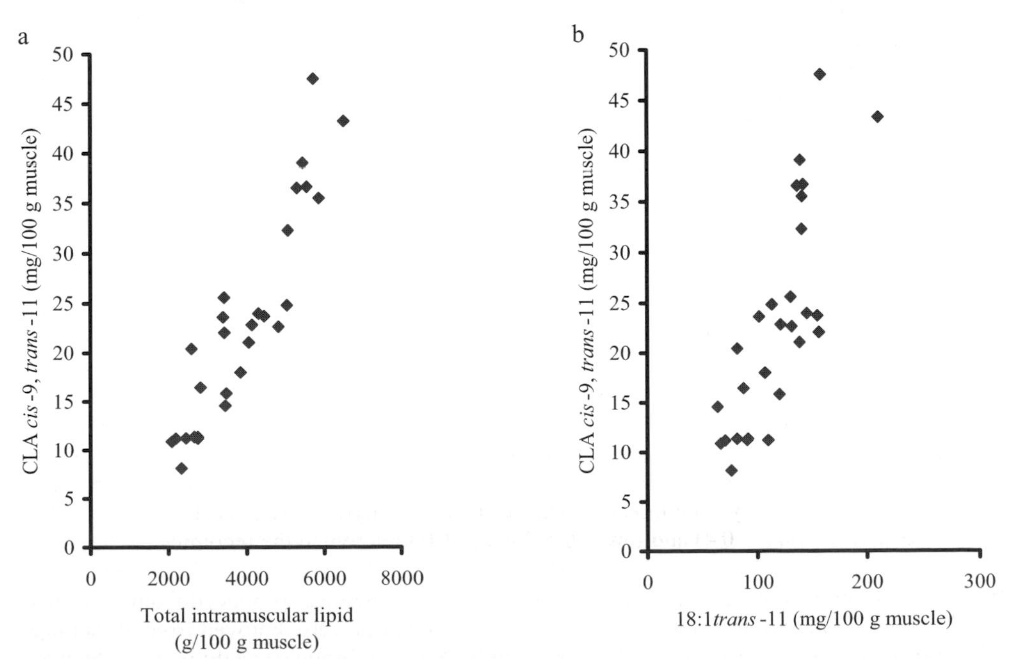

Figure 1. Typical relationship between (a) CLA cis-9, trans-11 and total lipid in muscle and (b) relationship between 18:1trans-11 and CLA cis-9, trans-11 in muscle.

Effect of genotype

Breed type influences fatness and as a result differences in fatty acid composition of beef are often confounded. Nevertheless, many studies have found differences in individual fatty acid concentrations between breeds and differences in the neutral lipid and phospholipid fractions. In a recent review by De Smet *et al.*, (2004) it was concluded that breed differences are generally small but nevertheless they reflect differences in underlying gene expression or enzymes involved in fatty acid synthesis, and hence merit attention. In our project of particular note was the low fat content (< 1%) and beneficially high P:S ratio (0.5-0.6) in the double-muscled Belgian Blue bulls. The double muscling is caused by a mutation in the myostatin gene. Raes *et al.*, (2001) examined the intramuscular fatty acid compostion of three myostatin genotypes (double muscling, heterozygous and normal) and noted that the high P:S ratio for double muscled animals may be explained by their lower fat content. These studies emphasised the relationships between fatness and fatty acid composition of beef. As the content of saturated and monounsaturated fatty acids increase faster with increasing fatness than does the content of PUFA, the relative proportion of PUFA and the P:S ratio decrease with increasing fatness (Figure 2). The *n*-6:*n*-3 ratio in meat of double-muscled Belgian Blue bulls is high (5-6), but may be improved by including grass and/or linseed in the diet (Raes *et al.*, 2003, 2004).

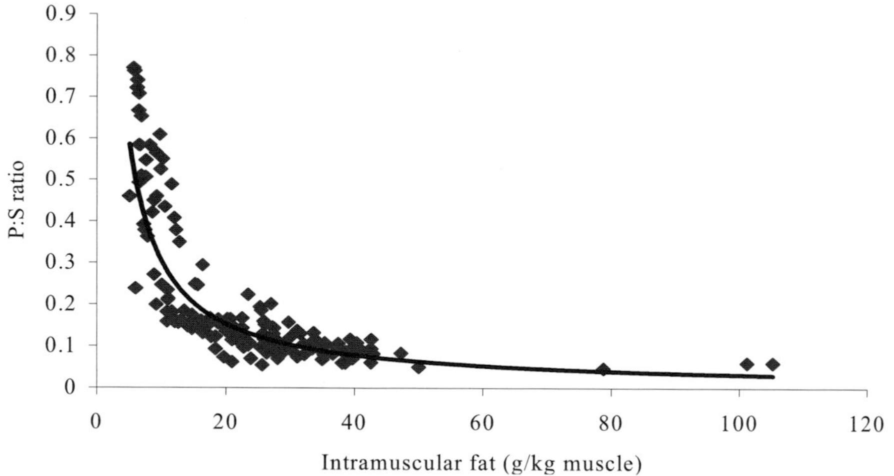

Figure 2. The relationship between intramuscular fat content and P:S ratio.

Beef fatty acids and meat quality

Altering the fatty acid composition of beef can alter meat quality by providing a different mix of reactive ingredients, which affect oxidative stability (shelf life), colour and flavour. As the number of double bonds is increased (i.e. moving from 18:1*n*-9 → 18:2*n*-6 → 18:3*n*-3) so their melting point and oxidative stability are reduced. This reduction in shelf life is the result of the oxidation of oxymyoglobin (bright red) to metmyoglobin (dark brown) reducing visual appeal and the oxidation of lipid, producing products which also cause rancidity in cooked meat (Wood *et al.*, 2003).

Lipid oxidation, colour stability and sensory attributes of the meat produced from all project partners were analysed by University of Bristol. Loin samples were removed from all animals, conditioned in vacuum packs for 10 days and then frozen for subsequent analysis. After thawing, samples were trimmed of discoloured surfaces, sliced into 20mm steaks, packed in a modified atmosphere (O_2:CO_2, 75:25), displayed under lights (600 lux, 16h on: 8h off, 4 ± 1°C) and colour measured daily for 8d. A second set of steaks was displayed for 5 and 10d respectively and then analysed for thiobarbituric acid reacting substances (TBARS) as a measure of rancidity. Sensory analysis of grilled steaks was conducted using a trained 10 member panel including the following descriptors: toughness, juiciness, beef, abnormal, greasy, bloody, livery, metallic, bitter, sweet, rancid, fishy, acidic, cardboard, vegetable/grassy, dairy and the hedonic, overall liking (Nute, 2003).

Lipid oxidation is increased in meat with a higher content of PUFA (Wood *et al.* 2003). This was particularly noticeable for studies involving ruminally protected lipid supplements that resulted in meat with an *n*-3 and *n*-6 PUFA of 3.5 and 7.0 compared to 1.6 and 3.6% in the control, respectively. After 5d of display little difference was noted in TBARS, but at 10 days values (Figure 3a) increased considerably above the level of 2 mg malonaldehyde per kg of meat at which rancidity may be detected by consumers (Campo *et al.*, 2004). These values were higher than normal for fresh meat as the samples had been previously frozen. Feeding protected fish oils at the highest concentration also produced higher TBARS (Richardson *et al.*, 2004). Colour saturation was also affected by feeding, declining fastest with higher levels of inclusion of the lipid supplement (Figure 3b). Grass silage fed animals produced *n*-3 rich meat with a greater stability than *n*-6 rich meat from those fed on concentrates (Warren *et al.*, 2002). This was related to the antioxidants in grass contributing to higher vitamin E content in the meat with benefits for lower lipid oxidation and better colour retention despite a greater potential for lipid oxidation. A number of papers have recognised that fatty acid composition and concentration of tissue vitamin E influence both lipid oxidation and colour stability (Liu *et al.*, 1996; Renerre, 2000; Richardson *et al.*, 2003; Wood *et al.*, 2003).

Increasing the content of *n*-3 PUFA increased sensory attributes such as "greasy" and "fishy" (for example with longer term grass feeding). On the other hand, when the levels of PUFA were further enhanced through either the provision of a ruminally protected lipid supplement rich in PUFA (Figure 3c) or by infusing PUFA directly into the duodenum the resultant meat was associated with higher sensory scores for factors such as "abnormal" and "rancid" (see Wood *et al.*, 2003). While some of these notes such as 'fishy' or 'rancid' appear negative, it should be emphasised that they contribute to the strength of an overall flavour and may increase overall attractiveness of the meat. Concentrated ingredients for perfumes often have unpleasant odours, which become pleasant when diluted and mixed with other fragrances. Further studies to optimise the amount of vitamin E (and other antioxidants) which need to be fed with different diets, especially those which challenge the oxidative capacity of the animals due to incorporation of higher concentrations of long chain PUFA, are required. This is particularly important when the raw meat is further challenged by processing such as size reduction, freezing or cooking (Richardson *et al.*, 2003).

Ruminal biohydrogenation

A major factor in attempting to manipulate the PUFA content of meat is the extensive transformation of fatty acids in dietary lipids in the rumen by the action of rumen microbial enzymes originating from different classes of bacteria and protozoa (Demeyer and Doreau, 1999). This lipolysis leads to the formation of free fatty acids and is generally very rapid but may be decreased by a low pH. Unsaturated fatty acids have relatively short half-lives in ruminal contents because they are rapidly hydrogenated by micro-organisms to more saturated end products. Project results confirm that the extent of biohydrogenation is very high for PUFA, averaging approximately 86 and 92% for 18:2*n*-6 and 18:3*n*-3, respectively. Biohydrogenation does give rise to a number of metabolically interesting intermediates such as the 18:1*trans* isomers and CLA and this research has extended considerably our knowledge of the range of isomers involved. It is now clear that the combined production of both 18:1 *trans*-11 (*trans*-vaccenic acid) and CLA *cis*-

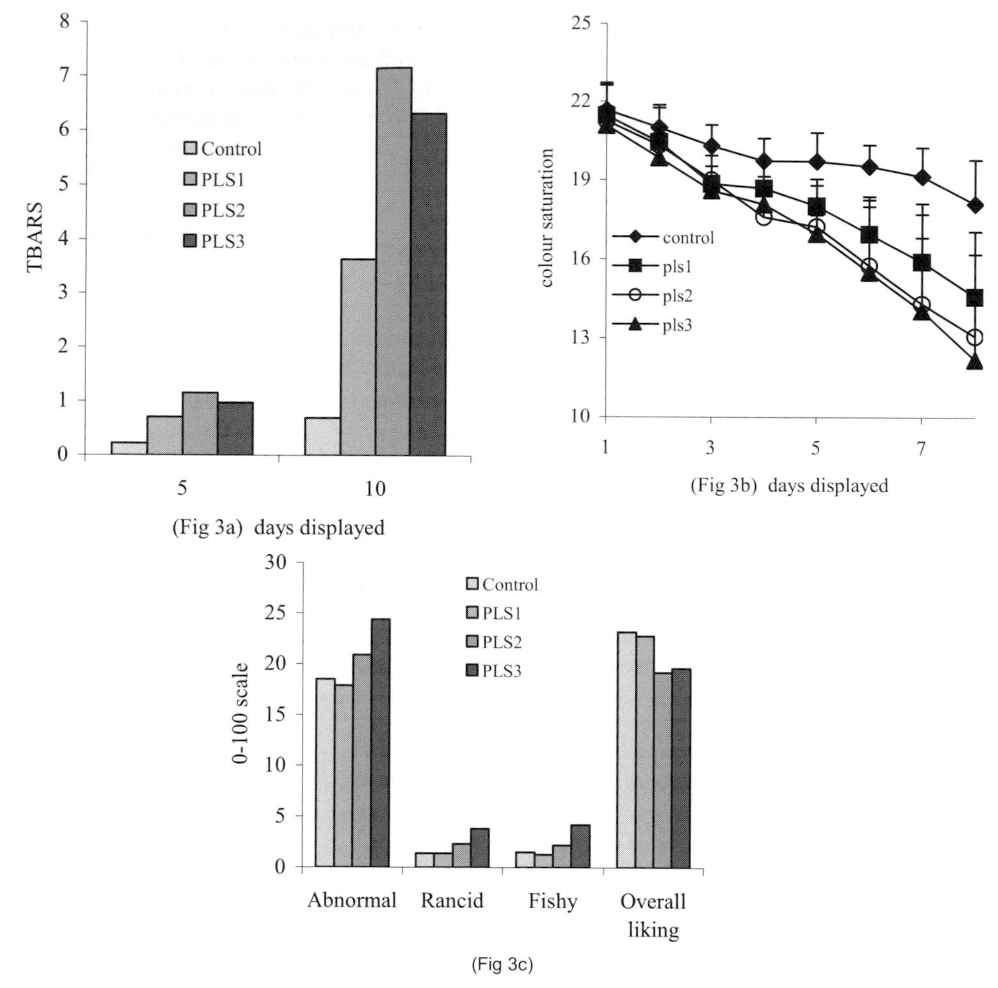

Figure 3. Effect of feeding a ruminally protected lipid supplement (PLS) at varying levels 0, 400, 800 and 1000 g/d (control, PLS1, PLS2 and PLS3, respectively) on meat quality in steers (a) Lipid oxidation in longissimus *after conditioning at 1°C and 5 and 10 days display in over wrapped packs at 4°C, (assessed as TBARS, (mg malonaldehyde/kg). (b) Colour saturation of loin steaks during retail display in modified atmosphere packs. (c) Sensory analysis for grilled* longissimus *steaks (0-100 line scales) (Scollan* et al.*, 2004).*

9, *trans*-11 in the rumen is responsible for CLA *cis*-9, *trans*-11 in the tissue. However, production of ruminal *trans*-vaccenic is the most important determinant of tissue CLA (Figure 1b) and hence factors regulating the production of *trans*-vaccenic in the rumen are important in promoting more CLA synthesis in the tissues.

Forages are a major source of dietary PUFA for ruminants. In general, achieving higher intakes of dietary PUFA result in higher flows of these fatty acids to the small intestine, despite high levels of biohydrogenation. Wilting grass for subsequent ensiling results in an important loss in total fatty acids but does not affect the fatty acid composition (Chow *et al.*, 2004a). The reduction in total fatty acids is most likely related to oxidation losses during wilting. Hay making was associated with losses in total fatty acids (approximately 20%) but not in fatty acid composition

(Doreau *et al.*, 2003). Biohydrogenation of dietary PUFA was lower in hay compared to fresh grass, with the result that net flow of 18:3*n*-3 to the small intestine was not different (Doreau *et al.*, 2003). This may relate to differences in the physical properties of the forage resulting in differences in microbial attachment and/or encouraging different microbial populations in the rumen. However, despite the high extent of biohydrogenation, differences in PUFA content of grass result in differences in duodenal PUFA and vaccenic acid flows, (Doreau *et al.*, 2005).

Plant enzymes may play a role in the initial breakdown of dietary lipids in the rumen (rather than rumen microbial mediated lipolysis; Lee *et al.*, 2003a). Different patterns and lower levels of lipolysis were noted on *Trifolium pratense* (red clover) compared to perennial ryegrass. It is thought that the enzyme polyphenol oxidase (PPO) in red clover may be involved in regulating the extent of lipolysis and biohydrogenation. Lee *et al.*, (2004) compared two lines of red clover with differing concentrations of polyphenol oxidase (Cv. Milvus, a genotypic mutant with reduced PPO activity, LowPPO and the wild type, NormalPPO) for the rate and extent of plant mediated lipolysis against red clover treated with ascorbate to inhibit PPO activity (AscPPO). Lipolysis, measured as the proportional decline in the membrane lipid polar fraction, was reduced with increasing PPO activity, with values after 12 h incubation of 0.12, 0.20 and 0.22 for NormalPPO, LowPPO and AscPPO, respectively. These results suggest that forages expressing high PPO activity may reduce lipid losses in the rumen. *In vivo* studies have demonstrated a lower degree of biohydrogenation on red clover compared with grass and beneficially higher flows of linolenic acid, 18:3*n*-3, to the duodenum (Lee *et al.*, 2003b).

Including additional PUFA rich oil such as linseed in the diet is an important method of enhancing beneficial PUFA supply to the animal (Loor *et al.*, 2004). However, even though biohydrogenation of PUFA is high, beneficial responses in PUFA deposition are noted in the tissues. Other sources of *n*-3 PUFA were investigated in the project (purslane, chia, camelina) but they offered little additional benefit compared to linseed. Fish oil inhibited complete biohydrogenation of 18:2*n*-6 and 18:3*n*-3 and caused an accumulation of 18:1*trans*-vaccenic acid (Chow *et al.*, 2004b). It is likely that fish oil caused an inhibition of the rumen micro-organisms involved in the final step of biohydrogenation from TVA to the saturated stearic acid (Chow *et al.*, 2004b).

It is evident from the studies involving ruminally protected lipids that evolving methods of reducing biohydrogenation to enhance PUFA deposition in tissues is necessary. Studies *in vitro* (Van Nevel and Demeyer, 1996) and *in vivo* (Kalscheur *et al.*, 1997) suggest that low ruminal pH reduces biohydrogenation. Using data from a number of studies, Ueda and Doreau (unpublished) have established a relationship between rumen pH and biohydrogenation (Figure 4). This suggests that in situations were rumen pH is low (< 6), biohydrogenation is markedly reduced. Feeding a high proportion of concentrate in the diet can reduce rates of biohydrogenation and alter profiles of 18:1, 18:2 and 18:3 biohydrogenation intermediates flowing to the small intestine (Loor *et al.*, 2004). However, in this study pH was unaffected by the proportion of concentrate in the diet, and it was hypothesised that changes in the extent of biohydrogenation when high concentrate diets are fed may be independent of pH, and may relate to changes in dietary starch content resulting in subtle changes in the microbial population (Ueda *et al.*, 2003; Loor *et al.*, 2004). There is a need for studies describing shifts in the microbial population linked to fatty acid metabolism in the rumen with a view to establishing natural methods of modifying extent of biohydrogenation and generation of important biohydrogenation intermediates.

Conclusions

The results of this project demonstrate that beef may be produced which is more compatible with current medical guidelines for the composition of the human diet. Studies have demonstrated that

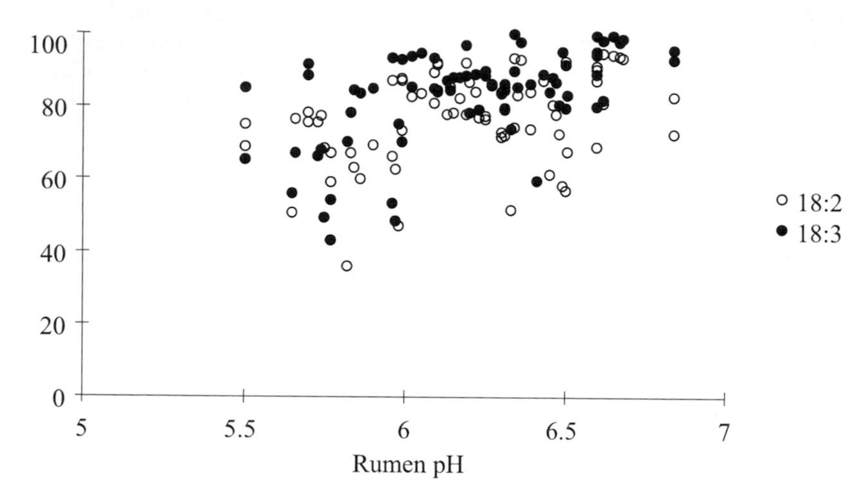

Figure 4. Relationship between pH and ruminal biohydrogenation of linolenic and linoleic acid (Ueda and Doreau, unpublished).

beef is low fat (less than 5%) and that nutritional opportunities exist to produce beef characterised by a lower content of atherogenic saturated fatty acids, higher content of more beneficial monounsaturated fatty acids and polyunsaturated fatty acids and lower *n*-6:*n*-3 PUFA ratio. It is difficult to have a large shift in the P:S ratio by nutritional means due to the high degree of biohydrogenation of dietary PUFA in the rumen. Studies which provided PUFA either post ruminally by direct infusion into small intestine or by using ruminally protected lipids confirmed that beef muscle does have a high capacity to incorporate beneficial PUFA in its lipids resulting in meat characterised by a high P:S ratio and low *n*-6:*n*-3 ratio. Important relationships exist between the fatty acid composition and sensory attributes and colour shelf life of the meat. As the content of *n*-3 PUFA in the meat are increased, sensory attributes such as "greasy" and "fishy" score higher and colour shelf life may be reduced. Under these situations high levels of dietary antioxidant are necessary to help stabilise the effects of incorporating high levels of long chain PUFA into meat.

The project has made a large contribution to our understanding of factors regulating fatty acid composition of beef and has helped to highlight areas which require further research. Consumers demand food that is safe, of consistent quality, and offers nutritional benefits. The benefits of the consumption of beef and beef products with a composition more in line with medical recommendations will contribute towards a reduction and/or prevention of diet-mediated illnesses and will lead to significant social and health care benefits. Addressing these issues is essential for the European consumer and the beef industry.

References

Bauman, D.E., L.H. Baumgard, B.A. Cori and J.M. Grinari, 2003. The biology of conjugated linoleic acids in ruminants. In: Advances in Conjugated Linoleic Research, Volume 2, J.L. Sebedio, W.W. Christie and R.O. Adlof, (editors). AOCS Press, Champaign, IL, p. 146-173.

British Nutrition Foundation, 1999. *Meat in the Diet*. London: British Nutrition Foundation.

Campo, M.M., G.R. Nute, S.I. Hughes, M. Enser, J.D. Wood and R.I. Richardson, 2004. Flavour perception of oxidation in beef. In: Proceedings of the 50th International Congress of Meat Science and Technology, Helsinki, Finland: in press.
Chow, T.T., V. Fievez, M. Ensberg, A. Elgersma and S. De Smet, 2004a. Fatty acid content, composition and lipolysis during wilting and ensiling of perennial ryegrass: preliminary findings. Grassland Science in Europe, Vol. IX, p. 981-983.
Chow, T.T., V. Fievez, A.P. Moloney, K. Raes, D. Demeyer and S. De Smet, 2004b. Effect of fish oil on *in vitro* rumen lipolysis, apparent biohydrogenation of linoleic and linolenic acid and accumulation of biohydrogenation intermediates. Anim. Feed Sci. and Tech.: in press.
De Smet, S., K. Raes and D. Demeyer, 2004. Meat fatty acid composition as affected by genetic factors. Anim. Res. 53: p. 81-88.
Dannenberger, D., G. Nuernberg, N.D. Scollan, W. Schabbel, H. Steinhart, K. Ender and K. Nuernberg, 2004. Effect of diet on the deposition of *n*-3 fatty acids, conjugated linoleic- and C18:1*trans* fatty acid isomers in muscle lipids of German Holstein bulls J. Agric. Food Chem. 52: p. 6607-6615
Demeyer, D. and M. Doreau, 1999. Targets and procedures for altering ruminant meat and milk lipids. Proc. Nutr. Soc. 58: p. 593-607.
Department of Health (1994). Report on health and social subjects No. 46. Nutritional aspects of cardiovascular disease. HMSO, London.
Doreau M., K. Ueda and C. Poncet, 2003. Fatty acid ruminal metabolism and intestinal digestibility in sheep fed ryegrass silage and hay. Trop. Subtrop. Agroec. 3: p. 289-293.
Doreau, M., J.L. Peyraud and D. Rearte, 2005. Relationship between soluble carbohydrate/ nitrogen ratio in grass and milk fatty acid composition in cows: role of ruminal metabolism. In: Indicators of milk and beef quality, J.F. Hocquette and S. Gigli (editors), Wageningen Academic Publishers, Wageningen, The Netherlands, this book.
Durand, D., V. Scislowski, D. Gruffat, Y. Chillard and D. Bauchart, 2005. High-fat rations and lipid peroxidation in ruminants: consequences on the health of animals and quality of their products. In: Indicators of milk and beef quality, J.F. Hocquette and S. Gigli (editors), Wageningen Academic Publishers, Wageningen, The Netherlands, this book.
Enser, M., 2001. The role of fats in human nutrition. In: Oils and fats, Volume 2. Animal carcass , R.I. Richardson and G.C. Mead (editors), Leatherhead Publishing, Leatherhead, Surrey, UK, p. 77-122.
Ip, C., S. Banni, E. Angioni, G. Carta, J. McGinley, H.J. Thompson, D. Barbano and D. Bauman, 1999. Conjugated linoleic acid-enriched butter fat alters mammary gland morphogenesis and reduces cancer risk in rats. J. Nutr. 129: p. 2135-2142.
Kalscheur, K.F., B.B. Teter, L.S. Piperova and R.A. Erdman, 1997. Effect of dietary forage concentration and buffer addition on duodenal flow of *trans*-C18:1 fatty acids and milk fat production in dairy cows. J. Dairy Sci. 80: p. 2104-2114.
Leaf, A., Y.F. Xiao, J.X. Kang and G.E. Billamn, 2003. Prevention of sudden cardiac death by *n*-3 polyunsaturated fatty acids. Pharmacol. Ther. 98: p. 355-377.
Lee, M.R.F., E.M. Martinez and N.D. Scollan, 2003a. Plant enzymatic mediated lipolysis of *Lolium perenne* and *Trifolium pratense* in an in vitro simulated rumen environment. Asp. Appl. Biol. 70: p. 115-120.
Lee, M.R.F., L.J. Harris, R.J. Dewhurst, R.J. Merry and N.D. Scollan, 2003b. The effect of clover silages on long chain fatty acid rumen transformations and digestion in beef steers. Anim. Sci. 76: p. 491-501.
Lee, M.R.F., A.L. Winters, N.D. Scollan, R.J. Dewhurst, M.K. Theodorou and F.R. Minchen, 2004. Plant-mediated lipolysis and proteolysis in red clover with different polyphenol oxidase activities. J. Sci. Food and Agric. 84: P. 1639-1645.
Liu, Q., K.K. Scheler, S.C. Arp, D.M. Schaefer and S.N. Williams, 1996. Titration of fresh meat stability and malondialdehyde development with Holstein steers fed vitamin E-supplemented diets. J. Anim. Sci. 74: p. 117-126.
Loor, J.J., K. Ueda, A. Ferlay, Y. Chilliard and M. Doreau, 2004. Biohydrogenation, duodenal flow, and intestinal digestibility of *trans* fatty acids and conjugated linoleic acids in response to dietary forage:concentrate ratio and linseed oil in dairy cows. J. Dairy Sci. 87: p. 2472-2485.

Mir, P.S., M. Ivan, M.L. Ha, B. Pink, E. Okine, L. Goonewardene, T.A. McAllister, R. Weselake and Z. Mir, 2003. Dietary manipulation to increase conjugated linoleic acids and other desirable fatty acids in beef: a review. Can. J.Anim. Sci. 83: p. 673-685.

Noci, F., A. P. Moloney, P. French and F.J. Monahan, 2003. Influence of duration of grazing on the fatty acid profile of *M.longissimus dorsi* from beef heifers. Proc. Brit. Soc. Anim. Sci. 23.

Nuernberg, K., D. Dannenberger, G. Nuernberg, K. Ender, J. Voigt, N.D. Scollan, J.D. Wood, G.R. Nute and R.I. Richardson, 2004. Effect of a grass-based and a concentrate feeding system on meat quality characteristics and fatty acid composition of *longissimus* muscle in different cattle breeds. Livest. Prod. Sci. in press.

Nute, G.R., 2003. Sensory analysis of meat. In: Meat processing: improving quality, J. Kerry, J. Kerry and D. Ledward (editors), Woodhead Publishing Limited, Cambridge, England, p. 175-192.

Raes, K., S. De Smet and D. Demeyer, 2001. Effect of double-muscling in Belgian Blue young bulls on the intramuscular fatty acid composition with emphasis on conjugated linoleic acid and polyunsaturated fatty acids. Anim. Sci. 73: p. 253-260.

Raes, K., S. De Smet, A. Balcaen, E. Claeys and D. Demeyer, 2003. Effect of diets rich in *n*-3 polunsaturated fatty acids on muscle lipids and fatty acids in Belgian Blue double-muscled young bulls. Reprod. Nutr. Dev. 43: p. 331-345.

Raes, K., L. Haak, A. Balcaen, E. Claeys, D. Demeyer and S. De Smet, 2004. Effect of feeding linseed at similar linoleic acid levels on the fatty acid composition of double-muscled Belgian Blue young bulls. Meat Sci. 66: p. 307-315.

Renerre, M., 2000. Oxidative processes and myoglobin. In: Antioxidants in muscle foods, E. Deker, C. Faustman and C.J. Lopez-Bote (editors), John Wiley, New York, 113-133.

Richardson, R.I., M. Enser, K. Hallett, J.D. Wood and N.D. Scollan, 2003. Effects of product type and fatty acid composition on shelf life of nutritionally modified beef. Proc. Brit. Anim. Sci. p. 41.

Richardson, R.I., K. Hallett, A.M. Robinson, G.R. Nute, M. Enser, J.D. Wood and N.D. Scollan, 2004. Effect of free and ruminally-protected fish oils on fatty acid composition, sensory and oxidative characteristics of beef loin muscle. In: Proceedings of the 50th International Congress of Meat Science and Technology, Helsinki, Finland, in press.

Ritzenthaler, K.L., M.K. McGuire, R. Falen, T.D. Schultz, N. Dasgupta and M.A. McGuire, 2001. Estimation of conjugated linoleic acid intake by written dietary assessment methodologies underestimates actual intake evaluated by food duplicate methodology. J. Nutr. 131: p. 1548-1554.

Scollan, N.D., N.J. Choi, E. Kurt, A.V. Fisher, M. Enser and J.D. Wood, 2001. Manipulating the fatty acid composition of muscle and adipose tissue in beef cattle. Brit. J. Nutr. 85: 115-124

Scollan, N.D., M. Enser, S. Gulati, I. Richardson and J.D. Wood, 2003. Effects of including a ruminally protected lipid supplement in the diet on the fatty acid composition of beef muscle in Charolais steers. Brit. J. Nutr. 90: p. 709-716.

Scollan, N.D., M. Enser, R.I. Richardson, S. Gulati, K.G. Hallett, G.R. Nute and J.D. Wood, 2004. The effects of ruminally protected dietary lipid on the fatty acid composition and quality of beef muscle. In: Proceedings of the 50th International Congress of Meat Science and Technology, Helsinki, Finland in press.

Simopoulos, A.P., 2001. *n*-3 fatty acids and human health: defining strategies for public policy. Lipids 36: p. S83-S89.

Ueda, K., A. Ferlay, J. Chabrot, J.J. Loor, Y. Chilliard, Y.and M. Doreau, 2003. Effect of linseed oil supplementation on ruminal digestion in dairy cows fed diets with different forage:concentrate ratios. J. Dairy Sci. 86, p. 3999-4007.

Van Nevel, C. and D. Demeyer, 1996. Influence of pH on lipolysis and biohydrogenation of soybean oil by rumen contents *in vitro*. Reprod. Nutr. Dev. 36: p. 53-63.

Warren, H.E., N.D. Scollan, K. Hallett, M. Enser, R.I. Richardson, G.R. Nute and J.D. Wood, 2002. The effects of breed and diet on the lipid composition and quality of bovine muscle. In: Proceedings of the 48th Congress of Meat Science and Technology 1: p. 370-371.

Wood, J.D., R.I. Richardson, G.R. Nute, A.V. Fisher, M.M. Campo, E. Kasapidou, P.R. Sheard and M. Enser, 2003. Effects of fatty acids on meat quality: a review. Meat Sci. 66: p. 21-32.

Nutritional quality of dairy products and human health

A. Lucas[1,2,], J.B. Coulon[1], P. Grolier[3], B. Martin[1] and E. Rock[3]*
[1]INRA-Centre de Clermont-Fd/Theix, Unité de Recherches sur les Herbivores (URH), 63122 Saint-Genès-Champanelle, France, [2]Institut Technique Français des Fromages, 419 route des Champs Laitiers, 74801 La Roche sur Foron, France, [3]INRA-Centre de Clermont-Fd/Theix, Unité de Maladies Métaboliques et Micronutriments (U3M), 63122 Saint Genès Champanelle, France.
**Correspondence: alucas@clermont.inra.fr*

Summary

The objective of that review is to make a statement on published data and to show the complex association between dairy products and human health. Whereas the group of food "milk and dairy products" present nutritional characteristics which distinguish it from the other food groups, its nutritional composition is likely to vary according to different parameters such as animal diet or parameters of dairy product manufacture. In particular, dairy products are richer in vitamins A and E, carotenoids, polyphenols and unsaturated fatty acids and lower in saturated fatty acids when animals are fed a pasture-based diet than when they consume a preserved forage-based one. In addition, the "negative" image of dairy products is mainly related to lipid hypothesis underlying cardiovascular diseases. The observational studies did not all conclude to deleterious effect of dairy products on cardiovascular diseases and on cancers. The questions under the data interpretation are as well on qualitative aspects (the real nature of the incriminated dairy product) as on quantitative ones. The dietary habits highly vary between the individuals within a given population. Deleterious associations of food items have been shown in certain population groups, such as an increase of animal products, including dairy products, and a decrease of plant-based food intake. Consequently, there remain many works to be defined and achieved for better assessing the nutritional quality of dairy products. However, in every case, the dietary recommendations concerning dairy product consumption must aim to include them in a balanced diet and not to exclude them from human diet.

Keywords: milk, dairy products, nutritional composition, human health

Introduction

To better understand the impact of milk and dairy product consumption on human health, it can be helpful to have an overview on the relationship between foods and pathologies, those defined as degenerative. The age-related diseases have summarily two potential origins: genetic and environmental. In this last factor, beside the parameters related to food safety, of which the effects usually appear in short and medium term, the more long term impacts of nutritional quality of the diet were highlighted in particular by epidemiologic approaches. A simple analysis of foods shows that they include energetic compounds (proteins, carbohydrates and lipids) and non energetic elements which are found in less amount and composed of micronutrients (vitamins, minerals and trace elements) and non essential microconstituents (carotenoids, polyphenols...). A limited intake of the macronutrients has negative impact on body growth leading to low body weight in undernourished people in developing countries or muscle loss (sarcopenia) in the elderly. Macronutrients by themselves are not sufficient to maintain a number of body functions. Experimentally-induced micronutrient deficiency on cellular models and laboratory animals largely confirmed the biological role of the micronutrients on targeted functions. For instance, deprivation of magnesium induced hyperactivation of immune system in the rat (Malpuech-Brugère *et al.*, 2000) and similarly, copper deficiency boosted systemic oxidative stress by decreasing the superoxide dismutase activity primarily involved in antioxidant defence system (Rock *et al.*, 1995). Questions remain as for the role of the microconstituents like polyphenols or carotenoids. Put aside pro-vitamin A activity of certain carotenoids, it seems that these microconstituents would act by supplementing endogenous antioxidant systems implied in the protection of the organism against the noxious effects of oxidative stress which is supposed to be at the origin of degenerative pathologies like cardiovascular diseases or certain cancers (Winklhofer-Roob *et al.*, 2003).

Within the Western diet, dairy products are likely to play a particular role on human health owing to their richness both in macro- and micronutrients and their high level of consumption. They supply a significant amount of energy, mainly in the form of proteins and lipids, but also vitamins (retinol) and minerals. Dairy products are an important source of calcium necessary for maintaining an optimal bone health (Prentice, 2004). Similarly, they also contain bioactive compounds such as carotenoids, bioactive peptides or conjugated linoleic acids which can positively influence human health. With the opposite, milk fat is high in saturated fatty acids and excessive intake relative to unsaturated fatty acids would raise the factor risk for cardiovascular diseases (Hu *et al.*, 2001). For this reason, dairy products are characterized by a "negative" image toward human health, although the observational studies did not all conclude to deleterious effect of dairy products on cardiovascular diseases. In addition, dairy product composition is subjected to relatively high variations, related to milk production conditions and milk processing. Consequently, according to pattern of dairy product consumption, which varies within population groups, dairy product consumption is likely to influence human health in a different way.

In the followings, some nutritional aspects will be considered including the variability of composition of dairy foods, the data available on impacts of dairy food intake on major degenerative diseases, and socio-economical aspects of dairy food consumption.

Nutritional composition of dairy products

Milk is secreted by mammary glands of mammalian females and it fulfils the nutritional needs of new-born mammals. Therefore, milk and milk-based products should contain essential macro- and micronutrients for the organism. However, nutritional needs of mammals are different according to specie, sex and age of the individuals. Consequently, from a quantitative point of view, the dairy products (bovine, caprine and ovine) and their nutrients usually consumed by humans in industrialised countries, may not always meet its physiological needs. In addition, milk production conditions, in particular animal diet, influence milk nutritional composition and the technological transformations to elaborate products (cheese, fermented milk, butter...) may further modify it. Consequently, nutritional quality of milk greatly varies according to different parameters.

The macronutrients (proteins, lipids and carbohydrates) correspond mainly to energy necessary for the anabolic processes i.e. for body growth and maintenance. The dairy products and more particularly milk, ripened cheeses and butter, because of their richness in macronutrients and their high level of consumption, supply a significant quantity of energy intake to the population, mainly in the form of proteins and lipids (Table 1). Moreover, the milk proteins are of high nutritional quality because of their digestibility and their composition in essential amino acids (Leonil, 2001). On the contrary, fatty acid composition of milk fat is not well balanced towards human nutritional needs. It is indeed well-known that beyond the consumed quantities of fatty acids, balance between the various groups of fatty acids (saturated, monounsaturated, polyunsaturated) is an important parameter in prevention of certain pathologies, including cardiovascular diseases. An excessive intake of saturated fatty acids has been significantly associated with a higher risk of cardiovascular diseases (Hu *et al.*, 1999; Kabagambe *et al.*, 2003), while a higher consumption of unsaturated fatty acids, in particular those of n-3 family, is associated with a lower risk (de Caterina and Zampolli, 2001). Milk fat is relatively rich in saturated fatty acids and low in polyunsaturated fatty acids (Table 1). In addition, because of the microbial transformation of unsaturated fatty acids in the rumen, milk fat can also be a significant source of *trans* fatty acids whose consumption in excess is associated with a higher risk of coronary arterial diseases (Clifton *et al.*, 2004). Among the different dairy products consumed, butter, cream and ripened cheeses are the main sources of saturated and *trans* fatty acids (Table 1).

Table 1. Nutritional composition of dairy products[1].

/Kg	Liquid milk[2]	Fresh products[2]	Cream	Butter	Ripened cheese[2]	Fresh cheese[2]	Concentrated milk[2]	Milk powder[4]
Energy (KJ)	1850	2970	13360	30910	15730	4980	5370	14780
Proteins (g)	31.9	42.0	22.7	7.0	294	77.0	64.0	355
Fatty acids (g)[3]								
SFA	9.5	22.2	212	526	173	50.6	46.6	0.0
MUFA	4.8	9.9	95.0	235	89.4	22.9	23.0	0.0
PUFA n-6	0.3	0.6	5.4	14.1	11.6	1.8	1.2	0.0
PUFA n-3	0.2	0.3	2.5	6.3	3.8	0.6	0.6	0.0
Vitamins								
C (mg)	11	10	8	0	0	12	10	50
B1 (mg)	0.5	0.4	0.3	0.1	0.5	0.2	0.8	3.8
B2 (mg)	1.7	1.8	1.4	0.2	3.4	1.6	3.3	18
PP (mg)	0.9	1	0.6	0.3	1.4	1.3	1.9	10
B5 (mg)	3.7	3.5	2.5	0.5	4	2.1	6.4	35
B6 (mg)	0.2	0.4	0.3	0.1	0.7	0.7	0.5	2.5
B8 (mg)	15	-	0.02	0	0.03	-	0.05	0.15
B9 (μg)	29	30	35	-	92	260	150	430
B12 (μg)	2.3	0	2	0	22	6.2	1.9	30
A (μg)	182	270	2500	7080	2130	740	700	0
D (μg)	-	0.5	-	16	16	1.3	1	0
E (mg)	0.8	0.9	8	20	3.5	1.9	2.7	0
Minerals								
Ca (mg)	1140	1550	750	150	11970	1110	2550	13010
P (mg)	845	1120	600	375	7590	1040	2010	11060
Mg (mg)	104	150	75.0	15.0	500	100	240	1120
Fe (mg)	0.5	0.5	1.0	1.2	7.8	1.4	2.0	5.2
Zn (mg)	3.7	3.8	2.5	0.2	40.0	3.7	10.6	47.5
Cu (mg)	0.1	0.1	0.8	0.2	1.5	0.3	0.5	2.0

[1]From data indicated in the tables of food composition (Feinberg *et al.*, 1987; USDA, 2003) ; [2] The data of composition used correspond to that of cow semi-skimmed milk (liquid milk), whole milk yogurt (fresh products), Emmental cheese (ripened cheese), 40% fat fresh cheese (fresh cheese), concentrated whole milk (concentrated milk) and skimmed milk powder (milk powder) ; [3]SFA: saturated fatty acids, MUFA: monounsaturated fatty acids, PUFA: polyunsaturated fatty acids.

The micronutrients (vitamins, minerals and trace elements) are non energetic compounds which take place as biocatalysts in many metabolic ways. Dairy products are an important source of vitamins A, B_2, B_5, B_{12} and D, and supply also significant amounts of vitamins B_1, B_6, B_9 and E to the population (Table 1). Among the different dairy product categories, milk and butter are respectively important sources of water-soluble vitamins (B) and fat-soluble vitamins (A, D, E), while cheese may supply both groups of vitamins (Table 1). Important quantities of minerals and trace elements are also supplied by dairy products, in particular by milk, cheese and fermented milk. They are indeed the main source of calcium for the French population and contain also high levels of magnesium and zinc. In addition, like much of other food, dairy products highly contribute to phosphorus intake but, while supplying calcium simultaneously, they may limit the phosphocalcic imbalance of the food. In the same way, the practice of salting during cheese manufacture contributes to high sodium intake observed in Western countries that can have deleterious impact on blood pressure (Nowson *et al.,* 2003).

Dairy products also contain bioactive compounds which, beyond their nutritional interest, are of potential functional interest. Thus, certain milk proteins have an antimicrobial activity (immunoglobulins, lactoferrin, lactoperoxydase...) (Korhonen *et al.*, 2000; Kussendrager and van Hooijdonk, 2000; Steijns and Hooijdonk, 2000). Moreover, the degradation of milk proteins during the fermentation of milk or its digestion leads to the formation of bioactive peptides with opioid-like, antihypertensive, antithrombotic, immunostimulating activities (Silva and Malcata, 2004). In addition, milk fat contains conjugated linoleic acids (CLA), a specific fatty acid of ruminant products (Chilliard *et al.*, 2001), which would have anticarcinogenic (Parodi, 1999) and immunostimulating (Albers *et al.*, 2003) activities and could reduce body fat mass (Blankson *et al.*, 2000). Dairy products also contribute to the consumption of antioxidant microconstituents such as carotenoids (β-carotene, lutein, zeaxanthin) (USDA, 2003) or some phenolic compounds (isoflavones, equol, enterolacton) (Antignac *et al.,* 2004). It is quite obvious that dairy products do not constitute an important source of these microconstituents, compared with plant-based foods, but the presence of these compounds in dairy products contributes to increase the endogenous antioxidant potential which may help to preserve the integrity of other nutrients against oxidation.

Whereas dairy products present nutritional characteristics which distinguish them from other foods (meat, eggs, and fruits and vegetables), their composition may also greatly vary within between the dairy products and within a given one. The main involved factors are milk production conditions, in particular animal diet, and processing. It is known that nutritional composition of dairy products varies according to season or geographical area of milk production. It is for example well established that milk contains more vitamins A and E, carotenoids and unsaturated fatty acids, including CLA, but also *trans* fatty acids and low amounts of saturated fatty acids in spring or in summer when animals are fed a pasture-based diet than in winter when they receive a diet rich in conserved forages or in concentrates as illustrated in Table 2 (Chilliard *et al.*, 2001; Kanno *et al.*, 1968; Martin *et al.*, 2002; Scott *et al.*, 1984; Thompson *et al.*, 1964; Timmen and Patton, 1988). Moreover, in a recent study Martin *et al.* (2002) have observed a higher content in phenolic compounds when cows received a pasture-based diet compared to winter feeding (Table 2). Similarly, milk fat is higher in vitamin A and β-carotene and lower in polyunsaturated fatty acids when cows graze an immature ray grass-based pasture compared to a mature ray grass-based one (Hawke, 1963; McDowall and McGillivray, 1963). Moreover, animal diet supplementation with lipids influences fatty acid composition of milk fat. The addition of plant oils rich in polyunsaturated acids to dairy cows' diet increases unsaturated fatty acid content of milk included CLA and *trans* fatty acids. As reviewed by Chilliard *et al.* (2000), the response to soybean oil is linear up to at least a level of 4% of oil in the diet (up to 2.0% of CLA in total milk fatty acids). Free oils are more effective than crude whole oilseeds, whereas extruded, micronized or heat-treated oilseeds have an intermediate effect (Chouinard *et al.*, 2001).

Table 2. Nutritional composition of milk according to season (from Martin et al.*, 2002).*

	Winter		Spring
	Concentrate-rich diet	Hay-rich diet	Pasture-rich diet
Vitamins and carotenoids (mg/L)			
vitamin A	0.17	0.13	0.20
β-carotene	0.12	0.09	0.19
vitamin E	0.46	0.47	0.63
Fatty acids (% total fatty acids)[1]			
SFA	70.4	66.3	56.8
MUFA	14.9	17.5	27.8
PUFA n-6	1.8	1.1	1.1
PUFA n-3	0.4	1.1	1.1
C18:1 *trans* 11	0.7	1.4	3.7
CLA	0.4	0.7	1.7
Phenolic compounds (mg/L)	2.0	3.7	9.0

[1]SFA: saturated fatty acids, MUFA: monounsaturated fatty acids; PUFA: polyunsaturated fatty acids, CLA: conjugated linoleic acids

According to geographical area, contents of dairy products in fatty acids, fat-soluble components (vitamins A, D, E, and β-carotene) and water-soluble ones (vitamins B and minerals) may also significantly vary (Dong and Oace, 1975; Mc Dowell and Mc Dowall, 1953; Smit *et al.*, 2000; Thompson *et al.*, 1964). Theses differences of milk composition according to geographical area could be due to differences of animal diet, but also to animal specie (Sağdlç *et al.*, 2003), breed (Thompson *et al.*, 1964) and stage of lactation (Larson *et al.*, 1983). For example, milk fat is know to be higher in β-carotene and lower in retinol when milk comes from Guernsey cows (17.7 and 4.8 μg/g) than when it comes from Holstein cows (4.9 and 5.8 μg/g) (Krukovsky *et al.*, 1950), while milk produced by Jersey cows would be significantly lower in CLA than Holstein milk (Sol Morales *et al.*, 2000). In addition, as altitude production site is high, milk fat content in polyunsaturated and *trans* fatty acids tend to increase whereas saturated fatty acids tend to decrease, likely due to differences of botanical composition of pasture (Bugaud *et al.*, 2001; Collomb *et al.*, 2002).

The variations of milk composition according to its production conditions suggest that nutritional quality of dairy products may be improved while modulating certain parameters, in particular animal diet. Thus, as previously shown in Table 2, a 15% decrease in saturated fatty acid content of milk fat may be observed while replacing a cereal-rich diet by a pasture-based one. In addition, it is interesting to notice that several components of nutritional interest vary in the same way according to animal diet. For instance, higher unsaturated fatty acid contents were associated with higher antioxidant contents (vitamin E, carotenoids and polyphenols) limiting thereby their oxidation. On the contrary, lower saturated fatty acid contents were associated with higher *trans* fatty acid contents in milk fat.

Beyond variations of nutritional composition within dairy products owing to milk production conditions, the processes applied to milk may also modify their nutritional composition. For instance, during transformation of milk into cheese, certain nutrients are concentrated (protein, fat, fat-soluble vitamins), while others are swept in the whey (water-soluble vitamins, minerals). Thus, cream, butter, cheese and milk powder are significantly richer in energy than milk, fresh products and fresh cheeses (Table 1). Moreover, micronutrient composition of dairy products also varies according to type of product. Thus, dairy products richest in fat (cream, butter and ripened cheese) contain higher fat-soluble vitamin amounts (A, D, E), while those containing high protein level also have high amounts of minerals (Table 1).

Nutritional properties of dairy products also vary between the different varieties belonging to a same type of dairy product because certain parameters of dairy product manufacture influence their nutritional composition. For example, there are great variations of composition between the different varieties of cheese as illustrated in Table 3.

Macronutrient content of cheese varies according to the technological process used to make it, in particular straining level of curd. For example, Gruyere, a hard cheese, is about two times higher in proteins than Cottage, a fresh cheese. The contents in B vitamins between the different cheese varieties also vary. Two process parameters are known to influence vitamins B composition of cheese. Firstly, during cheese making, vitamins B are swept in the whey because of their water-soluble property (Reif *et al.* 1976). Secondly, certain microorganisms synthesize vitamins B while others consume them during milk fermentation and cheese ripening (Reif *et al.*, 1976). The microorganism species used in cheese making may have an impact by synthesising B vitamins during milk fermentation and cheese ripening (Shahani *et al.*, 1962).

Table 3. Nutritional composition of cheese according to variety[1].

/ Kg	Cottage	Camembert	Roquefort	Cheddar	Gruyere
Energy (KJ)	3767	12557	15444	16868	17286
Protein (g)	137	198	215	249	298
Lipid (g)[2]	19	243	306	331	323
SFA	12.2	152	192	210	189
MUFA	5.50	70.2	84.7	93.9	100
PUFA	0.59	7.24	13.2	9.42	17.3
Vitamins					
B2 (mg)	1.85	4.88	5.86	3.75	2.79
B5 (mg)	2.42	13.6	17.3	4.13	5.62
B9 (μg)	130	620	490	180	100
B12 (μg)	7.10	13.0	6.40	8.30	16.0
A (μg)	210	2400	2900	2580	2680
E (mg)	0.20	2.10	-	2.90	2.80
β-carotene (μg)	50	120	0.0	850	330
Minerals					
Calcium (g)	0.69	3.88	6.62	7.21	10.1
Phosphorus (g)	1.51	3.47	3.92	5.12	6.05
Magnesium (g)	0.06	0.2	0.3	0.28	0.36
Potassium (g)	0.96	1.87	0.91	0.98	0.81
Sodium (g)	4.06	8.42	18.1	6.21	3.36
Zinc (mg)	4.20	23.8	20.8	31.1	39.0

[1]From data indicated in the tables of food composition (USDA, 2003)
[2]SFA: saturated fatty acids, MUFA: monounsaturated fatty acids, PUFA: polyunsaturated fatty acids.

In addition, mineral contents also vary in a great extent between cheese varieties. Mineral contents of cheese are all the more low that cheese dry matter are low and that level of acidification and of demineralization which curd undergoes during milk fermentation is high, because acidification of milk due to milk fermentation by lactic acid bacteria makes soluble minerals which are swept in the whey (Prieto *et al.*, 2002). Consequently, cheeses of which manufacture involve low straining and high acidification such as Cottage or Camembert are lower in mineral contents that cheeses of which making involves high straining and low acidification like Gruyere.

Altogether, these important variations of nutritional composition within the so-called food group "milk and dairy products" take a new look at pertinence to study the role that this food category plays on human health. Indeed, level and pattern of dairy product consumption vary within the population (Collet-Ribbing, 2001). For example, consumption of cheese is likely to be higher among subjects belonging to higher socioeconomic levels (Sanchez-Villegas *et al.*, 2003). Consequently, according to dietary patterns of individuals toward dairy products, human health is likely influenced in a different way. That perhaps explained why epidemiological works that studied the relation between dairy product consumption and human health led to contradictory results as we will illustrate it.

Relation between dairy product consumption and human health

To establish the role of a food on the health of a population, a first element to be realised is to find out the association of this food with defined pathologies. Correlation analysis allows such an association between the consumption of this food and the frequency of occurrence of pathologies in a population (case-control and cohort studies). The results of these studies establish a relationship of association which it is advisable to transform into relationship of causality by intervention studies consisting in supplementing healthy individuals with the targeted food (or the implied chemical component) and to study the effect on the incidence of pathologies. This step is difficult to realise because of the time necessary for the diseases to occur. Also, another possible way is to study the impact of the nutritional intervention on biomarkers or functional indicators of the physiopathological state of the individual. Thus, certain data existing in the literature show associations between the consumption of dairy products and the occurrence of degenerative pathologies like cardiovascular diseases, osteoporosis and cancers.

Dairy products and cardiovascular diseases

A great number of epidemiological observations showed a positive association between the incidence of cardiovascular diseases and lipid consumption, in particular those originating from animal foods such as cholesterol and saturated fatty acids (Mann *et al.*, 1997). Saturated fatty acids (58-82% of total lipids of milk) are generally recognised to increase the risk factor of atherosclerosis, in particular by increasing total cholesterolemia and cholesterol LDL, it is the case in particular for palmitic, myristic and lauric acids (Hu *et al.*, 2001, Figure 1).

The latter constitute 30-55% of total lipids and consequently the other saturated fatty acids (stearic acid and those with short chain) have only little effect on lipid profile. In the epidemiological study on the nurses (Nurses' Health Study), the results obtained from the food consumption surveys showed that dairy products (cheeses, skimmed milk, butter and unskimmed milk) contributed to 42% of myristic acid consumption, the more atherogenic fatty acid (Hu *et al.*, 1999). Dietary *trans* fatty acids also would negatively influence the risk of cardiovascular disease (Clifton *et al.*, 2004). Nevertheless, van de Vijver *et al.* (2000) have observed that at the currently intake levels of *trans* fatty acids in European countries, they are not associated with an unfavourable serum profile. Monounsaturated (oleic acid) and polyunsaturated fatty acids (including conjugated linoleic acids)

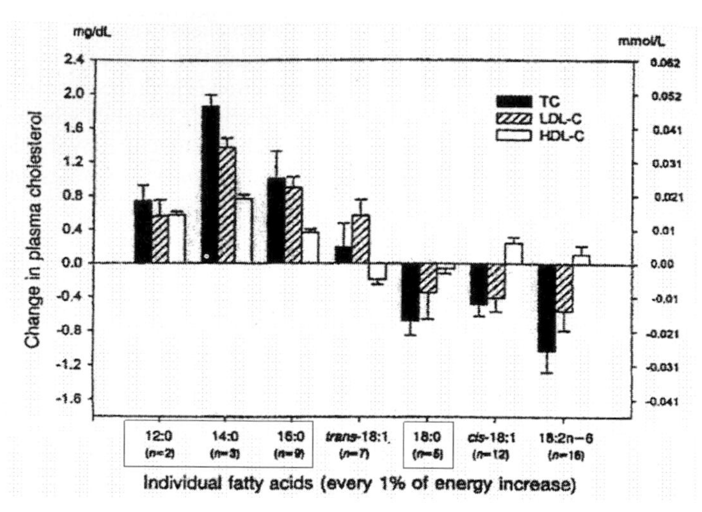

Figure 1. Effects of lauric (12:0), myristic (14:0), palmitic (16:0), elaidic (trans-18:1), stearic (18:0), oleic (cis-18:1), and linoleic (18:2n-6) acids on total cholesterol (TC), LDL cholesterol (LDL-C) and HDL cholesterol (HDL-C), (reproduced from Hu et al.*, 2001).*

could contribute to decrease the risk factors of atherosclerosis, in particular by increasing HDL level (Mensink and Katan, 1992). Lastly, the role of the cholesterol brought by dairy products (100-200 mg/L) is negligible on cholesterolemia which depends on the endogenous cholesterol synthesis. Despite, the observations of the effects of dairy product consumption on cholesterolemia remain incoherent (Pfeuffer and Schrezenmeir, 2000). In the Massaïs, the cholesterolemia is only marginally affected in spite of a daily consumption of approximately 4 litres of dairy products. The intervention studies do not show significant effects on cholesterolemia of consumption of 1-2 litres per day of milk. The epidemiological studies undertaken in Switzerland show a reduction in the plasmatic lipids in the milk consumers. However, other studies undertaken in Italy and Japan showed no effect or an increase of lipemia respectively. Similarly, data obtained in a cohort study carried out in United Kingdom provided no convincing evidence that milk consumption is associated with an increase in cardiovascular disease risk (Elwood *et al.*, 2004a). In addition, an analysis of cohort studies suggests that milk drinking may be associated with a small but worthwhile reduction in heart disease and stroke risk (Elwood *et al.*, 2004b).

As a whole, the results of epidemiological studies do not show a systematic effect (protective or noxious) of dairy product consumption on the cardiovascular system. Data on the specific role or the particular contribution of dairy products, like raw milk cheeses, on the cardiovascular diseases do not exist. A result recently obtained on healthy volunteers recruited in the area of Clermont-Ferrand (also a cheese production area) shows a significant increase in the plasmatic saturated fatty acid level compared to volunteers from Reus (Spain). However, the 2 groups of volunteers were recruited on the basis of criterion excluding particular biological or physical disorders; they were both in very good health (Cardinault, 2003).

Dairy products and osteoporosis

In nutrition research, calcium is one of the most studied elements for the prevention of osteoporosis. The epidemiological and intervention studies showed positive and significant effect of calcium intake consumption on gain or maintenance of bone mass (Dawson-Hughes, 1991; Recker *et al.*, 1996). Of these data, it appears that dairy products can play a protective role against osteoporosis in particular by the intermediary of calcium and proteins as well, in particular in Western countries (Dawson-Hughes, 1991; Recker *et al.* 1996) where dairy products contribute to 60% of the recommended daily allowance for calcium. A review of the studies carried out in 1990's concludes that the dairy product consumption benefit more to women of < 30 years old (Wiensier and Krumdieck, 2000). The calcium contribution during the period of growth could positively interfere on the risk of fracture at more advanced ages. This hypothesis was checked by experimental studies which showed a significant increase of bone mineral density in growing rats fed diet enriched with cheese (Kato *et al.*, 2002; Figure 2).

In addition, recent study carried out in the United States concluded that women with low milk intake during childhood and adolescence have less bone mass in adulthood and greater risk of fracture (Kalkwarf *et al.*, 2003). However, the above-cited study on the nurses' cohort (Nurses' Health Studies) did not show a significant protective effect of milk or high-calcium diet on the occurrence of osseous fractures. The authors noticed that an increased vitamin D intake was associated with a lower risk of osteoporotic hip fractures in postmenopausal women (Feskanich *et al.*, 2003). This is in agreement with a study showing that higher dairy product consumption in the elderly was associated with greater hip bone mineral density in men, even if the effect was found more effective in men than in women (McCabe *et al.*, 2004). From a quantitative point of view, recommended dietary allowance for calcium is 1000 to 1200 mg/d, whatever the age of individuals. Interestingly, 50 grams of Gruyere, Emmental, or Romano cheese can provide up to 500 mg of calcium.

Dairy products and cancers

Epidemiological studies were undertaken to study the effect of an increased consumption of dairy products on colorectal, breast and prostate cancers. A recent analysis of 10 cohort studies indicates significantly inverse association between milk consumption and risk of colorectal cancer (Cho *et al.*, 2004). Another study in which analysis of at least 30 case-control and cohort studies concluded that cohort studies consistently found a protective effect toward colorectal cancer of total dairy product intake, but the evidence is not supported by case-control studies (Norat and Riboli, 2003). Similarly, the available epidemiologic evidence does not support a strong association between the consumption of milk or other dairy products and breast cancer risk (Moorman and Terry, 2004). Such discrepancies may derive from the fact that some constituents of dairy products could be

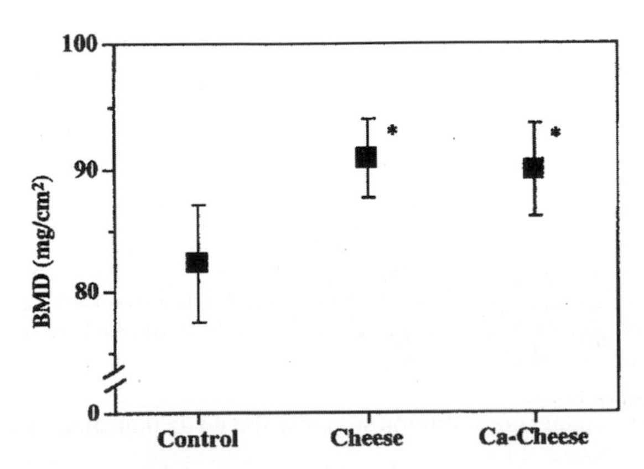

*Figure 2. Effects of experimental diets on bone mineral density (BMD) of the fourth lumbar spine in 12-week-old male rats. Datum points are means, and bars show 95% confidence intervals, n=8. * Significantly different from the control group (P<0.05), (reproduced from Kato* et al.*, 2002).*

able to reduce the risks of cancer (calcium, vitamin D, CLA, lactic bacteria...) and others, on the contrary, to increase the risks (saturated fatty acids but also pesticides and growth factors like IGF-1 known to be carcinogenic). For example, some studies clearly suggested that calcium is byself a protective element. The protective mechanisms of calcium in colorectal cancer results from its capacity to bind secondary bile acids and free fatty acids, primarly deoxycholic and lithocolic acids, thereby reducing their effective toxic dose to the colonic epithelial cells and preventing their stimulatory effects on proliferation of intestinal mucosa. A second hypothesis invokes an intracellular action of the calcium, which could inhibit the proliferation of epithelial cells of the colon by inducing their differenciation (Norat and Riboli, 2003). Other components of fermented dairy products could also protect from colorectal cancer by the intermediary of lactic bacteria by stabilising the colic flora, metabolising certain mutagen elements and stimulating immune response. With the opposite, calcium could have an indirect harmful effect on prostate cancer i.e. by reducing the production of parathyroidal hormone (Chan and Giovannucci, 2001), and resulting in a decreased bioavailability of vitamin D, necessary to cellular proliferation and differentiation. The milk lipids, and in general lipids from animal foods, also contribute to increase lipemia whose harmful role is recognised for colorectal cancer, in opposition with the protective role of lipids of vegetable products. However, the association between dairy product consumption and cancers are less obvious than that found between meat consumption and cancers. By definition, the lipids are energetic compounds and it has been suggested that high energy intake may modulate the process of carcinogenesis. It has been hypothesised that dietary fat may affect the metabolism of sex hormones implied in breast and prostate cancers. Among milk fatty acids, butyric acid and conjugated linoleic acids would have respectively anticarcinogenic and antitumor properties (Aro *et al.*, 2000).

In summary, the currently available studies on the relation between dairy product consumption and cancers are generally incoherent for colorectal and breast cancer, and show an increase in the risk of prostate cancer. Very few data exist as for the relationship to the consumption of specific dairy products like ripened cheese for instance.

A potential interpretation of the unfavourable image of dairy products: association with other food and practices of life

The population studies carried out to date do not allow distinguishing the various food categories, neither the associated population groups. Such an approach was carried out in a study on French population and allows distinguishing 6 different population groups from "small diversified eaters" until "big monotonous eaters". This differentiation has been based on 44 food categories consumed over a 7 days period (Collet-Ribbing, 2001). The extreme categories found above are characterised by:

1. Small and diversified eaters (SDE): this group accounts for 15% of the population studied and is composed of 80% women of which the two thirds are 45 years old and more. This group contains only 0.7% of obese subjects.
2. Big monotonous eaters (BME): this group accounts for 11% of the population and is composed of 90% men from 25 to 54 years old. In addition to low food diversity, alcohol intake was found significantly greater than in SDE group. This group is marked by a clear tendency to overweight with nearly 9% of obese subjects.

In these groups, the dairy product consumption was distributed as indicated in Table 4.

This table shows that milk is less consumed whereas cheeses and butter are more consumed by the individuals from BME group. For the other food items, there is a tendency to increase meat and alcohol intake together with low intake of fruits and vegetables. The epidemiological studies

Table 4. Daily consumption of dairy products according to dietary pattern.

	SDE[1] (g)	BME[1] (g)
Milk	103	57
Fresh dairy products	138	43
Cheeses	31	54
Butter	13	24
Meat + pork-butchery	79	164
Alcohol (wine + beer)	62	629
Fresh vegetables	199	95
Fruits	209	71

[1]SDE: small and diversified eaters, BME: big monotonous eaters

implying dairy product consumption do not always take into account such food practices of the various categories of the studied population. Therefore, it can be of interest to evaluate the published studies on that basis, knowing that dietary habits largely depend on lifestyle factors including smoking and use of alcohol. Such an effect was found in another study in which consumption of the various categories of food was analysed according to the daily alcohol consumption (Kesse *et al.*, 2001). Figure 3 shows that high alcohol consumption affects food intake and, in particular, high alcohol consumption is associated with higher cheese consumption and lower soup and fruit consumption.

In addition, two studies observed positive association between risk of cardiovascular disease and a dietary pattern characterised by high intakes of high-fat dairy products but also of meats, eggs

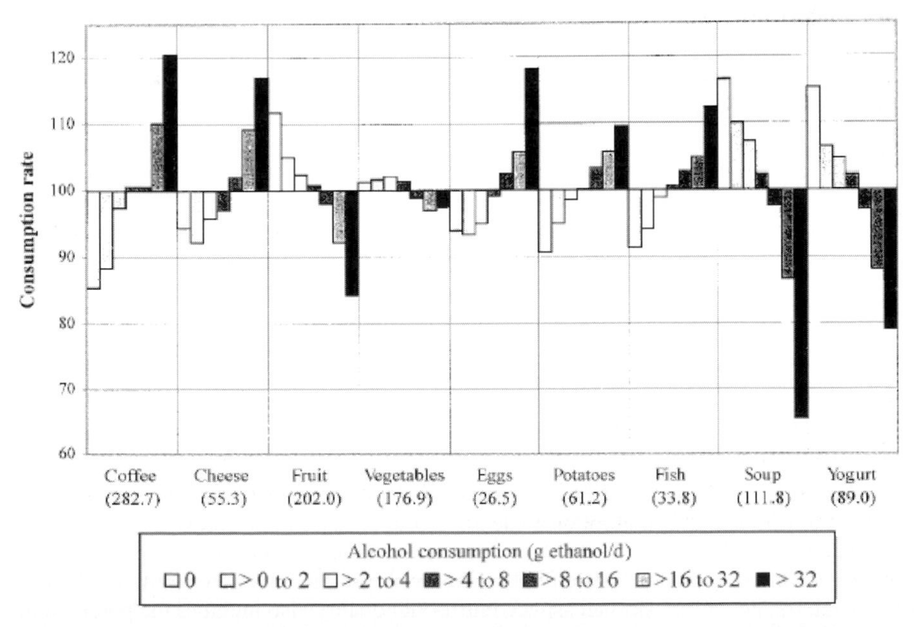

Figure 3. Consumption of selected food items (population mean in parentheses; arbitrary mean = 100) according to alcohol consumption in 72 904 French women (reproduced from Kesse et al.*, 2001).*

and red meat in the first study (Kerver *et al.*, 2003) and of red meat, processed meat, refined grains, sweets and dessert and French fries in the second study (Hu *et al.*, 2000).

The Table 5 illustrates the percentage of the recommended dietary allowances (RDA) for various elements and covered by dairy product consumption according to dietary pattern close to that found with SDE and BME. This table shows especially that a level of dairy product consumption close to that of BME, because the high saturated fatty acid intake from dairy products, should not be associated with high levels of consumption of the other saturated fatty acid-rich food categories (meat and meat products, oils and fats). Such an association could thus be at the origin of a food imbalance leading to the physical characteristics (overweight) of BME and to an increase in risk of pathologies. On the contrary, by taking into account the values obtained on the SDE, it would be advisable to adapt dairy product consumption of the population so as to satisfy its nutritional requirements.

Table 5. Contribution to adult man RDA[1] of dairy product consumption[2] according to dietary pattern[3].

	RDA	SDE (%)	BME (%)
Energy[4]	11400 Kj	13.1	16.0
Proteins[5]	56 g	32.6	35.1
Fatty acids[6]			
SFA	24.2 g	67.3	97.2
MUFA	60.7 g	12.7	18.4
PUFA n-6	12.4 g	5.3	8.1
PUFA n-3	2.5 g	10.6	15.4
Vitamins			
B2	1.6 mg	33.2	22.7
B12	2,4 µg	38.3	55.0
A	800 µg	26.8	38.4
D	5 µg	15.6	25.4
Minerals			
Ca	900 mg	78.3	86.8

[1]Recommended Dietary Allowances from Martin, 2001); [2]The data of composition used correspond to that of semi-skimmed milk (liquid milk), whole milk yogurt (fresh products) and Emmental cheese (ripened cheese) indicated in the tables of food composition (Feinberg *et al.*, 1987 ; USDA, 2003); [3]SDE: small and diversified eaters, BME: big monotonous eaters; [4]RDA for an adult man having usual activities of the majority of the population between 20-40 years and weighing 70 kg; [5]RDA for a adult man weighing 70 kg; [6]SFA: saturated fatty acids, MUFA: monounsaturated fatty acids, PUFA: polyunsaturated fatty acids.

Conclusion

Nutritional quality is still semantically too broad to be applied to dairy products. As stated before, it is essential to take into account not only the major but also the minor nutrients to better define the nutritional quality. The difficulty resides among others to better control their amount which can greatly vary according to factors related with either milk production and/or milk processing. On the other hand, besides compositional aspect, nutritional quality also includes the potential health benefits and risks. Globally, if dairy products are recognised as foods providing high quality proteins and calcium, it is also discussed as for the lipids that can contribute to increase risk factors for cardiovascular diseases. The observational studies are not consistent to show either a positive or a negative impact on the main pathologies of Western countries. However, it will be necessary in the future epidemiological studies to distinguish the individuals according to their pattern of consumption of dairy products and to study by intervention approach the specific role of certain dairy products such as cheeses on human health. There remain many works to be defined and achieved for better assessing nutritional quality of dairy products, and to test by targeted intervention studies the impact of dairy products on biomarkers of health effect of these products. In addition, beyond the nutritional role of dairy products, it will be of interest to study the functional role of their consumption in the human diet and in particular how nutrients of dairy products interact with the ones supplied by other foods. In every case, it will be necessary to accompany these studies by nutritional information on dairy products, and in particular on the place which they should occupy in a diversified and varied diet by taking for example the base of the food pyramid (see figure 4).

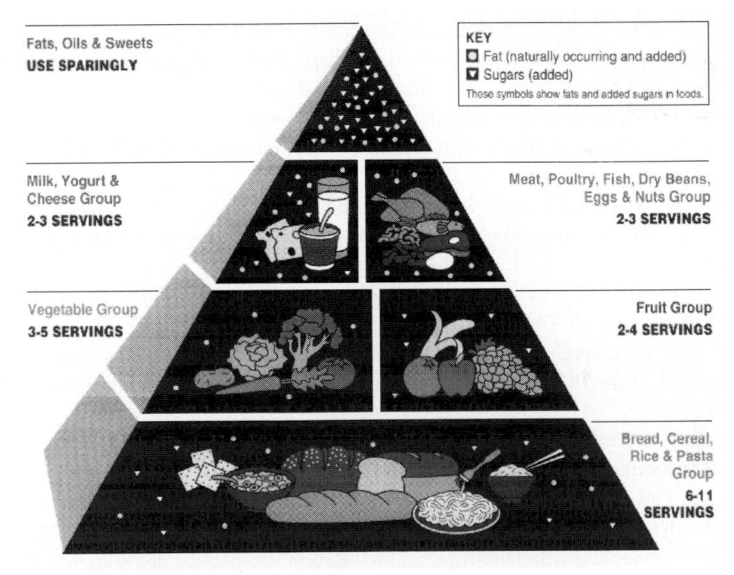

Figure 4. The Food Guide Pyramid (reproduced from USDA, 1992).

References

Albers, R., R.P. van der Wielen, E.J. Brink, H.F. Hendriks, V.N. Dorovska-Taran and I.C. Mohede, 2003. Effects of cis-9, trans-11 and trans-10, cis-12 conjugated linoleic acid (CLA) isomers on immune function in healthy men. Eur. J. Clin. Nutr. 57: p. 595-603.

Antignac, J-P., R. Cariou, B. Le Bizec and F. André, 2004. New data regarding phytoestrogens content in bovine milk. Food Chem. 87: p. 275-281.

Aro, A., S. Männistö, I. Salminen, M.L. Ovaskainen, V. Kataja and M. Uusitupa, 2000. Inverse association between dietary and serum conjugated linoleic acid and risk of breast cancer in postmenopausal women. Nutr.Cancer. 38: p. 1-157.

Blankson, H., J.A. Stakkestad, H. Fagertun, E. Thom, J. Wadstein and O. Gudmundsen, 2000. Conjugated linoleic acid reduces body fat mass in overweight and obese humans. J. Nutr. 130: p. 2943-2948.

Bugaud, C., S. Buchin, J-B. Coulon, A. Hauwuy and D. Dupont, 2001. Influence of the nature of alpine pastures on plasmin activity, fatty acid and volatile compound composition of milk. Lait 81: p. 401-414.

Cardinault, N., 2003. Contribution à l'étude multicentrique européenne VITAGE : Effets de l'âge sur le statut, l'absorption et le métabolisme du β-carotène, du lycopène et de la lutéine chez l'homme sain. PhD Thesis, Université Blaise Pascal et Université d'Auvergne, France.

Chan, J.A. and E.L. Giovannucci, 2001. Dairy products, calcium, and vitamin D and risk of prostate cancer. Epidemiol. Rev. 23: p. 87-92.

Chilliard, Y., A. Ferlay and M. Doreau, 2001. Effect of different types of forages, animal fat or marine oils in cow's diet on milk fat secretion and composition, especially conjugated linoleic acid (CLA) and polyunsaturated fatty acids. Livest. Prod. Sci. 70: p. 31-48.

Chilliard, Y., A. Ferlay, R.M. Mansbridge and M. Doreau, 2000. Ruminant milk fat plasticity: nutritional control of saturated, polyunsaturated, *trans* and conjugated fatty acids. Ann. Zootech. 49: p. 181-205.

Chouinard, P.Y., L. Corneau, W.R. Butler, Y. Chilliard, J.K. Drackley and D.E. Bauman, 2001. Effect of dietary lipid source on conjugated linoleic acid concentrations in milk fat. J. Dairy Sci. 84: p. 680-690.

Cho, E., S.A. Smith-Warner, D. Spiegelman, W.L. Beeson, P.A. van den Brandt, G.A. Colditz, A.R. Folsom, G.E. Fraser, J.L. Freudenheim, E. Giovannucci, R.A. Goldbohm, S. Graham, A.B. Miller, P. Pietinen, J.D. Potter, T.E. Rohan, P. Terry, P. Toniolo, M.J. Virtanen, W.C. Willett, A. Wolk, K. Wu, S.S. Yaun, A. Zeleniuch-Jacquotte and D.J. Hunter, 2004. Dairy foods, calcium, and colorectal cancer: a pooled analysis of 10 cohort studies. J. Natl. Cancer Inst. 96: p. 1015-1022.

Clifton, P.M., J.B. Keogh and M. Noakes, 2004. Trans fatty acids in adipose tissue and the food supply are associated with myocardial infarction. J Nutr. 134: p. 874-879.

Collet-Ribbing, C., 2001. La santé des Français et leurs consommations alimentaires. In: Apports nutritionnels conseillés pour la population française, A. Martin (Coordinnator), Editions Tec and Doc, France, p. 397-431.

Collomb, M., U. Bütikofer, R. Sieber, B. Jeangros and J.O. Bosset, 2002. Composition of fatty acids in cow's milk fat produced in the lowlands, mountains and highlands of Switzerland using high-resolution gas chromatography. Int. Dairy J. 12: p. 649-659.

Dawson-Hughes, B., 1991. Calcium supplementation and bone loss: a review of controlled clinical trials. Am. J. Clin. Nutr. 54: p. S274-280.

De Caterina, R. and A. Zampolli, 2001. n-3 fatty acids: antiatherosclerotic effects. Lipids. 36: p. S69-78.

Dong, F.M. and S.M. Oace, 1975. Folate concentration and pattern in bovine milk. J. Agric. Food Chem. 23: p. 534-538.

Elwood, P.C., J.E. Pickering, A.M. Fehily, J. Hughes and A.R. Ness, 2004a. Milk drinking, ischaemic heart disease and ischaemic stroke. I. Evidence from the Caerphilly cohort. Eur. J. Clin. Nutr. 58: p. 711-717.

Elwood, P.C., J.E. Pickering, J. Hughes, A.M. Fehily and A.R. Ness, 2004b. Milk drinking, ischaemic heart disease and ischaemic stroke. II. Evidence from cohort studies. Eur. J. Clin. Nutr. 58: p. 718-724.

Feinberg, M., J.C. Favier and J. Ireland-Ripert, 1987. Répertoire général des aliments. Tome 2 - Table de composition des produits laitiers. Editions Tec and Doc, France.

Feskanish, D., W.C. Willett and G.A. Colditz, 2003. Calcium, vitamin D, milk consumption, and hip fractures: a prospective study among postmenopausal women. Am. J. Clin. Nutr. 77: p. 504-511.

Hawke, J.C., 1963. Studies on the properties of New Zealand butterfat. J. Dairy Res. 30: p. 67-75.

Hu, F.B., J.E. Manson and W.C. Willett, 2001. Types of dietary fat and risk of coronary heart disease: a critical review. J. Am. Coll. Nutr. 20: p. 5-19.

Hu, F.B., E.B. Rimm, M.J. Stampfer, A. Ascherio, D. Spiegelman and W.C. Willett, 2000. Prospective study of major dietary patterns and risk of coronary heart disease in men. Am. J. Clin. Nutr. 72: p. 912-921.

Hu, F.B., M.J. Stampfer, J.E. Manson, A. Ascherio, G.A. Colditz, F.E. Speizer, C.H. Hennekens and W.C. Willett, 1999. Dietary saturated fats and their food sources in relation to the risk of coronary heart disease in women. Am. J. Clin. Nutr. 70: p. 1001-1008.

Kabagambe, E.K., A. Baylin, X. Siles and H. Campos, 2003. Individual saturated fatty acids and nonfatal acute myocardial infarction in Costa Rica. Eur. J. Clin. Nutr. 57: 1447-1457

Kalkwarf, H.J., J.C. Khoury and B.P. Lamphear, 2003. Milk intake during childhood and adolescence, adult bone density, and osteoporotic fractures in US women. Am. J. Clin. Nutr. 77: p. 257-265.

Kanno, C., K. Yamauchi and T. Tsugo, 1968. Occurence of γ-tocopherol and variation of α-tocopherol and γ-tocopherol in bovine milk fat. Dairy Sci. 51: p. 1713-1719.

Kato, K., Y. Takada, H. Matsuyama, Y. Kawasaki, S. Aoe, H. Yano and Y. Toba, 2002. Milk calcium taken with cheese increases bone mineral density and bone strength in growing rats. Biosci.Biotechnol. Biochem. 66: p. 2342-2346.

Kerver, J.M., E.J. Yang, L. Bianchi and W.O. Song, 2003. Dietary patterns associated with risk factors for cardiovascular disease in healthy US adults. Am. J. Clin. Nutr. 78: p. 1103-1110.

Kesse, E., F. Clavel-Chapelon, N. Slimani and M. van Liere, 2001. Do eating habits differ according to alcohol consumption? Results of a study of the French cohort of the European prospective investigation into cancer and nutrition (E3N-EPIC). Am. J. Clin. Nutr. 74: 322-327

Korhonen, H., P. Marnila and H.S. Gill, 2000. Bovine milk antibodies for health. Brit. J. Nutr. 84: p. S135-146.

Krukovsky, V.N., F. Whiting and J.K. Loosli, 1950. Tocopherol, carotenoid and vitamin A content of the milk fat and the resistance of milk to the development of oxilded flavors as influenced by breed and season. J. Dairy Sci. 33: p. 791-796.

Kussendrager, K.D. and A.C. van Hooijdonk, 2000. Lactoperoxidase: physico-chemical properties, occurrence, mechanism of action and applications. Brit. J. Nutr. 84: p. S19-25.

Larson, L.L., S.E. Wallen, F.G. Owen and S.R. Lowry, 1983. Relation of age, season, production, and health indices to iodine and beta-carotene concentrations in cow's milk. J. Dairy Sci. 66: p. 2557-2562.

Léonil, J., 2001. Protéines. In: Lait, nutrition et santé, G. Debry (Coordonnator), Editions Tec. and Doc., France, p. 45-76.

Malpuech-Brugère, C., W. Nowacki, M. Daveau, E. Gueux, C. Linard, E. Rock, J-P. Lebreton, A. Mazur and Y. Rayssiguier, 2000. Inflammatory response following acute magnesium deficiency in the rat. Biochim. Biophys. Acta. 1505: p. 91-98.

Mann, J.I., P.N. Appleby, T.J. Key and M. Thorogood, 1997. Dietary determinants of ischaemic heart disease in health conscious individuals. Heart 78: p. 450-455.

Martin, A. (Coordinator), 2001. Apports nutritionnels conseillés pour la population française. Editions Tec. and Doc., France, 612 p.

Martin, B., A. Ferlay, P. Pradel, E. Rock, P. Grolier, D. Dupont, D. Gruffat, J-M. Besle, N. Ballot, Y. Chilliard and J.B. Coulon, 2002. Variabilité de la teneur des laits en constituants d'intérêt nutritionnel selon la nature des fourrages consommés par les vaches laitières. Renc. Rech. Ruminants, 9: p. 347-350.

McCabe, L.D., B.R. Martin, G.P. McCabe, C.C. Johnston, C.M. Weaver and M. Peacock, 2004. Dairy intakes affect bone density in the elderly. Am. J. Clin. Nutr. 80: p. 1066-1074.

McDowall, F.H. and W.A. McGillivray, 1963. Studies on the properties of New Zealand butterfat. VII. Effect of stage of maturity of ryegrass fed to cows on the characteristics of butterfat and its carotene and vitamin A contents. J. Dairy Res. 30: p. 59-66.

McDowell, A.K. and F.H. McDowall, 1953. The vitamin A potency of New Zealand butter. J. Dairy Res. 20: p. 76-100.

Mensink, R.P. and M.B. Katan, 1992. Effect of dietary fatty acids on serum lipids and lipoproteins. A meta-analysis of 27 trials. Arterioscler. Thromb. 12: p. 911-919.

Moorman, P.G. and P.D. Terry, 2004. Consumption of dairy products and the risk of breast cancer: a review of the literature. Am. J. Clin. Nutr. 80: p. 5-14.

Norat, T. and E. Riboli, 2003. Dairy products and colorectal cancer. A review of possible mechanisms and epidemiological evidence. Eur. J. Clin. Nutr. 57: p. 1-17.

Nowson, C.A., T.O. Morgan and C. Gibbons, 2003. Decreasing dietary sodium while following a self-selected potassium-rich diet reduces blood pressure. J Nutr. 133: p. 4118-23.

Parodi, P.W., 1999. Conjugated linoleic acid and other anticarcinogenic agents of bovine milk fat. J. Dairy Sci. 82: p. 1339-49.

Pfeuffer, M. and J. Schrezenmeir, 2000. Bioactive substances in milk with properties decreasing risk of cardiovascular diseases. Brit. J. Nutr. 84: p. S155-159.

Prentice, A., 2004. Diet, nutrition and the prevention of the osteoporosis. Public Health Nutr. 7: p. 227-243.

Prieto, B., I. Franco, J.G. Prieto, A. Bernardo and J. Carballo, 2002. Compositional and physico-chemical modifications during the manufacture and ripening of León raw cow's milk cheese. J. Food Comp. Anal. 15: p. 725-735.

Reif, G.D., K.M. Shahani, J.R. Vakil and L.K. Crowe, 1976. Factors affecting B-complex vitamin content of Cottage cheese. J. Dairy Sci. 59: 410-415

Recker, R., S. Hinders, K.M. Davies, R.P. Heaney, M.R. Stegman, J.M. Lappe and D.B. Kimmel, 1996. Correcting calcium nutritional deficiency prevents spine fractures in elderly women. J. Bone. Miner. Res. 11: p. 1961-1966.

Rock, E., E. Gueux, A. Mazur, C. Motta and Y. Rayssiguier, 1995. Anemia in copper-deficient rats: role of alterations in erythrocyte membrane fluidity and oxidative damage. Am. J. Physiol. 269: p. 1245-1249.

Sağdıç, O., M. Döonmez and M. Demirci, 2004. Comparison of characteristics and fatty acid profiles of traditional Turkish yayik butters produced from goats', ewes' or cows' milk. Food Control. 15: p. 485-490.

Sanchez-Villegas, A., J.A. Martinez, R. Prättälä, E. Toledo, G. Roos and M.A. Martinez-Gonzalez, 2003. A systematic review of socioeconomic differences in food habits in Europe: consumption of cheese and milk. Eur. J. Clin. Nutr. 57: p. 917-929.

Scott, K.J., D.R. Bishop, A. Zechalko and J.D. Edwards-Webb, 1984. Nutrient content of liquid milk. I. Vitamins A, D_3, C and the B complex in pasteurized bulk liquid milk. J. Dairy Res. 51: p. 37-50.

Shahani, K.M., I.L. Hathaway and P.L. Kelly, 1962. B-complex vitamin content of cheese. II. Niacin, panthotenic acid, pyridoxine, biotin, and folic acid. J. Dairy Sci. 45: p. 833-841.

Silva, S.V. and F.X. Malcata, 2004. Caseins as source of bioactive peptides. Int. Dairy J. *in press*

Smit, L.E., H.C. Schönfeldt, W.H. de Beer and M.F. Smith, 2000. The effect of locality and season on the composition of South African whole milk. J. Food Comp. Anal. 13: p. 345-367.

Sol Morales, M., D.L. Palmquist and W.P. Weiss, 2000. Milk fat composition of Holstein and Jersey cows with control of depleted copper status and fed whole soybeans or tallow. J. Dairy Sci. 83: p. 2112-2119.

Steijns, J.M. and A.C. Hooijdonk, 2000. Occurrence, structure, biochemical properties and technological characteristics of lactoferrin. Brit. J. Nutr. 84: p. S11-17.

Thompson, S.Y., K.M. Henry and S.K. Kon, 1964. Factors affecting the concentration of vitamins in milk. I. Effect of breed, season, geographical location on fat-soluble vitamins. J. Dairy Res. 31: p. 1-25.

Timmen, H. and S. Patton, 1988. Milk fat globules: fatty acid composition, size and in vivo regulation of fat liquidity. Lipids. 23: p. 685-689.

U.S. Department of Agriculture (USDA) / U.S. Department of Health and Human Services, 1992. http://www.usda.gov/cnpp/pyramid.html

U.S. Department of Agriculture (USDA), Agricultural Research Service. 2003. USDA National Nutrient Database for Standard Reference, Release 16. Nutrient Data Laboratory Home Page, http://www.nal.usda.gov/fnic/foodcomp.

van de Vijver, L.P.L., A.F.M. Kardinaal, C. Couet, A. Aro, A. Kafatos, L. Steingrimsdottir, J.A. Amorim Cruz, O. Moreiras, W. Becker, J.M.M. van Amelsvoort, S. Vidal-Jessel, I. Salminen, J. Moschandreas, N. Sigfússon, I. Martins, A. Carbajal, A. Ytterfors and G. van Poppel, 2000. Association between *trans* fatty acid intake and cardiovascular risk factors in Europe: the TRANSFAIR study. Eur. J. Clin. Nutr. 54: p. 126-135.

Weinsier, R.L. and C.L. Krumdieck, 2000. Dairy foods and bone health: examination of the evidence. Am. J. Clin. Nutr. 72: p. 681-689.

Winklhofer-Roob, B.M., E. Rock, J. Ribalta, D.H. Shmerling and J.M. Roob, 2003. Effects of vitamin E and carotenoid status on oxidative stress in health and disease. Evidence obtained from human intervention studies. Mol. Aspects Med. 24: p. 391-402.

Current and emerging technologies for the prediction of meat quality

A.M. Mullen and D.J. Troy*
National Food Centre, Teagasc, Ashtown, Dublin 15, Ireland.
**Correspondence: amullen@nfc.teagasc.ie*

Summary

One of the greatest challenges to the meat industry, in the present time, is the ability to predict and guarantee the quality of their final product. While there is a need to predict quality at various stages during the life of the animal, the main focus of this review is directed to the *post-mortem* (in particular early *post-mortem*) prediction of quality. Many biochemical parameters and physical measurements demonstrate powerful predictive ability during this period. Other reviews in this session have focused more on the live animal and the way in which muscle biology or genetics can help to predict meat quality. In this review the more commonly used methods, namely pH, electrical properties and colour will be summarised initially. Emerging technologies such as NIR, image analysis, ultrasonics, NMR, immuno-based assays and biosensors are then discussed. Most of the emphasis is on tenderness, however, as technological aspects of quality are important to processors these are also referred to. Most references will be regarding bovine and porcine muscle. The application of proteomics and functional genomics to the prediction of meat quality is being dealt with in separate articles and, although highly relevant, will not be discussed here. Florescence spectroscopy is another emerging technology of interest which is also being discussed in a separate article.

Keywords: meat quality, prediction, new technologies, beef, pork

Introduction

Meat quality has been defined in many ways and will vary according to the end user (e.g. retail, processor, consumer). Currently, in many research fields including meat science, there is a strong focus on the consumer and addressing their needs. A common definition of quality is that it is the characteristics of a product "that are sought and valued by the consumer". Hoffman (1990) described meat quality as the 'sum of all quality factors of meat in terms of the sensoric, nutritive, hygienic and toxicological and technological properties.' The most important quality aspects for beef, rated by consumers in four EU countries were taste, tenderness, juiciness, freshness, leanness, healthiness and nutritional quality (Grunert, 1997). According to US studies, 50% of consumers rated tenderness to be the most important attribute of beef eating quality (Miller *et al.*, 1995). It is shown that consumers can distinguish between different levels of tenderness, and that they are willing to pay a higher price for tender beef (Boleman *et al.*, 1997; Lusk *et al.*, 2001).

In the majority of processing industries confidence in the quality of raw materials is a priority in the manufacture of the final product. In general, however, the meat industry does not have that advantage and hence is to a large extent unable to provide guarantees of product quality, especially in terms of palatability, to the end user. Variability of meat quality (Maher *et al.*, 2004a-d; Morgan *et al.*, 1991) prevents the meat industry marketing its produce according to quality. As a result meat produced today can not be guaranteed to possess the best quality attributes as its quality can only be truly assessed after consumption. The reasons for the variability of meat quality are numerous, but they emanate from the fact that these quality attributes are altered at many points along the meat chain from live animal through post-slaughter, in retail outlet and even in the purchaser's home. While there is a wealth of knowledge on factors which influence meat quality, and many significant improvements have been made in all aspects of meat quality, the prediction of this quality remains, to a large extent, elusive.

Ideally, the ultimate eating quality of meat needs to be predicted in the early *post-mortem* period (Mullen *et al.*, 1998a). By this, we mean within 24-48 hrs post-slaughter, during which time the carcass is within the confines of the meat factory. However, it may be also of interest for rearing or genetic purposes to predict meat quality earlier from muscle characteristics of live animals, however, this is outside of the scope of this review. There is also a need for a method of assessing meat quality at the point of sale. Presently, routine methods of measuring meat quality, within a typical European meat processing plant, revolve around a few measurements. However, a small number of meat processors have, reportedly, commissioned their own technology and are secretly reaping the awards (Swatland, 2002).

In recent years, new technologies have been developed which show promise for exploitation and use in the meat plant. These include both physical and biochemical techniques, many of which are described in more detail below. These show potential for use as indicators of meat quality and pave the way for further research into their predictive ability.

Indicators currently employed at factory level

pH

Generally, meat in the pH range 5.4 to 5.6 has the most desirable properties. Post-mortem glycolysis results in the accumulation of lactic acid which contributes to a decline in the pH of the muscle from about 7.2, at death, to roughly 5.5 after *rigor mortis* onset. In pork, rapid pH decline can result in pale, soft, exudative (PSE) meat, which presents a large problem to pork producers. To date an effective predictor of the occurrence of this condition is the measurement of the early *post-mortem* pH decline. In pork pH measurements at 45min *post-mortem* (pH_{45}) are used to detect the presence of PSE conditions (Somers *et al.*, 1985). Recently Kircheim *et al.*, (2001) showed pH_{45} displayed a high degree of reliability in correctly predicting PSE and RFN (reddish-pink, firm, non-exudative) meat. Garcia-Rey *et al.*, 2004 concluded that pH before salting is a good predictor for meat quality allowing the classification of the raw material in the first stage of manufacture. In beef higher ultimate pH (pH_u) values, which can reach pH 6.9, result in several defects, the most obvious being its colour, which becomes progressively darker as higher pH_u are reached. As a result, high pH meat is sometimes called 'dark-cutting' or 'DFD' (dark, firm, and dry) referring to the meat's physical properties. The microbiological stability of high pH meat is poor, tenderness is more variable, and cooked flavour is inferior (Gill and Newton, 1981; Simmons *et al.*, 2000).

There is some controversy over the usefulness of pH as a predictor of tenderness. Eikelenboom *et al.*, (1996) suggest pH_u is a good predictor of pork tenderness. Although Shackelford *et al.*, (1994) reported that pH at 3 hours *post-mortem* (pH_3) was not an effective predictor of meat tenderness, O'Halloran *et al.* (1997) concluded that both the early *post-mortem* and ultimate pH of muscle, has an important effect on the disruption of myofibrillar proteins and thus on beef tenderness. Indeed, Dransfield *et al.*, (2003) recently reported that pH_3 accounted for 0.52 of the variation in tenderness due to muscle and slaughter age in cattle. Lowe *et al.*, (2003) suggest that pHu is a good predictor of tenderness in bulls that are growing well on-farm, with only a short period of fasting prior to slaughter.

Recent research has focused on the manner in which even the temperature of the meat at the time of sampling will influence the pH (Jansen, 2001, Bruce *et al.*, 2001). In general glass or solid state electrodes are used to measure pH electrochemically. Significant variation in recorded pH values is introduced through the use of different types of meters and probes (O'Neill *et al.*, 2004). For industrial situations a solid state electrode would be the most suitable, however, some difficulties have been noted, such as the drift within the probe which may be associated with protein build up around the electrode (unpublished observations, The National Food Centre (NFC)). Andersen *et al.*, (1999) suggested that near infrared (NIR) and visual range spectrometric methods are comparable to the precision of the standard glass electrode pH meter.

An example of efforts being made to advance the prediction of meat quality is the application of a patented method of rapid pH determination by Young *et al.*, (2004a). The method is based on the rapid hydrolysis of muscle glycogen to glucose by the enzyme amyloglucosidase and subsequent measurement of the liberated glucose. The authors suggest this method may hold potential for transfer to the industry but recognise that it currently requires a degree of operator finesse and understanding (Young *et al.*, 2004b). Insofar as pH prediction can be a critical control point, the choice, training, and auditing of staff is important to the method's success. However, the current method is labour intensive and is best suited to slower processing lines where one operator is sufficient.

Electrical impedance and conductivity

The electrical conductivity of muscle changes when damage to differing degrees occurs to the membrane system of the muscle during *post-mortem* glycolysis. Conductivity is an indirect way of testing the intact cell membranes within a muscle tissue. Impedance, which is a combination of both resistance and capacitance, decreases when there is a disruption in membrane integrity (Kleibel *et al.*, 1983) and an increase in fluids within the muscle tissue (Pliquett *et al.*, 1995). Early *post-mortem* measurements of these electrical parameters have been acquired from trials within the NFC and have been compared with sensory and technological attributes in pork. Results have revealed that impedance and conductivity measurements were significantly correlated to the colour, drip loss and tenderness values (Byrne *et al.*, 2000 and personal communication). Schoeberlein *et al.*, (1999) ascertained that impedance measurements act as a good indicator of pork quality, while Lee *et al.*, (2000) suggest that conductivity (24hr *post-mortem*) may be a reliable predictor of water holding capacity in pork.

Segregating bovine carcasses on the basis of colour, drip loss and pH, a stronger correlation ($r = 0.84$) between conductivity 24 hr, and Warner Bratzler shear force (WBSF - mechanical measurement of tenderness) 14 days was observed (Mullen *et al.*, 2000) in one classification group. Some relationship exists between electrical measurements and tenderness, however, it would appear that this is not a simple, linear relationship. The relationship between electrical impedance and beef tenderness has been explored by Lepetit *et al.*, (2002), who suggest a relationship between the two measurements. Nevertheless, they conclude, the accuracy in determination of this parameter (impedance) must be improved for its potential to be able to be realised. The usefulness of conductivity and impedance measurements as quality predictors has been addressed by many researchers and more details, can be read in Swatland (1995a, 2002) and Pliquett and Pliquett (1998).

Colour

Colour measurements in the early *post-mortem* period have been investigated for their relationship with meat tenderness. In general it is accepted that this is a weak relationship. Jeremiah *et al.*, (1991) conclude segregation (tenderness) based on both pH and colour offered little in advantage over the use of pH alone. The evolution of colour in beef using samples which exhibited a wide range of ultimate pH values has been studied (Abril *et al.*, 2001). While more recent studies suggest a relationship between colour and tenderness (Wulf *et al.*, 1997), correlation coefficients were quite low (r=-0.38 correlation with WBSF; r=0.37 correlation with sensory tenderness). However, Wulf and Page (2000) suggested inclusion of muscle colour and pH would increase the accuracy and precision of the USDA (United States Department of Agriculture) quality grading standards for beef carcasses. A monochromatic fiber-optic probe was developed by MacDougall and Jones (1975) to measure the internal reflectance of meat. It has a peak sensitivity at 900nm, where absorbance by red heme pigments is minimal (Swatland, 1995b). The usefulness of the fiber-optic probe in segregating PSE and DFD classes has been demonstrated (MacDougal and Jones, 1980 and Garrido *et al.*, 1995). However, this probe has not been taken up by the industry to any great extent. Following analysis on a large number of topside pork muscles, we have shown both reflectance (using a hand held, non invasive probe, (Optostar)) and the CIE L* value of colour, performed well as objective methods of segregating the visual quality of pork topsides prior to processing into hams (McDonagh *et al.*, 2002). Tan *et al.*, (2000) concluded that a colour machine vision system was an accurate method of evaluating pork colour while van Oeckel *et al.*, (1999) found double density light transmission to be a useful method. Tan., (2004) also concluded that computer vision is a promising technology for objective meat quality grading. Results from a multichannel fibre optic probe indicate that adipose and collagenous connective tissue have high reflectance values (Swatland, 2000).

Emerging technologies for the prediction of meat quality

Ultrasound

Sound waves with frequencies above the audible range are called ultrasound waves. A number of variables characterising the propagation of an ultrasound signal, through a medium such as meat, can be measured. In addition the ultrasound measurements can also be made at a number of frequencies. Sound moves by compression waves, which are reflected and refracted when they move from one medium to the next. Sono-elastography is a method which uses ultrasonic pulses to track the internal displacements of small tissue elements in response to externally applied stress.

The mechanical properties of meat have been characterised by two sono-elastography parameters - the propagation velocity and the attenuation co-efficient of the mechanical wave. Optimisation of the measurement conditions considered temperature, sample dimensions, probe dimensions and frequency range aspects. Measurements taken from beef samples have indicated the usefulness of this technology as a non-intrusive method of visualising structural characteristics of beef (Ophir *et al.*, 1993; Cross and Belk, 1994). These investigations indicated that measurements may describe muscle structure at the muscle bundle level. The evolution of meat during *rigor* onset and ageing was followed by pH measurements and mechanical resistance of myofibrils at 20% of deformation. Comparisons were made between these parameters and the variables from the sono-elastrography analysis. Results indicate that sono-elastography can be used to follow *rigor mortis* onset and ageing in meat. Elastic deformation of intramuscular connective and adipose tissue caused by external stress is detected ultrasonically and has been proposed as a method of predicting beef quality (Swatland, 2001). Abouelkaram *et al.*, (2000) have analysed bovine muscle samples and investigated the influence of compositional and textural characteristics on ultrasonic measurements. The method used was based on the measurement of acoustic parameters (velocity, attenuation and backscattering). They conclude that the ultrasonic method used is robust enough to have real potential for muscle tissue characterisation. Ultrasonic strain image analysis has shown some relationship with WBSF values of tenderness in pork *M. semitendinosus* (Berg *et al.*, 1999).

A novel ultrasound technique has been developed with the potential to rapidly and accurately measure tenderness on small samples of meat (Allen *et al.*, 2001). This technique has much higher resolution than previously applied methods and is, therefore, very sensitive to meat texture. The method is also sensitive to differences in meat composition, particularly fat and protein content. It is possible, therefore, that composition and texture could be measured simultaneously to give an overall picture of the likely eating quality of meat. This could be particularly useful in processed meats where it is important to know the composition and textural properties of intermediate emulsions in order to control the eating quality of the final product. Anisotropy of ultrasonic velocity is capable of tracking structural changes in ageing beef and has potential as an indicator of eating quality in beef (Dwyer *et al.*, 2001a and b). However, further work on a broader spectrum of meat samples is needed to validate this approach for predicting beef eating quality.

Nuclear magnetic resonance (NMR)

NMR imaging measures energy differences between magnetic moments of naturally occurring, intrinsically magnetic atoms, when external fields are imposed. As water holding capacity is a commercially important attribute of porcine muscle the relationship between water dynamics and various factors (muscle type, chilling regime etc) has been explored using NMR. Various quantitative NMR parameters which are closely related to water dynamics, namely relaxation curves ($T_{20,}$ T_{21} and T_{22}) and magnetic transfer parameters, have been acquired and analysed. The T_{20} population is thought to correspond to water which is tightly associated with macromolecules, while the T_{21} reflects the intra-cellular water and the T_{22} reflects the extra-cellular/myofibrillar water. According to Beauvallet and Renou (1992), NMR has proven to be a powerful technique for the study of transformations in muscle tissue and is of value in assessing meat quality. Further measurements, recorded at 3hr and 14days *post-mortem* in bovine muscle, detected a redistribution of water over the ageing process. A study by Brondrum *et al.*, (2000) concluded NMR to be a better predictor of water holding capacity, intramuscular fat and total water content, than the use of a fibre optic probe or visible or NIR spectroscopy. Bertram *et al.*, (2004) showed *post-mortem* changes in the T_{22} water population were related to impedance characteristics and thereby membrane integrity. Moreover, it is suggested that changes in water distribution can be ascribed to the on-going disruption of membrane integrity and muscle contraction. A combination of NMR relaxation and diffusion techniques could be a potential tool for improving the understanding of the micro-structural properties associated with different meat quality (Bertram *et al.*, 2004). Laurent *et al.*, (2000) published a good overview of the characterisation of muscle by NMR imaging and spectroscopic techniques. These techniques give non-invasive access to structural, metabolic and chemical information as well as water dynamics and fat distribution. However, the cost of this analysis is prohibitive for on-line measurements.

Image analysis

Image analysis of meat has been investigated with the view to identifying features which are characteristic of the meat sample. Textural attributes of photographic images were analysed using several feature extraction methods to evaluate the possibility of identifying bovine muscle (Basset *et al.*, 2000). This involved acquiring photographic images of meat samples under both ultraviolet and visible light. Classifications experiments were performed to identify the bovine muscle according to muscle type, age and breed. Classification of muscles from animals within one age category led to high identification rates, while classifications based on age of the animal and breeds proved more difficult. Li *et al.*, (2001) analysed image texture features of beef from images. They concluded that useful beef tenderness information can be extracted by analyzing the muscle image texture, however this information seems to be mainly from the connective tissue component of the meat.

Near Infrared (NIR) spectroscopy

Many applications of NIR have dealt with the quantitative determination of compositional analysis of foods (Osborne and Fearn, 1986). NIR spectra reflect the organic constituents of food products and contain information about the conformation of components such as proteins and polysaccharides. NIR applications have also focused on direct estimation of quality parameters in foods. When using NIR spectra to predict the fat content it is advisable to mince or grind the meat into a homogenous mixture (Isaksson *et al.*, 1992; Rodbotten *et al.*, 2000). In this way a single NIR scan will probably detect the 'true' composition. Perhaps increasing the number of NIR scans substantially on intact meat samples may improve the prediction results for fat (Rodbotten *et al.*, 2000).

The ability of NIR to detect changes in the state of water and hydrogen bond interactions in foods has been observed (Hildrum *et al.*, 1995). Since such changes evidently occur in meat during tenderisation and ageing it was thought that a relationship may exist between NIR measurements and meat quality attributes. This was initially investigated by Mitsumoto (1991) who observed a strong relationship between NIR and both WBSF and compositional measurements of meat. Since then many studies concluded that NIR was a good predictor of meat tenderness (Byrne., 1998; Hildrum *et al.*, 1994; Park *et al.*, 1998). However, these prediction models were not always easily replicated (unpublished work by Mullen). Recently work at the NFC (Venel *et al.*, 2001) focused on avoidance of a number of potential experimental errors which may have mitigated against the development of satisfactory models. While NIR (750-1100nm) was unable to satisfactorily predict the selected organoleptic properties of bovine *M. semimembranosus* , better results were obtained for the WBSF values in the *M. longissimus dorsi* (r=0.51). While the predictive performance was improved, when the sample set was segregated according to animal grade, sex, ultimate pH or day of bone out (r= 0.54-0.72), the accuracies were no better than previously reported (unpublished work, NFC). NIR spectra (1100-2500nm) were collected pre-*rigor* in bovine LD (Rodbotten *et al.*, 2000). While these spectra had higher overall absorbancies compared to post-*rigor* measurements the results did not support that early *post-mortem* NIR spectra could be used as a predictor of beef tenderness. More recent research by Leroy *et al.*, (2004) concludes that NIR spectra collected on fresh meat showed good potential to predict CIE L^* and b^* parameters in reflectance mode. For the others parameters, the accuracy of the predictive models seemed to be weak, possibly due to reduced variability in quality of the sample set. The mincing of the meat sample before taking spectra could help to reduce heterogeneity.

Tenderness probe

World-wide there are many research projects focusing on the identifying or developing probes for the prediction of tenderness. As an overview of this field is beyond the scope of this review, a few examples will be documented to draw the readers attention to general area of research. Efforts have been made to develop physical methods to predict tenderness, as assessed by a sensory panel. For instance, in a study carried out with various French beef breeds, shear force (mechanical measure of tenderness) was shown to explain up to 48% of tenderness variability. However, the correlation between shear force and tenderness is thought to depend largely on the breed and the production system (Brouard *et al.*, 2001). As well as the shear force method, other instruments have been developed and tested. The MIRINZ (Meat Industry Research Institute of New Zealand) tenderness probe consists of two sets of pins on which meat samples are impaled (Jeremiah and Phillips, 2000). Tension is applied to the muscle fibres by one set of pins which rotate relative to a static set of pins. A torque signal is recorded against the angle of rotation. Results indicate it may be an alternative method to the Warner Bratlzer. However, this method would still require cooking of the meat. Ideally the industry requires a method which would allow measurements to be taken on raw meat, either on the carcass or on primal or retail cuts. A pressure based probe was recently developed at The National Food Centre, Ireland which measured both raw and cooked meat. However, a very low level of variation in ultimate tenderness was explained by this method.

Immunoassays

The application of immunoassay technology to the analysis of non-clinical samples in the food and agricultural sectors has been growing steadily for some time (Rittenburg and Grothaus, 1992). It is a powerful analytical tool that depends on the interaction between an antibody and the antigen being measured. The potential of this method for investigating endogenous food components such as vitamins, enzymes and structural proteins has been observed (Finglas *et al.*, 1990; Doumit *et al.*, 1996).

The myofibrillar component of meat tenderness provides a basis for the application of immuno-based assays to prediction tenderness. Proteolytic degradation of key myofibrillar proteins has been related to *post-mortem* tenderisation (Troy and Tarrant, 1987; Troy *et al.*, 1987; O'Halloran, 1996; Boyer-Berri and Greaser, 1998). Degradation of many structurally important myofibrillar proteins, during the *post-mortem* ageing of meat, has been observed by many research groups including NFC. Of these troponin T and its 30kDa myofibrillar proteolytic fragment have been related to meat tenderness (Buts *et al.*, 1986; Troy and Tarrant, 1987; Troy *et al.*, 1987). It has been suggested that the appearance of this 30kDa fragment could serve as an early *post-mortem* indicator of meat tenderness. A soluble 1734.8Da fragment of the troponin T molecule has recently been isolated (Nakai *et al.*, 1995; Mullen *et al.*, 1998b; Stoeva *et al.*, 2000;) which appears to be related to meat tenderness (Mullen *et al.*, 2000). Due to its solubility this fragment is more easily extracted from meat than the myofibrillar 30kDa and, therefore, it may be a more suitable candidate for routine factory analysis. An immuno-based assay was developed using polyclonal antibodies for the detection of this fragment. Initial results show a relationship between formation of the fragment and beef tenderness (Tsitsilonis *et al.*, 2002). The calpain/calpastatin proteolytic system which has been implicated in the tenderisation process has also been targeted for the development of an immunoassay test (Doumit *et al.*, 1996; Koohmaraie, 1996).

In pork, exoprotease activities can constitute a novel and adequate technique to predict early *post-mortem* technological quality (Toldra and Flores, 2000). Analysis of different peptide fractions from pork of various quality classes showed some differences between quality classes, over the ageing process (Moya *et al.*, 2001). Development of immuno-based assays to these and other important enzymes and proteolytic fragments are underway with the ultimate aim of predicting meat quality traits. There are some reservations about speed of immunoassays for use in a commercial situation (Koohmaraie, 1996). However, ultimately, and more ideally, a successful assay could be developed into a more rapid test, which would be more readily transferable to the meat industry.

Metabolites

Many metabolites exist in muscle, blood and urine which may contain valuable information regarding the variability of quality between different animals. Nucleosides and nucleotides also play an important role in the biochemical changes which occur during the early *post-mortem* period. Levels of hypoxanthine and inosine were different in exudative and normal pork meat up to 3 days *post-mortem* (Battle *et al.*, 2001). In the same study IMP and ATP only differed between 4 and 6hr *post-mortem* with the IMP:ATP ratio differing up to 2 hr *post-mortem*. The authors suggest that 2 hr is the optimum time for sampling when attempting to predict PSE. In another study the same authors suggest that nucleotide contents may serve as an index of some taste variations between pork quality classes but a direct link may be unlikely due to their rapid metabolism and the low degree of differences accounted for (Flores *et al.*, 1999).

Other novel metabolites of interest include nitric oxide (NO) which is a gaseous intercellular messenger and has recently been proposed in ubiquitous roles, including those within a normal functioning skeletal muscle. WBSF values for striploins incubated with NO enhancers were seen to decrease while those of striploins incubated with NO inhibitors increased at days 3 and 6 of conditioning (Cook *et al.*, 1997). The mechanism of NO in tenderisation is not yet known, however NO can mediate its effects by free radicals and/or calcium changes which can in turn affect proteolytic enzymes. The non-enzymatic tenderisation of meat proposed by Takahashi (Takahashi, 1996) and involving calcium ions may be partially explained by NO.

Biosensors

Biosensors have been defined in many ways. For example, they have been defined as analytical devices incorporating a biological material (eg. tissue, microorganisms, organelles, cell receptors, enzymes, antibodies, nucleic acids etc), a biologically derived material or biomimic intimately associated with or integrated within a physiochemical transducer or transducing microsystem, which may be optical, electrochemical, thermometric, piezoelectric or magnetic. Biosensors usually yield a digital electronic signal, which is proportional to the concentration of a specific analyte or group of analytes. While the signal may in principle be continuous, devices can be configured to yield single measurements to meet specific market requirements (http://www.biosensors-congress.com/about.htm). Following the identification of bioanalytes which are indicative of meat quality a further potential for development of diagnostic tests lies in the field of biosensors. These have many advantages including ease of use, sensitivity, minimum pre-treatment of sample, small sample volume. Jianrong *et al.*, (2004) discuss the increasingly important role nanotechnology is playing in the development of biosensors. Through the use of nanomaterials the sensitivity and performance of biosensors is being improved. In conclusion the authors recommend that nanotechnology-based biosensors be integrated within tiny biochips with on-board electronics, sample handling and analysis. This, the authors suggest, will greatly enhance functionality, by providing devices that are small, portable, easy to use, low cost, disposable, and highly versatile diagnostic instruments.

The future

Many innovative advances have been made in the area of measuring and predicting meat quality traits. It can be seen that in many cases the findings can be difficult to interpret and may be contradictory to others. Obviously factors such as experimental design, sampling methodology, sampling conditions, instrument type and data analysis are critical to the interpretation of results. The application of any of these techniques to an on-line situation will require large scale industrial based trials to verify and confirm their ultimate usefulness. As discussed in other articles, many advances have been made in the area of DNA and gene expression analysis. It is anticipated that this will contribute greatly to our understanding of meat quality traits. It is possible that early *post-mortem* prediction of meat quality will require recording more than one 'on-line' measurement. Ultimately predicting meat quality attributes may require a more holistic approach which sets criteria at a number of points along the chain of animal production through to consumption. Establishment of these criteria would enable implementation of a system such as the Meat Standards Australia (MSA). This system is based on the management of meat tenderness using a carcass grading scheme which utilizes the concept of total quality management of those factors which impact on beef palatability (See Thompson, 2002).

References

Abouelkaram, S., K. Suchorske, B. Buquet, P. Berge, J. Culioli, P. Delachartre and O. Basset, 2000. Effects of bovine muscle composition and structure on ultrasonic measurements. Food Chem. 69: p. 447-455.

Abril, M., M.M. Campo, A. Onenc, C. Sanudo, P. Alberti and A.I. Negueruela, 2001. Beef colour evolution as a function of ultimate pH. Meat Sci. 58: p. 69-78.

Allen, P., C. Dwyer, A.M. Mullen, V. Buckin, C. Smyth and S. Morrissey. 2001. Using ultrasound to measure beef tenderness and fat content. End of project report, ISBN 1-81470-202-1

Andersen, J.R., C. Borggaard, A.J. Rasmussen and L.P. Houmoller, 1999 Optical measurements of pH in meat. Meat Sci. 53: p. 135-141.

Basset, O., B. Buquet, S. Abouelkaram, P. Delachartre and J. Culioli, 2000. Application of tectural image analysis for the classification of bovine meat. Food Chem. 69: p. 437-445.

Battle, M., M.C. Aristoy and F. Toldra, 2001. ATP metabolites during aging of exudative and non-exudative pork meats. J. Food Sci. 66: p. 68-71.

Beauvallet, C., and J.P. Renou, 1992. Applications of NMR spectroscopy in meat research. Trends Food Sci. Tech. 3: p. 241-246.

Berg, E.P., F. Kallel, F. Hussain, R.K. Miller, J. Ophir and N. Kehtarnavaz, 1999. The use of elastography to measure quality characteristics of pork *semimembranosus* muscle. Meat Sci. 53: p. 31-35.

Bertram, H.C., A. Schafer, K. Rosenvold and H.J. Andersen, 2004. Physical changes of significance for early *postmortem* water distribution in porcine *M. longissimus*. Meat Sci. 66: p. 915-924.

Boleman, S.J., S.L. Boleman, R.K. Miller, J.F. Taylor, H.R. Cross, T.L. Wheeler, M. Koohmaraie, S.D. Shackelfors, M.F. Miller, R.L. Wets, D.D. Johnson and J.W. Savell, 1997. Consumer evaluation of beef of known categories of tenderness. J. Anim. Sci. 75: 1521-1524.

Boyer-Berri, C., and M.L. Greaser, 1998. Effect of portmortem storage on the Z-line region of titin bovine muscle. J. Anim. Sci. 76: p. 1034 -1044.

Brondrum, J., L. Munck, P. Henckel, A. Karlsson, E. Tornberg and S.B. Engelsen, 2000. Prediction of water holding capacity and composition of porcine meat by comparative spectroscopy. Meat Sci. 55: p. 177-185.

Brouard S., G. Renand and F. Turin, 2001. Relationships between *Longissimus lumborum* muscle characteristics and tenderness of three local beef cattle breeds. Renc. Rech. Ruminants 8: p. 49-52.

Bruce, H.L., J.R. Scott and J.M. Thompson, J.M. 2001 Application of an exponential model to early *postmortem* bovine muscle pH decline. Meat Sci. 58: p. 39-44.

Buts, B., E. Claeys and D. Deymeyer, 1986. Relation between concentration of troponin-T, 30,000 dalton and titin on SDS-PAGE and tenderness of bull *longissimus dorsi*. Proceedings of the 32nd European Meeting of Meat Research Workers, p. 175-178.

Byrne, C., 1998. Near infared reflectance spectroscopy and electrical measurements of beef muscle as indicators of meat quality, PhD, Thesis, National University of Ireland, Cork.

Byrne C.E., D.J. Troy and D.J. Buckley, 2000. Postmortem changes in muscle electrical properties of bovine *M. longissimus* dorsi and their relationship to meat quality attributes and pH fall. Meat Sci. 54: p. 23-34.

Cook, C.J., S.M. Scott and C.E. Devine, 1997.The effect of endogenous nitric oxide on tenderness changes of meat. ICOMST, G1-5, p. 558-559, New Zealand.

Cross, H.R., and K.E. Belk, 1994.Objective measurements of carcass and meat quality. Meat Sci. 36: p. 191-202.

Doumit, M. E., S.M. Lonergan, J.R. Arbona and J. Killefer and M. Koohmaraie, 1996. Development of an enzyme-linked immunosorbent assay (ELISA) for quantification of skeletal muscle calpastatin. J. Anim. Sci. 74: p. 2679-2686.

Dransfield E., J.F. Martin, D. Bauchart, S. Abouelkaram, J. Lepetit, J. Culioli, C. Jurie and B. Picard 2003. Meat quality and composition of three muscles from French cull cows and young bulls. Anim. Sci. 76: p. 387-399.

Dwyer, C, A.M. Mullen, P. Allen and V. Buckin, 2001a. Anisotropy of ultrasonic velocity as a method of tracking *postmortem* ageing in beef. Proceedings 47th ICoMST, Poland. 4 (P4), 250.

Dwyer, C, A.M. Mullen, P. Allen and V. Buckin, 2001b. Ultrasonic characterisation of dairy and meat systems. Proceedings, Food Science and Technology Research Conference, Cork. p. 30.

Eikelenboom, G., A.H. Hoving-Bolink and P.G. van der Wal 1996. The eating quality of pork. I. The influence of ultimate pH. Fleischwirtschaft 76: p. 392-393.

Finglas P.M., S.A. Alcock and M.R.A. Morgan, 1992. The biospecific analysis of vitamins in food and biological materials. In: Food safety and quality assurance, M.R.A. Morgan, C.J. Smith and P.A. Williams (editors), Elsevier Applied Science Publishers Ltd., London, p. 401-409.

Flores, M., E. Armero, M. C. Aristoy and F. Toldra, 1999. Sensory characteristics of cooked pork loin as affected by nucleotide content and *postmortem* meat quality. Meat Sci. 51: p. 53-59.

Garcia-Rey, R.M., J.A. Garcia-Garrido, R. Quiles-Zafra, J. Tapiador, and M.D. Lupue de Castro, 2004. Relationship between pH before salting and dry-cured ham quality. Meat Sci. 67: p. 625-632.

Garrido., M.D., J. Peduaye, S. Banon, M.B. Lopez and J. Laencina, (1995) On-line methods for pork quality detection. Food Control 6: p. 111-113.

Gill, C.O. and K.G. Newton, 1981. Microbiology of DFD meat. In: The Problem of Dark-Cutting in Beef, D.E. Hood and P.V. Tarrant, (editors), Martinus Nijhoff, The Hague, The Netherlands, p. 305-327.

Grunert, K.G. 1997. What's in a steak? A cross-cultural study on the quality perception of beef. Food Qual. Prefer. 8: p. 157-174.

Hildrum, K.I., B.N. Nilsen, M. Mielnik and T. Naes, 1994. Prediction of sensory characteristics of beef by near-infrared spectroscopy. Meat Sci. 38: p. 67-80.

Hildrum, K.I., T. Isaksson, T. Naes, B.N. Nilsen, M. Rodbotten and P. Lea, 1995. Near infared reflectance spectroscopy in the prediction of sensory properties of beef. J. Near Infared Spec. 3: p. 81-85.

Hoffman, K. 1990. Definition and measurement of meat quality. Proceedings of 36th Ann. Int. Cong. of Meat Sci. and Tech., Cuba, p. 941.

Isaksson, T., C.E. Miller and T. Naes, 1992. Nondestructive NIR and NIT determination of protein, fat and water in plastic wrapped, homogenised meat. App. Spec. 42: p. 1685-1694.

Jansen, M.L. 2001. Determination of meat pH - temperature relationship using ISFET and glass electrode instruments. Meat Sci. 58: p. 145-150.

Jeremiah, L.E., and D.M. Phillips, 2000. Evaluation of a probe for predicting beef tenderness. Meat Sci. 55: p. 493-502.

Jeremiah, L.E., A.K.W. Tong and L.L. Gibson, 1991. The usefulness of muscle colour and pH for segregating beef carcasses into tenderness groups. Meat Sci. 30: p. 97-114.

Jianrong, C., M. Yuqing, H. Nongyue, W. Xiahua and L. Sijiao, 2004. Nanotechnology and biosensors. Biotechnol. Adv. 22: 505-518.

Kircheim, U., C. Kinast and F. Schone, 2001. Early *postmortem* measurement as indicator of meat quality characteristics. Fleischwirtschaft 81: 98-90.

Kleibel, A., H. Psutzner and E. Krause, 1983. Measurement of dielectric loss factor routine method for recognising PSE muscle. Fleischwirschaft 63: p. 1183-1185.

Koohmaraie, M., 1996. ELISA test for calpastatin. Recip. Meat Conf. Proc. 49: p. 177-178.

Laurent, W., J.M. Bonny and J.P. Renou 2000. Muscle characterisation by NMR imaging and spectroscopic techniques. Food Chem. 69: p. 419-426.

Lee, S., J.M. Norman, S. Gunasekaran, R.L.J.M. van Laack, B.C. Kim and R.G. Kauffman, 2000. Use of electrical conductivity to predict water-holding capacity in post-*rigor* pork. Meat Sci. 55: 385-389.

Lepetit, J, P. Sale, R. Favier and R. Dalle, 2002. Electrical impedance and tenderisation in bovine meat. Meat Sci. 60: p. 51-62.

Leroy, B., S. Lambotte, O. Dotreppe, H. Lecocq, L. Istasse and A. Clinquart, 2004. Prediction of technological and organoleptic properites of beef *Longissimus thoracis* from near-infrared relectance and transmission spectra. Meat Sci. 66: p. 45-55.

Li, J., J. Tan and P. Shatadal, 2001. Classification of tough and tender beef by image texture analysis. Meat Sci. 57: p. 341-346.

Lowe, T.E., C.E. Devine, R.W. Wells and L.L. Lynch, 2003. The relationship between *postmortem* urinary catecholamines, meat ultimate pH, and shear force in bulls and cows. Meat Sci. 77: p. 251-260.

Lusk, J.L., J.A. Fox, T.C. Schroeder, J. Mintert and M. Koohmaraie, 2001. In-store valuation of beef tenderness. Amer. J. Agr. Econ. 83: p. 539-550.

MacDougall, D.B., and S.J. Jones, 1975. The use of a fibre optic probe for the detection of pale pork. Proceedings of the 21st European Meeting of Meat Research Workers; Berne, August 31st - September 5th, research notes, p. 113-5.

Maher, S.C., A.P. Moloney, A.M. Mullen, D.J. Buckley and J.P. Kerry, 2004a. Quantifying the extent of variation in the eating quality traits of the *M. longissimus dorsi* and *M. semimembranosus* of conventionally processed Irish beef. Meat Sci. 66: p. 351-360.

Maher, S.C., A.P. Moloney, A.M. Mullen, M.G. Keane, D.J. Buckley and J.P. Kerry, 2004b. Decreasing variation in tenderness of beef by controlling pre- and post slaughter management. Meat Sci. 67: p. 33-43.

Maher, S.C., A.M. Mullen, D.J. Buckley, J. Kerry, D. J. Troy and A.P. Moloney, 2005. The influence of biochemical differences of the variation in tenderness of *M. longissimus dorsi* of Belgian Blue steers managed homogenously pre- and post-slaughter. Meat Sci. 69: p. 215-224.

Maher, S.C., A. M. Mullen, A.P. Moloney, D.J. Buckley, J., Kerry, M.G. Keane and P. Dunne, 2004c. Variation in the eating quality of *M. longissimus dorsi* from light and heavy dairy or dairy cross beef steers and young bulls. Livest. Prod. Sci. 90: p. 271-177.

McDonagh, C., D.J. Troy, J.P. Kerry and A.M. Mullen, 2002. Objective and subjective methods for evaluating raw pork quality prior to processing. In: Proc. ICoMST, 268-269, Italy.

Miller, M.F., K.L. Huffman, S.Y. Gilbert, L.L. Hamman and C.B. Ramsey, 1995. Retail consumer acceptance of beef tenderized with calcium chloride. J. Anim. Sci. 73: 2308-2314

Mitsumoto, M., S. Maeda, T. Mitsuhashi and S. Ozawa, 1991. Near Infared Spectroscopy determination of physical and chemical characteristics in beef cuts. J. Food Sci. 56: p. 1493-1496.

Morgan, J. B., J.W. Savell, D.S. Hale, R.K. Miller, D.B. Griffin, H.R. Cross and S.D. Shackelford, 1991. National beef tenderness survey. J. Anim. Sci. 69: p. 3274-32831.

Moya, V.-J., M. Flores, M.-C. Arsitoy, and F. Toldra, 2001. Pork meat quality affects peptide and amino acid profiles during the ageing process. Meat Sci. 58: p. 197-206.

Mullen, A.M., Ú. Casserly, and D.J. Troy, 1998a. Predicting Beef Quality at the early *postmortem* period. Proceedings, International Conference, Challenges for the Meat Industry in the Next Millennium, May 1998, Malahide Castle, Ireland.

Mullen, A.M., S. Stoeva, W. Voelter, and D.J. Troy, 1998b. Identification of bovine protein fragments produced during the ageing process. ICOMST, Barcelona, Spain, 1998.

Mullen, A.M., B. Murray, and D.J. Troy, 2000. Predicting the eating quality of meat. End of project report. Teagasc, ISBN, 1-84170-157-2.

Nakai, Y., T. Nishimura, M. Shimizu and S. Arai, 1995. Effects of freezing on the proteolysis of beef during storage at 4°C. Biosci. Biotech. Biochem. 59: p. 2255-2258.

O'Halloran, G.R. 1996. Influence of pH on the endogenous proteolytic enzyme systems involved in tenderisation of beef. PhD thesis, The National University of Ireland.

O'Halloran, G.R., D.J. Troy and D.J. Buckley, 1997. The relationship between early *postmortem* pH measurements and the tenderisation process of beef muscle. Meat Sci. 45: p. 239-251.

O'Neill, D. J., D.J. Troy and A.M Mullen, 2004. Determination of potential inherent variability when measuring beef quality. Meat Sci. 66: p. 765-770.

van Oeckel, M.J., N. Warnants and C.V. Boucque, 1999. Measurement and prediction of pork colour. Meat Sci. 52: p. 347-354.

Ophir, J., R.K. Miller, H. Ponnekanti, I. Cespedes and A.D. Whittaker, 1993. Elastography of beef muscle. Meat Sci. 36: p. 239-250.

Osborne, B.G. and T. Fearn, 1986. Near Infared spectroscopy in food analysis. Longman Sci. and Tech. Harlow, Essex, UK.

Park, B., Y.R. Chen, W.R. Hruschka, S.D. Shackelford and M. Koohmaraie, 1998. Near-infared reflectance analysis for predicting beef longissimus tenderness. J. Anim. Sci. 76: p. 2115-2120.

Pliquett, F., U. Pliquett, L. Schoberleu and K. Freywald, 1995. Impedance measurements to characterise meat quality. Fleischwirschaft 75: 496

Pliquett, U. and F. Pliquett, 1998. Conductivity of meat as a quality parameter: critical remarks. Fleischwirtschaft 78: p. 1010-1012.

Rittenburg, J.H., and G.D. Grothaus 1992. Immunoassays: Formats and applications. In: Food safety and quality assurance, M.R.A Morgan, C.J. Smith C.J. and P.A. Williams P.A., (editors) Elsevier Applied Science Publishers Ltd., London. p. 3-12.

Rodbotten, R., B.N. Nilsen and K.I. Hildrum, 2000. Prediction of beef quality attributes from *post mortem* near infared reflectance spectra. Food Chem. 69: p. 427-436.

Shackelford, S.D., M. Koohmaraie and J.W. Savell, 1994. Evaluation of longissimus dorsi muscle pH at three hours *post mortem* as a predictor of beef tenderness. Meat Sci. 37: p. 195-204.

Schoeberlein, L., E. Scharner, K.O. Honikel, M. Altman and F. Pliquett, 1999. The Py value as a characteristic feature of meat quality. Fleischwirtschaft, 79: p. 116-120.

Simmons, N.J., M.M. Auld, B.C. Thomson, J.M. Cairney and C.C. Daly, 2000. Relationship between intermediate pH toughness in the striploin and other muscles of the beef carcass. Proceedings of New Zealand Society for Animal Production 60: p. 117-119.

Somers, C., P.V. Tarrant and J. Sherrington, 1985. Evaluation of some objective method for measuring pork quality. Meat Sci. 15: p. 63-76.

Stoeva, S., C. Byrne, A.M. Mullen, D.J. Troy and W. Voelter, 2000. Isolation and identification of proteolytic fragments from TCA soluble extracts of bovine *M. longissimus dorsi.* Food Chem. 69: p. 365-370.

Swatland, H.J. 1995a. On-line evaluation of meat. Swatland, H.J. Ed. Technomic Publishing Company, Pennsyvlania, USA.

Swatland, H.J. 1995b. Objective assessment of meat yield and quality. Trends Food Sci. Technol. 6: p. 117-119.

Swatland, H.J. 2000. Measurements with an on-line probe. Food Res. Int. 33: 749-757

Swatland, H.J. 2001. Elastic deformation in probe measurements on beef carcasses. J. Muscle Foods 12: p. 97-105.

Swatland H.J. 2002. On-line monitoring of meat quality. In: Meat Processing, Woodhead Publishing Limited, UK, p. 193-216.

Tan, F.J., M.T. Morgan, L.I. Ludas, J.C. Forrest and D.E. Gerrard, 2000. Assessment of fresh pork colour with machine vision. J. Anim. Sci. 78: p. 3078-3085.

Tan, J., 2004. Meat quality evaluation by computer vision. J. Food Eng. 61: p. 27-35.

Takahashi, K., 1996. Structural weakening of skeletal muscle tissue during *postmortem* ageing of meat: the non-enzymatic mechanism of meat tendrisation. Meat Sci. 43: p. S67-S80.

Thompson, J., 2002. Managing meat tenderness. Meat Sci. 62: p. 195-308.

Toldra, F. and M. Flores, 2000 The use of muscle enzymes as predictors of pork meat quality. Food Chem. 69: p. 371-377.

Troy, D.J. and P.V. Tarrant 1987. Changes in myofibrillar proteins from electrially stimulated beef. Biochem. Soc. Trans. 15: p. 297-298.

Troy, D.J., P.V.Tarrant and M.G. Harrington, 1987. Changes in myofibrillar proteins during conditioning of high pH beef. Biochem. Soc. Trans. 15: p. 299-300.

Tsitsilonis, O.E., S. Stoeva, H. Echner, A. Balafas, L. Margomenou, H.L. Katsoulas, D.J. Troy, W. Voelter, M. Papamichail and P. Lymberi, 2002. A skeletal muscle troponin T specific ELISA based on the use of an antibody against the soluble troponin T (16-31) fragment. J. Immunol. Methods 268: p. 141-148.

Venel, C., A.M Mullen, G. Downey and D.J. Troy, 2001. Prediction of tenderness and other quality attributes of beef by near infared reflectance spectroscopy between 750 and 1100nm; further studies. J. Near Infared Spec. 9: p. 185-198.

Wulf, D. M., S. F. O'Connor, J. D. Tatum and G. C. Smith, 1997. Using objective measurements of muscle color to predict beef longissimus tenderness. J. Anim. Sci. 75: p. 684-692.

Wulf, D.M. and J.K. Page, 2000. Using measurements of muscle colour, pH and electrical impedance to augment the current USDA beef quality standards and improve the accuracy and precision of sorting carcasses into palatability groups. J. Anim. Sci. 78: p. 2595-2607.

Young, O.A., J. West, A.L. Hart and F.F.H. van Otterdijk, 2004a. A method for early determination of meat ultimate pH. Meat Sci. 66: p. 493-498.

Young, O.A., R.D. Thomson, V.G. Merhtens and M.P.F. Loeffen, 2004b. Industrial application to cattle of a method for the early determination of meat ultimate pH. Meat Sci. 67: p. 107-112.

Determination of structure and textural quality of dairy and meat products using fluorescence and infrared spectroscopies coupled with chemometrics

E. Dufour
U.R. "Typicité des Produits Alimentaires", ENITA Clermont, 63370 Lempdes, France.
Correspondence: dufour@enitac.fr

Summary

Structure conditions both textural and flavour properties of cheeses and meats. Up to recently, delineating the structure of the food matrices and the interactions between components in food products has been impaired by the lack of techniques. More and more evidence shows that front face fluorescence spectroscopy and attenuated total reflection infrared spectroscopy may be valuable techniques to determine the structure (and hence the textural quality) of dairy and meat products. The large spectral collections recorded on food products have to be processed using multivariate analysis techniques such as principal component analysis, factorial discriminant analysis and canonical correlation analysis. This makes it possible to extract relevant information related to the molecular structures of proteins and fats, to the interactions between food product components and to the relation between structure and texture of cheeses and meats. These spectroscopic techniques remove methodological locks and allow the structure at molecular level and the dynamics of "real food products" to be investigated.

Keywords: structure, texture, meat, cheese, fluorescence, infrared, chemometry

Introduction

Although food safety becomes a major concern for consumers, the organoleptic characteristics of meat (e.g., meat tenderness) and cheese remain a major purchasing and loyalty factors of consumers. The management of the quality of meat and cheese requires the ability to anticipate their final properties. To reach this goal, on-line methods allowing real time measurements at the slaughterhouse or at the dairy plant; and prediction of the characteristics of final products will be useful tools. The possibilities of rapid and non-destructive measurements of quality related parameters in foods are advancing due to advances in spectroscopy and development of new sensor technology including cameras and spectrophotometers. Near-infrared (NIR) spectroscopy is one example of a method that is successfully used to quantify composition and as a tool for on-line prediction or classification of sensorial and physical product qualities (Isaksson *et al.*, 2001). Lately, fluorescence has also proven potential for rapid and non-destructive analysis of foods.

In recent years it has became more increasingly clear that the application of spectroscopic methods to food analysis can alleviate important problems in the processing and distribution of food and food products. Indeed, the traditional analysis methods for major food components are slow, relatively expensive, time-consuming, require highly skilled operators and not easily adapted to on-line monitoring. The chemical methods are not effective enough to cover the growing demand of the industry. To meet with this need, a great number of non-invasive and non-destructive instrumental techniques such as infrared have been developed to determine product composition. These new analytical techniques are relatively low-cost and can be applied in both fundamental research and in the factory as on line sensors for monitoring the food products. Results published in the last 5 years show that spectroscopic methods in combination with multivariate statistical analyses have broad applications in our understanding of food structure and properties. The goal of this article is to review the development and the application of infrared and fluorescence spectroscopies as research tools for bovine meat and dairy products and as sensors for on-line measurements in abattoirs and dairy plants.

Spectroscopic techniques for non-destructive study of food products

Infrared spectroscopy

Infrared radiation (IR) or the term infrared alone refers to energy in the region of the electromagnetic spectrum at wavelengths longer than those of visible light, but shorter than radio waves. The applications of this technique for animal nutrition, agricultural and food sciences have increased considerably in the last decade (Givens *et al.*, 1997).

Near infrared spectroscopy (NIR) is widely used for the determination of organic constituents in feeds, foods, pharmaceutical products and related materials. The technique is advantageous for many applications because it can provide rapid, non-destructive and multi-parametric measurements. The technique is also suitable for at-line and on-line process control and is based on the absorption of electromagnetic radiation at wavelength in the range 800-2500 nm. NIR spectra of food correspond mainly to overtones and combinations of vibrational modes involving C-H, N-H and O-H chemical bonds. Mid infrared (MIR) represents the spectrum of the absorption of all the chemical bonds having an infrared activity between 4000 and 400 cm^{-1}. The acyl-chain is mainly responsible for the absorption observed between 3000 and 2800 cm^{-1}, whereas the peptidic bond C-NH is mainly responsible of the absorption occurring between 1700 and 1500 cm^{-1}. Most of the absorption bands in the mid infrared region, but not in the near infrared region, have been identified and attributed to chemical groups. The absorbance of C-O (~ 1175 cm^{-1}) and C=O (~ 1750 cm-1) of the triacylglycerols ester bond and the acyl chain C-H (3000-2800 cm^{-1}) are commonly used to determine fat. The infrared bands appearing in the 3000-2800 cm-1 region are particularly useful because they are sensitive to the conformation and the packing of the phospholipid acyl chains (Casal and Mantsch, 1984). For example, the phase transition of phospholipids (sol to gel state transition) can be followed by mid infrared spectroscopy: increasing temperature results in a shift of the bands associated with C-H (~ 2850, 2880, 2935 and 2960 cm^{-1}) and carbonyl stretching mode of the phospholipids.

Water is a very strong infrared absorber with prominent bands centered at 3360 cm^{-1} (H-O stretching band), at 2130 cm^{-1} (water association band) and at 1640 cm^{-1} (the H-O-H bending vibration). Infrared spectroscopy can be used with proteins in aqueous solution. The subtraction of a large H_2O band from a large absorbance spectrum of protein in H_2O to get a small spectrum of protein was difficult considering older dispersive infrared spectrometers (Chittur, 1999). Precise subtractions of the H_2O band are now possible because of the frequency precision achievable with Fourier transform IR spectrometer. In the other hand, water exhibits a strong absorption band overlapping amide I and II protein bands. Due to the high absorbency at about 1640 cm^{-1} in the amide I and II region and to comply with the Beer-Lambert law, the path length of the cuvette has to be in the 10 μm range. The development of the attenuated total reflectance (ATR) device allows the sampling problems (water, opaque and viscous samples) encountered when collecting spectra from food products to be overcome.

Fluorescence spectroscopy

Absorption of light by a fluorescent molecule causes the excitation of an electron moving from a ground state to an excited state. After the electron has been excited, it relaxes rapidly from the higher vibrational states to the lowest vibrational state of the excited electronic state; after which, the excited state may decay to the ground state by the emission of a photon (fluorescence). Due to energy losses, the emitted fluorescence photon is always less energetic than the absorbed photon (Genot *et al.*, 1992). Fluorescence spectroscopy offers several inherent advantages for the characterizations of molecular interactions and reactions. Firstly, it is 100-1000 times more sensitive than other spectrophotometric techniques. Secondly, fluorescent compounds are extremely sensitive to their environment. For example, tryptophan residues that are buried in the hydrophobic interior of a protein have different fluorescent properties than residues that are on a hydrophilic surface. This environmental sensitivity enables to characterize conformational changes such as those attributable to the thermal, solvent or surface denaturation of proteins, as well as the interactions of proteins with other food components. Thirdly, most fluorescence methods are relatively rapid. If absorbance of the sample is less than 0.1, the intensity of the emitted light is proportional to fluorophore concentration and excitation and emission spectra are accurately recorded by classical right-angle fluorescence device. When the absorbance of the sample exceeds 0.1, emission and excitation spectra are both reduced and excitation spectra are distorted. To avoid these problems, a dilution of samples is currently performed so that their total absorbance would be less than 0.1. However, the results obtained on diluted solutions of food samples can not be extrapolated to native concentrated samples since the organisation of the food matrix is lost. To avoid these problems, the method of front-face fluorescence spectroscopy can be used to record the spectra of real food products. Fluorescence spectroscopy relies on fluorescence probes or fluorophores. Fluorophores can be divided broadly into two main classes: intrinsic and extrinsic. Intrinsic fluorophores are those which occur naturally. They include the aromatic amino-acids - tryptophan, tyrosine and phenylalanine in proteins, vitamin A and B_2, NADH derivatives of pyridoxal and chlorophyll, some nucleotides and numerous other compounds that can be found at low or very low concentration in food products.

As mentioned above, fluorescent molecules are extremely sensitive to their environment. For example, the emission of tryptophan is highly sensitive to its local environment, and is thus often used as a reporter group for protein conformational changes (Lakowicz, 1983). Spectral shifts have been observed as a result of several phenomena, such as tertiary structure change, binding of ligands and protein-protein association. In addition, the emission maxima of proteins reflect the average exposure of their tryptophan residues to the aqueous phase. Tryptophan is an essential amino acid with a well-characterised fluorescence response. The fluorescence properties of tryptophan, along with its chromophore moiety-indol ring, have been studied extensively due to its use as a standard optical probe for protein structure and dynamics and for the characterization of cheeses (Herbert *et al.*, 2000). Vitamin A located in the milk fat globule is another interesting probe. Solid fat content is an important quality control parameter in the edible fats and vitamin A located in the core of milk fat globules is a good fluorescent probe for the determination of this parameter. It has been shown that the shape of the vitamin A excitation spectrum is correlated with the physical state of the triglycerides in the fat globule (Dufour *et al.*, 1998). Indeed, F322nm/F295nm ratio is modified by increasing the temperature (Figure 1), allowing the melting point (28 °C) of the triglycerides to be derived. It was shown that the determination of the solid fat content from the fluorescence spectral data and from differential scanning calorimetry gave similar results. The two sets of data were highly correlated since the coefficient correlation was equal to 0.99. This result shows that the shape of the vitamin A fluorescence spectrum is sensitive to the physical state of the triglycerides in the fat globules of an emulsion.

Evaluation of the spectral data

The chemical information contained in the spectra are hidden in the band position, the band intensities and the band widths. Whereas the band positions give information about the appearance and the structure of certain chemical compounds in a mixture, the intensities of the bands are related to the yield of these compounds via the Beer Lambert law. The easiest way to determine

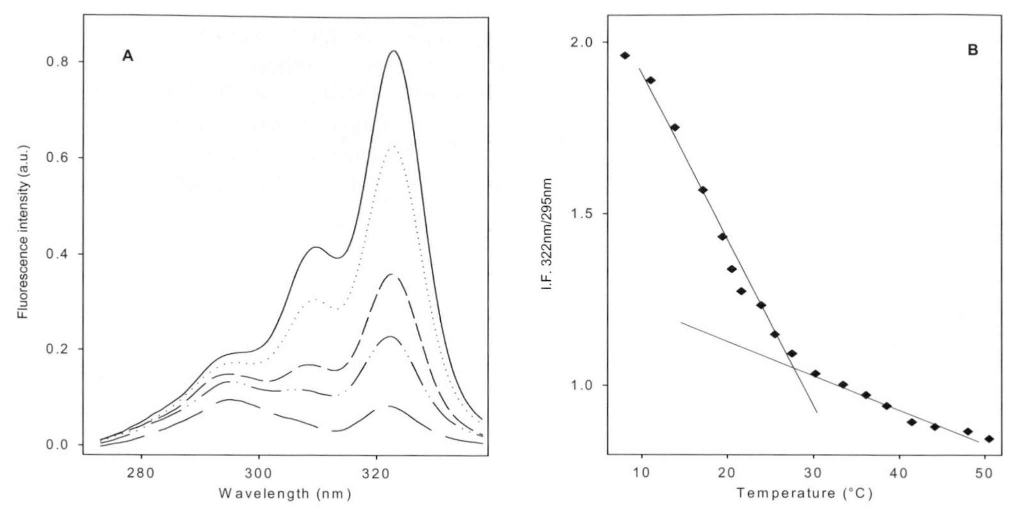

Figure 1. Fluorescence properties of an emulsion at different temperatures. (A) Excitation spectra of vitamin A of a milk fat-in- water emulsion (melting point of fats =28 °C) recorded at (——) 8.3 °C, (....) 13.8 °C, (–..–) 30.2°C, and (– –) 49.1 °C. (B) Changes in the ratio F $_{322\ nm/295\ nm}$ with temperature.

the content of a chemical compound is to measure the change of the intensity of a well-resolved band that clearly belongs to this compound. This is possible for pure component. But food products retain numerous components giving complex spectra with overlapping bands. The most valuable way to evaluate the data is to use chemometric tools to extract quantitative, qualitative or structural information from the spectra. These methods encompass descriptive techniques such as the Principal Component Analysis (PCA), Canonical Correlation Analysis (CCA), Common Components and Specific Weights Analysis (CCSWA) and predictive techniques such as Factorial Discriminant Analysis (FDA) and Principal Component Regression (PCR) (Figure 2).

Descriptive techniques

The most frequently used method with multivariate data is PCA. The purpose of this technique is to obtain an overview of all the information in the data set. In PCA, orthogonal directions in variable space describing the variation are found. In this way a new set of fewer coordinate axes called principal components (PCs) is generated. PCA allows the use of the entire spectrum for the quantitative analysis and it provides a synthetic description of large data sets with a minimum loss of information (Joliffe, 1986). The scaling factors associated with the PCs can be directly related to properties of the investigated systems, i.e., concentrations of components, protein-protein and protein-lipid interactions and can be "interpreted" as a spectrum. CCA makes it possible to identify and quantify the associations between two sets of variables recorded on the same samples. The goal is to find the maximal correlation between a chosen linear combination of the first set of variables and a chosen linear combination of the second set of variables. Maximally correlated pairs of variables may then be identified with linear combinations and are called the canonical variables (Saporta, 1990).

Predictive techniques

The aim of these techniques is to predict the membership of an individual to a qualitative group defined as a preliminary. The use of infrared and front-face fluorescence spectroscopy for quality

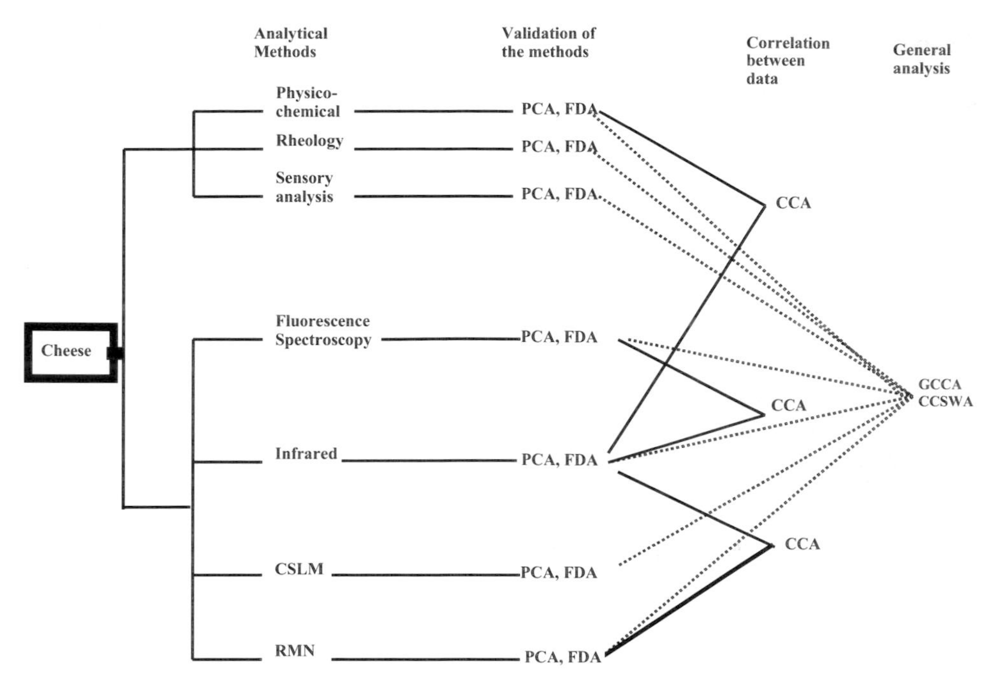

Figure 2. Evaluation of data sets recorded on food products using chemometric tools. CSLM: Confocal Scanning Laser Microscopy, RMN : Nuclear Magnetic Resonance, PCA: Principal Component Analysis, FDA: Factorial Discriminant Analysis, CCA: Canonical Correlation Analysis, GCCA: Generalized Canonical Correlation Analysis, CCSWA: Common Components and Specific Weight Analysis

control in the food industry require the development of prediction models for samples that have to be assigned to qualitative groups such as variates, species or quality classes (Robert *et al.*, 1996).

Such predictions can be handled by discriminant techniques such as FDA which involve building a model from a calibration set and choosing a rule to assign the calibration and validation samples to the qualitative groups. In a first step, a step-wise discriminant analysis is performed to select the PCs the most relevant for the discrimination of variables following the qualitative groups initially defined. From the selected variables, FDA assesses new synthetic variables which are linear combinations of the variables of origins called "discriminant factors" and which are not correlated and allow the best separation of the qualitative groups. The individuals can be reallocated within the various groups. For each individual, the distance separating it from the various centres of gravity from each group is calculated. So, the individual is affected to the one the distance is the shortest. The comparison of the assignment group to the real group is an indicator of the quality of the discrimination. The PCR method is a multiple regression applied from the PCs rather than from the raw data. Indeed, when the variables of origin are numerous and are strongly correlated between them, it is preferable to carry out the prediction on the PCs. All the PCs are not introduced in the regression model: the last components with small variances are discarded.

Characterization of cheeses and meat

Discrimination of cheeses from infrared and fluorescence spectra

Soft cheeses exhibit a wide range of texture resulting from different manufacturing processes. Herbert *et al.*, (2000) have investigated intrinsic fluorescence of cheeses in order to discriminate between 8 different marketed soft cheeses: four different mesophilic-processed cheeses (M1 and M2 at two ripening times (M1y, M1o, M2y, M2o), M3 and M4), one thermophilic-processed cheese (TH) and an ultrafiltered-processed cheese (UF). Protein tryptophan emission spectra and vitamin A excitation spectra were recorded directly on cheese samples using front-face fluorescence spectroscopy. The tryptophan emission spectra presented a maximum located at 322 nm, which location varied slightly from one cheese to another. Similarly, vitamin A spectra showed a maximum located at 322 nm and two shoulders at 295 and 305 nm. As well as for tryptophan spectra, vitamin A spectra of cheeses exhibited slight differences. The ability to discriminate the 8 groups from their fluorescence spectra was investigated using FDA. The map defined by the discriminant factors 1 and 2 showed that the tryptophan emission spectra were well suited for the soft cheese discrimination (Figure 3). Considering discriminant factor 1, M2y and M2o cheeses were observed on the far right, whereas M1o, M1y, UF and TH were located on the far left. The M3 and M4 red smear cheeses exhibited coordinates close to origin. The cluster formed by M1y, M1o, UF and TH included the cheeses with the highest fat yield (between 32.5 and 36 $g.100g^{-1}$ of cheese), whereas M2y and M2o exhibited about 23 $g.100g^{-1}$ of cheese (Herbert, 1999). In addition, M1y and M2y cheeses had negative scores according to discriminant factor 2, whereas M1o and M2o had positive scores. Fluorescence allowed to discriminate the changes of cheese structure during ripening. From the similarity map, it appears also that the fluorescent properties of a ripened cheese are close to that one of this cheese at a young stage. The tryptophan and the vitamin A fluorescence spectra of a cheese are fingerprints allowing its identification. These results also show that the different textures among investigated cheeses are related to the unique molecular structure of each cheese resulting from the process and the ripening time.

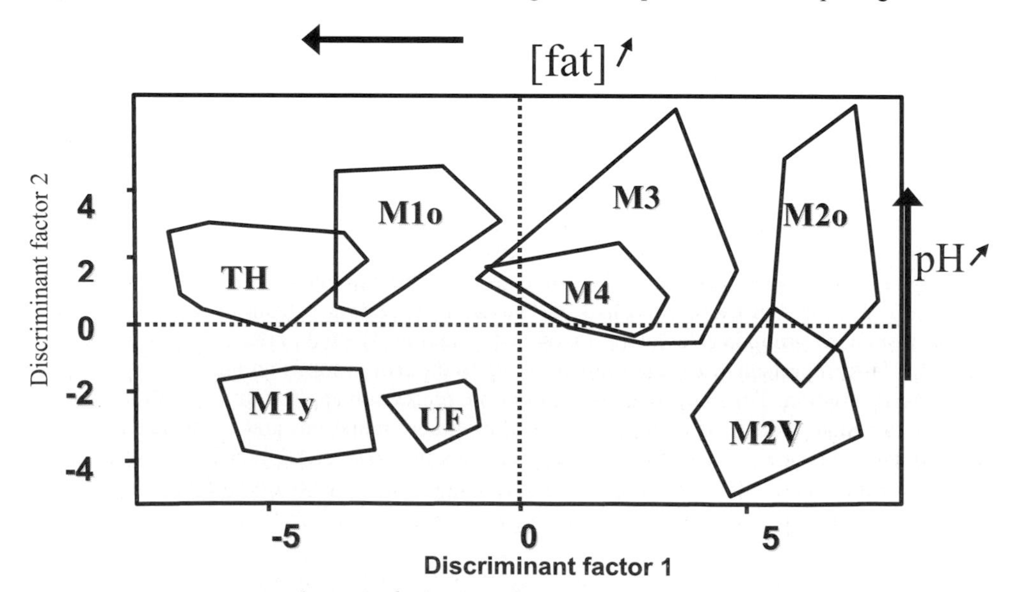

Figure 3. Discriminant analysis similarity map determined by discriminant functions 1 and 2 for the tryptophan spectra of the eight soft cheeses.
Mesophilic- (M1 to M4), thermophilic- (TH) and ultrafiltered- (UF) processed cheeses

In another study (Mazerolles *et al.*, 2001), 16 experimental semi-hard cheeses varying in moisture (42.1 to 49.8 $g.100g^{-1}$), protein (20.2 to 25.9 $g.100g^{-1}$) and fat (23.7 to 31.1 $g.100g^{-1}$) contents were studied by fluorescence and mid-infrared spectroscopies at 1, 21, 51 and 81 days of ripening. PCA was applied to the whole library of Amide I and II normalised infrared spectra. The results showed that the 4 ripening groups were well ordered according to ripening time on the factorial map defined by the principal components 3 and 6 (7% and 0.17% of the total inertia, respectively). The spectral pattern 3 showed an opposition between bands around 1650 cm^{-1} and 1620 cm^{-1}. These bands could be explained by the global modifications affecting the protein network during ripening. At the same time, opposition was observed between bands at 1580 cm^{-1} and bands below 1560 cm^{-1}. According to Byler and Farell (1989), the decrease in the intensity of the band at 1580 cm^{-1} could be attributed to a modification of the nature of the anion interacting with the carboxylic groups of aspartic and glutamic acids side chains in caseins. In addition, an opposition between bands at 1580 cm^{-1} and 1555 cm^{-1} was observed on spectral pattern 6, confirming that exchanges previously described happened during ripening.

Discrimination of muscles and meat from fluorescence spectra

Intrinsic fluorescence properties of composites of beef meat pieces have been evaluated by Skerjvold *et al.* (2003). Excitation and emission spectra were recorded by front-face fluorescence spectroscopy of fat, connective tissue and of meat samples to obtain specific univariate combinations of excitation and emission wavelengths. The combinations of excitation and emission wavelengths showing best ability to distinguish the three components are 290/332 nm for myofibrils, 322/440 nm or (322/405) nm for fat and 380/440 nm for connective tissue. The authors also reported that sample orientation and mincing meat samples affected fluorescence amplitudes but not the general shape of the spectra. Therefore, choice of characteristic wavelength combinations was not affected. The possibility of classification of muscle type by autofluorescence spectroscopy is proposed based on the different emission spectra (between 332 and 400 nm) obtained for *psoas major* and *semimembranosus* muscles. Using the above mentioned excitation/emission wavelengths, images were captured of tissue sample composed by 3 pieces of meat, 3 pieces of fat and 3 pieces of connective tissue. Image analysis showed that simple gray level thresholding in two dimensions based on two of the fluorescent images was sufficient in order to successfully segment the different tissue components (Skerjvold *et al.*, 2003).

Dufour and Frencia (2001) have reported that intrinsic fluorescence properties of meat can be used to develop a rapid method for the measurement of meat tenderness. In this study, emission spectra were recorded by front-face fluorescence spectroscopy between 305 and 400 nm (excitation: 290 nm) and 425 and 580 nm (excitation: 317 nm) directly on different muscles and at different ageing times. The possibility of classification of muscle type by autofluorescence spectroscopy is proposed based on the tryptophan emission spectra (305 - 400 nm range) obtained for *tensor fasciae lata*, *longissimus thoracis*, *semi-tendinosus*, *infraspinatus* and *triceps brachii*, 5 muscles exhibiting different collagen yields and different tenderness. In the same paper, the authors investigated at two ageing times (2 and 14 days *post-mortem*) the intrinsic fluorescence of *longissimus thoracis* (LT) and *infraspinatus* (IS), two muscles exhibiting different collagen contents. The emission spectra of protein tryptophan residues for the investigated muscle samples were recorded between 300-400 nm (excitation : 290 nm). Figure 4 shows the PCA similarity map defined by the principal components 1 and 2 for tryptophan spectra of LT and IS muscles.

The first two principal components took into account 67.9% and 15.5% of the total variance. The emission spectra of tryptophan residues of protein showed a good repeatability for a given sample and allowed to discriminate LT and IS muscles at the two ageing times. As previously shown for cheeses, the emission fluorescence spectrum of tryptophan residues of proteins is a fingerprint of

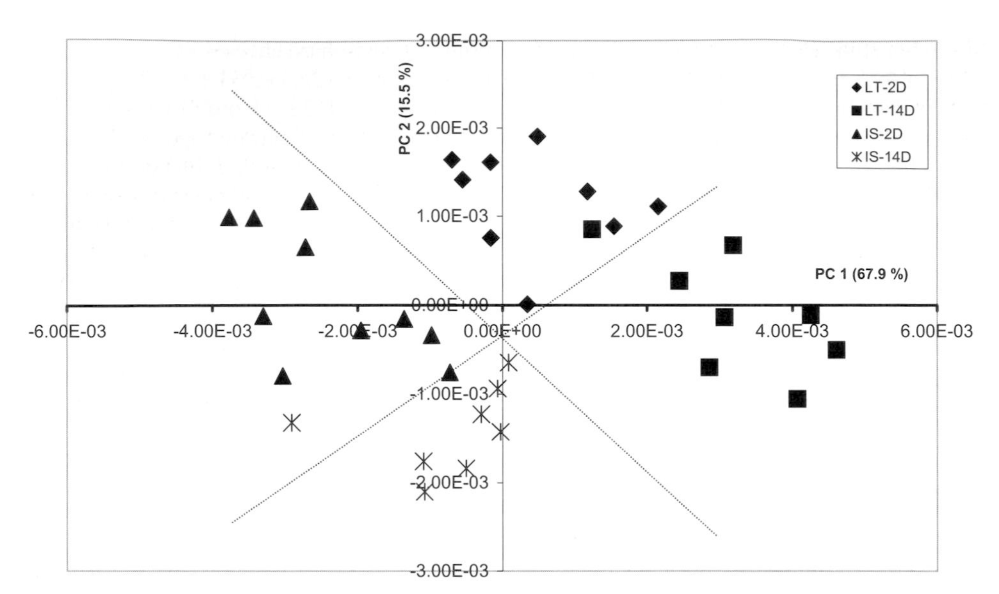

Figure 4. PCA similarity map defined by the principal components 1 and 2 for the tryptophan spectral data of longissimus thoracis (LT) and infraspinatus (IS) muscles at 2 (2D) and 14 days (14D) post-mortem.

the product resulting in this study from the fluorescence properties of the whole meat proteins (myofibrils, collagen, sarcoplasmic proteins) and of their changes during ageing time.

Joint analysis of fluorescence and infrared spectroscopic data

Coupling fluorescence and infrared techniques should contribute to a better understanding of structural modifications at a molecular level for a given product. In addition, a better discrimination could be observed with CCA applied to fluorescence and infrared data than with PCA applied to fluorescence or infrared data alone. The joint analysis of the data sets may allow to manage in a very efficient way the whole spectroscopic information collected, contributing to a better understanding of the main structural modifications affecting proteins and lipids in cheeses during ripening. Mazerolles *et al.*, (2001) applied CCA to the fluorescence and infrared data sets recorded on semi-hard cheeses at different stages of ripening (see part 2 of section "Discrimination of cheeses from infrared and fluorescence spectra" for investigated cheeses). The results show that the first two pairs of canonical variates are strongly correlated with the squared canonical correlation coefficients equal to 0.87 and 0.58, respectively. The similarity maps for fluorescence and infrared spectral data are similar and discriminated very clearly the four groups of samples. From these results, the authors conclude that the characteristic of the ripening stages of cheeses can be obtained by two rapid and non-invasive techniques.

In another paper, Dufour *et al.*, (2000) have investigated the vitamin A fluorescence and mid-infrared (2800-3000 cm^{-1} region) spectra of 16 different semi-hard cheeses at 1, 21, 51 and 81 days of ripening. The results of PCA performed on the data sets allow a rough discrimination of cheeses as a function of the storage time. As the methyl and methylene regions of the mid infrared spectra and the fluorescence of vitamin A are related to the lipid structure in the fat globule, the differences observed in the spectra suggest a phase transition of triglycerides in fat globule during ripening. In order to confirm this hypothesis, CCA was applied on the fluorescence and mid-infrared data sets. The results of CCA performed on the two data sets show that the first two pairs of canonical variates are correlated with squared canonical correlation coefficient of 0.89 and 0.64, and that the similarity maps obtained for the infrared and fluorescence spectra are similar. Considering the mid-infrared region, the spectral pattern associated with the second canonical variate shows a shift towards lower wave numbers of the 2850 and 2920 cm^{-1} bands from 21 to 81 days of storage. This pattern is characteristic of the liquid-to-gel phase transition (Casal and Mantsch, 1984). Consequently, it has been proposed that the shift of the bands at 310 and 322 nm observed by the authors for the fluorescence spectral pattern associated with the second canonical variate is characteristic of the changes in lipid viscosity following partial crystallization. This result strongly indicates that the two spectroscopic methods make it possible to observe the structure of lipids in cheeses at a molecular level.

The results of spectroscopic investigations of cheeses show that the intrinsic fluorescence spectra and the mid-infrared spectra recorded directly on cheeses provide fingerprints that allow cheeses to be identified. But a spectrum is more than a fingerprint since the analysis of spectral collections by multi-dimensional statistical methods makes it possible to derive information on the structure of the cheese matrix, i.e., on the protein/protein and protein/lipids interactions and on the structure of micelles and triglycerides in fat globules.

Relation between texture and structures at molecular and microscopic levels

It is generally accepted that structure conditions affect both texture and flavour properties of cheese and meat. However this theoretically appealing approach has been limited so far by lack of information on the relation between structure and texture.

Considering the 8 marketed soft cheeses mentioned in section "Discrimination of cheeses from infrared and fluorescence spectra", the description of their texture sensorial characteristics has been achieved using a sensory profile including 18 items (Herbert, 1999). From this point, it has been attempted to relate the sensory perception of texture analysed at the rate of the panel assessment to some structural characteristics of the same product. In the scope of this research project, soft cheese structures have also been considered at molecular (fluorescence of tryptophan and vitamin A) and microscopic scales. By comparing the data, some gross conclusions may be drawn. As shown by fluorescence (Figure 3) and sensory analysis, TH cheese exhibited a molecular structure and a texture closer to M1o than M1y (Herbert *et al.*, 2000). In addition, the investigation of the protein network of TH and M1o cheeses by Confocal Laser Scanning Microscopy indicated similar smooth protein network microstructures (Herbert, 1999). On the other hand, the fluorescence images of M1y and M1o showed a different protein network, i.e., granular for M1y and smooth for M1o. As large data sets including fluorescence spectra, microscopic images, rheological data and sensory data have been collected on the same samples, multivariate analysis techniques such as CCA make it possible to provide a global measure of the links between the groups of variables and to extract relevant information on the relation between structure and texture of cheeses. Considering the tryptophan fluorescence spectra and the sensory data collected on the 8 soft cheeses, four pairs of canonical variates were assessed to describe the correlation between the two data sets. The authors report that the squared canonical coefficients for canonical variates 1 and 2 are 0.93 and 0.80, respectively. The two canonical variates were correlated significantly and showed that the structure at the molecular level is correlated to the sensory attributes. Considering the similarity map of the canonical variates 1 and 2 assessed from the fluorescence spectra, M1y, M2y and UF cheeses have negative scores according to canonical variate 1, whereas M1o, M2o, TH, M3 and M4 have positive scores. In fact CCA results indicate that the sensory attributes characterising M1y, M2y and UF are firmness, disintegration, pastiness, solubility, graininess, crumbliness, grains and surface state, whereas M1o, M2o, TH, M3 and M4 cheeses are well described by melting quality, springiness and oval holes.

Several others papers have been published on the correlation between spectroscopic data and sensory data for cheeses. In 1998, Sørenson and Jepsen reported that a good assessment of sensory attributes of semi-hard cheeses was obtained from NIR spectroscopy (1110-2490 nm) after 5, 7, 9 and 11 weeks of ripening using Partial Least Square (PLS). In addition, Lebecque *et al.*, (2001) report that high correlations are observed between texture sensory attributes, rheological data and fluorescence data recorded on Salers cheeses. Considering fluorescence of vitamin A and sensory profiles, the first two pairs of canonical variates have been found correlated with squared canonical correlation coefficients equal to 0.85 and 0.69.

Fluorescence spectroscopy has also been investigated as a tool for quantification of composites of meat (Wold *et al.*, 1999) and of meat properties such as lipid oxidation (Wold and Mielnik, 2000) and water holding capacity (Brøndum *et al.*, 2000). Intrinsic fluorescence properties of meat has also been evaluated by Dufour and Frencia (2001) with the aim to develop a rapid method for the measurement of meat tenderness. In this study, canonical correlation analysis was used to investigate the correlations between sensory (texture attributes) and rheology data, sensory and spectral data, and rheology and spectral data recorded on *longissimus thoracis*, *semi-tendinosus* and *triceps brachii* muscles at 2, 6 and 11 days *post-mortem*. Strong correlations were observed between the different data tables (Table 1). Considering sensory and spectral data, the canonical coefficient for the canonical variates 1 was 0.95. This result indicates that the texture attributes of meat may be derived from the fluorescence spectra of protein tryptophan residues. Considering the strong correlations between rheology and spectral data, the authors also suggest that fluorescence spectra of muscles retain information on the 2 components of tenderness: 1 - basic hardness related to the different contents in collagen varying with the type of muscle, and 2 - myofibril component related to the proteolysis during meat ageing. Similar correlations between rheology and spectral data have been found in other food products such as frankfurters (Allais *et al.*, 2004). From these results, it appears that fluorescence spectra recorded directly on meat allow prediction of its mechanical characteristics and texture.

Table 1. Canonical correlation coefficients (R) for the first canonical variates of CCA performed on the sensory, rheology and spectral data.

	R
Sensory analysis / Spectroscopy with optic fibre	0.93
Sensory analysis / Spectroscopy with front face device	0.95
Texturometer / Spectroscopy with front face device	0.95
Spectroscopy with front face device / with optic fibre	0.96

Conclusion

Simple and rapid spectroscopic methods such as front-face fluorescence and mid-infrared spectroscopies have a great potential for the investigation of the structure in meat and dairy products and for the development of rapid methods for the prediction of texture quality of food products.

References

Allais, I., C. Viaud, A. Pierre and E. Dufour, 2004. A rapid method based on front-face fluorescence spectroscopy for the monitoring of the texture of meat emulsions and frankfurters. Meat Sci. 67: p. 219-229.

Brøndum, J., L. Munck, P. Henckel, A. Karlsson, E. Tornberg and S.B. Engelsen, 2000. Prediction of water-holding capacity and composition of porcine meat by comparative spectroscopy. Meat Sci. 55: p. 177-185.

Byler, D.M. and H.M. Farell, 1989. Infrared spectroscopic evidence for calcium ion interaction with carboxylate groups of casein. J. Dairy Sci. 72: p. 1719-1723.

Casal, H.L. and H.H. Mantsch, 1984. Polymorphic phase behavior of phopholipid membranes studied by infrared spectroscopy. Biochim. Biophys. Acta 779: p. 382-401.

Chittur, K.K. 1999. FTIR and protein structure at interfaces. Bull. BMES 23: p. 3-9.

Dufour, E., C. Lopez, A. Riaublanc and N. Mouhous Riou, 1998. La spectroscopie de fluorescence frontale: une approche non invasive de la structure et des interactions entre les constituants des aliments. Agoral 10: p. 209-215.

Dufour, E., G. Mazerolles, M.F. Devaux, G. Duboz, M.H. Duployer and N. Mouhous Riou, 2000. Phase transition of triglycerides during semi-hard cheese ripening. J. Dairy Sci. 10: p. 81-93.

Dufour, E. and J.P. Frencia, 2001. Les spectres de fluorescence frontale - Une empreinte digitale de la viande. Viandes Prod. Carnés 22: p. 9-14.

Genot C., F. Tonetti, T. Montenay-Garestier, D. Marion and R. Drapon, 1992. Front-face fluorescence applied to structural studies of proteins and lipid-protein interactions of visco-elastic food products. 1- designation of front-face adaptor and validity of front-face fluorescence measurements. Sci. Aliments 12: p. 199-212.

Givens, D.I., J.L. de Boever and E.R. Deaville, 1997. The principles, practices and some future applications of near infrared reflectance spectroscopy for predicting the nutritive value of foods for animals and humans. Nutr. Res. Rev. 10: p. 83-114.

Herbert, S., 1999. Caractérisation de la structure moléculaire et microscopique de fromages à pâte molle. Analyse multivariée des données structurales en relation avec la texture. Ph.D. Thesis. Ecole Doctorale Chimie Biologie de l'Université de Nantes, France, 118 p.

Herbert, S., N. Mouhous Riou, M.F. Devaux, A. Riaublanc, B. Bouchet, J.D. Gallant and E. Dufour, 2000. Monitoring the identity and the structure of soft cheeses by fluorescence spectroscopy. Le Lait, 80: p. 621-634.

Isaksson, T., B.N. Nilsen, G. Tøgersen, R.P. Hammond and Hildrum K.I., 1996. On-line, proximate analysis of ground beef directly at a meat grinder outlet. Meat Sci. 43: p. 245-253.

Isaksson T., L.P. Swendsen, R. Taylor, S.O. Fjæra and P.O. Skjervold, 2001. Non-destructive texture measurements of farmed Atlantic salmon, using Near Infrared Reflectance spectroscopy. J. Sci. Food Agric. 82: p. 53-60.

Jollife, I.T. (editor) 1986. Principal Component Analysis. Springer, New York, USA, 271 pp.

Lakowicz, J. R., 1983. Fluorophores. In: Principles of fluorescence spectroscopy, J. R. Lakowicz (editor), Pelenum Press, New York, USA, p. 63-93.

Lebecque, A., A. Laguet, M.F. Devaux and E. Dufour, 2001. Delineation of the texture of Salers cheese by sensory analysis and physical methods. Le Lait, 81: p. 609-623.

Mazerolles, G., M.F. Devaux, G. Duboz, M.H. Duployer, N. Mouhous Riou and E. Dufour, 2001. Infrared and fluorescence spectroscopy for monitoring protein structure and interaction changes during cheese ripening. Le Lait 81: p. 509-527.

Robert, P., M.F. Devaux and D. Bertrand, 1996. Beyond prediction: extracting relevant information from near infrared spectra. J. Near Infrared spectroscopy, 4: 75-84

Saporta, G. (editor) 1990. Probabilités - Analyse des données et statistique, Technip Ed, Paris, France, 493 p.

Skjervold, P.-O., J.P. Wold, R.G. Taylor, Ph. Berge, S. Abouelkaram, J. Culioli and É. Dufour, 2003. Development of intrinsic fluorescence multispectral imagery specific for fat, connective tissue and myofibers in meat. J. Food Sci. 68: p. 1161-1168.

Sørenson, L.K. and R. Jepsen, 1998. Assessment of sensory properties of cheese by near-infrared spectroscopy. Int. Dairy J. 8: p. 863-871.
Wold, J.P., F. Lundby and B. Egelandsdal, 1999. Quantification of connective tissue (Hydroxyproline) in ground beef by autofluoresence spectroscopy. J. Food Sci. 64: 377-385
Wold, J.P. and M. Mielnik, 2000. Non-destructive assessment of lipid oxidation in minced poultry meat by autofluoresence spectroscopy. J.Food Sci. 65: p. 87-95.

The influence of nutrition, genotype and stage of lactation on milk casein composition

D.G. Barber[1], A.V. Houlihan[2], F.C. Lynch [1] and D.P. Poppi[1]
[1]Schools of Animal Studies and Veterinary Science, University of Queensland, St Lucia. Qld. 4072, Australia, [2]Emerging Technologies, Queensland Department of Primary Industries and Fisheries, 19 Hercules St, Hamilton. Qld. 4007, Australia.
**Correspondence: David.Barber@dpi.qld.gov.au*

Summary

Variation in milk protein content and the casein composition is observed in many production systems. This is associated with changes in the level of nutrition, with stage of lactation and with cows having different milk protein variants, all of which affect cheese yield and properties to variable extents. Casein composition is difficult to change experimentally but, in those cases where it has altered, this has been associated with cows at sub-optimal levels of nutrition, a situation which occurs throughout the annual production cycle of cows at pasture. Changes of up to 25% (2.8 to 3.5% (m/v) range) are commonly seen in milk protein content and milk casein fractions can vary within a 10% (casein fraction/total casein) range. Milk protein variants have a profound influence on casein concentration and processability, however the influence on casein composition is not clearly defined. The interaction of milk protein variants, stage of lactation and level of nutrition has not been adequately studied.

Keywords: milk protein, casein composition, nutrition, genotype, stage of lactation

Introduction

Milk composition is an important issue in dairy production systems, particularly the seasonal variation seen in pasture based systems, eg the large variation seen in the north Australian dairy industry. This fluctuation often results in a decline in milk protein concentration below levels set by the processing factories and the standard (3.0 % (m/m) crude protein (approximately 2.95 % true protein (m/v)) set by Food Standards Australia New Zealand (FSANZ). The decline in milk protein concentration is also associated with the apparent decline in cheese yield, quality and functionality observed in factories in Queensland, Australia.

The lowest concentrations of milk protein are predominantly seen in the summer months (December and January) in northern Australia. However, the decline in milk protein concentration begins in most cases during the spring months (September and October). Similar seasonal trends in milk protein concentration have been reported in Canada (Sargeant *et al.,* 1998), Ireland (Murphy and O'Mara, 1993; Mehra *et al.,* 1999), USA (Bruhn and Franke, 1976) and New Zealand (Auldist *et al.,* 1998). On-farm monitoring in northern Australia suggests that one of the main factors is the effect of the pasture transition period from temperate to tropical pasture systems on milk protein synthesis (Barber *et al.,* unpublished data). The nutritive value of tropical pastures (C4) is of lower quality compared to temperate pastures (C3) and concentrates. It has been well established in the literature that reduced forage quality tends to result in lower levels of milk protein (Murphy and O'Mara, 1993; Beever *et al.,* 2001; Erasmus *et al.,* 2001). Hence it may be suggested that one of the major causes of the decline in milk protein concentration in northern Australia is nutrition, with nutrition also being identified as a major factor influencing milk protein in other dairy regions of the world (DePeters and Cant, 1992; Murphy and O'Mara, 1993). Other important factors associated with the seasonal variation in milk protein concentration include photoperiod (Coulon, 1994) and environmental temperature (DePeters and Cant, 1992), however the relative importance of these factors is primarily dependant on the latitude of a specific dairy region.

Another aspect of relative importance to the dairy production chain, particularly cheese production, is milk protein composition and the seasonal variation seen in the casein fractions, alpha S1 (α_{S1}-), alpha S2 (α_{S2}-), beta (β-) and kappa (κ-) casein. The most important factor in determining cheese yield is the compositional quality of the milk, in particular the casein and fat components (Barbano and Sherbon, 1984; Ng Kwai Hang and Grosclaude, 1992; Banks, 2000). Of these components, casein is considered the most important. While fat contributes its own weight to cheese yield, casein contributes its weight together with the weight of the associated moisture, milk salts and minerals (Lawrence, 1991). Therefore, to increase cheese yield, the casein content (in terms of the casein: true protein ratio) and the efficiency of fat retention in the curd need to be increased. The composition and structure of the casein micelles are also important factors in determining cheese yield and functionality, as these influence the renneting properties of milk (eg rennet coagulation time, curd firmness, syneresis or moisture expulsion by the curd) which, in turn, affect the extent of losses of protein and fat in the whey (Puhan and Jakob, 1994).

Data from an on-farm monitoring program (Barber *et al.* unpublished data) suggests that the proportions of the casein fractions in total casein of milk are altered at specific times of the year associated with periods of pasture transition and declining pasture quality, thus resulting in an overall lower plane of nutrition for the animal. The main period of seasonal variation in the proportions of the casein fractions is seen during the summer-autumn months (January to May), which coincides with the period of decline in Mozzarella cheese yield and functionality in northern Australia. This suggests that casein composition is under the influence of nutritional changes in the animal and can be manipulated through nutrition to improve the processing efficiency and functionality of dairy products.

The alteration of milk protein composition is linked with the processability of milk through a direct effect on cheese yield and functionality. Pasture based systems vary in nutrient supply which can be ameliorated to some extent by supplements. The transition of C3, C4 pastures, silage and grain throughout the year results in a changing nutrient supply and nutrient balance with poorly defined effects on milk composition and processability. The underlying effects of stage of lactation and milk protein genetic variant also contribute to variations in casein composition, however their influence is generally counteracted by the year round calving patterns and mixed genotypes seen within herds in northern Australia. The extent to which they affect casein composition is also poorly defined, particularly the effect of stage of lactation. Kefford *et al.* (1995), Auldist *et al.* (1996) and Coulon *et al.* (1998) clearly showed that stage of lactation affects cheese yield and/or properties, which in some cases was linked to protein and casein content and also to diet quality. However, in some cases (Auldist *et al.*, 1996, Coulon *et al.*, 1998), late lactation milk had detrimental effects whilst Kefford *et al.* (1995) showed positive effects.

This review will outline the areas associated with the nutritional manipulation of milk protein composition, in particular the manipulation of the casein fractions. Reviews by (DePeters and Cant, 1992; Murphy and O'Mara, 1993) comprehensively outline the effect of nutrition on milk protein concentration. The underlying effects of genotype and stage of lactation on the casein concentration and composition will also be discussed.

Milk protein composition and synthesis

Milk protein composition

True protein in milk consists of 2 main protein types, the casein (ca. 80%) and whey proteins. The casein proteins are the major component of milk true protein and are comprised of alpha S1 (α_{S1}-), alpha S2 (α_{S2}-), beta (β-), kappa (κ-) and gamma (γ-) caseins. γ-Casein is a product of the proteolytic degradation of β-casein. Casein exists in the form of micelles (Rollema, 1992; Swaisgood, 1992), which are spherical complexes ranging in size from 50 - 500 nm (average 120 nm) in diameter (Fox and McSweeney, 1998). Casein micelles contain the individual casein components, α_{S1}-, α_{S2}-, β- and κ-casein, in a structure which is held together by colloidal calcium phosphate, with other salts such as magnesium, sodium, potassium and citrate also present (Rollema, 1992). Casein micelles are highly hydrated, binding approximately 2.0 g H_2O per gram of protein (Fox and McSweeney, 1998).

The relative proportions of the individual caseins in the micelles vary with micelle size. α_S-Casein and β-casein decrease with decreasing micelle size, while κ-casein increases (Donnelly *et al.*, 1983; McGann *et al.*, 1980). In studies on micelles of different sizes, it was shown that β-casein was located predominantly in the interior while the surface contained equimolar amounts of α_S- and κ-caseins (Dalgleish *et al.*, 1989). The relative proportion of glycosylated κ-casein was found to be inversely related to micelle size (Zbikowska *et al.*, 1992).

The individual caseins are phosphorylated to varying extents, with α_{S2}-caseins showing the greatest variability, containing 10 - 13 phosphate groups (Fox and McSweeney, 1998). Calcium binds to the caseins through these phosphate groups stabilising the micelle structure through the formation of calcium phosphate bridges. κ-Casein stabilises the highly phosphorylated calcium-sensitive caseins, α_{S1}-, α_{S2}- and β-, in the micelle structure (Fox and McSweeney, 1998).

κ-Casein is the only casein component that is glycosylated and there is extensive heterogeneity in this fraction, although the non-glycosylated form is the major form (Swaisgood, 1992). This variability in glycosylation results in 9 - 10 forms of κ-casein (Fox and McSweeney, 1998).

The main whey proteins include β- lactoglobulins and α- lactalbumin, with other proteins such as the immunoglobulins and serum albumin that are transferred from the periphery in a preformed state (DePeters and Cant, 1992). Whey proteins are not phosphorylated.

The non-protein N fraction of milk contributes about 2-6% of N to the total N content of milk (DePeters and Cant, 1992; Kaufmann and Hagemeister, 1987). The major component of the NPN fraction is urea derived by diffusion from plasma.

Milk protein synthesis and secretion

Milk proteins are synthesised within the epithelial cell of the lactating mammary gland from blood metabolites and secreted into the alveolus for incorporation with other milk constituents. Following the synthesis of the individual proteins within the endoplasmic reticulum in the mammary gland, the casein fractions are transported to the Golgi apparatus where they undergo phosphorylation (Wilde *et al.*, 1995). Casein phosphorylation is required on multiple sites to allow for Ca^{2+} binding and the aggregation of the casein to form micelles (Wilde *et al.*, 1995). The micelles are then packed in secretory vesicles with lactose molecules and passed onto the apical plasma membrane where they are secreted into the lumen of the alveolus.

The genes encoding the 4 casein proteins are contained on a short genomic DNA fragment in the order, α_{S1}-, β-, α_{S2}- and κ-, and are found on chromosome 6. The expression of these genes is a result of complex hormonal regulation at both the transcriptional and post-transcriptional level by prolactin and glucocorticoids (Martin and Grosclaude, 1993). The whey proteins, α-lactalbumin and β-lactoglobulin, are located on chromosomes 5 and 11 respectively (Martin and Grosclaude, 1993). The relative proportions of these individual proteins, particularly the casein fractions, are likely to be under the control of a number of factors. These include the availability of amino acids and other nutrients as substrates for protein synthesis, hormone concentrations, initiation rate of peptide chain formation, rate of chain elongation and post-translational changes to the protein structure prior to casein micelle formation. Recently, Toerien *et al.* (2004) have shown that glucose stimulates milk protein synthesis through its effect on the energy status of the cell whereas leucine increases ribosomal RNA and histidine increases the rate of protein translation most probably by increasing initiation rate. These studies at the level of individual casein proteins have not been done.

It is generally assumed that the proportions of the casein fractions tend to be relatively constant (Beever *et al.,* 2001; Davies and Law, 1980; Mariani, 1975). There are however many reported cases of differences in the proportions of casein components due to the influence of genetic polymorphisms (Auldist *et al.,* 1997; Creamer and Harris, 1997; FitzGerald and Hill, 1997; Horne *et al.,* 1997; Johnson *et al.,* 2000; Kennelly and Glimm, 1998; Lodes *et al.,* 1997; Macheboeuf *et al.,* 1993; Ng-Kwai-Hang, 1997; Ostersen *et al.,* 1997; Puhan, 1997).

Non-nutritional factors affecting milk protein composition

Effect of milk protein genetic variants on milk protein composition

Genetic variants of milk proteins are an important issue with respect to the composition and technological properties of milk. A number of milk protein polymorphisms exist naturally in animals and have been shown to have an influence on the overall composition of milk. Generally, relationships between a number of single protein variants and the total protein content of milk have been demonstrated, in particular the positive influence of a genetic variant of β-lactoglobulin has been extensively noted (Auldist *et al.,* 1997; Ng-Kwai-Hang, 1997; Puhan, 1997). The relationship refers to the increase in total protein content with the β-Lg AA variant compared to β-Lg BB, which relates to an increase in the whey protein content and often a reduction in casein content (Auldist *et al.,* 1997). However, β-Lg BB milk has been shown to have a higher casein content than β-Lg AA (Coulon *et al.,* 1998; Puhan, 1997), which has been related to an improvement in cheese manufacturing properties (Auldist *et al.,* 1997).

Investigations of milk protein polymorphisms have revealed a total of 30 genetic variants for the four casein and two whey proteins seen in bovine milk. There are five (A, B, C, D, E) variants for α_{S1}-casein, four variants (A, B, C, D) for α_{S2}-casein, seven variants (A^1, A^2, A^3, B, C, D, E) for β-casein and four variants (A, B, C, D) for κ-casein. The whey protein, β-lactoglobulin, has seven variants (A, B, C, D, E, F, G) and α-lactalbumin has three (A, B, C) variants (Ng-Kwai-Hang, 1997). A summary of the relationship with milk yield and composition and the relationship with casein composition is outlined in Table 1.

Overall, the effect of α_{s1}-casein variant on total milk protein concentration follows the trend: BC>AB>BB and CC>AB>BB (Aleandri *et al.,* 1990; Puhan, 1997). For casein content, it tends to act in order of BC>BB>AB (Puhan, 1997). Other studies have reported no effect of α_{s1}-casein on total protein or casein concentrations (McLean *et al.,* 1984). The C variant is associated with a

Table 1. Genetic polymorphisms of the milk protein fractions of bovine milk and their relationship to aspects of milk production and composition. The relationships between milk production and α_{S2}-casein and α-lactalbumin are not included, as they are generally monomorphic in dairy cattle breeds (Creamer and Harris, 1997; Ng-Kwai-Hang, 1997; Puhan, 1997).

Protein fraction	Common variants	Phenotypes	Relationship to:			
			Milk yield	Total protein	Casein	Whey protein
α_{S1}-Casein	A, B, C	AB, BB, BC	BB>AB>BC	BC>BB>AB	BC>BB>AB	AB>BB>BC
β-Casein	A^1, A^2, A^3, B	A^1A^1 A^1A^2 A^1A^3 A^1B A^2A^2 A^2A^3 A^2B	$A^1A^3>A^2A^3>A^2A^2>A^2B>A^1A^2>A^1A^1>A^1B$	$A^1B>A^2B=A^1A^1>A^2A^2$	$A^1B>A^2B>A^1A^1>A^2A^2$	$A^1B>A^2B=A^1A^1=A^2A^2$
κ-Casein	A, B	AA AB BB	AB>AA>BB	BB>AB=AA	BB>AB>AA	AA>AB>BB
β-Lactoglobulin	A, B	AA AB BB	AA>BB>AB	AA=AB>BB	BB>AB>AA	AA>AB>BB

higher concentration and proportion of α_{s1}-casein and a lower concentration and proportion of κ-casein in the milk when compared with the B variant (McLean *et al.,* 1984).

The effect of β-casein variant on total protein content of milk appears to be $A^1A^1>A^2A^2$ (Lodes *et al.,* 1997; Lundén *et al.,* 1997; Puhan, 1997). In terms of variant effect on casein content, β-casein B would appear to be associated with the highest value, followed by A^1 and then A^2 (Puhan, 1997). However, in the study of Lodes and colleagues, the B variant was associated with the lowest content of casein and total protein (Lodes *et al.,* 1997). There are no differences between the $A^1$and A^2 alleles in terms of the concentrations or proportions of the various casein fractions (McLean *et al.,* 1984). The B variant had significantly lower α_{s1}-casein concentration and higher κ-casein concentration than the A variants (McLean *et al.,* 1984). In practical terms, however, it must be noted that the β-casein B variant is relatively rare among Western dairy herds (Lodes *et al.,* 1997).

Although there are four known genetic variants of κ-casein, almost all studies into the association of κ-casein with milk composition traits limit their scope to two variants - A and B (Ng-Kwai-Hang, 1997). The majority of researchers agree that κ-casein B is associated with a higher concentration of protein and casein in milk (Ng-Kwai-Hang, 1997). A few studies, however, have shown no significant effect of κ-casein variant on either protein or casein content (Bobe *et al.,* 1999; Lodes *et al.,* 1997; Lundén *et al.,* 1997). The different phenotypes of κ-casein affect total protein content in the order BB>AB>AA and casein content in the order BB>AB>AA (Puhan, 1997). B-variant κ-casein has been demonstrated to have a higher proportion of total casein present as κ-casein (Horne, *et al.* 1997; McLean. *et al.* 1984). This genotype favours the expression of κ-casein at the expense of other milk proteins, in particular of α_{s1}-casein (Bobe *et al.,* 1999; McLean *et al.,* 1984).

Most research concerning β-lactoglobulin genotype has focussed on its association with milk protein. Some studies have concluded that β-lactoglobulin phenotype (A or B) exerts little or no effect on total milk protein (Bobe *et al.,* 1999; McLean *et al.,* 1984). However β-Lactoglobulin AA has been associated with higher total protein than β-lactoglobulin BB (Ng-Kwai-Hang, 1997; Winkelman, 1997), due to higher yields of whey protein (McLean *et al.* 1984; Puhan, 1997), particularly of β-lactoglobulin (Bobe *et al.* 1999; McLean *et al.* 1984; Ng-Kwai-Hang, 1997). Casein content therefore tends to be lower (McLean *et al.,* 1984; Ng-Kwai-Hang, 1997; Puhan, 1997), particularly the concentration of α_{s1}- and β-casein (Bobe *et al.,* 1999).

The polymorphic variations in milk proteins have been associated with milk composition and production traits. However, the causal relationship between the two is not yet understood, and the link is one of correlation only (Ng-Kwai-Hang, 1997). Nevertheless, the possibility of using these genetic variations as markers for selection offers potential for research and application in industry. Milk protein genetic variants offer the possibility of genetically selecting for milk that not only yields more product, but also provides proteins that are more suitable for product manufacture. Distinction should also be made between genetic polymorphism and non-genetic polymorphism, of which the latter is due to post-transcriptional modifications such as different degrees of phosphorylation and/or glycosylation of the protein (Ng-Kwai-Hang and Grosclaude, 2003).

Australian Holstein herds vary in the distribution of protein variants. Results of a study (1500 cows) from the Colac region in Victoria (Ng Kwai Hang and Bennett, 2002) showed that the major variants for each of the milk proteins were α_{S1}-casein B, β-casein A^2 (the frequency of β-casein A^1 was only slightly lower), κ-casein A, and β-lactoglobulin B (the frequency of β-lactoglobulin A was only slightly lower). These results were similar to those obtained for New Zealand Holsteins. In Canadian herds, β-casein A^1 was slightly higher than the A^2 variant and the frequency of β-lactoglobulin B was almost twice that of β-lactoglobulin A. Netherlands Holsteins showed a lower frequency of β-casein A^2 (Ng Kwai Hang and Bennett, 2002).

Effect of stage of lactation on milk protein composition

As lactation progresses, the morphology and physiology of the mammary gland changes. Secretory cell number, measured as udder capacity, is significantly affected by stage of lactation (SOL), causing an effect on milk production (Eichler and McFadden, 1996). Although the changes in the milk protein distribution over lactation have been studied, many of these papers have shown contradictory results and are based on cows of unknown feeding practices, bulk herd milk samples, infrequent sampling events, or outdated analysis techniques, such as electrophoresis.

One of the first investigations into the production of specific milk proteins across lactation was a study of eight cows by Larson and Kendall (1957). At this time, the only casein fractions yet identified were 'α-casein', β-casein and γ-casein, and electrophoresis was used to analyse their presence in milk. While this means that the specific values cannot be compared under the current understanding of detailed milk protein composition, it is interesting to note the general trend of decreasing individual casein yields, and the large changes in individual casein production during the first month of lactation.

The relative percentages of various casein fractions varied with stage of lactation in two studies assessing monthly samples of individual cows across 62 Canadian herds (Kroeker *et al.,* 1985;

Ng-Kwai-Hang *et al.,* 1987). The concentration of α_s-casein in milk fell quickly in the first month after calving, and then gradually rose again for the remainder of lactation (Ng-Kwai-Hang *et al.* 1987). In terms of relative percentage of casein, α_s-casein fell initially after parturition, and then remained steady for the rest of lactation (Kroeker *et al.,* 1985). Conversely, the relative percentage of β-casein rose in the first two months of lactation, remaining relatively steady thereafter (Kroeker *et al.,* 1985), but the concentration of β-casein showed little variation, other than a slight decline in early lactation, followed by a gradual increase for the remainder of lactation (Ng-Kwai-Hang *et al.,* 1987). The concentration of κ-casein fell in the initial two to three months after parturition and, like the concentrations of the other caseins, gradually rose again until the end of lactation (Ng-Kwai-Hang *et al.,* 1987). Relative percentage of κ-casein appeared to be unaffected by stage of lactation (Kroeker *et al.,* 1985).

Analysis of bulked herd milk samples (monthly) by Barry and Donnelly (1980) compared stage of lactation and seasonal influences on the casein composition of milks of autumn- and winter/spring-calving herds. Casein composition was determined using the column chromatography method. α_s-Casein, as a proportion of total casein, rose in early lactation, and then fell for the reminder of the lactation, in contrast to the findings of Kroeker *et al.* (1985), above. β-casein and γ-casein proportions varied in a directly inverse relationship. The proportions of each varied greatly during early and mid lactation, but the levels of β- and γ-casein experienced a dramatic drop and increase, respectively, in late lactation. This is consistent with a higher milk proteinase activity seen in late lactation, particularly plasmin and plasminogen. (Davies and Law, 1977, Politis *et al.,* 1989). When β-and γ-caseins were considered together (i.e. β-casein observed + 10/6 γ-casein observed), the proportions of total β-casein showed considerable variation, but the highest values were recorded in late lactation, and the lowest in early lactation. Uncorrected κ-casein proportion was high in late lactation, but when κ-casein was corrected for contaminant proteins in the κ-fraction, the proportion was relatively consistent across lactation.

A small study measured the relative values of the casein components in Egyptian Friesian cows at the first, third, fifth, seventh and ninth months of lactation, and noted a consistent fall in α_s-casein content as lactation progressed, and a consistent increase in the relative proportions of the β- and κ-casein fractions (Fahmy, 1995). Ostersen *et al.* (1997) measured casein fractions four times throughout lactation (at weeks 2, 9, 24 and 46 of lactation), and found small although significant changes in the casein distribution. The proportions of α_s- and κ-caseins declined, while the proportion of β-casein increased as lactation progressed. The proportion of γ-casein was highest at the beginning and end of lactation. These authors have ascribed the consistently decreasing proportional content of κ-casein, compared to the studies of Barry and Donnelly (1980), Kroeker *et al.* (1985), Ng-Kwai-Hang *et al.* (1987) and Fahmy (1995) to a low proteolysis of α_s- and β-casein resulting from a high nutritional state. The cows in this study were dried off at production levels in excess of ten kg per day, and so were not likely to be in advanced mammary involution.

Groen *et al.* (1994) analysed monthly individual milk samples from ten heifers using High Pressure Liquid Chromatography (HPLC). α_{s1}-casein was significantly affected by SOL, registering a significantly higher mass fraction of total casein in the first month of lactation. This declined to a relatively constant level in mid-lactation. The α_{s2}, β- and κ-casein mass fractions were unaffected by stage of lactation. They noted a significant effect of calendar month of test on the mass fraction of β- and κ-casein, but there was no evidence of a pattern of variation.

Coulon *et al.* (1998) found an effect of stage of lactation on protein content but not on the casein:protein ratio. Protein content and cheese properties varied with stage of lactation with Auldist *et al.* (1996) and Coulon *et al.* (1998) showing detrimental effects of late lactation milk versus earlier lactation milk but Kefford *et al.* (1995) found late lactation milk from high quality diets had higher cheese yields.

Nutritional factors affecting milk protein composition

Nutrition or management are not thought to alter casein composition (Mariani, 1975; Davies and Law, 1980; Beever *et al.,* 2001). However there are some reports that suggest that milk casein composition can be altered with nutritional manipulation (Auldist *et al.,* 1998a; Christian *et al.,* 1999a; Christian *et al.,* 1999b; Grandison *et al.,* 1985; Mackle *et al.,* 1999b). Generally, an increase in energy intake has been associated with changes in the proportions of casein fractions.

Effect of dietary energy intake

Changes in the proportions of the casein fractions in milk have mainly been reported with an increase in the concentrate level of the diet (DePeters and Cant, 1992). Generally, total protein and casein concentration both increased with an increased dietary energy intake but an alteration in the proportions of the casein fractions was not always seen.

Yousef *et al.* (1969) increased grain intake and increased milk protein and casein concentration by 0.14%, and 0.27% units respectively. α- casein increased by 5.3% and γ- casein decreased by 5.7% (Yousef *et al.,* 1969). Christian *et al.* (1999a,b) found an alteration in the concentrations of α_{S1}- and α_{S2}- casein in milk from cows fed different types of energy and protein supplements (lupin-wheat, canola-wheat, oats-sunflower, wheat-sunflower or spring pasture) in conjunction with a control diet of silage and pasture hay. The changes seen in the proportions of casein components were quite marked in some cases. This would suggest that the type of energy source is important together with an increase in ME intake. Mackle *et al.* (1999b) demonstrated significant changes to β- casein concentrations with different feeding regimes involving pasture, grain and silage feeding. No significant differences were seen between treatments for protein, casein, whey protein, or the α_S-, κ- and γ-casein concentrations (Mackle *et al.,* 1999b). Coulon *et al.* (2001) showed that increasing energy intake above requirements had a significant effect on the protein and casein concentrations but there was no change in the proportion of casein fractions. κ-Casein did show minor increases in concentration with an increase in energy supply (Coulon *et al.,* 2001). Similar results were also seen in a study by Grant and Patel (1980) who found no significant dietary effects on the casein fractions, casein or whey protein concentrations. A recent experiment increased ME intake, either with silage or with grain, and found no change in the casein fractions of the treatment groups although there were substantial within animal period effects (Figure 1, Barber *et al.,* unpublished data).

Effect of level of pasture intake

Pasture intake is another factor that may affect the proportions of casein fractions in milk through its direct effect on energy and protein intake. Pasture quality, composition and availability are the main factors that will control the level of intake, especially in tropical and sub-tropical environments where pasture transition periods occur. These pasture transition periods result through a changeover from temperate to tropical pasture species in late spring, and from tropical to temperate pasture species in autumn. Milk protein concentration declines quite markedly during the spring pasture transition period from September to November, with the lowest concentrations often seen in December to January (summer). Milk protein levels tend to peak at the same time that the temperate pastures are at their highest quality and availability (September). These same issues arise in production systems throughout the world especially those that use pasture as a large part of the annual feed supply or utilise it as part of a designated and quality assurance product as seen in parts of Europe.

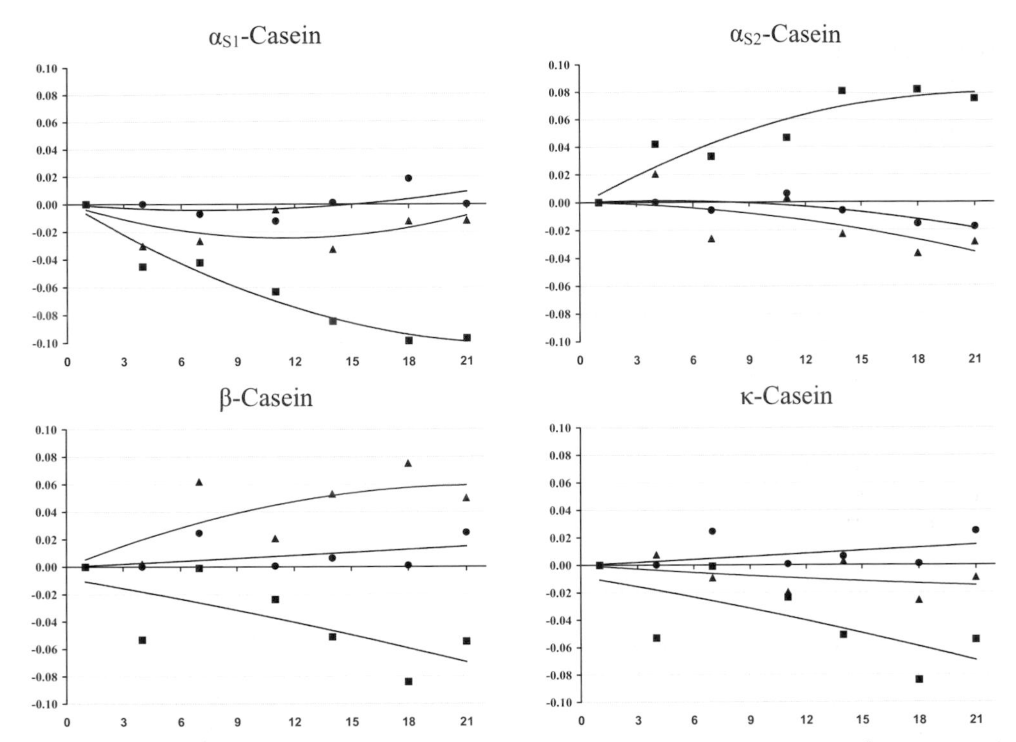

Figure 1. Time related responses in the proportions of αs1-casein, αs2-casein, β-casein and κ-casein fractions of total casein of an individual cow from day 1 to day 21 for the control (triangle), fibre-based (circle) and starch-based (square) diets. The day 1 value is used as the base value (covariate) and the response is measured as a deviation from the day 1 (zero) value.

Auldist *et al.* (1998a) investigated the influence of temperate pasture allowance (and hence energy intake) on milk composition under *ad libitum* and restricted grazing regimes across two seasons, with subsequent increases in the concentrations of protein (spring), casein, whey protein (spring), casein: whey protein ratio (summer), α-casein, β-casein and urea with cows having *ad libitum* access to pasture. κ-casein was increased in the spring period and decreased in the summer period with *ad libitum* pasture allowance. Seasonal changes in temperate pasture quality are the likely reason for this alteration in κ-casein (Auldist *et al.,* 1998a). Mackle *et al.* (1999a) examined the nutritional influences on milk composition of cows with different protein phenotypes. They suggested that changes in milk composition caused by nutrition were different for cows with different β-lactoglobulin phenotypes, particularly with restricted pasture allowance. They concluded that ME intake affected casein composition, with a further interaction with the protein phenotype of the animal (Mackle *et al.,* 1999a).

Other studies have also investigated the effect of varying levels of pasture allowance on milk composition with similar results. O'Brien *et al.* (1997) noted significantly higher concentrations of total protein, casein and lactose with increasing levels of pasture allowance. Petch *et al.* (1997) found that restricted feeding reduced crude protein, true protein and casein in early and mid, but not in late lactation. Grandison *et al.* (1985) suggested that pasture type was important, with cows grazing white clover showing an increase in the concentration of total casein, α_s- and β-caseins compared to cows grazing ryegrass. No significant difference was seen in the concentration of κ-casein between pasture types (Grandison *et al.,* 1985).

Effect of dietary protein intake

Dietary protein supply has been thoroughly reviewed with respect to milk protein concentration, with the general consensus that the inclusion of protein in the diet above animal requirements has variable results (DePeters and Cant, 1992). Therefore it could be assumed that milk protein composition, particularly casein composition, would also show a variable response to dietary protein intake.

Christian *et al.* (1999a) studied the effects of different dietary protein types on milk composition and found an increase in α_{S2}-casein and a decline in α_{S1}-casein concentration from a lupins-wheat diet when compared to a control diet of silage and hay. The canola-wheat diet also increased α_{S2}-casein concentrations above the control diet. Hermansen *et al.* (1999) investigated the effects of dietary protein supply on caseins and suggested that there was an interaction with ME intake. The study looked at the effect of the type of protein supply in terms of rye/clover versus rye/ N fertilised pasture, high rumen degradable protein (RDP) versus low RDP and high bypass protein versus low bypass protein. No significant difference was seen between any of the comparisons tested. However, there was a significant increase in κ- casein proportions when a high level of supplement was fed in conjunction with clover pastures. Coulon *et al.* (2001) investigated the effects of feeding three levels of protein and energy and the effect of genetic potential on milk protein composition. The feeding levels ranged from 85 to 125% of requirements and a significant increase in milk yield and protein concentration was seen with an increase in feeding level. There were no significant effects noted on the proportions of casein fractions in total protein from genetics or level of nutrition. However, there was a positive correlation between the proportion of κ-casein and milk protein concentration (R^2=0.28, P<0.01), which was attributed to the effects of either genetics or nutrition (Coulon *et al.,* 2001). This suggested that the proportions of casein fractions may not be as responsive to nutritional manipulation when dietary protein and energy levels are close to the animals requirements.

Other studies have also been conducted to investigate the effects of supplying individual amino acids on milk protein composition, particularly the supply of essential amino acids such as methionine and lysine with inconsistent results (Donkin *et al.,* 1989; Pacheco Rios *et al.,* 1999). However, Coulon *et al.* (1998) concluded in a review of casein in milk that amino acid supplementation had no significant effect on the proportion of casein within milk protein, but did have a positive significant effect on milk protein concentration.

Conclusions

The scope to alter milk protein composition using nutritional manipulations and, in turn, improve milk-processing qualities is an area of increasing interest for the worldwide dairy industry. The literature to-date is limited in terms of investigating some of the specific nutritional implications on casein composition, particularly on the proportions of individual casein fractions. The composition of the casein could have significant effects on cheese production, through enhanced coagulation properties, increased cheese yield and improved product quality and functionality.

The influence of milk protein genotype on milk yield and composition has been clearly defined by a number of authors. However, there is limited research data of the effect on the concentrations and proportions of the casein fractions. Responses of the casein fractions to milk protein genotype is not clearly defined and needs further research.

Research into the stage of lactation effects on casein composition is relatively limited and variable with respect to analytical methods used and the timing of analysis throughout lactation. Further study is needed to definitively establish the detailed shape of the stage of lactation curves for the casein fractions under the influence of different feeding systems and using modern analytical techniques.

The effect of nutrition on the expression of the casein fractions in milk is also relatively variable. However the level of ME intake appears to be the most likely means of manipulating the concentration and proportion of the casein fractions, especially when ME intake is sub-optimal. This may be achieved by increasing pasture allowance or quality, or level and type of supplement. Milk protein genotype has a marked effect on casein composition and needs to be accounted for in studies on casein composition.

This review has outlined the important factors associated with the manipulation of casein composition. The ability to manipulate casein composition on-farm would have considerable benefits along the dairy processing chain, particularly an improvement in the processing efficiency and functionality of dairy products.

Acknowledgements

This work was funded by Dairy Australia.

References

Aleandri, R., L. Buttazzoni, J. Schneider, A. Caroli and R. Davoli, 1990. The effects of milk protein polymorphisms on milk components and cheese-producing ability. J. Dairy Sci. 73: p. 241-255.

Auldist, M.J., S.Coats, B.J.Sutherland, J.J.Mayes, G.H.McDowell and G.L.Rogers, 1996. Effects of somatic cell count and stage of lactation on raw milk composition and the yield and quality of Cheddar cheese. J. Dairy Res. 63: p. 269-280.

Auldist, M. J., T. R. Mackle, J. P. Hill and C. G. Prosser, 1997. Effects of plane of nutrition on the composition of milk from cows of different beta-lactogloulin phenotype. In: Milk Protein Polymorphism, L.K. Creamer (editor), International Dairy Federation, Brussels, Belgium, p. 77-79.

Auldist, M. J., W. van der Poel, P. Laboyrie and C. G. Prosser, 1998a. Influence of pasture allowance on the composition and cheese-yielding potential of milk. Proc. NZ. Grsslds Assoc. 60: 199-202.

Auldist, M. J., B.J. Walsh and N.A. Thomson, 1998. Seasonal and lactational influences on bovine milk composition in New Zealand. J. Dairy Res. 65: p. 401-411.

Banks, J. M., 2000. Milk composition. In: Practical guide for control of cheese yield, D. B. Emmons (editor), International Dairy Federation, Brussels, 14-18.

Barbano, D.M. and J. Sherbon, 1984. Cheedar cheese yield in New York. J. Dairy Sci. 67: p. 1873-1883.

Barry, J. G. and W. J. Donnelly, 1980. Casein compositional studies. 1. The composition of casein from Friesian herd milks. J. Dairy Res. 47: p. 71-82.

Beever, D.E., J.D. Sutton and C.K. Reynolds, 2001. Increasing the protein content of cow's milk. Aust. J. Dairy Technol. 56: p. 138-149.

Bobe, G., D.C. Beitz, A.E. Freeman and G.L. Lindberg, 1999. Associations among individual proteins and fatty acids in bovine milk as determined by correlations and factor analyses. . J. Dairy Res. 66: p. 523-536.

Bruhn, J.C. and A.A. Franke, 1976. Monthly variations in gross composition of California herd milks. J Dairy Sci. 60: p. 696-700.

Christian, M P., C. Grainger, B J. Sutherland, J J. Mayes, M C. Hannah and B. Kefford, 1999a. Managing diet quality for Cheddar cheese manufacturing milk. 1. The influence of protein and energy supplements. . J. Dairy Res. 66: p. 341-355.

Christian, M.P., C. Grainger, B.J. Sutherland, J.J. Mayes, M.C. Hannah and B. Kefford, 1999b. Managing diet quality for Cheddar cheese manufacturing milk. 2. Pasture v. grain supplements. J. Dairy Res. 66: p. 357-363.

Coulon, J. B., 1994. Effect of physiological stage and season on the chemical composition of milk and its technological properties. Rec. Méd. Vet., 170: 367-374.
Coulon, J. B., D. Dupont, S. Pochet, P. Pradel and H. Duployer, 2001. Effect of genetic potential and level of feeding on milk protein composition. . J. Dairy Res. 68: p. 569-577.
Coulon, J. B., C. Hurtaud, B. Rémond and R. Vérite, 1998. Factors contributing to variation in the proportion of casein in cow's milk true protein: a review of recent INRA experiments. J. Dairy Res. 65: p. 375-387.
Creamer, L. K. and D. P. Harris, 1997. Relationship between milk protein polymorphism and physico-chemical properties. In: Milk Protein Polymorphism, L.K. Creamer (editor), International Dairy Federation, Brussels, Belgium, p. 110-123.
Dalgleish, D.G., D.S. Horne and A. Law, 1989. Size-related differences in bovine casein micelles. Biochim. Biophys. Acta 991: p. 383-387.
Davies, D.R. and A. J.R. Law, 1980. The content and composition of protein in creamery milks in south-west Scotland. J. Dairy Res. 47: p. 83-90.
Davies, D.T. and A.J.R. Law, 1977. The composition of whole casein from the milk of Ayrshire cows. J. Dairy Res. 44: p. 447-454.
DePeters, E.J. and J.P. Cant, 1992. Nutritional factors influencing the nitrogen composition of bovine milk: a review. J Dairy Sci., 75: p. 2043-2070.
Donkin, S.S., G.A. Varga, T.F. Sweeney and L.D. Muller, 1989. Rumen-protected methionine and lysine: Effects on animal performance, milk protein yield and physiological measures. J. Dairy Sci. 72: p. 1484-1491.
Donnelly, W. J., D. S. Horne and J. Barry, 1983. Casein compositional studies. III. Changes in Irish milk for manufacturing and the role of milk proteinase. J. Dairy Res., 50: p. 433-441.
Eichler, S. J. and T. B. McFadden, 1996. Effects of stage of lactation and season on udder development and milk yield in pasture-fed cows. Proc. NZ. Soc. Anim. Prod. 56: p. 58-60.
Erasmus, L., J. E. Hermansen and H. Rulquin, 2001. Nutritional and management factors affecting milk protein content and composition. In: Influence of feed on major components of milk, Bulletin of the International Dairy Federation, E. Hopkin (editor), International Dairy Federation, Brussels, Belgium, 366: p. 49-61.
Fahmy, M.A., 1995. Casein components of Friesian cow's milk as affected by the stage of lactation. Egyptian J. Dairy Sci. 23: p. 59-68.
FitzGerald, R. J. and J. P. Hill, 1997. The relationship between milk protein polymorphism and the manufacture and functionality of dairy products. In: Milk Protein Polymorphism, L.K. Creamer (editor), International Dairy Federation, Brussels, Belgium, p. 355-371.
Fox, P. F. and P. L. H. McSweeney (editors) 1998. Dairy Chemistry and Biochemistry. Blackie Academic and Professional, 478 p.
Grandison, A.S., D.J. Manning, D.J. Thomson and M. Anderson, 1985. Chemical composition, rennet coagulation properties and flavour of milks from cows grazing ryegrass or white clover. J. Dairy Res. 52: p. 33-39.
Grant, D.R. and P.R. Patel, 1980. Changes of protein composition of milk by ratio of roughage to concentrate. J. Dairy Sci. 63: p. 756-761.
Groen, A. F., R. van der Vegt, M. A. J. S. van Boekel, O. L. A. M. de Rouw, and H. Vos, 1994. Case study on individual animal variation in milk protein composition as estimated by high-pressure liquid chromatography. Netherlands Milk and Dairy Journal 48: p. 201-212.
Hermansen, J.E., S. Ostersen, N.C. Justesen and O. Aaes, 1999. Effects of dietary protein supply on caseins, whey proteins, proteolysis and renneting properties in milk from cows grazing clover or N fertilized grass. J. Dairy Res. 66: p. 193-205.
Horne, D. S., J. M. Banks and D. D. Muir, 1997. Genetic polymorphism of bovine k-casein: effects on renneting and cheese yield. In: Milk Protein Polymorphism, L.K. Creamer (editor), International Dairy Federation, Brussels, Belgium, p. 162-171.
Johnson, D. L., S. F. Petch, A. M. Winkelman and M. Bryant, 2000. Genetics of milk characteristics in New Zealand dairy cattle. Proc. NZ. Soc. Anim. Prod. 60: p. 318-319.

Kaufmann, W. and H. Hagemeister, 1987. Composition of milk. In: Dairy-Cattle Production, H.O. Gravert (editor), Elsevier Science Publishers, Amsterdam, p. 107-171.

Kefford, B., M. P. Christian, B. J. Sutherland, J. J. Mayes, and C. Grainger, 1995. Seasonal influences on cheddar cheese manufacture: Influence of diet quality and stage of lactation. J. Dairy Res. 62: p. 529-537.

Kennelly, J.J. and D.R. Glimm, 1998. The biological potential for altering the composition of milk. Can. J. Anim. Sci. 78: p. 23-56.

Kroeker, E.M., K.F. Ng Kwai Hang, J.F. Hayes and L.J. Moxley, 1985. Effects of environmental factors and milk protein polymorphism on composition of casein fraction in bovine milk. J. Dairy Sci. 68: p. 1752-1757.

Larson, B.L., and K.A. Kendall, 1957. Protein production in the bovine. Daily production of the specific milk proteins during the lactation period. J. Dairy Sci. 40: p. 377.

Lawrence, R. C., 1993. Cheese yield potential of milk. In: Factors Affecting the Yield of Cheese, D.B. Emmons (editor), Special issue of the International Dairy Federation, International Dairy Federation, Brussels, Belgium, p. 109-120.

Lodes, A., J. Buchberger, I. Krause and H. Klostermeyer, 1997. The influence of rare genetic variants of milk proteins on compositional properties of milk, rennetability, casein micelle size and content of non-glycosylated k-casein. In: Milk Protein Polymorphism, L.K. Creamer (editor), International Dairy Federation, Brussels, Belgium, p. 158-161.

Lundén, A., M. Nilsson and L. Janson, 1997. Marked effect of beta-lactoglobulin polymorphism on the ratio of casein to total protein in milk. J. Dairy Sci. 80: p. 2996-3005.

Macheboeuf, D., J.-B. Coulin and P. D'Hour, 1993. Effect of breed, protein genetic variants and feeding on cows' milk coagulation properties. J. Dairy Res. 60: p. 43-54.

Mackle, T.R., A.M. Bryant, S.F. Petch, J.P. Hill and M.J. Auldist, 1999a. Nutritional influences on the composition of milk from cows of different protein phenotypes in New Zealand. J. Dairy Sci. 82: p. 172-180.

Mackle, T.R., A.M. Bryant, S.F. Petch, R.J. Hooper and M J. Auldist, 1999b. Variation in the composition of milk protein from pasture-fed dairy cows in late lactation and the effect of grain and silage supplementation. New Zeal. J. Agr. Res. 42: p. 147-154.

Mariani, P., 1975. Protein distribution in the milk of Friesian, Brown Swiss, Reggio and Modena cows. Dairy Science Abstracts, 38: p. 46-47.

Martin, P. A. and F. Grosclaude, 1993. Improvement of milk protein quality by gene technology. Livest. Prod. Sci. 35: p. 95-115.

McGann, T., W.J. Donnelly, R. Kearney and W. Buchheim, 1980. Composition and size distribution of bovine casein micelles. Biochim. Biophys. Acta, 630: p. 261-270.

McLean, D. M., E. R. B. Graham, R. W. Ponzoni, and H. A. McKenzie, 1984. Effects of milk protein genetic variants on milk yield and composition. J. Dairy Res. 51: p. 531-546.

Mehra, R. K., B. O'Brien, J.F. Connolly and D. Harrington, 1999. Seasonal variation in the composition of Irish manufacturing and retail milks 2. Nitrogen fractions. Irish J. Agr. Food Res. 38: p. 65-74.

Murphy, J. J. and F. O'Mara, 1993. Nutritional manipulational of milk protein concentration and its impact on the dairy industry. Livest. Prod. Sci. 35: p. 117-134.

Ng-Kwai-Hang, K.F., J.F. Hayes, J.E. Moxley and H.G. Monardes, 1987. Variation in milk protein concentrations associated with genetic polymorphism and environmental factors. J. Dairy Sci. 70: p. 563-570.

Ng Kwai Hang, K. F. and F. Grosclaude, 1992. Genetic polymorphism of milk proteins. In: Advanced Dairy Chemistry Volume 1 - Proteins, P.F. Fox (editor), Elsevier Applied Science, p. 405-455.

Ng-Kwai-Hang, K. F., 1997. A review of the relationship between milk protein polymorphism and milk composition/milk production. In: Milk Protein Polymorphism, L.K. Creamer (editor), International Dairy Federation, Brussels, Belgium, p. 22-79.

Ng Kwai Hang, K.F. and L. Bennett, 2002. Improving profits for the Australian dairy industry through genetic selection. Personal communication.

Ng-Kwai-Hang, K. F. and F. Grosclaude, 2003. Genetic polymorphism of milk proteins. In: Advanced Dairy Chemistry, Volume 1: Proteins, P.F. Fox and P.L.H. McSweeney (editors), Kluwer Academic/Plenum, New York, USA, p. 739-816.

O'Brien, B., J.J. Murphy, J.F. Connolly, R. Mehra, T.P. Guinee and G. Stakelum, 1997. Effect of altering the daily herbage allowance in mid lactation on the composition and processing characteristics of bovine milk. J. Dairy Res. 64: p. 621-626.

Ostersen, S., J. Foldager and J.E. Hermansen, 1997. Effects of stage of lactation, milk protein genotype and body condition at calving on protein composition and renneting properties of bovine milk. J. Dairy Res. 64: p. 207-219.

Pacheco Rios, D., W.C. McNabb, J. P. Hill, T.N. Barry and D.D.S. Mackenzie, 1999. The effects of methionine supplementation upon milk composition and production of forage-fed dairy cows. Can. J. Anim. Sci. 79: p. 235-241.

Petch, S. F., A. M. Bryant and A. R. Napper, 1997. Effects of pasture intake and grain supplementation on milk nitrogen fractions. Proc. NZ. Soc. Anim. Prod. 57: p. 154-156.

Politis, I., E. Lachance, E. Block and J.D. Turner, 1989. Plasmin and plasminigen in bovine milk: a relationship with involution? J. Dairy Sci. 72: p. 900-906

Puhan, Z., 1997. Introduction to the subject. In: Milk Protein Polymorphism, L.K. Creamer (editor), International Dairy Federation, Brussels, Belgium, p. 12-21.

Puhan, Z. and E. Jakob, 1994. Genetic variants of milk proteins and cheese yield. In: Cheese Yield and Factors Affecting its Control, D.B. Emmons (editor), International Dairy Federation, p. 111-122.

Rollema, H.S., 1992. Casein association and micelle formation. In: Advanced Dairy Chemistry Volume 1 Proteins, P. Fox (editor). Elsevier Applied Science, London, UK, p. 111-140.

Sargeant, J.M., M.M. Shoukri, S.W. Martin, K.E. Leslie and K.D. Lissemore, 1998. Investigating potential risk factors for seasonal variation: an example using graphical and spectral analysis methods based on the production of milk components in dairy cattle. Prev. Vet. Med. 36: p. 167-178.

Swaisgood, H., 1992. Chemistry of the caseins. In: Advanced Dairy Chemistry Volume 1 Proteins, P. Fox (editor). Elsevier Applied Science, London, UK, p. 63-110.

Toerien, C.A., D.R. Trout and J.P. Cant, 2004. Effects of nutrients on p70S6K activation in the bovine mammary gland. Journal of Animal and Feed Sciences. 13: p. 449-452

Wilde, C. J., C. V. P. Addey, L. M. Boddy-Finch and M. Peaker, 1995. Autocrine control of milk secretion: From concept to application. In: Intercellular signalling in the mammary gland, C.J. Wilde, M. Peaker and C.H. Knight (editors), Plenum Press, New York, p. 227-237.

Winkelman, A. M. 1997. Associations of beta-lactoglobulin A, B and C variants with production traits in New Zealand dairy cattle. In: Milk protein polymorphism, L.K. Creamer (editor), International Dairy Federation, Brussels, Belgium, p. 83-86.

Yousef, I.M., J.T. Huber and R.S. Emery, 1969. Milk protein synthesis as affected by high-grain, low-fibre rations. J. Dairy Sci. 53: p. 734-739.

Zbikowska, A., J. Dziuba, H. Jaworska and A. Zaborniak, 1992. The influence of casein micelle size on selected functional properties of bulk milk proteins. Pol. J. Food Nutr. Sci. 1/42: p. 23-32.

Milk for Protected Denomination of Origin (PDO) cheeses: I. The main required features

G. Bertoni[1], L. Calamari[1], M.G. Maianti[1] and B. Battistotti[2]*
[1]Istituto di Zootecnica, Facoltà di Agraria, Università Cattolica del Sacro Cuore, Via Emilia Parmense 84, 29100 Piacenza, Italy, [2]Istituto di Microbiologia, Facoltà di Agraria, Università Cattolica del Sacro Cuore, Via Emilia Parmense 84, 29100 Piacenza, Italy.
**Correspondence: giuseppe.bertoni@unicatt.it*

Summary

The value of PDO cheeses has greatly increased, not only in economic terms but also as a cultural value, in particular in some European countries. They can fulfil particular organoleptic expectations, and at the same time represent the product of a traditional and more friendly exploitation of the land. Quite often the PDO cheese-making process requires milk with specific traits that can guarantee:

- organoleptic features linked to milk origin (species, breed, feeding system, geographical area, feedstuffs and feeding system, etc.);
- a good cheese-making aptitude to optimize the "natural" dairy process resulting in a high-quality cheese and lower risk of defects. This aptitude in milk means good renneting properties and good microbiological population.

If the origin is a self-evident objective, the better milk traits are justified considering that PDO cheese production does not allow "heavy" chemical or biological emendations of milk.

Keywords: raw milk, PDO cheese, chemical analysis, microbiological analysis, dairy

Introduction

Protected Denomination of Origin cheeses (PDO) are a great source of income for many European countries; not only in economic terms, but also because they are vehicles of tradition and culture. The objectives to pursue in their case are not high levels of safety but standardized organoleptic traits, as is the case with industrially-produced cheeses; on the contrary, the emphasis is primarily on the specific and then the genuine and particular sensorial aspects (for this reason there is no perfect standardization), without of course ignoring safety.

The differentiation between these two wide-ranging categories - which by no means cover the whole dairy system- has come about in more or less the last 100 years since important technological means became available (centrifuge, selected lactic acid bacteria (LAB) cultures, pasteurisation, continuous cheese-making etc.) all of which offer the producer guarantees against frequent defects and reduce processing costs (Blanc, 1981). In fact, whilst in some countries according to area, socio-economic conditions, type of traditional production, etc., the main aim was the guarantee of technological results, in other countries was of foremost importance the maintenance of the best traits of typical cheeses. Nevertheless, it would be a narrow view, which would not correspond to the truth, to claim that in the last century nothing has changed in the production of PDO cheeses. Let us take as an example the evolution of Parmigiano-Reggiano, in "Le Cuisinier et le Médecin" (Lombard, 1860 cit. by Alberini, 1993). Parmesan is defined a "Cheese originally from the Duchy of Parma, in Italy. It is sold in cylindrical forms weighing 30 kilos. It has a dry consistency and is salty; saffron powder is used to give it a sulphur colour". Today we know that Parmigiano-Reggiano is no longer either very salty or coloured; just as we know that innovations have been introduced to improve hygiene in dairies, but also to improve the process (natural creaming, natural whey starter, etc.). There has been a cautious balance between tradition and innovation, trying to take into account technological progress and the evolution of consumer needs without losing the specificity, originality and authenticity of a PDO cheese; similar suggestions were made by Morand-Fehr *et al.* (1998).

What is necessary for the production of a real PDO cheese? It is difficult to generalize, but probably: raw specific milk, natural LAB culture, discontinuous cheese-making, slow salting and ripening. We must not disregard other needs too, and, of these, we highlight hygienic-sanitary conditions and economic tenability. As far as the first is concerned, there is no real incompatibility if the cows are healthy, and if milking and processing takes properly place; furthermore these cheeses are generally consumed after more than 60 days. As far as the economic tenability is concerned, it is also essential that the cheese-making process should be successful: with a high degree of sensorial features and a reduced number of defects. Both aspects are the consequence of the appropriate combination of milk traits and of dairy technology. However, when the application of technology is limited (no pasteurisation, no selected microbial cultures, no chemical improvements of the raw milk), the milk becomes a more important element, without however ignoring other aspects (plant features and the dairyman's skill).
In general, milk used for this purpose should guarantee 3 main features (Figure 1):

- specific differentiation of cheese by means of some particular traits: i.e. animal species, breed etc., breeding system (extensive, mountain grazing, feeding techniques and especially forage type as pasture, hay, silage etc.);
- good renneting properties because these can condition the cheese yield, the chemico-physical and microbiological processes occurring in the cheese-making and ripening;
- good microbial-fermenting properties because these can affect cheese hygiene as well as the more or less favourable consequences of the whole cheese process.

These three milk features are consequence of many factors (Figure 1) and the latter two are particularly important when technological tools must be used sparingly as in the case of PDO cheese production.

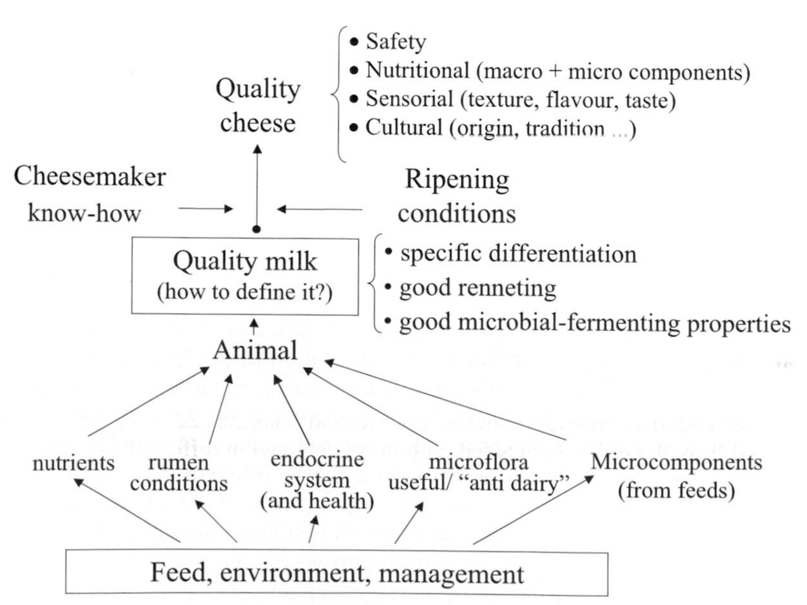

Figure 1. Milk quality, factors that influence it and consequences on the quality of PDO cheese.

Specific differentiation

This possibility is first of all aimed at ensuring genuine and real quality to consumers, but it can be also useful for the technological reasons. The specificity of a PDO cheese can depend on the species (for example goat, sheep, cow), and the breed (for example, Valdostana for Fontina cheese); the breeding system can also affect the milk yield of cows, but even more important is the feeding technique (grazing or not, forage/concentrates ratio, forage type, etc).

Animal species and breeds

While species can be taken for granted, breed differences are not so evident except when part of the traditional system (Valdostana for Fontina, Podolian for Caciocavallo etc.). In this case breed may be specifically adapted to the environment (mountains, extensive breeding, etc.); however, this too, independently of the breeding system, can be a source of positive difference in the cheese making process. This is the case with Reggiana or Italian Brown breeds that show better renneting properties (Pecorari *et al.*, 1987); it has also been shown that some French breeds improve the sensorial characteristics of the cheese (Coulon, 2004). In this respect, from the review of Martin *et al.*, (2003), it appears that the fat/protein ratio is responsible for the texture modification of the Saint-Nectaire cheese; therefore the Holstein Friesian's milk - with a higher fat content - produces more melting cheese. Nevertheless, the cheese obtained from milk of similar fat content has a more intense taste in Montbelliarde and Tarentaise breeds (Verdier-Metz *et al.*, 1998).

Breeding and feeding systems

The breeding system, no matter whether different species and breeds are involved, can first of all affect the feeding and life-style of the animals. Moio *et al.* (1996) have already shown a significant content of sesquiterpenes in the milk of sheep grazing on pastures; these compounds have a minimal effect on the sensorial traits, but they are very important as marker of the breeding system. Positive effects on the aroma of Sicilian cheeses (Carpino *et al.*, 2003) have been attributed to several chemical compounds originating from specific plants in those pastures: namely L-carvone and geranil acetate. In a wider study undertaken to compare the chemical and sensory characteristics of Abondance cheeses made with milk from animals grazing in areas within the same highland pasture, but with different predominant plants (Buchin *et al.*, 1999), significant sensorial differences were observed, particularly regarding texture and aroma. Differences in texture noted between the cheeses from milk produced in the two areas may come from differences in primary proteolysis, partly due to different amounts of plasmin and plasminogen in the milk and in cheeses. The aroma differences were related to differences in volatile compounds. Some compounds had a microbial origin, while some others may have originated from the pasture.

In a similar comparison of some highland and lowland Gruyere type cheeses from Switzerland, the results clearly highlight numerous differences in the composition (heavy metals, fatty acids, triglycerides, volatile components including terpenoids, polyaromatic hydrocarbons and flavour) between highland and lowland milk products" (Bosset and Jeangros, 2000). Even for Abondance cheese produced from the milk of cows fed with grass obtained in different mountain areas, few changes of texture and aroma have been observed (Bugaud *et al.*, 2002). As well as by the activity of the plasmin, the texture is influenced by unsaturated fatty acids - much more abundant in mountain milk - but the acidification rate during cheese-making is also important.

It can therefore be concluded that many changes occurring in cheeses obtained from different breeding systems or feeding techniques could be attributed to some volatile flavouring compounds; however a large part of them have a microbial origin, therefore several differences among cheeses can derive from different microbial activity in milk obtained from different areas or in different ways. It is noteworthy that the majority of microbes come from raw milk, which is in turn also contaminated by the wild micro-population occurring in feeds and stalls (Mordenti and Pecorari, 1999). Nevertheless, their activity can be positively or negatively affected by the milk itself; for example, terpenoids, which come from some plants, inhibit the formation of sulphur compounds (Bugaud *et al.*, 2002).

Apart from special mountain conditions, with or without grazing cows, an effect can be attributed to the techniques of forage storage: i.e. silage vs. hay. Verdier-Metz *et al.* (1998) have already observed that grass from natural pastures, fed as silage as opposed to hay, yields a more yellow cheese (higher in carotenoids) and with a spicier taste, no matter whether *C. tyrobutyricum* grew.

Renneting properties

These properties concern the most important of milk proteins and major components of cheese: caseins. In fact the real importance of proteins, particularly for PDO cheeses, is in their renneting aptitude and more proteins in milk do often but not always mean better coagulation. The major features of milk that explain renneting properties are protein content, casein fractions and their Ca and P contents that modify the micelle size and the pH (Remeuf, 1994). In addition to these traits, other important traits involved in milk renneting are (Figure 2): (a) the casein fractions, because high κ-casein (Losi and Mariani, 1984), as well as β-casein (Remeuf and Hurtaud, 1991) and perhaps α_{s2}-casein (Christian *et al.*, 1999) improve the renneting property; (b) the polymorphisms of k-casein and β-lactoglobulin because their BB genotype has a positive influence on renneting property (Russo and Mariani, 1978); (c) t he high Cl^- content which tends to slow κ-casein hydrolysis, and therefore renneting. Moreover, the latter is an index of mastitis that causes a reduction in normal casein available for the proteolytic activity associated with the higher plasmin content (Politis *et al.*, 1989); as a consequence there is an increase in γ-casein and an

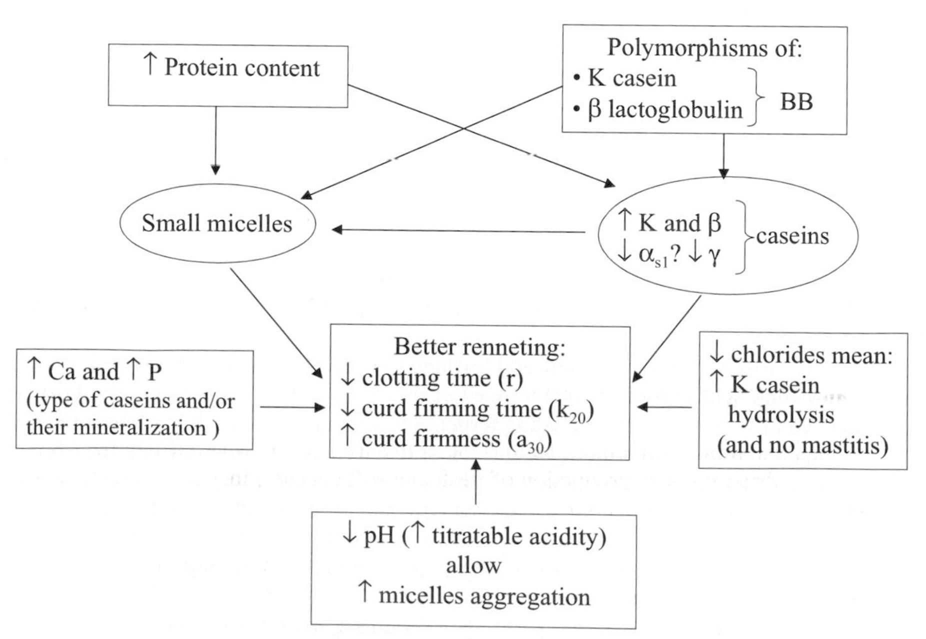

Figure 2. Factors and milk traits that influence the coagulation aptitude (Bertoni and Calamari, 2000).

impairment of the renneting properties. However a more exhaustive review on milk rennetability and its protein fractions, as well as on the causes of their changes has been discussed in the companion paper of Barber *et al.* (2005).

There is no doubt that a good renneting property in milk is of major importance for cheese yield (Losi and Mariani, 1984; Mariani and Pecorari, 1987; Remeuf *et al.*, 1991; Fossa *et al.*, 1994) and also for cheese quality, both for texture and colour, and for organoleptic properties such as flavour and taste (Pecorari *et al.*, 1995; Martin *et al.*, 1997).

The renneting aptitude of milk can be influenced by many different factors, both of genetic and environmental origin. Unfortunately they cannot be modified to a great extent; except for protein polymorphisms (Russo and Mariani, 1978) and cow health; our experience (Bertoni and Calamari, 2000) suggests that appropriate management of feeding and of general welfare (low stress and good health conditions) can be of great help for optimal milk renneting properties. Likewise, renneting properties was modified by a different sun exposure of highland pasture areas (Buchin *et al.*, 1999), suggesting that different forage could change them.

Bacteriological and fermentability traits

These traits are of major importance when raw milk is used; the presence of useful or dangerous microbes and their capability of growth in specific milk are of paramount importance for the cheese-making success. In fact, according to Martin and Coulon (1995), variations in the organoleptic qualities of cheese are mainly due to technological factors and among them the possibility to modify type of microbes and their acidification rate. A good example of this can be the influence of the degree of acidification on the process technology as described by Mariani and Pecorari (1987) for Parmigiano-Reggiano:

- when renneting is mainly of the enzymatic type, the micelle complex maintains its properties, the losses of calcium are moderate and consequently the curd has optimal rheological properties;
- when renneting is mainly of the acidic type (lactic acid), the micelle complex is modified and the calcium losses are more substantial. Thus, the curd is less elastic and firm, so whey drainage is reduced.

This means that milk must be processed before any substantial acidification could occur (often caused by a strong contamination of milk); nevertheless, it is worthwhile remembering that a high rate of acidification after the coagulation phase during cheese-making plays an important part in the control of gas-producing microflora (mainly propionic, citrate and ethero fermentative bacteria) and could be a useful way of reducing the blowing problems in the few months after cheesemaking with milk obtained from silage-fed cows (Bottazzi *et al.*, 1992).

Useful vs. antidairy bacteria

In general terms, microbial contamination must be as low as possible when milk is destined, after pasteurisation, to drinking or to production of fresh and soft cheeses; instead it can constitute - if appropriate for type and number - a reason for specificity when milk is destined to cheese production, obtained from raw milk, especially if they are of "Protected Denomination of Origin". In that last case, favourable microflora characterises the environment in which the cheese is made and reflects the connection with the territory, and constitute the initial starter that can direct and determine physico-chemical and biological processes of cheesemaking and cheese ripening (Mariani and Battistotti, 1999).

The microflora and its management (Figure 3) is therefore of great importance and has been well-known since the beginning of the twentieth century when Gorini (1907; cited by Mucchetti *et al.*, 1998), suggested that the differences between the most famous Italian cheeses (Parmigiano-Reggiano and Grana Padano) were due to the effects of forage, climate and technology on microbes. This supports the idea that typical cheeses require indigenous microflora (Bouton and Grappin, 1995; Demarigny *et al.*, 1997; Freitas and Malcata, 2000) or, preferably, a natural whey culture such as that used for Parmigiano-Reggiano (Mucchetti *et al.*, 1998). The importance of the natural starter is particularly relevant when milk lacks useful bacteria or has too many undesirable (and sometimes pathogenic) microbes; it modulates the fermentation and reduces cheese defects (Mucchetti *et al.*, 1998; Bottazzi *et al.*, 1999).

Montel *et al.* (2003) have also recently confirmed that early microbial contamination of milk is influenced by environmental farm conditions and that it considerably affects milk processing. In fact, for PDO cheeses, the bacterial quality can only be optimised according to mild tools; reduction of active bacteria and spores (through the physical effect of fat globules rising during the natural creaming), inoculation of natural starters, addition of anticlostridia substances (i.e. lysozyme in the Grana Padano technology). Of major importance is therefore any useful attempt to reduce contamination, particularly by the spores of *Clostridum* spp. (*C. sporogenes*, *C. butyricum* and particularly *C. tyrobutyricum*), which can cause the blowing of cheese (Bottazzi *et al.*, 1982). The most studied of these species is *C. tyrobutyricum,* which causes late blowing and is particularly influenced by the feeding regimen: i.e. the type and amount of silage in the diet (Baraton, 1985). Depending on the cheese, the deleterious effects of spores can be avoided by the reduction of the number of spores in milk to a safe level of less than 200 spores per litre. Spores occur in faeces, and any means of achieving a clean udder (buildings, bedding, available area, season, etc.) and high milking hygiene (parlour, washing and drying teats, etc.; Rasmussen *et al.*, 1991) is of prime importance (Pecorari, 1984). Nevertheless, the number of spores in faeces can be modified by a number of factors, including silage in the diet, since the spores occur in the soil and can germinate and grow in silage (Leibensperger and Pitt, 1987; Spoelstra, 1990); however, the digestive process pattern (i.e. the faeces pH as showed in Figure 4) can affect the faecal spore output.

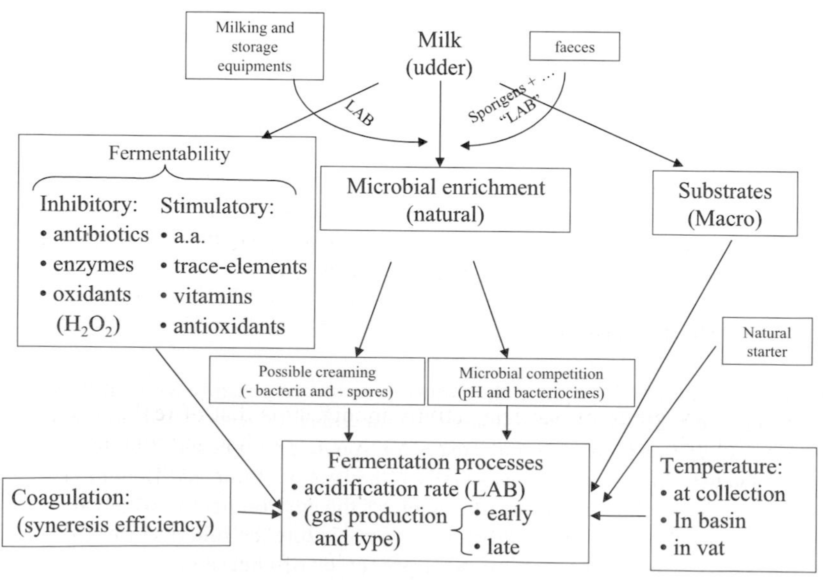

Figure 3. Main factors that influence the microbe contamination of milk and their activity during the cheese-making process.

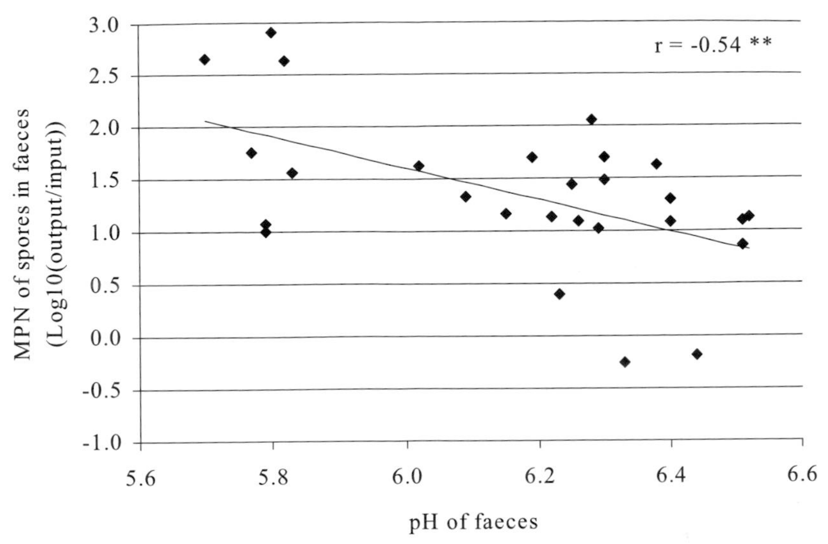

Figure 4. Relationship between pH and the ratio between most probable number (MPN) of Clostridium thyrobutyricum spores in dairy cows faeces and MPN of spores in feeds (Bani et al., 2001).

Fermentability traits

The milk bacteriological traits are of paramount importance, but the whole cheese process is greatly affected by other factors and some of them can modify the microbial activity; again according to Montel *et al.* (2003), they correspond to different technologies (cooking T°, size of curd, salting, etc.), but also to the acidification rate. In turn, this acidification rate could be influenced by the natural whey starter supply, as mentioned before, but it can also depend on the natural fermentability of milk as shown in some field conditions (Vecchia *et al.*, 2003). This trait, the so-called bacteria fermentation ability or acidification rate (Marth, 1974), is particularly important in the case of raw milk. Unfortunately, despite long scientific interest (since 30 years ago we have demonstrated a lower acidification rate of milk obtained from cows fed very large amounts of corn silage, Battistotti and Bertoni, 1974), it is still difficult to explain the phenomenon. Nutrient availability, particularly of small nitrogen compounds (amino acids, nucleic acid derivatives), essential fatty acids, trace elements and vitamins, could be of some importance (Rašic and Kurmann, 1978). This availability could in fact explain the positive effect of yeast extract on lactic fermentation of raw or pasteurised milk (Chazaud and Larpent, 1980), and also the positive effect of Fe^{3+} and Zn^{2+} supply (Maianti *et al.*, 1995). However, in trials with an amino acid mixture or with cysteine supplements, there was no increase in the acidification rate of *Lactobacillus delbrueckii subsp. bulgaricus* and *Streptococcus thermophilus* (Maianti M.G., personal communication).

A second important group of bacteria activity modifiers is that of real antimicrobial agents: lysozyme, lactoferrin, lactoperoxidase system, immunoglobulins and vitamin binding proteins. The effects of these agents on the inactivation of pathogens or saprophytic micro-organisms, as well as on the LAB, need further studies. Lactoperoxidase (LP) in bovine is involved in natural defence system from bacteria through peroxidation of thiocyanate ions by hydrogen peroxide, both present in biological fluids in a so-called lactoperoxidase system. It is quite thermo-tolerant, because a complete inactivation of LP in milk requires 15s/78°C (Griffiths, 1986). The antimicrobial effect of the LP system differs according to organism (Pruitt and Reiter, 1985).

Valdez *et al.* (1988) observed that starter cultures for cheese-making appeared to be strongly reduced by the LP system. In our trial (Calamari *et al.*, 2003), a significant but weak correlation (r= -0.21; P<0.05) between acidification rate and LP was observed. It would appear that, in bovine milk, lactoperoxidase (LP) is always present in excess and that the factors limiting its activity could be thiocyanate (SCN^-) and hydrogen peroxide availability. The SCN^- concentration of bovine milk varies with breed, species, udder health and type of feed and lies between 1 and 15 ppm (Kussendrager and Van Hoijdonk, 2000) and might be too low. Hydrogen peroxide (H_2O_2) may be formed endogenously, e.g. by polymorfonuclear leucocytes in the process of phagocytosis. Many lactobacilli, lactococci and streptococci may also produce sufficient H_2O_2, under aerobic conditions, to activate the LP system (De Wit and Van Hooydonk, 1996); furthermore, H_2O_2 may be generated by a number of H_2O_2 systems such as the oxidation of hypoxanthine by xanthine oxidase (XO). A significant correlation was in fact observed between XO activity in milk and acidification rate (r= -0.33, P<0.01), that seems to indicate an involvement of this enzyme in the inhibitory systems that affect growth of *Streptococcus thermophilus* and *Lactobacillus delbreuckii* ssp. *bulgaricus* in raw milk (Calamari *et al.*, 2003).

Milk fermentescibility (acidification rate) and the factors that can modify it are therefore an interesting issue and our studies can help to increase knowledge about them. According to these studies, antioxidants (vitamin C and E) improve the acidifying activity of raw milk while heat treatments show different effects according to T° (Bertoni *et al.*, 1992; Maianti *et al.*, 1993). This could mean that high temperature treatment is able to destroy any antibacterial mechanism and/or allows the release of some essential nutrient, furthermore the expulsion of oxygen could be effective.

It should however be underlined that some complex interference could occur because:

- colostrum, very rich in Ig and lactoferrin (Ye and Yoshida, 1995), acidifies in fact more quickly than milk (Maianti *et al.*, 1992);
- mastitic milk, with a high leukocyte content, is also better acidified (Maianti M.G., unpublished results); maybe because, according to Okello and Marshall (1986), the increase of somatic cells results in a stimulation of *S. thermophilus* - unable to degrade protein - due to a higher proteolysis (plasmine effect);
- the lactation phase modifies the acidification rate with the lowest values at the 2nd-4th month (Calamari *et al.*, 1986; Maianti *et al.*, 1992);
- milk from small farms generally performs better (Calamari *et al.*, 2003).

Considering these results it seems that evaluation of the milk acidification rate on raw milk and on pasteurised milk, together with evaluation of the parameters related to mammary epithelial functionality and of the enzyme activity involved in the main mammary antimicrobial systems, could improve our knowledge of the factors that modify the milk acidification rate. However our recent data (Vecchia *et al.*, 2003) confirm that both nutrients and antimicrobial enzymes could be involved.

To conclude, a variable acidification rate has been observed several times (Limsowtin and Powell, 1996; Masle and Morgan 2001; Bugaud *et al.*, 2002; Montel *et al.*, 2003; Coulon *et al.*, 2003) and explained with the above-mentioned mechanisms, but there is a lack of definitive knowledge which could allow any attempt to improve it.

Conclusions

The value of PDO cheeses has greatly increased, not only in economic terms but also as a cultural value, in particular in some European countries. They can fulfil particular organoleptic expectations, and at the same time represent the product of a traditional and more friendly exploitation of the land.

In spite of the wide variety of situations, ranging from real traditional to semi-industrial, the PDO cheese-making process requires milk with specific characteristics that could guarantee:

- organoleptic features, linked to milk origin (species, breed, feeding system, geographical area, feedstuffs and feeding system, etc.), as well as to the cheese-making process and the ripening conditions (namely the microbial population) that follow;
- a good cheese-making aptitude, to optimize the "natural" dairy process resulting in a high-quality cheese and lower risk of defects. This aptitude in milk means:
- good renneting properties, to maximize yield and to drive the chemico-physical and microbiological processes toward their optimal conditions;
- good microbiological population (proper amount and proper types) and good milk fermentability, because microbes are the cause of many organoleptic traits but are also responsible for many cheese defects.

This means that milk for PDO cheeses should be monitored to guarantee its origin, but also its suitability to a successful cheesemaking; the reasons to ensure an origin guaranty is self-evident, while for cheese-making it can be remembered that only the industrial procedures can emend some defects (low renneting, improper microflora). Therefore appropriate analyses of milk (and perhaps of cheese) would allow milk traceability as well as ensure proper coagulation and microbiological traits, besides the traditional composition (i.e. fat and protein contents); all together very helpful for cheese-making and therefore a guaranty for high levels of cheese yield and quality.

References

Alberini, M., 1993. The fascinating and homemade story of Parmigiano-Reggiano. In: Parmigiano Reggiano. A symbol of culture and civilization, F. Bonilauri (editor), Leonardo Arte srl, Segrate, Milano, Italy, p. 10-16.

Bani, P., L. Calamari, E. Bettinelli and G. Bertoni, 2001. Composizione della razione e contenuto in spore di clostridi nelle feci di bovine in lattazione. Sci. Tecn. Latt-Cas. 52 : p. 369-386.

Baraton, Y. 1985. Review on the spore contamination. ITEB, Paris, France.

Barber, D.G., A.V. Houlihan, F.C. Lynch and D.P. Poppi, 2005. The influence of nutrition, genotype and stage of lactation on milk casein composition. In: Indicators of milk and beef quality, J.F. Hocquette and S. Gigli (editors), Wageningen Academic Publishers, The Netherlands, in press.

Battistotti, B. and G. Bertoni, 1974. The milk from cows fed with corn silage as medium for lactic acid bacteria growth. Sci. Tecn. Latt-Cas. 25 Notiziario Tecnico: p. 7-10.

Bertoni, G. and L. Calamari, 2000. Fattori alimentari ed ambientali che agiscono sull'acidità del latte e sulla caseificazione. Atti del V Convivio Formaggi d'alpeggio: il pascolo, l'animale, la razza, il prodotto. Cavalese (TN), 15 settembre, p. 19-42.

Bertoni, G., L. Calamari, M.G. Maianti and A. Azzoni, 1992. Preliminary research into factors that modify the milk acidification rate. Proceedings of the Progetto Finalizzato Moderne Strategie Lattiero-Casearie, Tecniche Nuove, Milano, Italy, p. 35-37.

Blanc, B., 1981. Typical cheese varieties in the technological development of the dairy industry. Sci. Tecn. Latt-Cas. 32: p. 377-406.

Bosset, J.O. and B. Jeangros, 2000. Comparison of some highland and lowland Gruyere type cheeses from Switzerland: a study of their potential PDO-characteristics. In: Livestock farming systems. Integrating animal science advances into the search for sustainability. D. Gagnaux and J.R. Poffet (editors). Proceedings of the fifth international symposium on livestock farming systems. Posieux (Fribourg), Switzerland, 19-20 August 1999, p. 337-339.

Bottazzi, V., B. Battistotti and A. Rebecchi, 1999. Ambiente e microflora del latte: una relazione che cambia. Latte 24: p. 84-88.

Bottazzi, V., F. Bodini, B. Battistotti, C. Corradini and M. Lauritano, 1982. Removal of clostridia from milk by bactofugation, and production of Grana cheese. Sci. Tecn. Latt-Cas. 33: 123-165

Bottazzi, V., G.L. Scolari, F. Cappa, B. Battistotti, F. Bosi and E. Brambilla, 1992. Lactic acid bacteria for Grana cheese production. Part III: Acidification rate and blowing. Sci. Tecn. Latt-Cas. 43: p. 71-93.

Bouton, Y. and R. Grappin, 1995. Comparison of the quality of scalded-curd cheeses manufactured from raw or microfiltered milk. Lait 75: p. 31-44.

Buchin, S., B. Martin, D. Dupont, A. Bornard and C. Achilleos, 1999. Influence of the composition of Alpine highland pasture on the chemical, rheological and sensory properties of cheese. J. Dairy Res. 66: p. 579-588.

Bugaud, C., S. Buchin, A. Hauwuy and J.B. Coulon, 2002. Texture et flaveur du fromage selon la nature du pâturage: cas du fromage d'abondance. INRA Prod. Anim. 15: p. 31-36.

Calamari, L., P. Bani, G. Bertoni and V. Cappa, 1986. Changes of milk composition and characteristics during lactation. In: Quality control of the milk and dairy products. Proceedings XXI International Simposium on Animal Production, [CN Baldissera, editor], Società Italiana per il progresso per la zootecnia, Milano, Italy, p. 223-229.

Calamari, L., P. Bani, M.G. Maianti and G. Bertoni, 2003. New researches on the factors affecting milk acidification rate. 38° Simposio Internazionale di Zootecnia, Milk and Research, Lodi 30 maggio 2003, G.F. Greppi and G. Enne (editors): p. 137-144.

Carpino, S., G. Licitra and P.J. Van Soest, 2003. Selection of forage species by dairy cattle on complex Sicilian pasture. Anim. Feed Sci. Tecn. 105: p. 205-214.

Chazaud, M.T. and M.J.-P. Larpent, 1980. Stimulation of the activity of starter cultures by yeast extract. Rev. lait. fr. 383(19): p. 21-24.

Christian, M.P., C. Grainger, B.J. Sutherland, J.J. Mayes, M.C. Hannah and B. Kefford, 1999. Managing diet quality for cheddar cheese manufacturing milk. 1. The influence of protein and energy supplements. J. Dairy Sci. 66: p. 341-355.

Coulon, J.B., 2004. Qualità del latte e produzione casearia: l'effetto della genetica e dell'alimentazione sulle proprietà coagulanti del latte e sulle caratteristiche sensoriali del formaggio. 7ma Conferenza mondiale allevatori razza bruna, Verona, 3-7 marzo 2004, p. 55-65.

Coulon, J.B., E. Rock and Y. Noël, 2003. Caractéristiques nutritionnelles des produits laitiers et variations selon leur origine. INRA Prod. Anim. 16: 275-278

De Wit, J.N. and A.C.M. Van Hooydonk, 1996. Structure, functions and applications of lactoperoxidase in natural antimicrobial systems. Neth. Milk Dairy J. 50: p. 227-244.

Demarigny, Y., E. Beuvier, S. Buchin, S. Pochet and R. Grappin, 1997. Influence of raw milk microflora on the characteristics of Swiss-type cheeses: II. Biochemical and sensory characteristics. Lait 77: p. 151-167.

Fossa, E., M. Pecorari, S. Sandri, F. Tosi and P. Mariani, 1994. The role of milk casein content in the Parmigiano Reggiano cheese production: chemical composition, rennet coagulation properties and dairy-technological behaviour of milk. Sci. Tecn. Latt-Cas. 45: p. 519-535.

Freitas, A.C. and F.X. Malcata, 2000. Microbiology and biochemistry of cheeses with appélation d'origine protegée and manufactured in the Iberian peninsula from ovine and caprine milks. J. Dairy Sci. 83: p. 584-602.

Griffiths, M.W., 1986. Use of milk enzymes as indices of heat treatment. J. Food Prot. 49: p. 896-905.

Kussendrager, K.D. and A.C.M. Van Hoijdonk, 2000. Lactoperoxidase: physico-chemical properties, occurrence, mechanism of action and application. Brit. J. Nutr. 84 suppl. 1: p. S19-S25.

Leibensperger, R.Y. and R.E. Pitt, 1987. A model of clostridial dominance in ensilage. Grass Forage Sci. 42: p. 297-317.

Limsowtin, G.K.Y. and I.B. Powell, 1996. Milk quality for cheesemaking. Aust. J. Dairy Technol. 51: p. 98-100.

Losi, G. and P. Mariani, 1984. Technological importance of milk protein polymorphism in manufacturing grana cheese. L'industria del latte 20: 23-54
Maianti, M.G., L. Calamari, G. Bertoni and V. Cappa, 1993. Researches on some factors interfering with the acidification rate of milk. In Moderne strategie lattiero-casearie. 2° anno, S Carini (editor), Tecniche Nuove, Milano, Italy, p. 30-33.
Maianti, M.G., L. Calamari, G. Bertoni G and V. Cappa, 1995. Researches into some factors interfering with the acidification rate of milk. Il latte 7: p. 718-721
Maianti, M.G., L. Calamari, V. Cappa and A. Soressi, 1992. The effect of the phase of lactation on some milk characteristics. In Moderne strategie lattiero-casearie. 1° anno, S. Carini (editor), Tecniche Nuove, Milano, Italy, p. 28-30.
Mariani, P. and B. Battistotti, 1999. Milk quality for cheesemaking. Proceedings of the A.S.P.A. XIII Congress, Piacenza, June 21-24, 499-516.
Mariani, P. and M. Pecorari, 1987. Genetic factor, cheesemaking aptitude and cheese yield of milk. Sci. Tecn. Latt-Cas. 38: p. 286-326.
Marth, E.H., 1974. Fermentations. In Fundamentals of dairy chemistry, 2nd ed., B.H. Webb, A.H. Johnson and J.A. Alford (editors), The Avi Publishing Company Inc., Westport, Connecticut, p. 772-872.
Martin, B. and J.B. Coulon, 1995. Milk production and cheese characteristics. II. Influence of bulk milk characteristics and cheesemaking practices on the characteristics of farmhouse Reblochon cheese from Savoie. Lait 75: p. 133-149.
Martin, B., J.F. Chamba, J.B. Coulon and E. Perreard, 1997. Effect of milk chemical composition and clotting characteristics on chemical and sensory properties of Reblochon de Savoie cheese. J. Dairy Res. 64: p. 157-162.
Martin, B., S. Buchin and C. Hurtaud, 2003. Conditions de production du lait et qualités sensorielles des fromages. INRA Prod. Anim. 16: p. 283-288.
Masle, I. and F. Morgan, 2001. Aptitude du lait de chèvre à l'acidification par les ferments lactiques - Facteurs de variations liés à la composition du lait. Lait 81: p. 561-569.
Moio, L., L. Rillo, A. Ledda and F. Addeo, 1996. Odorous constituents of ovine milk in relathionship to diet. J. Dairy Sci. 79: p. 1322-1331.
Montel, M.C., E. Beuvier and A. Hauwuy, 2003. Pratiques d'élévage, microflore du lait et qualités des produits laitiers. INRA Prod. Anim. 16: p. 279-282.
Morand-Fehr, P., R. Rubino, J. Boyazoglu and J.C. Le Jaouen, 1998. Reflections on the history, on the present situation and on the evolution of the typical animal products. In: Basis of the quality of typical Mediterranean animal products. EAAP Publication n° 90, J.C. Flamant, D. Gabiña and M. Espejo Díaz (editors), Wageningen Pers., Wageningen, The Netherlands, p. 17-29.
Mordenti, A. and M. Pecorari, 1999. L'alimentazione delle bovine nell'espressione di tipicità dei formaggi. Sci. Tecn. Latt-Cas. 50: p. 57-75.
Mucchetti, G., F. Addeo and E. Neviani, 1998. History evolution of PDO cheeses production. 1. Production, use and composition of whey starters in the Grana Padano and Parmigiano-Reggiano cheeses making: statements on the relationship between whey starter and PDO Sci. Tecn. Latt-Cas. 49: 281-311
Okello, U. and V.M.O. Marshall, 1986. Influence of mastitis on growth of starter organism used for the manufacture of fermented milks. J. Dairy Res. 53: p. 631-637.
Pecorari, M., 1984. Observation on the relationship between feeding and Clostridia blowing of cheese. Proceedings Moderni sistemi di fienagione e qualità casearie del latte, Reggio Emilia, Italia, p. 35-51.
Pecorari, M., E. Fossa, S. Sandri, G. Tedeschi, L. Pellegrino and P. Mariani, 1995. The role of casein of milk on Parmigiano-Reggiano cheese production: yield, chemical, composition, proteolysis, lipolysis and organoleptic properties of the aged cheese. Sci. Tecn. Latt-Cas. 46: p. 211-232.
Pecorari, M., S. Sandri and P. Mariani, 1987. The renneting properties of milk of Italian Friesian, Italian Brown, Reggiana and Modenese breeds. Proceedings. La qualità casearia del latte: fattori genetici ed ambientali (Quale latte per il miglior formaggio?), Ed. Regione Veneto, p. 97-105.
Politis, I., E. Lachance, E. Block and I.D. Turner, 1989. Plasmin and plasminogen in bovine milk: a relationship with involution? J. Dairy Sci. 72: 900-906

Pruitt, K.M. and B. Reiter, 1985. Biochemistry of peroxidase system: antimicrobial effects. Immunology Series 27: The lactoperoxidase system: p. 143-178.
Rašic, J.L. and J.A. Kurmann, 1978. Fermented fresh milk products. Yoghurt. Scientific grounds, technology, manufacture and preparations. Volume 1. Copenhagen, Denmark: Technical Dairy Publishing House.
Rasmussen, M.D., D.M. Galton and L.G. Petersson, 1991. Effects of premilking teat preparation on spores of anaerobes, bacteria, and lodine residues in milk. J. Dairy Sci. 74: p. 2472-2478.
Remeuf, F., 1994. Relations entre caractéristiques physico-chimiques et aptitudes fromagères des laits. Rec. Méd. Vét. 170: p. 359-365.
Remeuf, F., V. Cossin, C. Dervin, J. Lenoir and R. Tomassone, 1991. Relationship between milk physico-chemical traits and cheesemaking property. Lait 71: p. 397-421.
Remeuf, F. and C. Hurtaud, 1991. Relationship between physico-chemical traits of milk and renneting property. In: Qualité des laits à la production et aptitude fromagère, M. Journet, A. Hoden and G Brule (editors), INRA, ENSAR, Rennes, France, p. 1-7.
Russo, V. and P. Mariani, 1978. Milk polymorphism and relationship between the genetic variants and the characteristics of zootechnical, technological, dairy-farming interest. Rivista di Zootecnia e Veterinaria 6: p. 289-304.
Spoelstra, S.F., 1990. Comparison of the content of clostridial spores in wilted grass silage ensiled in either laboratory, pilot-scale or farm silos. Neth. J. Agr. Sci. 38: p. 423-434.
Valdez, G.R. de, W. Bibi and M.R. Bachman, 1988. Antibacterial effect of the lactoperoxidase/thiocyanate/hydrogen peroxide (LP) system on the activity of thermophylic starter culture. Milchwissenschaft 43: p. 350-352.
Vecchia, P. M. Intini, L. Calamari and B. Battistotti, 2003. Le principali cause delle fermentazioni anomale nel latte. Inf. Agr. 59: p. 49-53.
Verdier-Metz, I., J.B. Coulon, P. Pradel, C. Viallon and J.L. Berdagué, 1998. Effect of forage conservation (hay or silage) and cow breed on the coagulation properties of milks and on the characteristics of ripened cheeses. J. Dairy Res. 65: p. 9-21.
Ye, X. and S. Yoshida, 1995. Lactoperoxidase and lactoferrin: changes in post partum milk during bovine lactational disorders. Milchwissenschaft 50: p. 67-71.

Milk for Protected Denomination of Origin (PDO) cheeses: II. The evaluation techniques of milk suitability

L. Calamari[1], G. Bertoni[1], M.G. Maianti[1] and B. Battistotti[2]*
[1]Istituto di Zootecnica, Facoltà di Agraria, Università Cattolica del Sacro Cuore, Via Emilia Parmense 84, 29100 Piacenza, Italy, [2]Istituto di Microbiologia, Facoltà di Agraria, Università Cattolica del Sacro Cuore, Via Emilia Parmense 84, 29100 Piacenza, Italy.
**Correspondence: luigi.calamari@unicatt.it*

Summary

PDO cheeses promise organoleptic peculiarities and traditional production systems that involve higher technological risks and costs. This suggests that milk for PDO cheeses would be monitored to guarantee its origin, but also its suitability for a successful cheese-making. In other words, suitable analyses of milk (and perhaps of cheese) would allow its traceability as well as proper evaluation of coagulation and microbiological traits. Many well known methods exist to accomplish these tasks, but our effort has concentrated on new techniques, of which some are on-line, now available to recognize whether milk is obtained from healthy cows and its origin corresponds to PDO rules; furthermore whether milk shows good renneting properties, while its microbial contamination is at an acceptable degree and finally it represents an optimal substrate for the specific LAB development. All these parameters are costly and are justified when they guarantee the customers, but also contribute to the improvement of cheese quality (farmers aware and obtaining higher prices, cheese-maker helped in his job) maintaining the cheese peculiarities.

Keywords: PDO cheese, milk traits, milk evaluation, raw milk

Introduction

It is widely recognized that PDO cheeses have very desirable and peculiar traits, but it is also known that these features are strictly influenced by milk, namely his specificity, composition, cheese making aptitude and microbiological traits. This means that milk for PDO cheeses should be monitored to guarantee its origin, but also its suitability to a successful cheesemaking (Bertoni *et al.*, 2005).

In late 1500 the mistress of the house knew how to recognize good milk: "...from its whiteness, pleasing odour, sweet taste, a substance of medium thickness, dripped from a nail to see whether it tends to drip immediately or to keep the round shape of a drop for a long time" (Stefano, 1609). A couple of centuries later, with reference to the Lodigiano (similar to Grana Padano), the good dairyman wishing to optimise cheese production, after having checked with the herdsman that the cows had received good forage, and had not been treated badly and that the milking had been done in the best conditions, "went to examine the milk with all his senses...." (it can be supposed sight, taste, smell and touch). In fact, in "The safest rules to make the longest-lasting and the best Lodigiano cheese" by Castelli (1789), it is conveniently affirmed that: "To know therefore when the milk is more or less healthy to suggest the use of more or less heat and rennet is part of the art and skill which distinguishes one dairyman from another". In other words, the dairyman would recognize the milk traits in order to optimize the process.

Nowadays the milk suitability for PDO cheese production is based on more objective tools and must regard special features that could differentiate cheese from a sensorial point of view, but that can also allow its traceability. Moreover they must demonstrate its good renneting properties, as well as its proper microbial population (a high level of useful ones and low level of dangerous ones) with optimal growth conditions in the milk itself.

The traceability of milk can be useful in milk itself, but many aspects can be also recognised (and are sometimes better) in cheese; on the contrary, renneting and microbiological traits should be recognized in milk to facilitate the dairyman's "mission" which is that of good and safe cheese.

Previously referred historical knowledge allow us to understand that milk evaluation is not new for the dairyman. More than 2 centuries ago, the dairyman was already aware of the need to adapt the cheese-making process to milk features; nowadays still seems sound the Quintilianus' (III book) statement, "Long, hard study is meaningless if it does not provide the possibility to improve existing knowledge". So far we can in fact evaluate the quality of milk thanks to its major components and to several parameters: chemical, physical and microbiological (Figure 1); we can also recognize the factors responsible for the quality, which are those quoted as territory/environment in Figure 1.

This knowledge is extremely important; in fact, although the quality of cheese depends on the dairy technology too, the variable quality of the milk could suggest technological changes. Nevertheless, the range of dairyman adaptability is not endless; therefore, milk traits can change but within a quite well-defined reference range in many parameters. However we need to refine continuously this knowledge to allow a better expectation of milk response to cheese-processing and to facilitate the job of the dairyman, resulting in a lower number of defects and in a higher quality level of cheese.

Therefore milk evaluation for PDO cheese production would regard, as shown in the "holistic" view of Figure 2:

- specificity and traceability, that is the possibility of verifying the origin of milk from a certain area (and/or production system). This can mean - depending on the situation - a different breed (or species) and also different specific feedstuffs from the geographical area that could modify the micropopulation, as well as the presence of different chemical substances (fatty acids, terpenoids etc.);
- the proper renneting aptitude which is of major importance for the rheological characteristics of curd and then for the chemico-physical and structural properties of cheese. They influence whey drainage, which affects the moisture content of the curd, which is fundamental for the

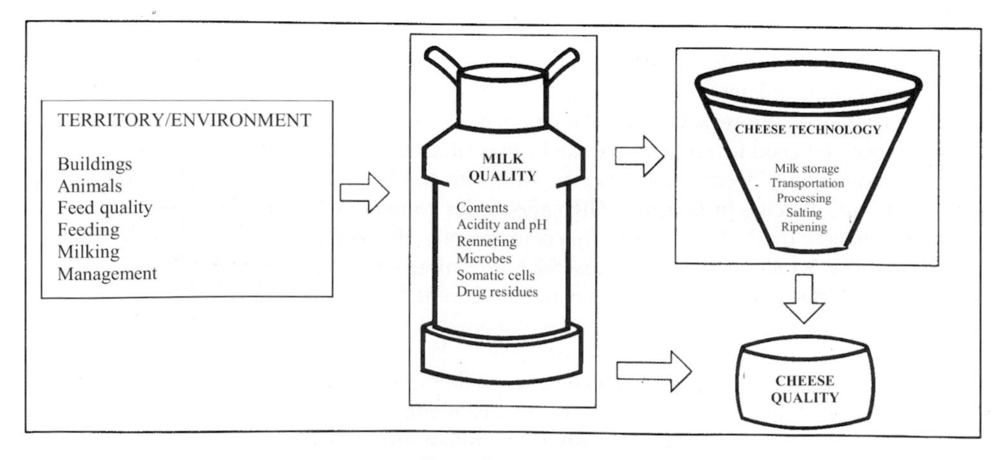

Figure 1. Cheese quality depends on milk quality and technology (Anonymous, 1992).

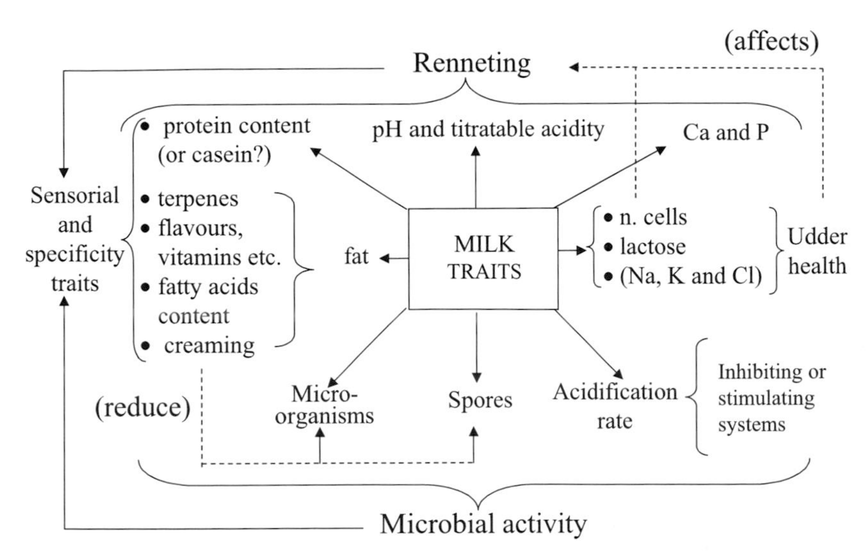

Figure 2. Milk traits affecting PDO cheese success (and parameters useful to monitor them).

appropriate start of the cheese-ripening process which is in turn linked to fermentation activity (Pecorari *et al.*, 1988);

- the microbial-fermentative properties that can first of all guarantee the hygiene and furthermore the majority of the organoleptic characteristics of the cheese (Martin and Coulon, 1995). These traits, however, cannot only be measured in terms of the "absence" of undesirable micro-organisms (Listeria, clostridia etc.), but also in terms of the presence of local, useful LAB strains specific to the territory (Bouton and Grappin, 1995), as well as in terms of fermentation suitability (Bertoni, 1996).

Parameters checked at milking

Before discussing which parameters can allow proper differentiation of milk peculiarities (traceability), its cheese-making aptitude and its microbial properties, it appears useful to discuss shortly the first tests carried out on the milk. It is the first thing the milker does, and its aim is both to ascertain whether the milk is normal, and to evaluate the general health status of the cow and particularly of the udder. It is an old practice mainly based on the senses: colour, consistency of the liquid, presence of clots, smell etc.; unfortunately it is becoming progressively less accurate because the milker has less and less time, and also because there is a larger use of more or less automated milking systems (even the robot) for which this first test would become an automatic one.

Otherwise there is a E.U. mandatory rule according to which some hygiene conditions must be guaranteed, namely "before milking the milker must check the appearance of the first few milk spurts. The milk of any cow must be excluded from commercialisation if it shows any physical anomaly" (Council directive EEC, 1992a; Council directive EEC, 1992b).

As stated before, it is becoming more and more difficult for milkers to carry out this first monitoring, particularly in the case of automatic milking; thus it becomes even more useful to use online analytical methods, in particular to detect infectious mastitis, which requires quick therapy. This need is currently met by a combination of human inspection of animals, by electrodes to measure in the electrical conductivity of milk, and by the analysis of data in herd management software to detect changes in milk yield and other factors (Hogeveen and Meijering, 2000). However, De Mol and Ouweltjes (2000) reported results of single and combined measurements from 29,033 milkings to detect clinical mastitis and concluded that early warning with sensors and software then on the market was not reliable.

Nevertheless, there is a relatively high genetic correlation between electrical conductivity and mastitis (Rogers, 2002, cit. by Norberg *et al.*, 2004a). This is quite promising for more reliable monitoring of udder health, and it is likely it will change due to the use of more automated sensor-based systems, especially in automatic milking systems (Norberg *et al.*, 2004a). In fact, according to Pyorala (2003), the most important promising parameters for monitoring sub-clinical mastitis are milk N-acetyl-beta-D-glucosaminidase (NAG ase) activity, lactose, but also electrical conductivity as confirmed by Norberg *et al.* (2004b), along with some other indicators such as optical and milk flow measurements, preferably with an inter-quarter evaluation included in the test. This hypothesis could be reliable considering that lactose evaluation via biosensors is available (Tkac *et al.*, 2000) as well as the evaluation of NAGase by an electro-chemical sensor (Mottram *et al.*, 2000), while Wiedemann and Wendl (2003) have suggested the use of colour measurement for the on-line determination of milk quality during milking. The on-line colour measurement has in fact allowed the detection of milk from diseased udders and the method disregarded the strong influence of milk fat content on milk colour. Therefore a better detection rate of sub-clinically diseased quarters and, hence, altered milk quality, could be obtained by combining these colour values with other parameters (i.e. the physico-chemical and biological sensors previously reviewed).

These coming possibilities are of great interest for milk producers, but they could be of further interest if new quick sensors could allow the evaluation of other physiological, nutritional and health indices: progesterone (Delwiche *et al.*, 2001; De Mol, 2001), urea (Jenkins *et al.*, 1999; Jenkins *et al.*, 2002a), mastitis, as shown before, as well as antibiotic residues (Gaudin *et al.*, 2001). On the same direction, although a little more complex, are the possibilities foreseen with infrared technology, namely Fourier transform Infrared technology (FT-IR). This has been used to analyse bulk and individual milk for urea content thus allowing a judgement on protein (and partly energy) imbalance (Van den Bijgaart, 2003). The direct comparison between individual milk urea values obtained by FT-IR and an automated flow injection method (FIA) showed a high correlation (r=0.931) (Baugmartner *et al.*, 2003a). Furthermore the determination of acetone by FT-IR could be used for the detection of sub-clinical ketosis (Heuer *et al.*, 2001) and Hansen (1999) indicated that the accuracy of the acetone analysis by FT-IR was enough to classify cows into 2 groups: healthy and possibly ketotic (energy deficiency).

Many of these last parameters are not directly regarding milk as cheese raw matter, but it is worth remembering the motto: “healthy milk in a healthy cow”.

Parameters of specificity and traceability

Animal species and breeds

The tracing of the origin of raw material (McKean, 2001) is particularly important in PDO products linked to specific animal species or breed within species (i.e buffalo-milk for Mozzarella di bufala or Valdostana cow for Fontina cheese, etc.). DNA markers have already proven useful for evaluating raw material quality, i.e. through the detection of aplotypes of milk casein genes that influence the efficiency of cheese-making (Coulon, 2004), and can also find an application in animal product traceability along the food chain.

A wealth of literature and molecular methods is available on the identification of species, ranging from the use of centromeric DNA probes (Buntjer, 1997) to PCR on specific nuclear or mitochondrial genes (Tartaglia *et al.*, 1998). The targeting of mitochondrial DNA is particularly interesting, since the stability of this circular genome and the thousands of copies per cell permit the implementation of successful diagnostic attempts also in materials that have undergone severe technological treatment (Frezza *et al.*, 2003).

In addition, molecular methods have the potential to trace breed origin. This is somehow easier in products derived from single animals, since DNA fingerprinting is unique in each individual, with the exception of twins or highly inbred lines. In this case a "*farm to fork*" traceability system can be organised, collecting samples (e.g. hair) on the farm and verifying the identity between these and samples collected along the production and distribution chain. Microsatellite analysis is typically performed to accomplish this task. If the complex and expensive systematic collection of on-farm samples is to be avoided, two alternative "*fork to farm*" molecular traceability systems may be designed. The first relies on the fact that allele frequencies at polymorphic markers differ between breeds. Upon availability of a database of marker allele frequencies, the multilocus genotype of a product obtained from a single animal can be assigned to a specific breed using a probabilistic approach (Pritchard *et al.* 2000). Alternatively, breed-specific markers are to be identified and used to assign a product to a breed.

The molecular tracing of milk-derived products is more complex, since milk from several animals is first pooled and then processed. Each animal contributes DNA to the pool through milk somatic cells. This contribution is not constant, depending on fluctuations in somatic cell count in milk and production level, compared to that of all other animals contributing milk to the pool. Lactation stage, season, climate, other environmental factors, including management, and the onset of sub-clinical mastitis are all sources of variation. In addition, new animals (about 30% per year) regularly enter the farm to replace those removed for various reasons and contribute an additional source of variation in pools. DNA contribution to cheese of single animals, single farms and even groups of farms is therefore highly variable and makes most of the strategies previously described inapplicable to milk or cheese traceability. Indeed, the only strategy that can be pursued to assign a cheese to a specific breed is the use of breed specific markers or gene alleles.

These are not easily found. Domestic animal history is quite recent (domestication occurred somewhere 10,000 years ago) and breed formation is very recent (only a handful of generations ago). In these conditions, it is very unlikely that different alleles have been fixed in different breeds by genetic drift. However, fixation of different alleles might have happened by human or natural directional selection. In this view, candidate genes to be investigated are those controlling phenotypic traits and production aptitude and, among those, genes controlling coat colour.

Indeed, polymorphisms at the melanocortin receptor 1 (MC1R) gene have been proposed as tools for verifying animal products produced by specific cattle breeds to be free from "Holstein contamination" (Chung *et al.*, 2000; Maudet and Taberlet, 2002; Russo and Fontanesi, 2004). MC1R maps at the locus Extension and is involved in molecular mechanisms regulating eumelanin/pheomelanin production in melanocytes. Two functional mutations have been identified at this locus, that modify MC1R either to a completely inactive state (recessive *e* allele, producing pheomelanic red coat) or to a constitutively active state (dominant E^d allele, producing eumelanic black coat; Klungland *et al.*, 1995). Molecular methods may therefore distinguish milk derived from red pheomelanic (homozygous for the *e* allele), black eumelanic (homozygous or heterozygous for the E^d allele) and brown (carrying the wild type *E* allele) breeds. The investigation of other loci controlling coat colour (e.g. *agouti*, *kit*, etc.,) may allow the identification, in the near future, of breed specific genomic formulae univocally characterising breeds, and permitting their traceability.

Breeding and feeding systems

Traceability of a specific cheese, namely PDO, can also mean the possibility to know the pattern of the chain: area of animal breeding production system, feeding rules etc. This can be obtained for instance by the evaluation of some compounds common in plants which are specific in some areas. This seems to be the case of 1-carvone and many other chemical compounds identified in the plants of Sicilian pastures and that could characterize the locally obtained cheese (Carpino *et al.*, 2004a; Carpino *et al.*, 2004b). It is also the case of terpens evaluation, in milk and cheese, which is an index of natural mixed pastures (Viallon *et al.*, 2000; Rubino and Claps, 2000; Cornu *et al.*, 2002); furthermore the fatty acid composition, namely the long chain unsaturated fatty acids, can be useful to differentiate milk from cows grazing on lowland or highland pastures (Collomb *et al.*, 1999; Bugaud *et al.*, 2002).

The possibility to trace milk (and cheese) according to the use of fresh forages rather than hay or silage can also be of great importance (Martin *et al.*, 2002). According to Verdier-Metz *et al.* (1998), the cheese obtained from the milk of cows fed on grass silage is yellower than from grass hay; unfortunately the difference cannot be easily interpreted as an indication of hay or silage feeding, also because fresh grass causes a higher intensity of yellow colour (Coulon and Priolo, 2002) and there are often mixed situations. Of course the use of many different parameters can help in this effort (Coulon *et al.*, 2004).

So far it remains difficult to say whether the analysis of milk (or cheese) can allow a precise fingerprinting of its origin, nevertheless it seems promising for the future; the levels in casein of some stable isotopes like $^{13}C/^{12}C$ and $^{15}N/^{14}N$ (Versini *et al.*, 2000) could be of great interest. In fact, these stable isotopes seem to be modified by the type (C_3 or C_4) of photosynthetic cycle of plants, by the use of chemical fertilizers, by grass vs. legumes forages, by water stress etc. These different factors could allow the geographical localization of cheese origin. Very promising results have been obtained with multi-element stable isotope analysis of butter for C, N, O, S and Sr, together with discrimination analysis of the results and their evaluation by comparison with data for certified authentic samples; the authors (Rossmann *et al.*, 2000) suggest that it can be a very potent tool with which to solve the problem of butter origin assignment. More recently the potential value of this technique for cheeses has been confirmed by Chiaccherini *et al.* (2002) who were able to characterize sheep and cow cheeses according to geographical area and feeding techniques.

An index of silage feeding (at least corn silage) could be the chromatographic evaluation of 1-Methyl-β-carboline-3-carbozilic Acid (Ia), an unusual metabolite observed in urine and milk of corn-fed silage (Bertoni *et al.*, 1980). Average values of Ia in the rumen, milk and urine of control animals were respectively 0.2, 0.08, and 0.54 μg/mL. The concentration of Ia increased steadily in fluids of animals fed with corn silage, reaching an average value of 30 times the control value in rumen and 100 times the control value in urine after 20-25 days from the beginning of the diet. The concentration in milk (approx. five times the control value) reached its highest point after 2 months of feeding with corn silage. Because Ia is not present in corn silage, it is possible that the silage feeding conditions either induce different fermentation processes in the rumen, with a higher production of Ia, or hinder the normal catabolic activity, with an increasing absorption of the compound, which is then excreted in milk and urine (Bertoni *et al.*, 1980). Further studies are obviously needed for a better comprehension of its usefulness.

Heat milk treatments

In the end, the evaluation of milk heat treatments could also be useful; it is in fact known that they can improve the technical results of cheese production, but as a consequence there are some losses of cheese peculiarities. In other words it can be important to know whether a cheese has been produced with real raw milk. Because some milk enzymes are inactivated by heat treatments, their evaluation in milk or cheese can be an indicator of heat-treatments or thermization; this is the case of alkaline phosphatase (Pellegrino *et al.*, 1995) and of α-L-fucosidase (McKellar and Piyasena, 2000).

Parameters of cheese-making aptitude

The cheese-making aptitude of milk is tremendously important for success in the cheese-making process and the possibility to evaluate it in milk can be of great interest: in order to adapt the technology, but also to select suitable vs unsuitable milk and therefore to pay for milk also on the basis of this trait. However there are many factors affecting this trait (Remeuf *et al.*, 1991): casein content, polymorphisms of k-casein and β-lactoglobulin, size and T° tolerance of casein micelles, pH (or titratable acidity), calcium and chlorine contents etc.. Many of these factors have been discussed in other companion papers (Barber *et al.*, 2005; Bertoni *et al.*, 2005). Moreover, the interactions between these factors are not yet well clarified; also for these reasons the best way to evaluate the renneting properties of milk has been the Formagraph (Annibaldi *et al.*, 1977), a parameter that tries to mimic the first steps of a real cheese-making process and include rennet clotting time, curd firming rate and curd firmness.

Unfortunately it is a time-consuming method; furthermore the rheological aspects of renneting are well-known but a complete set of the parameters affecting it would be worse in terms of costs and interpretation. However, it may be worthwhile evaluating the various parameters affecting the renneting (i.e. caseins, titratable acidity etc.) or rheological properties with quick single multi-purpose equipment.

Infra-red spectrometry is by far the most frequently applied methodology for the compositional analysis of milk and milk products at present. The technique is successfully applied to provide analytical data for milk payment based on composition, for farm management and breeding purposes in the framework of dairy herd improvement and for the testing of raw, intermediate and finished products in dairy processing. The determination of fat, protein and lactose, has been described during recent decades (Barbano and Clark, 1989). In recent years, the new application of infra-red methodology has been launched and successfully implemented for dairy analysis. This was partly triggered by the increased interest in cost-effective collection of data on the composition and quality of milk and milk products. At the same time, the introduction of Fourier Transform Infra-red (FT-IR) technology in combination with the application of multi-dimensional procedures such as the classic least square, principal component regression or partial least squares, has improved this methodology. The spectral information provided by these instruments makes it possible to obtain more detailed information on milk composition; despite the increased data information, rapid measurements can still be obtained, especially when the technique is based on interferometry rather than on dispersive spectrometry (Agnet, 1998). Technological developments have now made full spectrometers available for routine analysis of milk and have opened perspectives for the simultaneous and routine determination of many parameters such as casein, urea, freezing point, pH, specific sugars, lactic acid, free fatty acids and citric acid, as well as fat, lactose and protein.

A relevant parameter for the renneting property of milk is the casein content, but the reference method for determination in milk, Kjeldahl, is a very time-consuming procedure involving precipitation and nitrogen determinations, while several studies have shown that casein/protein ratio - a simpler calculation method - is not fixed, but may be influenced by genetic and physiological factors (Coulon *et al.*, 1998). Therefore, for routine quality-control and process-control purposes there is a need for a cheaper and faster method, and IR spectroscopy in combination with multivariate calibration is a potential candidate as FT-IR has been the most widespread method used for compositional analysis in the dairy industry (Pedersen, 2003). For the determination of casein by FT-IR in milk samples with a large variation in casein number, two possibilities exist: first the double IR protein measurement before and after precipitation of casein, second the direct FT-IR method. The former measures with an accuracy of approximately 1.4% and the latter measures with an accuracy of approximately 4% (Pedersen, 2003). Nevertheless poor correlation was observed by Sørensen *et al.* (2003) between casein numbers obtained by reference analysis and direct FT-IR spectrometry: R^2 of 0.73 and a standard error of prediction of 0.89 for casein numbers in the range 70.8-81.0. Better results were obtained by Hewavitharana and Van Brakel (1997); these authors observed a good correlation coefficient (r=0.976) between direct casein measurement by FT-IR and reference method on 300 samples obtained from several dairy farms. Further studies would be welcomed.

Although commercial FT-IR equipments could evaluate the titratable acidity - another relevant parameter of renneting properties - limited scientific literature is available only for milk pH. Baugmartner *et al.* (2003b) studied the possibility to measure pH by mid-IR FT-IR spectrometry and concluded that, although based only on few available data, this is a very promising tool. Accuracy data, obtained by using the potentiometric pH measurements as a reference method, have shown a standard error of prediction of 0.064 pH units.

In conclusion, the FT-IR technology could help the herdsman and dairyman with a quicker and cheaper evaluation of milk also regarding cheese-making aptitude. At present this is available only through some parameters like casein, titratable acidity, pH etc., but the possibility to have from FT-IR technology a response similar to the Formagraph one is at least a hope for the future.

Parameters of microbial-fermentative properties

Microbial traits

First of all the amount and type of microbes are important because they could affect cheese-production success and safety. The total microbial population is evaluated (Battistotti, 1990) by indirect methods (i.e. lactic acid, methilene blue reduction etc.) or direct ones (i.e. plate count technique). At present there are automated methods based on the impedimetric technique that measure the electrical changes occurring in a substrate as evidence of bacterial metabolism (Pirovano *et al.*, 1995; Curda and Plockova, 1995).

For cheeses obtained from raw milk the total number of bacteria would be important to ensure an autochthonous population which in turn could improve the PDO cheese specific features (Mordenti and Pecorari, 1999). Nevertheless, to evaluate this microbial specificity, the total bacterial count is of very poor interest; the microbiological classification based on the species is useful but the major differences could be those regarding the biotypes of the same species. For this reason the new methods use DNA characterization to obtain a fingerprinting of the specific strains, after a PCR which utilize species-specific primers (Cavazza *et al.*, 2000), that can allow a differentiation of *crus* of the same cheese (Bertheir *et al.*, 2001; Andrighetto *et al.*, 2002).

These new methods can be useful also for cheese-processing based on the use of natural cultures (i.e. whey starter for Parmigiano-Reggiano or for Grana Padano) because the naturally selected bacteria can be modified according to milk contamination (forages, stalls) but also its micro constituents composition (Coulon *et al.*, 2003) that can modify microbe activity (Bugaud *et al.*, 2002) as will be suggested later (milk fermentability).

A fairly new possibility has been proposed by Oberreuter *et al.* (2003) who used the FT-IR spectroscopy for monitoring the population dynamics of microorganisms in foodstuff. Application of this method will enable a food producer to obtain a detailed overview on the microbiological status of a product down to the species level, which can suggest correction measures to be taken early in the process. Amiel *et al.* (2000) used the FT-IR for the discrimination and identification of dairy Lactic Acid Bacteria (LAB).

Besides a specification of the useful LAB, the evaluation of milk micro-population would regard the "antidairy" bacteria (i.e. clostridia) and potentially pathogenic bacteria (i.e. coliform). The clostridia (namely *C. tyrobutyricum*) are particularly dangerous for hard long-ripening cheeses (Emmenthal type, Parmigiano-Reggiano etc.) providing a late swelling. They can contaminate milk through the faeces where they arrive after a long journey starting from the soil with a growing stage in the silages (Demarquilly, 1998) and perhaps in the digestive tract of the cow (Bani *et al.*, 2001), as well described by Bottazzi *et al.* (1999). So far spore evaluation has been based on the More Probable Number (MPN) after incubation in a liquid medium (Bergère and Sivelä, 1989); nevertheless, new methods based on the use of specific antibodies were described by the same authors. Unfortunately, although the MPN method is time- and labour-consuming, it remains widely used and only recently has some attention been paid to immuno-enzymatic methods that are labour-intensive too (milk has to be filtered or the MPN incubation phase must be carried out), but they allow the specific identification of *C. tyrobutyricum* or that of *C. sporogenes* and *C. butyricum* which are the cause of early swelling in cheeses (Lodi and Brasca, 2000). For similar reasons, to distinguish between different species as well as alive or dead cells, detection based on molecular methods has also been tried (Herrewegh and Nieuwenhof, 2000) using DNA by PCR or RNA after amplification. A fairly new method for total bacteria and spore counting with automatic techniques is based on impedance technology; Fontana *et al.* (2002) have in fact obtained results in less than 48 hrs and the data obtained from the rapid impedance method were statistically comparable to those obtained using the reference method (MPN).

The coliform bacteria should be evaluated as indicators of milking hygiene because they come from faeces; however, they are of very great importance in the case of cheese-processing without added LAB cultures. The present and future evaluation methods are based on traditional methods (MPN or plate count with selective substrates).

In some cheese technologies, like Gruyère, Parmigiano-Reggiano and Grana Padano, the milk is submitted to a natural creaming process; in this case the final microbial population could be very different for the reduction in active bacteria and spores through their agglutination and the physical effect of fat globules rising during natural creaming. This means that the creaming activity should be measured too, and a fairly simple method has been proposed by one of us (Speroni and Bertoni, 1984); however in recent studies (Vecchia *et al.*, 2003) we have shown a close relationship between creaming activity and Aspartate amino transferase (AST-GOT). It has been suggested that the measurement of the enzyme (index of increased citoplasmatic material in the fat globule) could be an indirect parameter of creaming (Bertoni, 1996).

Fermentability traits

Knowledge of the microbial population of milk is important but it does not exhaust the topic; milk can in fact reduce the bacteria fermentation ability (Marth, 1974). This aspect has not received enough research attention, although a different LAB activity, and therefore acidification rate, has been suggested for a long time as a factor of hard cheese defects, corn silage being a possible cause (Battistotti and Bertoni, 1974). Unfortunately, a major difficulty is a proper and repeatable evaluation method; in fact it has been previously observed that the acidification rate can be reduced by antimicrobial systems - many of which can be neutralized by more or less severe heat treatments- or by nutrient deprivation. Therefore its evaluation has to be performed with a direct method, monitoring the growth rate of bacteria involved in the process. Namely during milk fermentation, organic acids, mainly lactic acid, are produced thereby lowering the pH. Then a direct measurement of pH or the total acid production can be used to characterize milk fermentation rate. Battistotti and Bertoni (1974) evaluated the acidification rate throughout the measurement of the increase of the titratable acidity on pasteurised milk (80°C x 10 min) inoculated with *Streptococcus thermophilus* and *Lactobacillus delbrueckii ssp. bulgaricus* and incubated at 45°C for a maximum time of 6 hours. Later on the incubation temperature was reduced to 42°C and the method was simplified, measuring the pH reduction (Calamari *et al.*, 1986) or measuring the increase of titratable acidity after 3.5 hours of incubation (Calamari *et al.*, 2003). Then Bertoni *et al.* (1992) and Calamari *et al.* (2003) suggested an evaluation of the acidification rate both on raw milk and on pasteurised milk to better evaluate the two aspects involved: the presence of antimicrobial agents (reduced after pasteurisation) and the availability of nutrients for bacteria. After milk has been pasteurised, the acidification rate is usually faster than on raw milk, but the highest values (3-5 times) are obtained after sterilization (Maianti *et al.*, 1995).

To determine the initial milk acidification rate, Jenkins *et al.* (2002b) use the slope of the regression line of the pH values measured during the first 2 hrs of incubation. Clementi *et al.* (1998) used a fourth degree polynomial to determine the maximum instantaneous acidification rate and the time and the pH at which it occurred. Demarigny *et al.* (1994) observed, using milk cultures with different strains of *Lactococcus lactis*, a nearly linear pH decrease in the pH range 6.2-4.8, regardless of the strains or the milk in which the bacteria were grown. They proposed a mean acidification rate (pH unit/h) calculated by linear regression in the pH range 6.2-4.8. The mean acidification rate did not depend on the level of milk inoculation, and enabled significant differences to be observed between strains and/or milks. Fonseca *et al.* (2000) used the continuous titration of the acid produced with the aid of alkali addition during fermentation. Evaluation of the acidification rate could be carried out throughout the measurement of the acid production and the FT-IR offers the possibility to measure lactic acid in fermented dairy products (Bendtsen, 2003). No matter the detection method, the best results would be those obtained with the natural autochthonous bacteria.

Conclusions

The value of PDO cheeses has greatly increased, not only in economic terms but also as a cultural value, in particular in some European countries. They can fulfil particular organoleptic expectations, and at the same time represent the product of a traditional exploitation of the land. Nevertheless their production needs traditional technology that include milk with some specific traits and naturally suitable for cheese-making. This means that milk for PDO cheeses should be monitored to guarantee its origin, but also its suitability to a successful cheesemaking. In other words, a suitable analysis of milk (and perhaps of cheese) would allow milk traceability as well as proper evaluation of coagulation, microbiological traits, besides the traditional composition (i.e. fat and protein contents). Many well known methods exist to accomplish these tasks, but our effort has concentrated on new techniques, some of which are on-line, now available to recognize whether:

- milk is obtained from healthy cows (electrical conductivity, NAGase, urea, ketone bodies, color etc.);
- milk origin corresponds to PDO rules (breed of origin, by DNA analysis; breeding and feeding systems, by microcomponents like terpenes, stable isotopes evaluation, fatty acid composition etc.);
- milk could properly coagulate (new possibilities are made available by FT-IR: casein content, titratable acidity are indirect index of the rheological traits which in future might be directly evaluated in milk with the same methodology);
- the level of milk contamination by antidairy microbes is acceptable while the useful LAB are well represented (DNA and immunological analysis will improve the precision and decrease the time required for the assay, but impedimetric methods will be also useful);
- milk represents an optimal substrate only for the specific LAB development (acidification rate evaluation needs a better setting).

Before concluding, it is important to remember that all these analyses are very useful, but expensive too; therefore they should be selected according to real needs. Furthermore, some of

Table 1. Pattern for the evaluation of milk quality [bonus in Euro (€)] established by Parmigiano-Reggiano since 1983 (Zannoni and Mora, 1993).

Parameter	Rules for scoring		
	Values	Score (€/1000l)	
		Evening milking	Morning milking
Titratable acidity	3.20-3.80	5.16	5.16
	3.00-3.15	2.58	2.58
	<3.00->3.80	0	0
Somatic cell count	≤ 300.000	7.75	7.75
	301.000-600.000	5.16	5.16
	601.000-1.000.000	2.58	2.58
	> 1.000.000	0	0
Coliforms	≤ 5.000 cfu/ml	5.16	5.16
	6-20.000 cfu/ml	2.58	2.58
	> 20.000 cfu/ml	0	0
Clostridia	Absent	2.58	2.58
	Present	0	0
Clotting features[1]	Type A-B-C	-	10.33
	Type D-E-EA-EF	-	5.16
	Type F-FF-DD	-	0
Fat %	*According to percentage*		
Casein %	*According to percentage*		
Inhibitors	*At least once a month*		

[1] Cheesemaking aptitutde in relation to clotting time, clot firming rate and curd firmness (Zannoni and Annibaldi, 1981):
A = optimal; B = good; C, D = fairly good; E, EA = middling; DD, EF = poor; F, FF = not suitable

them can be useful to demonstrate extra value in milk - following from the possibility of producing higher price cheeses - which would represent a bonus for the breeder. This is particularly important, because milk for PDO cheeses is often more expensive, but the higher price must be justified by objective better traits. A good example of a convenient cost/benefit proportion of milk analyses can be found in the procedure to evaluate milk for Parmigiano-Reggiano (Table 1).

To conclude, an easier possibility to define the specific traits of milk for PDO cheese would represent a guarantee either for the consumers as well as for the farmers. Nevertheless, these new opportunities must be a chance to improve PDO cheeses, but never a risk for their traditional traits. It is in fact wise to use new technological tools (innovation) to obtain better results:

- improving milk quality by an advisory activity for the farmers, but also encouraging them with milk price according to its quality (properly evaluated);
- improving the cheese-making process with better equipments, but also adapting the process conditions (rennet, natural culture, temperature pathway etc.) to the actual milk traits (still properly evaluated).

On the contrary it would be dangerous to distort the milk production system or the cheese process according to these new technology, with a strong risk to loose the cheese peculiarities that are the aim of any change and must be used to discriminate what useful or not.

References

Agnet, Y., 1998. Fourier transform infrared spectrometry. A new concept for milk and milk product analysis. Bulletin of the IDF 332: p. 58-68.

Amiel, C., L. Mariey, M.C. Curk-Daubie, P. Pichon and J. Trevert, 2000. Potentiality of Fourier transform infrared spectroscopy (FT-IR) for discrimination and identification of dairy lactic acid bacteria. Lait 80: p. 445-459.

Andrighetto, C., F. Borney, A. Barmaz, B. Stefanon and A. Lombardi, 2002. Genetic diversity of *Streptococcus thermophilus* strains isolated from Italian traditional cheeses. Int. Dairy J. 12: p. 141-144.

Annibaldi, S., G. Ferri and R. Mora, 1977. Nuovi orientamenti nella valutazione tecnica del latte: tipizzazione lattodinamografica. Sci. Tecn. Latt-cas., 28:115-126

Anonymous, 1992. Il nuovo metodo di pagamento del latte secondo la qualità. Il Parmigiano-Reggiano, 22(2): p. 1-4.

Bani, P., L. Calamari, E. Bettinelli and G. Bertoni, 2001. Composizione della razione e contenuto in spore di clostridi nelle feci di bovine in lattazione. Sci. Tecn. Latt-Cas., 52 : p. 369-386.

Barbano, D.M. and J.L. Clark, 1989. Infrared Milk analysis - Challenges for the future. Symposium: instrumental methods for measuring components of milk. J. Dairy Sci. 72: p. 1627-1636.

Barber, D.G., A.V. Houlihan, F.C. Lynch and D.P. Poppi, 2005. The influence of nutrition, genotype and stage of lactation on milk casein composition. In: Indicators of milk and beef quality, J.F. Hocquette and S. Gigli (editors), Wageningen Academic Publishers, The Netherlands, in press.

Battistotti, B., 1990. Recenti acquisizioni sul valore di alcuni parametri microbiologici del latte. In Atti del Convegno "Qualità del latte, proprietà tecnologiche e resa casearia. Aggiornamenti e prospettive", Verona, 13 marzo 1990, p. 43-52.

Battistotti, B. and G. Bertoni, 1974. The milk from cows fed with corn silage as medium for lactic acid bacteria growth. Scienza e Tecnica Lattiero-Casearia 25, Notiziario Tecnico, p. 7-10.

Baugmartner, M., M. Flock, P. Winter, W. Luf and W. Baugmartner, 2003a. Evaluation of Fourier Transform Infrared spectrometry for the routine determination of urea in milk. Milchwissenschaft 58: p. 599-602.

Baugmartner, C., A. Landgraf and J. Buermeyer, 2003b. pH determination. Bulletin of the IDF 383: p. 23-28.

Bendtsen, A.B., 2003. Lactic acid and specific sugars. Bulletin of the IDF 383: 29-32

Bergère, J.L. and S. Sivelä, 1989. Detection and enumeration of clostridial spores related to cheese quality. Classical and new methods. A-Doc 112, FIL-IDF: p. 2-7.

Bertheir, F., E. Beuvier, A. Dasen and R. Grappin, 2001. Origin and diversity of mesophilic lactobacilli in Comté cheese, as revealed by PCR with repetitive and species-specific primers. Int. Dairy J. 11: p. 293-305.

Bertoni, G., 1996. Environment, feeding and milk quality. L'Informatore Agrario 52 Suppl.: p. 5-41.

Bertoni, G., L. Calamari, M.G. Maianti and A. Azzoni, 1992. Preliminary research into factors that modify the milk acidification rate. Proceedings of the Progetto Finalizzato Moderne Strategie Lattiero-Casearie, Tecniche Nuove, Milano, Italy, p. 35-37.

Bertoni, G., L. Calamari, M.G. Maianti and B. Battistotti, 2005. Milk for Protected Denomination of Origin (PDO) cheeses: I. The main required features. In: Indicators of milk and beef quality, J.F. Hocquette and S. Gigli (editors), Wageningen Academic Publishers, The Netherlands, in press.

Bertoni, G., L. Merlini, G. Nasini and U. Abenaim, 1980. 1-Methyl-β-carboline-3-carboxylic Acid, an unusual metabolite from cows fed with corn silage. J. Agric. Food Chem., 28: p. 672-673.

Bottazzi, V., B. Battistotti and A. Rebecchi, 1999. Ambiente e microflora del latte: una relazione che cambia. Latte 24: p. 84-88.

Bouton, Y. and R. Grappin, 1995. Comparison of the quality of scalded-curd cheeses manufactured from raw or microfiltered milk. Lait 75: p. 31-44.

Bugaud, C., S. Buchin, A. Hauwuy and J.B. Coulon, 2002. Texture et flaveur du fromage selon la nature du pâturage: cas du fromage d'abondance. INRA Prod. Anim. 15: p. 31-36.

Buntjer, J., 1997. Ph. D. Thesis "Dna repeats in the vertebrate genome as probes in phylogeny and species identification, University of Utrecht, NL.

Calamari, L., P. Bani, G. Bertoni and V. Cappa, 1986. Changes of milk composition and characteristics during lactation. In Quality control of the milk and dairy products. Proceedings XXI International Simposium on Animal Production, C.N. Baldissera (editor), Società Italiana per il progresso per la zootecnia, Milano, Italy, p. 223-229.

Calamari, L., P. Bani, M.G. Maianti and G. Bertoni, 2003. New researches on the factors affecting milk acidification rate. 38° Simposio Internazionale di Zootecnia, Milk and Research, Lodi 30 maggio 2003 (edited by G.F. Greppi. G. Enne): p. 137-144.

Carpino, S., J. Horne, C. Melilli, G. Licitra, D.M. Barbano and P.J. Van Soest, 2004a. Contribution of native pasture to the sensory properties of Ragusano cheese. J. Dairy Sci. 87: p. 308-315.

Carpino, S., S. Mallia, S. La Terra, C. Melilli, G. Licitra, T.E. Acree, D.M. Barbano and P.J. Van Soest, 2004b. Composition and aroma compounds of Ragusano cheese: native pasture and total mixed rations. J. Dairy Sci. 87: p. 816-830.

Castelli, C., 1789. Regole più sicure di fare il migliore, e più durevole formaggio nel lodigiano. A cura di G. Bologna, Grana Padano, Milano, 1996.

Cavazza, A., E. Poznanski, M.S. Tedeschi, F. Gasperi, P. Cocconcelli and F. Cappa, 2000. Caratterizzazione e tutela delle produzioni tipiche di montagna: accertamenti microbiologici innovativi. Atti del V Convivio Formaggi d'alpeggio: il pascolo, l'animale, la razza, il prodotto. Cavalese (TN), 15 settembre, p. 135-144.

Chiaccherini, E., P. Bogoni, M.A. Franco, M. Giaccio and G. Versini (Coautori: F. Camin, L., Ziller), 2002. Characterisation of the regional origin of sheep and cow cheeses by casein stable isotope $^{13}C/^{12}C$ and $^{15}N/^{14}N$) rations. J. Commodity Sci. 41(IV): p. 303-315

Chung, E.R., W.T. Kim, Y.S. Kim and S.K. Han, 2000. Identification of Hanwoo meat using PCR-RFLP marker of MC1R gene associated with bovine coat colour. Korean J. Anim. Sci 42: p. 379-390.

Clementi, F., B.T.C. Goga, M.T. Marinucci and E. Di Antonio, 1998. Use of selected starter cultures in the production of farm manufactured goat cheese from thermized milk. Italian Journal of Food Science 10: p. 41-56.

Collomb, M., U. Bütikofer, M. Spahni, B. Jeangros and J.O. Bosset, 1999. Composition en acides gras et en glycérides de la matière grasse du lait de vache en zone de montagne et de plaine. Sci. Aliments 19: p. 97-110.

Cornu, A., B. Martin, I. Verdier-Metz, P. Pradel, J.B. Coulon and J.L. Berdagué, 2002. Use of terpene profile in dairy produce to trace the diet of dairy cows. In: Multi-function grasslands: quality forages, animal products and landscapes, J.L. Durand, J.C. Emile, C. Huyghe and G. Lemaire (editors), p. 520-551. Proc. 19th General Meeting of the European Grassland Federation, La Rochelle, France, 27-30 May 2002. Diffusion British Grassland Society, Reading, UK.

Coulon, J.B., 2004. Qualità del latte e produzione casearia: l'effetto della genetica e dell'alimentazione sulle proprietà coagulanti del latte e sulle caratteristiche sensoriali del formaggio. 7ma Conferenza mondiale allevatori razza bruna, Verona, 3-7 marzo 2004, p. 55-65.

Coulon, J.B., A. Delacroix-Buchet, B. Martin and A. Pirisi, 2004. Relationships between ruminant management and sensory characteristics of cheeses: a review. Lait 84: p. 221-241.

Coulon, J.B., C. Hurtaud, B. Rémond and R. Vérité, 1998. Facteurs de variation de la proportion de caséines dans les protéines du lait de vache. INRA Prod. Anim. 11: p. 299-310.

Coulon, J.B. and A. Priolo, 2002. La qualité sensorielle des produits laitiers et de la viande dépend des fourrages consommés par les animaux. INRA Prod. Anim 15: p. 333-342.

Coulon J.B., E. Rock and Y. Noël, 2003. Caractèristiques nutritionnelle des produits laitiers et variations selon leur origine. INRA Prod. Anim 16: p. 275-278.

Council directive EEC, 1992a. Laying down the health rules for the production and placing on the market of row milk, heat-treated milk and milk-based products. 92/46/EEC. Official Journal EEC L 268, 14-09-1992, p. 1-32.

Council directive EEC, 1992b. On the conditions for granting temporary and limited derogations from specific community health rules on production and placing on the market of milk and milk-based products. 92/47/EEC. Official Journal EEC L 268, 14-09-1992, p. 33-34.

Curda, L. and M. Plockova, 1995. Impedance measurement of growth of lactic acid bacteria in dairy cultures with honey addition. Int. Dairy J. 5: p. 727-733.

De Mol, R.M., 2001. Automated detection of oestrus and mastitis in diary cows. Tijdschrift voor Diergeneeskunde 126: p. 99-103.

De Mol, R.M. and W. Ouweltjes, 2000. Detection model for mastitis in cows milked in an automatic milking system. In Robotic Milking, proceedings of the international symposium held in Lelystad, The Netherlands, 17-19 August 2000, Wageningen Press, The Netherlands, p. 97-107.

Delwiche, M., X. Tang, R. BonDurant and C. Munri, 2001. Improved biosensor for measurement of progesterone in bovine milk. Transactions of the ASAE 44(6): p. 1997-2002.

Demarigny, Y., V. Juillard, N. Deschamps and J. Richard, 1994. Comaparaison, sur le plan pratique, de 3 modèles d'étude de la cinétique d'abaissement du pH du lait cultivé par *Lactococcus lactis*. Proposition du concept "Vmar". Lait 74: p. 23-32.

Demarquilly, C., 1998. Ensilage et contamination du lait par les spores butyriques. INRA Prod. Anim. 11: p. 359-364.

Fonseca, F., C. Béal and G. Corrieu, 2000. Method of quantifying the loss of acidification activity of lactic acid starters during freezing and frozen storage. J. Dairy Res. 67: p. 83-90.

Fontana, M., S. Busiello, S. Bisotti, G. Dallorto, G. Unger, G. Schwameis, U. Juterschnig, B. Unger, H. Masaniger and M. Schinkinger, 2002. Rapid enumeration of clostridial spores in raw milk sample using an impedimetric method. J. Rapid Meth. Aut. Mic. 10: p. 107-116.

Frezza, D., M. Favaro, G. Vaccari, C. von-Holst, V. Giambra, E. Anklam, D. Bove, P.A. Battaglia, U. Agrimi, G. Brambilla, P. Ajmone-Marsan and M. Tartaglia, 2003. A competitive PCR-based approach for the identification and quantification of bovine MBM in feeds: Analysis of DNA degradation in relation to different heat treatments. J. Food Prot. 1: p. 103-109.

Gaudin, V., J. Fontaine and P. Maris, 2001. Screening of penicillin residues in milk by a surface plasmon resonance-based biosensor assay: comparison of chemical and enzymatic sample pre-treatment. Anal. Chim. Acta, 436: p. 191-198.

Hansen, P.W., 1999. Screening of dairy cows for ketosis by use of infrared spectroscopy and multivariate calibration. J. Dairy Sci. 82: p. 2005-2010.

Herrewegh, A.P.M. and F.J. Nieuwenhof, 2000. Molecular methods for detection of clostridia. In: New developments in detection and identification of spore-forming bacteria in milk and milk products. Monograph of IDF Group A19. Bulletin IDF 357: p. 36-39.

Heuer, C., H.J. Luinge, E.T.G. Lutz, Y.H. Schukken, J.H. van der Maas, H. Wilmink and J.P.T.M. Noordhuizen, 2001. Determination of acetone in cow milk by Fourier transform infrared spectroscopy for the detection of subclinical ketosis. J. Dairy Sci. 84: p. 575-582.

Hewavitharana, A.K. and B. van Brakel, 1997. Fourier transform infrared spectrometric method for the rapid determination of casein in raw milk. Analyst. 122: p. 701-704.

Hogeveen, H. and A. Meijering, 2000. Robotic Milking, proceedings of the international symposium held in Lelystad, The Netherlands, 17-19 August 2000, Wageningen Press, The Netherlands.

Jenkins, D.M., M.J. Delwiche, E.J. DePeters and R.H. BonDurant, 1999. Chemical assay of urea for automated sensing in milk. J. Dairy Sci 82: p. 1999-2004.

Jenkins, D.M., M.J. Delwiche, E.J. DePeters and R.H. BonDurant, 2002a. Factors affecting the application of on-line milk urea sensing. Transactions of the ASAE 45: p. 1687-1695.

Jenkins, J.K., W.J. Harper and P.D. Courtney, 2002b. Genetic diversity in Swiss cheese starter cultures assessed by pulsed field gel electrophoresis and arbitrarily primed PCR. Letters in Applied Microbiology 35: p. 423-427.

Klungland, H., D.I. Vage, L. Gomez-Raya, S. Adalsteisson and S. Lien, 1995. The role of melanocyte-stimulating hormone (MSH) receptor in bovine coat colour determination. Mamm. Genome 6: p. 636-639.

Lodi, R. and M. Brasca, 2000. Detection and enumeration of clostridia, of importance to cheese, with immunological methods. In: New developments in detection and identification of spore-forming bacteria in milk and milk products. Monograph of IDF Group A19. Bulletin IDF 357: p. 33-36.

Maianti, M.G., L. Calamari, G. Bertoni G and V. Cappa, 1995. Researches into some factors interfering with the acidification rate of milk. Il latte 7: p. 718-721.

Marth, E.H., 1974. Fermentations. In: Fundamentals of dairy chemistry, 2nd ed., B.H. Webb, A.H. Johnson and J.A. Alford (editors), The Avi Publishing Company Inc., Westport, Connecticut, p. 772-872.

Martin, B. and J.B. Coulon, 1995. Milk production and cheese characteristics. II. Influence of bulk milk characteristics and cheesemaking practices on the characteristics of farmhouse Reblochon cheese from Savoie. Lait 75: p. 133-149.

Martin, B., C. Hurtaud and D. Micol, 2002. Le rôle des fourrages dans la qualité des produits animaux: comment répondre aux attentes du consommateur? Fourrages 171: p. 253-264.

Maudet, C. and P. Taberlet, 2002. Holstein's milk detection in cheeses inferred from melanocortin receptor 1 (MCR1) gene polymorphism. J. Dairy Sci. 85: p. 707-715.

McKean, J.D., 2001. The importance of traceability for public health and consumer protection. Rev. Sci. Tech Off. Int. Epiz. 20(2): p. 363-371.

McKellar, R.C. and P. Piyasena, 2000. Predictive modelling of inactivation of bovine milk alpha-L-fucosidase in a high-temperature short-time pasteurizer. Int. Dairy J. 10: p. 1-6.

Mordenti, A. and M. Pecorari, 1999. L'alimentazione delle bovine nell'espressione di tipicità dei formaggi. Sci. Tecn. Latt-cas. 50: p. 57-75.

Mottram, T., J. Hart, R. Pemberton, H. Hogeveen and A. Meijering, 2000. Robotic milking. Proceedings of the International Symposium held in Lelystad, The Netherlands, 17-19 August, p. 108-113.

Norberg, E., H. Hogeveen, I.R. Korsgaard, N.C. Friggens, K.H.M.N. Sloth and P. Løvendahl, 2004b. Electrical conductivity of milk: ability to predict mastitis status. J. Dairy Sci. 87: p. 1099-1107.

Norberg, E., G.W. Rogers, R.C. Goodling, J.B. Coope and P. Madsen, 2004a. Genetic parameters for Test-Day electrical conductivity of milk for first-lactation cows from random regression models. J. Dairy Sci. 87: p. 1917-1924.

Oberreuter, H., A. Brodbeck, S. von Stetten, S.D. Goerges and S. Scherer, 2003. Fourier transform infrared (FT-IR) spectroscopy is a promising tool for monitoring the population dynamics of microorganisms in food stuff. Eur. Food Res. Technol. 216: p. 434-439.

Pecorari, M., E. Fossa and G. Avanzini, 1988. The milk with bad coagulation: technological behaviour in the Parmigiano-Reggiano productions. Sci. Tecn. Latt-cas. 39: p. 319-337.

Pedersen, D.K., 2003. Determination of casein and free fatty acids in milk by means of FT-IR techniques. Bulletin of the IDF 383: p. 48-51.

Pellegrino, L., P. Resmini and U. Fantuzzi, 1995. Valutazione ed interpretazione della attività della fosfatasi alcalina nel controllo dei formaggi Grana Padano e Parmigiano Reggiano. L'industria del latte, XXXI, 1, fasc. 2: p. 17-30.

Pirovano, F., I. Piazza, F. Brambilla and T. Sozzi, 1995. Impedimetric method for selective enumeration of specific yoghurt bacteria with milk-based culture media. Lait 75: p. 285-293.

Pritchard, J.K., M. Stephens and P. J. Donnelly, 2000. Inference of population structure using multilocus genotype data. Genetics 155: p. 945-959.

Pyorala, S., 2003. Indicators of inflammation in the diagnosis of mastitis. Vet. Res. 34: p. 565-578.

Remeuf, F., V. Cossin, C. Dervin, J. Lenoir and R. Tomassone, 1991. Relationship between milk physico-chemical traits and cheesemaking property. Lait 71: p. 397-421.

Rossmann, A., G. Haberhauer, S. Hölzl, P. Horn, F. Pichelmayer and S. Voerkelius, 2000. The potential of multielement stable isotope analysis for regional origin assignment of butter. Eur Food Res Technol 211: p. 32-40.

Rubino, R. and S. Claps., 2000. Relazione fra pascolo e qualità del formaggio. Atti del V Convivio Formaggi d'alpeggio: il pascolo, l'animale, la razza, il prodotto. Cavalese (TN), 15 settembre, p. 43-56.

Russo, V. and L. Fontanesi, 2004. Analisi dei geni del colore del mantello: tracciabilità di razza. Atti 7ma Conferenza mondiale allevatori razza bruna, Verona, 3-7 marzo 2004, p. 95-100.

Sørensen, L.K., M. Lund and B. Juul, 2003. Accuracy of Fourier transform infrared spectrometry in determination of casein in dairy cows'milk. J. Dairy Res. 70: p. 445-452.

Speroni, A. and G. Bertoni, 1984. L'affioramento del grasso del latte: nuove proposte per la valutazione e l'interpretazione del fenomeno. Sci. Tecn. Latt.-cas. 35: p. 97-108.

Stefano, C., 1609. Agricoltura et casa di villa, Per Gio, Domenico Tarino, Torino, p. 64-65.

Tartaglia, M., E. Saulle, S. Pestalozza, L. Morelli, G. Antonucci and P.A. Battaglia., 1998. Detection of bovine mitochondrial DNA in ruminant feeds: a molecular approach to test for the presence of bovine-derived materials. J. Food Prot. 61: p. 513-8.

Tkac, J., E. Sturdik and P. Gemeiner, 2000. Novel glucose non-interference biosensor for lactose detection based on galactose oxidase-peroxidase with and without co-immobilised beta-galactosidase. Analyst 125: p. 1285-1289.

Van den Bijgaart, H., 2003. Urea. Bullettin of the IDF 383: p. 5-15.

Vecchia, P., M. Intini, L. Calamari and B. Battistotti, 2003. Le principali cause delle fermentazioni anomale nel latte. L'Informatore Agrario 59: p. 49-53.

Verdier-Metz, I., J.B. Coulon, P. Pradel, C. Viallon and J.L. Berdagué, 1998. Effect of forage conservation (hay or silage) and cow breed on the coagulation properties of milks and on the characteristics of ripened cheeses. J. Dairy Res. 65: p. 9-21.

Versini, G., F. Camin, D. Carlin, D. Depentori, F. Gasperi and L. Ziller, 2000. Accertamenti innovativi per la caratterizzazione e tutela delle produzioni tipiche di montagna. L'analisi chimica isotopica e dell'aroma. Atti del V Convivio Formaggi d'alpeggio: il pascolo, l'animale, la razza, il prodotto. Cavalese (TN), 15 settembre, p. 145-158.

Viallon, C., B. Martin, I. Verdier-Metz, P. Pradel, J.P. Garel, J.B. Coulon and J.L. Berdagué, 2000. Transfer of monoterpenes and sesquiterpenes from forages into milk fat. Lait 80: p. 6354-341.

Wiedemann, M. and G. Wendl, 2003. The use of colour measurement for the on-line determination of milk quality during milking. Landtechnik 58: p. 272-273.

Zannoni, M. and S. Annibaldi, 1981. Standardization of the renneting ability of milk by Formagraph. Sci. Tecn. Latt-Cas. 32: p. 79-94.

Zannoni, M. and R. Mora, 1993. Evolution of the milk quality program for the Parmigiano-Reggiano cheese. Il latte 18: p. 572-581.

Key success factors of competitive position for some Protected Designation of Origin (PDO) cheeses

D. Barjolle, J.-M. Chappuis and M. Dufour*
ETHZ - Institute of Agricultural Economics, Lausanne, Switzerland.
**Correspondence: d.barjolle@srva.ch*

Summary

This paper presents a method to assess the success of some PDO (Protected Designation of Origin) cheeses in a comparative way. The first part is a review of different meanings of the concept of performance according to the main theoretical approaches. We give a definition of the *"filière"* and select some criteria to assess the competitive position and the attractiveness of the reference markets in order to get an evaluation of the success in terms of prices and market shares. We try then to identify the main factors which influence the competitive position of the product.

Our research highlights that the success of the PDO cheeses is determined mainly by four determinants: the specificity of the product (its typicality, price, convenience, taste, symbolic component and its representation by the consumers), the effectiveness of the co-ordination (code of practice, governance structure, variety and quality management, promotional and research policies, lobbying ability), the market attractiveness, the public supports. In conclusion, the concept of *filière* has to be studied more in depth at the theoretical level. If it is possible to highlight the role played by these production channels (several small-scaled firms) on the rural employment, it will be necessary to set up a better legal framework to encourage their activities and common strategies.

Keywords: AOC, PDO cheeses, performance, competitive position, specificity, co-ordination, Industrial Organisational economy, New Industrial Organisation theory, market attractiveness

Introduction

The purpose of this paper is to identify the factors of success of PDO (Protected Designation of Origin) products in a comparative way. The aim of the PDO/PGI (Protected Geographical Indication) policy goes much beyond a strict economic performance. Its objectives are not only to protect traditional denominations and to allow products to compete effectively against cheaper lines, but also to have value-added products in decentralised regions often experiencing low economic outputs. A number of social, environmental and cultural benefits should thus emerge from such an approach which would be the stability of rural populations in remote areas, the protection of the landscape, the heritage and the tradition. This paper will only consider the economic performance of a PDO product.

This analysis will show us that the co-ordination within the *filière* and the producers' understanding to valorise their product through its specificity are among the most important factors to explain the marketing success of a PDO cheese.

The notion of "performance" in various theoretical approaches

We want to identify factors of success in order to provide some best-practice recommendations to the professionals involved in PDO products management as well as to the policy makers at regional, national and European level (the recommendations themselves are not part of this paper, but part of the final report of the European Research by Barjolle *et al.*, 2000). To identify the factors of success, we decided to elaborate a specific method after having examined several theoretical approaches and established that the performance is a word subject to competing meanings in the economic literature.

General equilibrium and market competition: the welfare as a performance

The neo-classical approach (NCA) defines the main components of the firm's behaviour as the cost minimisation and the profit maximisation. The firm's performance is not studied for itself.

From the NCA point of view, the performance of one firm is not isolated from the others. NCA considers the **efficiency of competitive markets** as a whole. The general equilibrium and the economic efficiency are the results of an economic process which is the adjustment between supply and demand. For any initial allocation of resources, a competitive process of exchange among individuals through exchange, input markets or output markets, would lead to an economically efficient outcome. A competitive market, built on the individual goals of consumers and producers and on the ability of market prices to convey information to both parties, will achieve an efficient allocation of resource thanks to Adam Smith's invisible hand. That is the first paradigm of welfare economics (Debreu, 1959; Arrow and Hahn,1970). Taking into account that the consumer preferences curve is convex, any efficient allocation of resources can be achieved by a competitive process with a suitable redistribution of those resources. These theorems of welfare economics depend crucially on the assumption that markets are competitive (Pindyck and Rubinfeld, 1998).

This approach does not suit for our purpose because its method is not strategy oriented (i.e. it gives indications at the end of a selection process of an optimal equilibrium). Moreover it does not offer any answer to the question of why a strategy succeeds when another does not.

Industry's performance related to market's structure and conduct

Mason (1939, 1949), who used inferences from micro-economic analysis to discuss industrial organisations, introduced the traditional Structure-Conduct-Performance approach (SCP). Following the mainstream approach, the Industrial Organisation Theory (IO) associates the industry's performance with the structure of the market and the strategies of the firms composing the industry. Three different measures, reflecting directly or not the profits or the relationship between the price and the costs, are commonly used to appreciate how close the performance of an industry is to the compe-titive benchmark (Carlton and Perloff, 1994): (1) the rate of return, which is based upon profits, (2) the price-cost margin, which should be based upon the difference between price and marginal cost, (3) the ratio of the market value of a firm to its value based upon the replacement cost of its assets (*Tobin's q*).

To describe the structure of a market, several measures are used. They are all considered as having some relation with the degree of competitiveness in an industry, for example: industry concentration (market shares of the different firms present on the market), barriers to entry (i.e. the ability of the firms to enter the industry).

SCP takes into account the conduct of sellers and buyers because it assumes that their strategic choices can explain the relationships between the market structure and the performance of the industry. We face difficulties in using this approach for our research because that the conduct analysis is adapted to a well-defined industry as a whole when we want to compare different strategies conducted by competitors on the same market.

The analysis of the performance of a firm

Coase's analysis of the transaction and the related costs highlighted that the firm is an alternative to the market as governance structure of the exchange process (Coase, 1937). Simon's contribution related to bounded rationality (Simon, 1955) was also decisive for a new generation of economists. Transaction Cost Analysis (TCA) put away one assumption of the mainstream economy that is the perfect rationality (Williamson, 1975, 1985). TCA is also based on institutional economy which deeply considers the transaction at the agent level in order to better explain the economic phenomena (Commons, 1934).

One point of neo-institutional economy is to observe how the structure, the strategy and the performance of a firm are influenced by several internal and external factors. The opportunist behaviour of agents as well as the costs of the seller's assets or the frequency of the transactions are the three main factors explaining the governance structure for a given transaction.

This way to consider a firm allows a new field of investigation for the economists: the firm itself. Not only the practitioners could also discuss the question of "firm management" but also the academic corpus. For example, Liebenstein (1976, 1982) introduced the notion of X-efficiency related to individual firm performance. He argued that the quality of the internal organisation of the firm (i.e. its ability to combine in an efficient way the different production factors) plays a determinant role in explaining its success vis-à-vis its competitors. What he called X-factor was the way of combining the internal and external resources within a firm. He took into account more factors than the two traditional factors (labour and capital).

Organisational economy (OE) considers the firm as the main subject matter. There are currently two main theoretical points of view in considering the firm. On the one hand, there is the Agency Theory, which tries to elaborate some mathematical representation in a normative way. The organisation is considered as a nexus of contracts. In this contractual system, the agents (single persons) are maximising their own, diverse and conflicting objectives (which are not necessary the cost minimisation or the profit maximisation). In that view, the behaviour of the organisation is like the equilibrium behaviour of a market (Jensen, 1998). This first point of view agrees with some of the main assumptions of NCA.

On the other hand, there is the New Industrial Organisation theory (NIO) (Tirole, 1988). It is based on the welfare economic paradigms but it rejects some assumptions like the lack of externalities, the ignorance of the public goods existence as well as the consumer's perfect information (Tirole takes into account the product quality problem). NIO also deals with the existence of market power when few firms serve the market. Tirole (1988) considers the firm as the main subject matter. Following several authors, he gives four definitions of the firm: the firm as (1) a loophole for the exercise of the monopoly power, (2) a static synergy, (3) a long-run relationship, (4) an incomplete contract. He does not propose himself an analysis of the performance of the firm. However he wishes that the conclusions of NIO will remain valid (at least at a descriptive level) when the profit-maximisation postulate will be abandoned for a full-fledged model of internal organisation as described by Liebenstein (Tirole, 1988, p. 51). This postulate has been until now nothing more than a wish. As it has been explained by Barney and al (1996), TCA, Agency Theory and NIO can be used to explain why firms exist. They cannot be used to explain why some firms outperform others. In effect, these approaches assume that all the firms, facing similar market context, will develop similar governance solutions. They do not take into account the heterogeneity of the firms (internal resources).

Firms which outperform others

OE second way of thinking progresses in a less normative manner, closer to the strategic management literature. The two questions "Why do some organisation outperform others ?" and "How can organisations co-operate ?" are definitively in the middle of this kind of investigations. Rumelt (1984), Wernerfelt (1984) and Barney (1991) following Teece (1982) and Prahalad and Bettis (1986) have introduced the resource-based view of the firm. For these authors, any analysis of a firm must be based on its resources and capabilities, which are divided into four types: (1) financial, (2) physical, (3) human, (4) organisational resources. Two main assumptions are used to explore the heterogeneity among several firms belonging to the same industry: first, the assumption of firm heterogeneity and second, the resource immobility. For Barney (1991), the resources which give a firm its best chance to outperform others are: (1) valuable (i.e. resources which enable the firm to exploit environmental opportunities and help it to avoid threats), (2) rare among its competitors, (3) costly to imitate and (4) without close strategic substitutes. This conception of the performance helps us better identify and analyse the different strategic choices and product performances. Methodology remains difficult to be used as an explanatory tool to explore collective strategies. The resource-based view of the firm considers co-operation and strategic alliances mainly in the case of vertical integration or joint ventures. Other forms of collective management are not yet analysed.

From industry and market to "*filière*" and Reference Market

The above approaches are not very helpful to consider co-operative strategies between firms along a supply chain. Indeed, an industry is defined as a group of firms, all producing the same good and addressing it to the same consumers.

To analyse the performance of a system of firms[1] producing a same good, we have to consider the whole *filière* as a production and marketing entity in competition with other entities (*filières* or organisations) producing very close substitutes composing a Reference Market.

We define the *filière* as a system composed by several different firms at different production levels in the supply chain (for example: milk producers, cheese producers, ripening houses). This system might act like one single organisation because (1) all the firms are producing either raw material or intermediary products with the same standards[2] and (2) they have interest to share expenses in advertising or research (Barjolle *et al.*, 1998a). The strategy of the whole *filière* on the market of reference is generally more efficient than addition of the individual ones due to the synergies. Supply Chain definition[3] usually includes all the stages from raw material producers to consumers. Our conception of the *filière* is different because only the production side is included. In extension, several organisations and *filières*, all producing very close substitutes and addressing the same consumer segment, can compose a supply chain.

The *Reference Market* (Day *et al.*, 1979) is defined by the ***substitute*** products[4]. Clear definition of what is a market is easy to give from a theoretical point of view, but difficult to transpose into the

[1] In the case of PDO products, this system can be composed by a lot of small-scaled firms. It can be as well composed by very diverse kind of firms. But generally, these firms have common goals and close relationships.

[2] Some differentiation is possible, but the product specifications are the same.

[3] Saporta B.,1989, p. 4 et 6 : "Une chaîne verticale d'activités successives".

[4] 1.- Two goods are substitutes if they can satisfy the same need (Dictionnaire d'économie et de sciences sociales, Nathan, 1996). 2.- Substitute : what can replace something else (Larousse Dictionary).

praxis (Tirole, 1997). Due to the difficulties to be right (neither narrow nor too broad) when defining the substitute products of a market, we decided to define case by case what we considered as the *Reference Market*.

In the following chapter, we will try to explain the relations between the **competitive position of the product** and the **attractiveness of the reference market** with the **effectiveness of the strategic choices** made by the *filière* (more exactly by one body representative of it, like a union). We will also take into account **any other factor** which might influence the competitive position of the cheese.

Methodology

Our aim is to better understand the reasons of the success of some typical products in terms of price or sales increasing. We also wanted to establish in which way the PDO Regulation could play a role in the performance of the product. Firstly, and in order to measure their success, we took into consideration two elements which are the **attractiveness of the reference market** and the **competitive position** of the product. We characterised these two elements for each product in a comparative way. The indicators chosen to assess the attractiveness and the competitive position follow a free adaptation of several methodologies (La Rue T. Hormer, 1982; Hinterhuber, 1989; Kotler, 1994; Lombriser and Ablanalp, 1998).

Secondly, we discussed the results in order to identify the factors which seem to have an influence of the observed success. We found out that the most important factor to explain the success of the product is the degree of effectiveness of the system of firms within the *filière* (Barjolle *et al.*, 1998c), but that it is not the only one. Some other factors are to be found in the institutional framework and in the own resources of the *filières*.

For each criteria, we identify in the bibliography mentioned in the theoretical part (see above) a complete list of indicators, mentioned as useful and significant for the evaluation on basis of the available data. The data have been collected through empirical studies, and the assessments of the scores have been done using a comparison through the authors, who have studied in depth and visited all the initiatives.

The attractiveness of the various reference markets

To assess the attractiveness of the various reference markets, we should ideally look at different determinants (see Table 1).

The competitive position of the product

To assess the competitive position of the various products, we should also ideally look at different determinants which are mentioned in Table 2.

[5] If the average age is lower than the optimal time span, the product is rare and very demanded on the market; restrictions and production crises also have direct effects on the scarcity of the product.

[6] The partners in the programme are: Fearne A. and Wilson, N., Wye College (GB), De Roest K and al, CRPA (IT), Galanopoulos K. and al, University of Thessaloniki, FotopoulosC., Vakrou A. and al, NAGREF (GR), Sylvander B. and Lassaut B., INRA-UREQUA, Leusie M., Chrysalide (F), Van Ittersum K. and al., Wageningen (NL), Barjolle D, Chappuis JM, Dufour M, IER-EPFZ (CH).

Table 1. Indicators for the attractiveness of the Reference Markets.

Ideal indicator	Chosen indicator[1]
• The size of the market of reference	• The size of the market of reference, which has been the only criteria for which quantitative data could be collected for each studied supply chain
• The growth rate of the market of reference	
• The barriers to entry on the reference market	
• The access conditions to the market (tariffs or non-tariffs barriers)	
• The economic stability on the reference market (sensitiveness to inflation in particular, evolution of the purchasing power)	• The socio-economic context of the countries of production and of the countries importing the cheeses (using the same way)
• The legal, political and social frame	• The region's image (through a qualitative appreciation conducted in a comparative way)
• The image of the sector (impacts on environment and public health in particular)	• The sector's image (using the same way)
• The structure of the partners downstream in the supply chain	
• The benefit margins realised in the past	
• The intensity of competition	

[1]based on available data, collected during the European research program PDO-PGI, market, supply chain and institutions

Table 2. Indicators for the competitive position of the product.

Ideal indicator	Chosen indicator [1]
• The evolution of the volumes (average annual growth)	• Evolution of sales
• The evolution of the part in the consumption	
• The consistency of the product's sensorial qualities (quality problems)	
• The presence of similar products on the market or direct substitutes which could disturb the PDO pro-duct's sales, by disrupting the balance volume/price	• Threats through imitations (direct substitutes)
• The resistance of the product to a hypothetical withdrawal of public support which strengthens the product position	• Dependency to public support / measures
• The evolution of the price (price tendency in relation to similar products)	• Price tendency
• The effect of scarcity, measured differently according to the product (for cheeses, this could be appreciated by checking the average age of stored cheeses and comparing it with the optimal length of storage[5])	
• The notoriety of the product in its country of origin (do the consumers know the product and do they associate the product with its production area ?)	• Notoriety of the product. Image of the product
• The diffusion of the product outside the country of origin ; assessment of the potential for further exportation (expansion to larger markets)	• Geographical spreading of the sales

[1] based on available data, collected during the European research program PDO-PGI, market, supply chain and institutions

Results

Nine different cheeses were studied in the frame of a European research programme[6]: the Comté and the Cantal in France, the Parmigiano Reggiano and the Fontina in Italy, the Feta in Greece, the Boeren Leidse met Sleutels and the Noord Hollandse Edammer in the Netherlands, the West Country Farmhouse Cheddar Cheese in Great Britain and the Gruyère in Switzerland. All these cheeses are very traditional cheeses produced only in a delimited area. The context and the results of the global research program are published in the final synthesis report (Barjolle and Sylvander., 2000). The characteristics of their reference markets are described in Tables 3 and 4.

Reference markets

We could identify two types of reference markets in the different countries (Table 5).
* The market of reference is growing. Here we can rank the cheese market in general (with a growth of 7% to 15% between 1990 and 1996), the PDO cheese market (with a growth of 10% to 30% between 1990 and 1996), the soft cheeses market in general and the mature cheeses markets.
* The market of reference is stable or decreasing. The semi-hard cheeses market in general, the hard cheeses market in Switzerland, the Edammer cheese market in the Netherlands and the Cheddar cheese market in the UK are either stable or decreasing.

The sector's image

With all the problems relative to meat production and consumption ("mad cow", "chicken disease"), the cheese market is attractive. The nutritional quality of the dairy products does not need to be proved anymore. The cheese consumption in Europe increases according to the *Fédération Internationale Laitière* (FIL). We can really say that the cheese sector has a very positive image.

The region

Several factors can contribute to the image and potential of a region. First there are the landscapes, the notoriety and the perception that consumers have about this region which might be under the influence of a promotional strategy. For example, the province of Noord-Holland has such a reputation to produce high quality cheeses that the consumers agree to pay more for the cheeses produced in this part of the Netherlands (Van Ittersum and Candel, 1999).

The socio-economic situation

High costs of production, like in Switzerland, are of course a negative economic factor. It weakens the opportunities of exports and threatens the products' sale inside the native country itself if imported products tend to be cheaper.

Competitive position (Table 6)

Consumption

The habits of consumption of the studied PDO cheeses are very different. It depends on the average quantity of cheeses consumed by inhabitant in the country of production, on the cheese production volume, on its notoriety inside and outside the production area, on the general consumption of cheese. The general trend in the consumption is appreciated by the development of national and exported sales. The exported quantities also vary a lot according to each considered cheese.

Table 3. Market attractiveness.

PDO products	Size of the markets of reference*	Growth of the markets of reference	Image of the sector**	Intensity of the competition from the substitute products	Barriers to entry/ barriers to exit	Market attractiveness		
						Weak	Middle	High
Boeren-Leidse met Sleutels	Small (<15000 tons)	↑	Traditional product	strong	Weak		x	
Cantal	Large, PDO cheeses (172000 tons) (+10%) PPPNC (200000 tons) (+8,7%)	↑	Traditional product	middle	Middle		x	
Comté	Large, PDO cheeses (172000 tons) (+10%) pressed and cooked cheeses (270000 tons)	↑	Traditional product	strong	Middle (ripening time)		x	
Feta	Middle, (98000 tons) (+28%)	↑↑↑	Traditional product	Greece: middle, export: strong	Middle (sheep and goat milk)			x
Fontina	Middle, semi hard cheeses (75000 tons) (-3,8%)	↓	Artisanal and traditional product	strong	weak	x		
Gruyere	Middle, Swiss hard cheese (80000 tons) (-9,6%)	↓	Artisanal and traditional product	strong	Middle (ripening		x	
NH Edammer	Large (>150000 tons)	↓	Industrial product	strong	Weak	x		
Parmigiano Reggiano	Large, hard cheese out of cow milk (240000 tons) (+15%)	↑↑	Traditional product	strong	High (ripening time)			x
WCF Cheddar	Large, Cheddar sold in the UK (153000 tons)	↓	Industrial product	strong	Middle (ripening time)	x		

*Size of the market of reference; more than 150000 tons: large, between 50000 and 150000 tons: middle, below 50000 tons:small.
**A family of products can be considered as traditional but not artisanal (traditional does not mean artisanal or industrial)

Table 4. Production, exportation and consumption of the PDO cheeses.

	Consumption	Production in 1996	Production in 1991	Exported sales in 1996	Exported sales in 1991	Evolution between 1990 and 1966	Trend
Boeren-Leidse met Sleutels		425 tons		Hardly anything	Hardly anything	Strong competition from PanPan cheese, a factory made Leidse cheese and from other large cheese brands. Some quality problems for the Boeren-Leidse met Sleutels.	→
Cantal		16'500 tons[7]	16'100 tons			Production decreased of 1.5%.	↓
Comte		41'000 tons	33'400 tons	~ 500 tons (higher than in 1991)		Comté is still the first AOC cheese. Production increased greatly between 1990 and 1994 leading to a decline in prices. The production was limited in 1995. Quotas were imposed to the cheese processors in order to preserve the price level. The stability of the market remains fragile	→
Feta	8,5 kg	76'000 tons	54'000 tons	9'600 tons	7'000 tons	Production increased of 39.4%. Exports increased of 25.5%.	↑↑↑
Fontina	0.2 kg[8]	3'500 tons	3'300 tons	7-8%	2%	Production increased of 32.3%	↑↑
Gruyere	3,2 kg	24'000 tons	22'400 tons	7'200 tons	8'200 tons	Increase of 4.9% between 1991 and 1996. Market share of the Gruyère raised from 21.6% to 25.1%. Sales of Gruyère on the Swiss market increased of 3.4% between 1991 and 1996. The market share decreased from 15.5% in 1992 to 14.3% in 1995. Exports increased of 12.9% between 1991 and 1995.	↑
Noord-Hollandse Edammer		Confidential between 5'000 and 15'000 tons	Higher than in 1996	Confidential information	Confidential information	Some wholesalers and retailers switched from NH Edammer to Friese Edammer cheese. The production and supply of NH Edammer to the largest wholesaler decreased from 40,000 cheeses per week in 1992 to 17,000 cheeses per week in 1996.	↓↓
Parmigiano Reggiano	1.55 kg	104'900	109'000 tons (1990)	6%	4%	Production decreased of 4.2% Nevertheless volumes can be considered as stable around a fluctuating trend.	→
WCF Cheddar	1,55 kg	22'000 tons	24'000 tons	400 tons		Production volumes have fallen since 1994. Several producers have left the industry, others have cut back their production. Higher prices obtained do not lead to an increase of the production. High milk prices paid to the farmers	↓

[7] in 1995
[8] 9 kg in the production area

Table 5. Evolution of the volumes on the various reference markets. Comparisons with the studied PDO cheeses.

Product	Market of reference	Evolution of the production volumes	Trend
Boeren-Leidse met Sleutels	Leidse cheeses	Cheese consumption increased of 6.3% between 1990 and 1994 in the NL. Meanwhile the domestic mature cheeses win 74% of market share (from 12.3% in 1990 to 21.4% in 1994).	↑
Cantal	Farmhouse cheeses (France)	Production increased of 7.4% between 1992 and 1995.	↑
	French PDO cheeses	Production increased of 10% between 1992 and 1995.	↑
	Auvergne PDO cheeses	Production increased of 5.9% between 1992 and 1995.	↑
	Pressed and uncooked paste cheeses	Production increased of 8.7% between 1992 and 1995.	↑
	All French cheeses	Production increased of 6.9% between 1992 and 1995.	↑
Comté	French PDO cheeses	Production increased of 10% between 1992 and 1995.	↑
	Pressed and cooked paste cheeses	Production increased in the last years (+ 7% in 1995; libre service).	↑
	All French cheeses	Production increased of 6.9% between 1992 and 1995.	↑
Feta	Sheep/Goat Greek cheeses	Production increased of 27.8% between 1990 and 1994.	↑
	All Greek cheeses	Production increased of 15.2% between 1990 and 1994.	↑
Fontina	Semi hard cheeses	Production decreased of 3.8% between 1990 and 1996.	↓
	Fontal		
	PDO cheeses in Italy	Production increased of 29.1% between 1990 and 1996.	↑
	All Italian cheeses	Production increased of 9.5% between 1990 and 1996.	↑
Gruyère	Swiss hard and semi-hard cheeses	Production decreased of 9.6% between 1991 and 1996. Market share decreased from 77% in 1991 to 72% in 1996.	↓
	All Swiss cheeses	Production decreased of 4.7% between 1991 and 1996. Production of soft cheeses is increasing though.	↓
	Cheeses (in and outside Switzerland)	Consumption has increased both in Switzerland (+8.9% between 1992 and 1995) and in the rest of Europe.	↑
Noord-Hollandse-Edammer	Edammer cheeses	Cheese consumption increased of 6.3% between 1990 and 1994 in the NL. Meanwhile the Edammer cheeses lost 30.6% of their market share (from 4.9% in 1990 to 3.4% in 1994).	↓
Parmigiano-Reggiano	Hard cheeses out of cow milk (Italy)	Production increased of 14.1% between 1990 and 1996.	↑
	PDO cheeses in Italy	Production increased of 29.1% between 1990 and 1996.	↑
	Grana Padano	Production increased of 35.2% between 1990 and 1996.	↑
	All Italian cheeses	Production increased of 9.5% between 1990 and 1996.	↑
West Country Farmhouse Cheddar	Cheddar cheese	The most important cheese (57% of sales in 1996) although losing market shares. Consumption of cheese in the UK has stabilised. The soft cheeses are increasing their share steadily.	↓
	Mature Cheddar	Stronger flavours are becoming more important and the higher value mature Cheddar sector has been gaining share since 1991.	→

Table 6. Competitive position of the product.

PDO product	Evolution of sales (volumes)	Price tendency	Notoriety of the product	Image of the product	Imitations (direct substitutes)	Geographic spreading of the sales	Dependency on public support/measures	Competitive position		
								Weak	middle	High
Boeren Leidse met sleutels	→	→	Well known in the western part of the NL	++	Pan Pan ++	No export	-		X	
Cantal	↓ (-1.5%)	↓	Well known in France	+	-	Little export	Direct payment to farmers	X		
Comté	→	→	Well known in France	++	-	Little export	Direct payment to farmers		X	
Feta	↑↑↑ (+40%)		Very high	++	+++	15% export (increasing)	EU subsidies			X
Fontina	↑↑ (+32%)	→	Well known in Italy	++	Fontal +/-	Little export	Very high (regional and EU subsidies)			X
Gruyere	↑ (+5%)	↓	Well known in Switzerland, confusion in France	++	Etivaz +/-	20% export all over Europe	Very high		X	
NH Edammer	↓↓	→	Well known in the Netherlands	++	Edammer +++	Marketed in the NL	-	X		
Parmigiano Reggiano	→ (stable around a fluctuating trend)	Variations (→↑↓)	Well known in Europe	++	Grana Padano +/-	6% export in 1996; all over Europe	Direct payments to the farmers/subsidies to the ripeners		X	
WCF Cheddar	↓		Little known	+	UK Cheddar, Canadian Cheddar ++		-	X		

Evolution of sales and price tendency

Except for the West Country Farmhouse Cheddar, the Noord-Hollandse Edammer and the Parmigiano Reggiano, we can see that the production of the studied PDO cheeses increased. The most significant growth was for the Comté and the Feta. Exportations contributed to their development but it was mainly in the country of production that sales increased.

Notoriety of the product and image of the product

We give an assessment of the notoriety for each studied product. We assessed that the image of the product was quite similar for all the products except for Cantal and West Country Farmhouse Cheddar because only few people recognise them as a must.

Imitations (direct substitutes)

When direct substitutes are in the stores, it creates a weakness for the competitive position of the product. Some products have specific substitutes but they are not very dangerous (Fontal and Etivaz) or not at the same position in the mind of the consumer (Grana Padano is less prestigious than Parmigiano Reggiano). Some have few direct substitutes and are more unique on their market, like Cantal or Comté.

Geographical diffusion of the products

It is also important to look at if the PDO cheese is mainly consumed within the PDO area or if it is also consumed outside of the PDO area (in the surroundings, in the whole country or even at an international level).

Dependency to public support

Public subsidies and support to farming activities are positive for the production of a PDO product and for the performance of its *filière*, as long as it is not a "poisonous gift" contributing to develop a kind of passivity among the actors. A general interest in preserving a regional identity and therefore local products can help the actors cooperate and involve themselves in a *filière* with collective objectives.

Discussion

The research allows us to speak about products whose competitive position is weak, medium or strong. We try to explain hereafter which are the determinant factors which make us rank for each studied PDO cheese in one of these three categories. We position each product in a diagram in relation to their competitive position and the attractiveness of their main reference market.

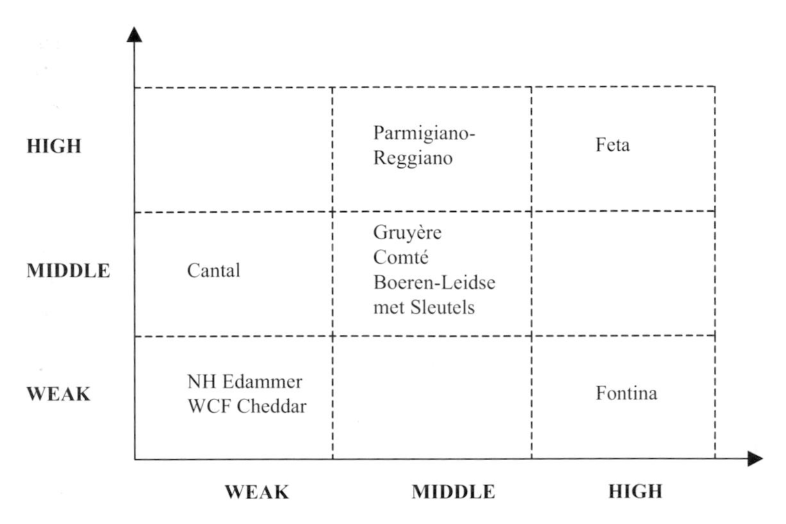

Figure 1. Competitive position of the products in relation to the attractiveness of their main reference market.

Influence of the PDO on the cheese competitive position

We cannot consider today the PDO as a well-known sign which would lead the consumers to buy a PDO cheese just because it has a PDO sign. Consumers do most often ignore the PDO existence or its signification[9]. The perspective of the PDO as a label which would be famous enough to become a sufficient purchase argument was yet a motivation for some *filières* (for instance the CONO: Company for the Noord-Hollandse Edammer cheese or the West Country Farmhouse Cheddar producers).

A PDO can have a major influence on the product competitive position for three other reasons. The first one is the fact that a PDO strategy chosen by a *filière* will force the professionals to define the quality of their product in a code of practice and to stick to it. It can thus help a *filière* to preserve and to strengthen the quality of its product, and to prevent the cheese from becoming more and more standardised.

The second indirect effect of the PDO is on the structure of the whole *filière* itself and on the co-ordination of the actors. The necessity to work together in order to define the specificity of their product, to improve its quality, to promote the cheese, to find solutions to manage the quantities might encourage the professionals to organise their *filière* in a better way. This improvement may help the *filière* to reduce for instance the transaction costs and to be more efficient and effective in general. This could be a positive effect of the PDO on the Gruyère *filière*, because the perspective of a PDO was for the professionals a strong incentive to create quickly a professional

[9] The Eurobaromètre 50.1 requested in 1998 by the European Commission shows that only 6,3% of the consumers knows the three letters "PDO", and 13,5% the full denomination "protected designation of origin". Moreover, consumers do not always know what guarantees a PDO. A third of them knows that it means that the product has a well-defined geographical origin but only a quarter can say that the main ingredients must all come from the production area. Many consumers only consider the PDO as a quality label without highlighting the importance of the origin.

organisation, a body which would help them overcome the important difficulties arising with the actual system being deregulated.

The third potential effect of the PDO is of course the protection of the designation which can give new opportunities of development, in particular on foreign markets.

Main factors of success

We believe that the performance of a *filière* is its capacity to give a good and high specificity to the product and to influence the public institutions in order to get financial subsidies and supports. It supposes first an effective co-ordination between all the firms involved in the *filière*.

The studied cheeses show us that the competitive position of a PDO product does not depend neither on its **country of origin** (the Cantal and the Comté are for instance both French products but their competitive position is different) nor on its **volume** (the Fontina, 3,500 tons, and the Feta, 76,000 tons, have both a good competitive position), nor on the **extend of the consumption** outside the production area (the Fontina is mainly consumed around the production area when the Feta is well exported), nor on the degree of the **vertical integration** in the *filière* (both the Boeren Leidse met Sleutels and the West Country Farmhouse Cheddar have a strong vertical integration but they have a different competitive position).

Our research highlight indeed four main determinants which will contribute to the product competitive position (Figure 2):

If one of these determinants is weak, it can be compensated by another strong factor. We can also understand that if a *filière* is not successful in differentiating its product from its substitutes, the competition will take place at the price level. We believe that such a competition will lead to a decreasing quality of the product, a loss of its *typicity* since quality and *typicity* require some extra production costs. *Typicity* has to be understood as an intrinsic component of the product, rooted in

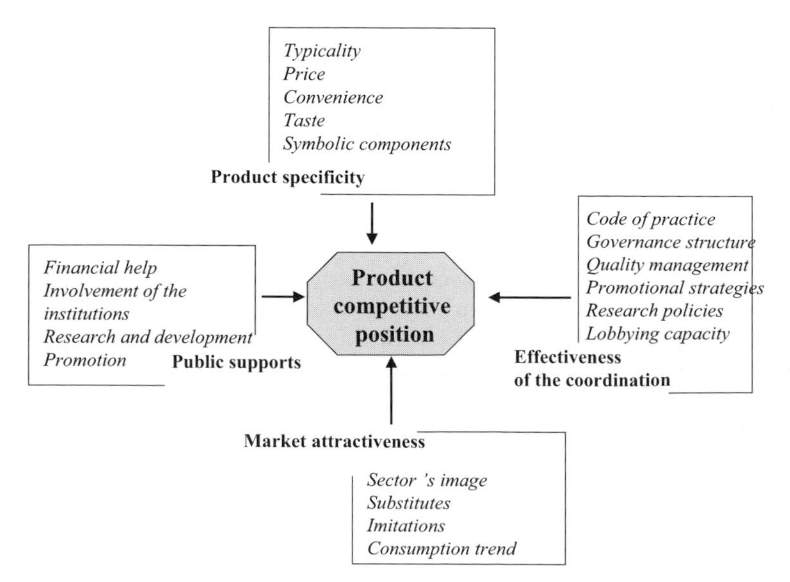

Figure 2. The four determinants of the product competitive position.

a historical and geographical context specific to the region of origin. The *typicity* will help the product performance, if it is well transcribed in the code of practice (crucial points of the technology and of the product definition). For the cheeses, the *typicity* is possible only if some process conditions which give evidences between the product and the geographical constraints are respected.

A lot of researches (see Barjolle *et al.*, 1998b for a review) give indications of what these conditions are.

Based on that review, the Federal Ministry for Agriculture in Switzerland defines the criteria to be fulfilled in order to be registered as PDO or PGI as follow (see Table 7).

In the particular case of PDO-cheeses, the registration principles for "cheese" and "Alpine pasture cheese" PDOs to assure a sufficient link to the territory are the following:
1. Milk from "non-silage" feeds;
2. Milk processed within less than 24 hours;
3. Sowing with typical cultures (only indirect sowing), no additives or technological auxiliaries other than water, salt and rennet;
4. Raw milk for hard and semi-hard cheeses, thermised for soft cheeses;
5. Ripening period that permits the organoleptic potential of the cheese be expressed.

For "alpine pasture cheese" the additional conditions are the following:
1. Milking and first processing should happen on the same alpine pasture range;
2. Natural skimming (creaming off);
3. Curd extraction with a cloth;
4. Use of the cloth for the shaping of the cheese (Järb);
5. "Traditional" mechanical press.

Conclusion

Our aim was to identify how some typical products could have some success on their own reference market, in terms of price or sales. In order to set up an appropriate method, we did a review of several theoretical approaches of the concept of performance. It made sense for us to assess the success of the product in terms of competitive position. We took then the **attractiveness of the market of reference** as the first factor of success, already identified by several authors in strategic management. We found out other explanatory factors coming both from the institutional framework and the internal resources of the *filières*[10].

These results are similar to those of other researches in organisational economics (Liebenstein, 1976, 1982, Barney, 1991 and more generally authors dealing with the resource-based view of the firm). Those highlight that a lot of internal and external factors have to be taken into consideration to assess the performance of several organisations in a comparative way.

We found out that the factor which might influence the success is mainly the co-operation within a *filière* in the pursuit of the competitive advantage. Only a good co-ordination will make possible an efficient strategy to valorise the PDO specificity, to promote the product and to face the market power that the retailers may have. Other factors are the specificity of the product itself and all the public supports given to the *filière*. The concept of *filière* has to be studied more in depth at the

[10] We had to introduce this specific element to be able to identify the factors of success. It is the system of the firms acting together as one single organisation.

Table 7. Recognition of a product as a PDO - Retained criteria and observation points.

Retained criteria	Observation points
The History	Historical existence of the product Historical reasons for the localised development of the product and of its characteristics (justify the product by its history)
The tradition	Continuity of collective practices Local consumption
The delimitation of the area	Characteristics that are common to the whole zone. The elements found on the whole delimited area (the area can not be homogeneous in all its aspects, but should be for some) Conformity between the production, processing and finalising of the product (are these three zones identical?) Matching of the proposed zone with the current zone of production, processing and finalising Proposed criteria to justify the delimitation of the zone
Plant varieties and cattle breeds	Existence of a contribution of the varieties and breeds to the typicity of the product Breeds and varieties should be adapted to the local conditions
Local fodder	Local origin of basic fodder
A good definition of the product, of its typicity and its differentiating elements	Identification and inscription in a code of practice of the outstanding points, the founding elements of the identity amd of the typicity of the product. Description of the characteristics of the raw material, the production, processing and finalising methods of the product These characteristics and/or the production methods must be very specific, distinctive and be linked to the delimited area, to the determined geographical zone Scientific characterisation of the product. Sensory profile Recognition by an « informed » consumer, supply chain specific sensory jury Mention of similar products on the market Points of differentiation between these products and the PDO Final characteristics Production, processing and finalising methods. Designation. Reputation, local tradition. Perceived quality.
Local usages, loyal and constant	Unity of usage and non contested usage Collective or potentially collective usage Traditional usage (see criteria of tradition).
Traceability	Use of indelible traceability markers required by the code of practice.
Representativeness of the petitioning group	Number of people belonging to each stage of production Distribution of these people in the delimited area Status of these people in the organisation
The existence of a trade mark	The existence of a trade mark using the designation and whose owner(s) intend to submit an opposition.
The reputation	On the basis of a notoriety survey on a significant sample of consumers (in particular for PGIs).

theoretical level. If it were possible to highlight the role played by these production channels (several small-scaled firms) on the rural employment, it would be necessary to set up a better legal framework to encourage their activities and common strategies.

References

Arrow, K. and F. Hahn, 1970. General Competitive Analysis. Holden Way, San Franscisco.

Barjolle, D., J.M. Chappuis and B. Sylvander, 1998a. From individual competitiveness to collective effectiveness: a study on cheese with Protected Designation of Origin. In proceedings of the 59th EAAE Seminar "Does economic theory contribute to a better understanding of competitiveness ?", EAAE, Appeldoorn, The Netherlands.

Barjolle, D., S. Boisseaux and M. Dufour, 1998b. Le lien au terroir. Bilan des travaux de recherche. Institut d'économie rurale de l'EPFZ, Antenne romande, Lausanne.

Barjolle, D., J.M. Chappuis and M. Dufour, 1998c. Cheese Synthesis, in final report European Project FAIR 95-306 "PDO-PGI Products: Market, Supply Chain and Institutions", IER-EPFZ.

Barjolle, D. and B. Sylvander, 2000, Protected designations of origin and protected geographical indications in Europe: regulation or policy ?, in final report of the FAIR Project "PDO-PGI: markets, supply chains and institutions.

Barney, J.B., 1991. Firm Resources and Sustained Competitive Advantage. Journal of Management, no. 17, p. 99-120.

Barney J.B. and W. Hesterly, 1996, "Organizational Economics : Understanding the relationship between Organizations and Economic Analysis", in *Handbook of Organizational Studies*, S.R. Clegg, C. Hardy and W.R. Nord Ed., SAGE Publications.

Carlton, D.W. and J.M. Perloff, 1994. Modern Industrial Organization. 2nd ed., HarperCollins College Publishers, New York.

Coase, R.H., 1937. The nature of the Firm. Economica, no. 4, p. 386-405.

Commons, J. R., 1934. Institutions: Its Place in Political Economy. University of Wisconsin Press, Madison.

Day, G.S., A.D. Shocker and R.K. Srivastava, 1979. Customer-oriented approaches to identifying product-markets. Journal of Marketing, Vol 43, p. 8-19.

Debreu, G., 1959. The Theory of Value. Wiley, New York.

Hinterhuber, 1989. Strategische Unternehmensführung. Band II. Strategisches Handeln. 4. Auflage, Berlin/New York.

Jensen, M.C., 1998. Foundations of Organizational Strategy. Havard University Press, Cambridge and London.

Kotler, P., 1994. Marketing Management, Prentice Hall New York.

La Rue T. Hormer, 1982. Strategic Management. Englewood Cliffs N.J., Prentice Hall, New York.

Liebenstein, H., 1976. Beyond Economic Man. Havard University Press, Cambridge, Mass.

Liebenstein, H., 1982. The Prisoners Dilema in the Invisible Hand: An Analysis of Intra-Firm Productivity. American Economic Review, no. 72 (2), p. 92-97.

Lombriser, Ablanalp, 1998. Strategisches Management: Visionen entwickeln, Strategien umsetzen, Erfolgspotentiale aufbauen. Versus Verlag AG, Zürich.

Mason, E.S., 1939. Price and Production Policies of Large-Scale Enterprise. American Economic Review, no. 29, supp., p. 61-74.

Mason, E.S., 1949. The Current State of the Monopoly Problem in the United States. Harvard Law Review, no. 62, p. 1265-85.

Pindyck, R.S. and D.L. Rubinfeld, 1998. Microeconomics. 4th ed., Prentice Hall, New York.

Prahalad, C.K. and R.A. Bettis, 1986. The dominant logic: a new linkage between diversity and performance Strategic Management Journal, no. 7, pp. 484-502.

Rumelt, R.P., 1984. Toward a strategic theory of the firm. Strategic Management Journal, no. 12, p. 167: 86.

Saporta, B., 1989. "Une chaîne verticale d'activités successives", Journal de Marketing industriel. Eyrolles management, issue, p. 4 and 6.

Simon, H.A., 1955. A Behavioral Model of Rational Choice. Quaterly Journal of Economics, no. 69, p. 99-118.

Teece, D.J., 1982. Toward an economic theory of the multiproduct firm. Journal of Economic Behavior and Organization, no. 3, p. 39-63.

Tirole, J. 1988. The Theory of Industrial Organization, The MIT Press, Cambridge, Massachusetts and London, England.

Tirole, J. 1997. "Formal and Real Authority in Organizations," (with P. Aghion), Journal of Political Economy, 105: 1-29. Reprinted in The Theory of the Firm: Critical Perspectives, ed. by Nicolai J. Foss, Routledge (1998), and in The International Library of the New Institutional Economics, ed. by Claude Ménard, Edward Elgar (2004).

Van Ittersum, K. and M. Candel, 1999. The European Consumer of PDO/PGI Protected Regional Products, Wageningen. Rapport rédigé dans le cadre du projet européen FAIR CT 905-306 "PDO-PGI Products: Market, Supply Chain and Institutions", INRA-UREQUA, Le Mans.

Wernerfelt, B., 1984. A resource-based view of the firm. Strategic Management Journal, no. 5, p. 171-80.

Williamson, O.E., 1975. Market and Hierarchies, Analysis and Anti-trust Implications: A Study in the Economics of Internal Organization. Free Press, New York.

Williamson, O.E., 1985. The Economic Institutions of Capitalism: Firms, Markets, Relational Contracting. Macmillan, London.

Qualification of milk producers in some Protected Designations of Origin: expert assessment conditions.

P. Parguel[1], F. Arnaud[1], D. Renard[2] and G. Risoud[3]*
[1]Institut de l'Elevage, CRA Valparc, 25048 Besançon, France, [2]CIGC, 39801 Poligny, France, [3]Syndicat de Défense de l'Epoisses, 21460 Epoisses, France.
**Correspondence: pierre.parguel@inst-elevage.asso.fr*

Summary

European dairy products can demand protection as regards origin. Economic operators establish the links between product characteristics and production conditions expressing the 'terroir' or locality of origin. If this link is validated in a procedure set up by each of the European Union member states, the area and the production conditions are written into national regulations. Europe can then register the product as Protected Designation of Origin (PDO). The product then becomes an element of the collective heritage. The milk producer, the first 'terroir' processor, is at the heart of this approach. Beyond simple respect for rules, the question arises as to the support and guidance for these producers in projects based on authenticity (respect for the 'terroir') and quality (response to customer demands including their societal demands). The initiation, by the French Inter-professional Committee for Comté Cheese and by the Association for the Defence of Epoisses Cheese, into a livestock farming qualification procedure, is the opportunity to strengthen the involvement of milk producers in these AOC (Appellation d'Origine Contrôlée), national equivalent of PDO. The methods take their inspiration from socio-economic and quality control procedures -territorial governance, cost of transaction, process control- ; they allow a progressive and durable appropriation by livestock farmers of dairy production rules (breed, feedstuffs...), and also the acquisition of real professional competence in inspection conditions. By the means of internal audits, they are invited to establish progress plans, aimed at strengthening the typicity of the product. The regulatory authorities are studying the recognition of such a qualification method.

Keywords: AOC, Protected Designation of Origin, Audits, Comté cheese, Epoisses cheese

Introduction

In France, a Protected Designation of Origin (Appellation d'Origine Contrôlée -AOC-) is a product whose characteristics are attributed to environmental elements in its geographical, historical and human dimensions. Links with the 'terroir' or locality are spoken of.

This distinction of products by origin is also recognised in Europe by the Protected Designation of Origin (PDO). Europe has extended this protective measure by setting up the Protected Geographic Indication (PGI) (rule 2081/92 of 14 July 1992). This sign concerns products with one specific step of the production system, and at least an established reputation links them to a place of production.[1]

As a general rule, these products are examined by an independent examining body (satisfying the requirements of ISO 45011 standard) and approved by the member State. For France, it is the INAO (the French National Institute for Protected Designations of Origin) which has been mandated by the Ministry of Agriculture for the inspection of AOC and IGP (equivalent of PGI in France).

[1] In France, only products that have obtained or requested a CCP (Certification of Product Conformity) or a Label Rouge (brand name of the French Ministry of Agriculture distinguishing products of superior quality) can apply for the PGI.

In this context, the INAO, on proposals from professionals in the sector - cheese-makers and milk producers - defines the production rules attached to their activity, and in this way supports and guides the development of the designation.

Many AOC are now wondering about a more formalised recognition of the work of milk producers, for several fundamental reasons:

At the commercial level, the buyers of Large and Medium-sized Supermarkets are introducing inspections at milk production level into their product listing[2] requirements. Purchasing procedures have incorporated audits of dairy farms. These new commercial requirements can result in segmentation within the product. Moreover, as for all other products claiming a distinctive sign (Label Rouge, organic farming), the AOC have to have an inspection structure for milk and cheese production conditions with consumer credibility. So the challenge is to ensure respect for production conditions by inspections that remain within the legal and philosophical framework of the AOC (control by the professionals), that are understood by the outside world and recognised by supervisory bodies (INAO, fraud services).

At the legal level, protection of product origin at the French and European levels is written into the regulations of the AOC and PDO in France and Europe respectively. Nevertheless, and in particular in international debates at the World Trade Organisation[3] (WTO), on the protection of geographical indications as intellectual property, validated justification of the link with the locality, particularly at the level of milk production, should offer more assets to the products concerned and their defenders.

At the technical level, the debate between tradition and modernity is still going on. Greater involvement by milk producers and better recognition of their practices are just some of the means of directing research and development actions, integrating more cultural or environmental elements.

It is in this context that some AOC Associations have wondered about an assessment of dairy production that can satisfy these three commercial, legal and technical objectives. This article sets out to study how these approaches restate the problem of the recognition of milk producers and more generally the product with commerce and institutions, how the livestock technician is positioned, and how the procedure modifies the social structure charged with managing the product.

The methodological principles of the procedure

The proposed qualification came from a combination of socio-economic and quality procedures.

The structural justification for qualification

Among the socio-economic procedures, the natural strategy of the AOC is very much directed towards territorial governance as defined by B. Sylvander (Sylvander, 1998). In fact, the production conditions of these AOCs define breeds, depend on the flora of the grazing land and supervise their use. What is more, quality is controlled within a collective action at the level of product management bodies. The purpose of qualification is then to determine heterogeneity of biodiversity that cannot be reduced, at the level of production and of the products marketed. In

[2] The listing is the procedure implemented by the purchasing services of LMS central buying offices.

[3] More particularly in ADPIC agreements: Aspect of Rights to Intellectual Ownership affecting Trade.

both cases, collective management emerges as a means of reducing costs (in particular by combining the means for research and development), and of joining forces to confront distribution requirements when they are judged to be inappropriate to the dairy farmers' real situations, or even in contradiction with the maintenance of the link with the locality. Focused as it is on self-inspection procedures right from production level, and beyond its technical content, qualification participates in consolidating the position of producers in the AOC (Letablier, 2000).

The economic justification of qualification

The commercial structures have taken the calculated decision that products from an organisation that cares for quality and the continuous progress of this quality are more interesting in terms of costs than a situation of the "spot market"[4] type, which supposes a large organisation for the purchase and quality control of goods. This hypothesis, studied under the term of "cost of transactions" (Barjolle and Chappuis, 2000), showed how farmhouse cheese sectors managed to maintain a price positioning on the market. The qualification of dairy farms comes within this logic.

The territorial justification of qualification

The AOC is a protection given to a product linked with its locality. In other words, the essential characteristics of the product are considered to be due to elements of the geographical environment where this product was created (Barjolle *et al.*, 1998). These characteristics can only be expressed by human know-how. The first level of this know-how, in the context of dairy products, is the dairy farm, since the animal is the first processor of the environment. So the dairy farmer is at the centre of an AOC issue which includes, more than the feed and housing of the animals : the appearance of the place, more than genetic choices : the biological potential of the herds, more than historical reflection : the durability of the production and finally, more than the quality of the milk produced : thc typicity and the reputation of the AOC cheese (Parguel *et al.*, 2002).

The input of process control procedures

These different socio-economic procedures are to be combined with quality procedures and tools. Livestock farming has been thc subject of research into the applicability of quality procedures. In the 1980s, the HACCP (Hazard Analysis and Critical Control Point) was born from the meeting of process analysis and quality circles. This approach, at the initiative of the NASA, became the subject of a CODEX recommendation (CODEX Alinorm 93/13 A). Its principles were tested in dairy farms, firstly on the health control of milk production (Parguel, 1997), then more globally for the definition and management of production conditions in AOC (Risoud and Parguel, 2002). With these tools, farmer know-how comes within a programme of reflection and development that can be received by all.

[4] The spot market (or free market) is the whole of sales that are not the subject of medium term contracts between producers and buyers. Prices are fixed instantly or can be revised in the short term.

The input of quality assurance

The standards of the ISO 9 000 series[5] formalise the requirements of universal recognition of the quality organisation. In their "2000" versions, these standards include requirements concerning in particular the notion of continuous progress. Even if the strict application of the HACCP and ISO 9000 standards remains limited in dairy farms, their methodological spin-offs are considerable. In the case of qualification, these standards have established the obligation of a "quality policy"[6], the commitment of operators and the setting up of inspection procedures.

The audit procedure

Quality procedures have made a significant contribution to thoughts on inspection and on the setting up of quality audits[7]. The audit is the observation of a situation in relation to a system of reference. For a same situation, the measurement of discrepancies is supposedly identical, whoever the auditor may be. It must therefore be competent regarding the situation observed, and independent of the customer-supplier relationship. In the case of the AOC, the audit must equally be capable of making a reliable assessment of the progress made by the dairy farmers.

The measurement of Total Quality

This measurement of progress in quality organisations is the subject of reference systems used for TQM (Total Quality Management). In France, the French Prize for Quality (PFQ) is a permanently evolving model, adapted to small and medium-sized enterprises (Europe has a Quality Price reference system: the European Foundation for Quality Management (EFQM), less targeted on small businesses).

For a given objective, this model proposes the classification of each point of the situation observed according to four criteria:

- Existence: the objective is understood and there is the beginning of a response;
- Method: a reactive procedure is implemented to deal with the objective;
- Systematism: the procedure is used as often as necessary;
- Exemplarity: the method, its application or the results obtained deserve to be communicated externally (Ministère de l'Industrie, 1995).

This classification is also designated under the acronym EMSE, taking the initials of the different criteria. The notion of systematism has to be understood as the capacity of the entity, for a given objective, to apply the most suitable solution, the solution that is known in the technical world. Exemplarity also considers the potential for disseminating the solution and the capacity and will of

[5] The ISO 9000 standards deal with Quality Assurance in companies. Until 2000, these standards were composed of requirement standards (ISO 9001, 9002, 9003) enabling the certification by an independent organisation (usually in France the French Association for Quality Assurance (AFAQ)) and a guide standard (ISO 9004). Since 2000 there has only been one requirement standard (ISO 9001) whose field of application has been reviewed, including among other things the notions of listening to the customer and of continuous progress, which are close to the concept of Total Quality Management (TQM).

[6] Quality policy: general directions and objectives of a company concerning quality, as formally expressed by the General Management (NF X 50-120).

[7] Audit: methodical and independent examination with a view to determining if the activities and results relating to quality satisfy pre-established provisions, and if these provisions are implemented effectively and likely to attain the objectives sought. One of the aims of the quality audit is to evaluate the needs for improvement or corrective actions (NF X 50-120).

the enterprise to carry out this dissemination. It is on this notion of exemplarity that tools are developed to support the exchange of experiences between enterprises on their points of excellence.

As the link with the locality in AOC is constructed every day in the evolution of farmers' practices, the qualification of dairy farms must adopt an audit procedure that can assess not only errors of understanding, or the non-implementation of compliant practices (existence and method), but also the expertise that can enrich and strengthen collective know-how (systematism and exemplarity).

Application in two French AOC: Comté and Epoisses

Motivations for the qualification of dairy farms

The cheese AOC have differences in their size (the number of producers in particular), in their history, the nature of their quality requirements, and the sensitivity of the product to health accidents. This is why we have chosen for our analysis two AOC with very different characteristics, to observe how a qualification approach could be set up with milk producers.

Comté is a cooked pressed cheese with long maturing process, which makes it relatively insensitive to health risks. For the most part, it is produced in cheese dairies (small cooperatives at commune or canton level) managed directly by the milk producers. This situation has resulted in a long-standing and very real involvement of the farmers in managing the appellation. The milk is produced in about 3 300 dairy farms representing a large proportion of the AOC area. The French Inter-professional Committee of Comté Cheese (CIGC) is a Committee whose members are designated by the French Ministry of Agriculture on proposals by the professional organisations. It also has Inter-professional status[8].

Epoisses is a soft washed-rind cheese, sensitive to health accidents. It is made in small and medium-sized enterprises in Burgundy, which do not control the whole of the milk collection. 58 milk producers are identified for this production. Their involvement in the AOC sector is recent (Risoud and Parguel, 2002). They only represent about 20 % of the milk collected over the appellation area. The AOC Epoisses is run by the Association for the Defence of Epoisses Cheese, which has the legal status of an Association under the law of 1901.

In Comté, the first argument for the qualification of dairy farms[9] is setting up a system that guarantees the consumer respect for regulatory aspects and certain environmental aspects (including the treatment of waste). Technical assistance to producers appears in the second argument, in continual progress towards more quality and typicity. Qualification is initially situated alongside of official inspection procedures (Veterinary Services, Fraud Squad, National Institute for Protected Designation of Origin). It places the stress on a generalised self- inspection, of which the CIGC is the project manager (Renard, 2003).

In Epoisses, qualification is above all presented as the logical outcome of the involvement of milk producers in strengthening links with the locality and improving the health and organoleptic qualities of the milk (Chevalier, 2002). The field covered is limited to AOC and health aspects of the milk, basically as regards regulations. The qualification is based on a system of self-inspection

[8] The interprofessional status enables the promulgation of rules applicable to everybody.

[9] The term qualification must be taken here as an evaluation of a level of quality with the aim of working out a progress plan. It must not be confused with the qualification of livestock farms implemented in other quality signs, such as Label Rouge, where it corresponds to an audit carried out by an agent approved by an authorised and accredited Certification body.

Table 1. Comparison of the principal structural characteristics between Comté and Epoisses.

	Comté	Epoisses
Region of origin	Franche-Comté	Burgundy
Date of AOC and PDO recognition	French Decrees of 14 January 1958 and 30 March 1976, Decree of 30 December 1998 modified by the Decree of 10 January 2000, Community Regulation L 148/1 of 21/6/96, EC Regulation N° 1107/96 of the commission of 12 June 96	French Decree of 14 May 1991, Modified by the Decree of 16 July 2004, Community Regulation 1107/96 of 12/06/1996, Modified by EC Regulation 828/2003 of 14 May 2003
Nature of product	Pressed cooked cheese	Soft cheese with a washed rind
Sensitivity of the technology to health accidents	Low	High
Volume produced	46 000 tonnes	800 tonnes
Number of milk producers	3 300	58
Processing structures	188 small cooperative cheese dairies	3 small and medium-sized private enterprises, 1 farmer producer
Involvement of dairy farmers	Long-standing	Recent
Structure of the AOC grouping	Inter-professional committee	Association under law of 1901

under the project management of the Production Conditions Approval Commission[10]. This situation eventually results in a formal link between farm qualification and official INAO inspections.

In both cases, qualification is intended to support and recognise the technical progress made by milk producers, to encourage customer loyalty (by providing better information), to keep control of distribution and anticipate any developments in regulations concerning products.

The commitment

Due to the dairy production structure of the two products studied, the notion of voluntary participation is applied in a different way. In Comté, where almost all the milk of a village is processed at the cheese dairy, producers are considered to be voluntary participants who, after receiving information about the qualification, accept its challenges and the audit procedure. The other producers continue to have their milk collected and processed in Comté. The dynamic of the system lies in the procedure's advantages as perceived by the farmers, the mobilisation of the Association and the way in which the cheese dairy board adheres to the qualification. Everything

[10] Regulations on the AOC (1992) provide for the existence of two commissions composed of agents from the INAO and professionals, named by the INAO, and charged with monitoring respect for production conditions (Production Conditions Commission) and the special features and sensorial qualities of the product (Product Approval Commission).

Table 2. The principal stages in setting up a qualification action for dairy farms in AOC.

Stage	Who?	How?
1 - the commitment	Prepared by a Steering Committee, Validated and communicated by the Association	From elements of the AOC Decree, quality results and information coming from the market, Individual (letter) and collective (meetings) communication
2 - the technical objectives	Prepared by the Steering Committee, Validated by the Association	The objectives are translated into requirements then into criteria of appreciation of individual situations
3 - the audits	Carried out by recognised technicians, Analysed and validated by the Association	Drawing up audit grids, Training of technicians, Possibly, inspection (super-audits), Setting up individual and collective progress plans, Revision once a year

is done to ensure that enough producers belong to the qualification to develop the quality and credibility of the sector[11].

In Epoisses, where a small part of the area's production is processed in AOC, producer commitment is necessary. The rest of the milk is processed into other added value products.

In both cases, the terms of the commitment were prepared by a producers' group (CIGC technical commission, the Epoisses Association production commission), and then validated by the Association.

In Comté, beyond the AOC decree, the commitments concern the monitoring of cows, the appearance of buildings, animal well-being, respect for grassland flora, training young people and reception of visitors to the farms. The farmers' commitments have been combined in the same document with those of the cheese makers and maturers.

In Epoisses, producer commitment is in the form of a reciprocal contract with the processor. It concerns health control and respect for the AOC decree as well as the conditions for differentiating ways of adding value to the appellation milk.

For these two products, farmers are approached collectively at local meetings organised by the Association and the processor, and individually by documents in the form of a diary for Comté and a producer's guide for Epoisses.

The technical objectives

Each commitment is translated into technical objectives with the help of local technicians (Institut de l'Elevage, Chambers of Agriculture, cheese dairies, Milk Recording Associations, veterinary officers). Table 3 presents the technical criteria discussed in professional groups in the framework of the AOC project. Some of these criteria are taken up in the AOC decree. During qualification

[11] The CIGC proposes that in the next decree, the qualification of dairy farmers should become obligatory every 5 years.

Table 3. Technical elements specific to the AOC Comté and Epoisses taken into account for the qualification of dairy farms.

	Comté	Epoisses
Animal management	Breeds authorised : Montbéliarde, French Simmental[1] Knowledge of ancestry and blood lines Herd qualified for health and health record up to date Display of farm characteristics Organised storage of medications Dairy production according to the objectives of the sector	Breeds authorised : Brown Swiss, Montbéliarde, French Simmental[1] Insemination practices aimed at strengthening the cheese-making potential of dairy cows[1] National Charter of good practices validated and up to date Dairy production according to the objectives of the sector[1]
Feed	At least one hectare of grass per cow[1] Self-sufficiency in forage Records of feed Access to different qualities of forage No more than 30% of concentrates in the total diet[1] No silage[1] Traceability of supplements Easy access to grazing land	85% self-sufficiency in feed on the AOC area[1] Spring grazing (20 ares/dairy cow) [1] Supplements < 30% total diet[1] Dry forage in winter > 30 % of the basic diet[1] Positive list of authorised feed and supplements[1] Records of feed practices[1]
Milking and milk quality	Milking at regular hours[1] Monitoring of herd results Monitoring the cow/cow quality Access to the clean dairy	Renewed plan of risk control (Haccp) Training in the ecology of pathogenic germs
Building and animal welfare	Clean animals Well-ventilated building, in good condition and of sufficient size Comfortable observation of the herd Farm surrounding well cared for and planted with flowers Integration of the building into the landscape	Clean animals Building to the standards of the Institut de l'Elevage
Monitoring the grasslands and crops	Development of natural grasslands Sowing of grassland with appropriate mixtures Respect for diversity of flora Sustainable use of plant health products Sustainable use of mineral fertilisers	Spreading record [1] Sustainable use of plant health products
Respect for the environment	Knowledge of the holding and its environment Respect for spreading recommendations Economy of natural resources (water) Non polluting elimination of waste	Control of organic spreading[1] Sustainable use of mineral fertilisers[1]
Communication	Concern for professional exchanges Seeking data and technical support Organisation of internal work Welcome for young people Promotion of the work of the farmer	Producer's information board Farm surroundings well cared-for Guide *Milk Producer's guide in Epoisses* updated and available for consultation

[1]regulatory obligation for the AOC

audits, all these criteria are completed by other elements arising from the general regulations. For each of the objectives, elements of proof are also defined, from which the auditor analyses the situation encountered. The expected results are classified according to the four-box grid of the PFQ : Existence, Method, Systematism, Exemplarity. The procedure is strictly identical for both products, Comté and Epoisses. Everything is presented in the form of a grid. This presentation was chosen because it is similar to documents used in audits of reference systems carried out in livestock farms by buyers of the large and medium-sized supermarkets. These grids have been completed with regard to the requirements of the Charter of Good Livestock Farming Practices set up by the National Livestock Confederation[12] and to elements that may be considered in the Farm Holding Territorial Contract (now the Contract of Sustainable Agriculture).

The audits

First of all, it is the farmer himself who is invited to assess his situation against the established audit grid. To do this, he uses record documents already at his disposal for regulatory (e.g. farm record) or technical (e.g. Milk Recording results) aspects. This phase is the opportunity for the producer to bring together all the documentation relative to the qualification. Then, in Comté and in Epoisses, regular audits are carried out at producer level (every 2 years in Epoisses and every 5 years in Comté) by the technicians who usually give technical support to the farmers. This solution has the advantage of the farmer's situation being analysed by staff already familiar with the farm environment. In this way producer membership is favoured and the skills required for setting up a progress plan are immediately available. Substantial savings are also made for carrying out this inspection. This situation amounts to an internal inspection procedure, compliant with the spirit of the AOC, which involves farming development in the objectives of the sectors. The auditors receive training in audits, provided by agents recognised by the regional inter-professional authorities.

The results of these audits are summarised at product Association level, which, beyond simply analysing producers' individual situations, make it possible to envisage the setting up of collective actions.

To ensure that the audit results are reliable, Epoisses, with the Production Conditions Approval Commission, brings all the auditors together twice a year and reviews the difficulties encountered from the combined audits. The INAO is then the body which can apply individual sanctions in the case of repeated, unjustified breaches of AOC rules. In addition, it was decided, as of 2003, that the INAO would once a year accompany each technician in his auditing work (upstream inspection). In Comté, the large number of auditors does not allow for such direct consultation. So an inspection procedure has been put in place, called super-audit, which makes it possible to ensure that the information provided by the auditors is correct, the producers have understood the qualification procedure and the qualification does culminate in a progress plan.

An agent of the Institut de l'Elevage on behalf of the CIGC carries out this super-audit. Each auditor is thus inspected twice a year. The first super-audit is carried out at the same time as the audit by the auditor technician, and the second time, in another farm some months after his visit. As for the audit, the super-audit is only implemented with volunteer auditors and only collective results are available. They are mainly used to improve the performance of auditors by training actions.

[12] It has in fact been considered, after discussion in each AOC, that the Charter of Good Livestock Practices should be integrated into the qualification for the following reasons : the Charter contains the regulatory requirements relative to livestock and it participates in the promotion of French livestock in general, which includes livestock farming producing in AOC. It participates in controlling the dissipation of productive practices, which could weaken the credibility of the product (Giraud-Heraut et al., 1998).

The annual meetings of the auditors in Epoisses and the super-audit procedure in Comté serve as a basis for revising the qualification. This revision is operated once a year in each Association, and can concentrate as much on the procedure as on the content of the qualification. In this way, the content of the audit plan and certain record documents were modified in 2004 and serve as a basis for a version II of the Milk Producer's Guide in Epoisses.

Results

Achievements

Table 4. Audits in Comté and Epoisses.

	Number of milk producers	Period	Number of producers audited	% of producers audited	Number of auditors	Number of auditors inspected
Comté[1]	3 300	2001/2003	1 600	48 %	84	61
Epoisses[2]	58	Year 2002	23	40 %	4	0
		Year 2003	35	60%	4	4

[1]in Comté, objective to cover the whole of production in 5 years
[2]in Epoisses, objective to cover the whole of production in 2 years

In Comté, 90 % of farmers have accepted the audit. 30 % of refusals concern producers who will shortly be stopping work. In Epoisses, all the dairy farms have accepted the procedure, which appears in their commitment.

Qualification and technical support to milk producers

In Comté, and considering the 61 super audits carried out in 2002 as representative, the results are presented in the following way: 9 qualification audits do not cause modifications on the farm, 25 encourage the setting up of better traceability (recordings concerning animals, crops or fertilisation), 26 have initiated real progress plans and a respect for deadlines for planned actions, 1 producer have launched procedures for continuous improvement (improvement actions will be or have been implemented more often that the frequency of audits).

The principal improvement actions noted concern the establishment of a register or health record, identification of animals under treatment, classification of milk quality results, preservation of labels and feed delivery notes, and registering harvests and fertiliser spreading. Other actions concern equipment: improvement of ventilation and lighting in the milking parlours, renovation of the dairy, modification to the building for the comfort of the animals, layout of access tracks to the grassland.

In Epoisses, the Qualification has made it possible to increase knowledge of AOC rules that are being drawn up, and still insufficiently understood by 3 farmers (7 % of the farmers audited). All the shortfalls noted in the farms have been the subject of a progress plan. These plans focus first of all on setting up records, something that is not too often to be found on farms which do not belong to a technical or technical and economic monitoring system, usually carried out by the Milk Recording. As for animal nutrition, the planned progress centres on self-sufficiency in feed (43 % of farmers audited) and the distribution of dry forage (26 %). For the health aspect, 6 farmers (30 %) have reviewed the definition of their Critical Control Points (CCP).

Recognition by the INAO

The INAO, via the National Dairy Produce Committee (CNPL) has not yet taken an official position as regards the qualification of dairy farms as presented here. Nevertheless thought is being given to inspections of production conditions implemented by the AOC Associations. In this context, a group composed of Heads of Centres and INAO technicians has studied the qualification procedures of Epoisses and Comté from the documents used in each sector : Milk Producer's Guide for Epoisses, the Agenda for Comté.
The criteria for their analysis were as follows:

- the objectives stated - links with the AOC decree or a future decree -,
- the controls - initiative, principal partners, financiers, technical experts, involvement of milk producers -,
- the means - number of players involved, number of operators for the implementation, communication structure, consumer credibility structure -,
- the positioning in relation to quality procedures - tools of reference, application, relevance to the AOC -,
- the results and effects expected or possible - effect on the diversity of products and know-how, effect on traditional know-how, effect on the natural flora of milk, effect on the consistency of the sector, communication to the consumer, enhanced status for the work of the dairy farmer -,
- the positioning in relation to public inspections - partnership with the INAO, links with the DSV (or DGAL), links with the DGCCRF (General Bureau for Competition, Consumption and the Prevention of Fraud), the place of producer commitment -.

For Comté, the structure has appeared credible because of the number of auditors and the establishment of super-audits under the authority of the CIGC board. Even if there were no formal link with official inspections (DSV, DGCCRF), the INAO could have been a partner in this qualification.

For Epoisses, the major objective is the involvement of milk producers in the draft revision of the decree. The Association has initiated this qualification and has been able to mobilise technical skills and regional and national financing in its project. It is the Production Conditions Commission that has responsibility for building credibility for the approach inside and outside the sector. The partnership with the INAO is established at this level and in exchanges with members of the CNPL Commission of Enquiry charged with the study of the draft decree.

Recognition by buyers and consumers

It is difficult to measure the recognition of dairy farm qualification by the buyers of the large and medium-sized supermarkets. In Comté, the chosen indicator is the communication by these buyers of quality requirements concerning milk producers. Since the beginning of the year 2000, the CIGC has recorded no requests either during direct contacts with the large/medium supermarkets or via the cheese maturers.

In Epoisses, the number of marketing companies is low. Naturally, dialogue is becoming established between companies within the Association, on demands from the large and medium supermarkets. Qualification has not brought about any increase in quality requirements from the supermarkets, nor any segmenting by Distributor Brand (MDD), satisfying a specific set of specifications established between the cheese dairy and the distributor.

At consumer level, it is risky to establish a link between volumes sold and prices and the setting up of the qualification to deduce any effects, direct (improvement of confidence) or indirect (improvement of quality). Nevertheless, it can be noted that qualification goes along with the growing demands of consumers who, beyond a place that makes " sense " for the product, are increasingly interested in understanding production methods defined and controlled by the producers (Valcescini, 2000). For this reason, it is legitimate to include in the criteria of dairy farm qualification, aspects which are associated with communication with the public (welcome to the farm, surroundings, communication about the farms).

The use of qualification to anticipate AOC rules

The use of dairy farm qualification in the context of AOC decree revision is not a declared objective of the Comté and Epoisses Associations. The position of Epoisses is justified by the fact that the production conditions taken into account in the qualification reference system have only been applied since 2001, and recently recognised by the authorities (decree of 16 July 2004). They still have to prove they are sustainable. As for Comté, an internal debate is in place for the revision of the decree. In this case it is obvious that information from the qualification procedure come into play in this internal negotiation, if only to validate the feasibility of the options under discussion.

Discussion

Qualification is a special tool of milk producers' involvement in the AOC, whether this involvement is of long date as in the case of Comté, or recent as in Epoisses. Given that the link with the locality is established (and not proven on scientific bases), it becomes indispensable to set up a support framework for production conditions, such that technical innovation is controlled to be mastered. Paradoxically this is the condition necessary for maintaining the typicity characteristics of the product over the long term.

Qualification is distinguished from certification as defined in specifications, within which it is supposed that requirements are at the origin of the desired quality. In this case, these requirements represent a series of constraints for the operator, justified by a commercial demand in return for money. So it is normal and justified, for simple reasons of commercial honesty, that the application of these rules should be checked in the most independent way possible. But in the AOC, the product is more a proposal of quality made by the operators themselves and recognised as original by the legislator and the market. Specifications are more a statement of rules distinguishing operators who are respectful of the concept. Qualification is then a means of assessing the capacity of operators to progress towards more "local identity " in the framework of rules thus defined.

The technician, whose work includes attention to the needs of farmers, helps them to build their own know-how with the help of their technical, sensorial and mental resources. This exposure to a different expertise from his own instructs him on his own technical knowledge, and his limitations as to the expectations of AOC dairy production. This process has already been described by Muriel Faure in her study of cheese sectors in the North Alps (Faure, 2000). At the level of bodies charged with research and development in livestock farming, farm qualification leads to a political positioning regarding the commercial reality of local products.

Qualification leads to progress plans. These plans can be reduced to action plans defined jointly between the farmer and the technician, in a concern for obtaining better positioning at the next audit. In other cases, it concerns real farm progress plans, including a definition of the farm's quality policy and personalised technical objectives. With this qualification, the AOC partly takes charge of the direction of the farm in its technical and structural aspects. This situation is not abnormal insofar as the AOC is an element of local heritage whose management is entrusted to professional structures. If the farmers' progress plans are limited to action plans, without a minimal formalisation of the farmers' objectives, then the AOC takes the risk of cutting itself off from the objectives of its most basic operators, the milk producers. This phenomenon would be aggravated if action plans were limited to technical reference systems whose link with the typicity of AOC products is not durably established. If it were admitted that the best-established links are those between the farming technique and elements of product quality (and not elements of link with the locality), qualification could then be taxed with drifting from quality standardisation. The strong involvement of professionals in the assessment procedures of the two cases studied, makes it possible to avoid such drift, and refocus the debate around the product in its original environment and its specific features.

Conclusion

The qualification of dairy farms in AOC is a tool at the service of Associations charges with the management of these products. This procedure is a place for arbitration between the notions of quality (responding to consumer expectations in the matter of taste, price and use of the product) and notions of locality (respect for links with the place). By the definition of farming practice judgement criteria, the qualification comes on the one hand from a sectional logic (seeking a standard quality from industrial standards), and on the other hand from a territorial logic (organising a variety of supply among craftsmen). It is a tool of regulation by the flexible model described by B. Sylvander (Sylvander, 1998).

This AOC management model cannot only be made by active participation of the different economic players concerned. By credible recording of farmer practices and the initiation of a system of permanent revision, qualification gives farmers a power of decision on the management of their product. Qualification therefore serves the social construction of the product by improving confidence among the different players.

By integrating into the AOC project major cross-disciplinary procedures such as the Charter of Good Practices (and tomorrow perhaps sustainable agriculture), qualification can preserve - whilst controlling it - a certain heterogeneity of the AOC product, that is necessary for its image and its successful value enhancement. It is a way of making the AOC credible in the eyes of consumers. If this aim is achieved, collective communication becomes stronger than what individual economic players would have had the means to implement.

Consumers and citizens are entitled to require that statements in AOC regulatory texts should be respected by all operators. So, as far as the fundamental allegations are concerned, strict inspection has to be put in place either by independent bodies (the general case in Europe) or by public bodies appointed by states (the case of France). But fundamentally the AOC carries the promise of a product associated with a place, and this link cannot be formalised as a whole by technical criteria capable of being tested. So it is preferable for the inspection to be based on evolutionary procedures directly involving all the players concerned. Local technicians are an integral part of this structure and there is no doubt that their positioning still needs to be specified. The question remains of the recognition of the principles of such a qualification by the supervisory bodies. The debate has to begin with inspection procedures that respect the AOC concept and whose modalities certainly differ from those implemented for other signs distinguishing the quality of products.

References

Barjolle, D., S. Boisseaux and M. Dufour, 1998. Le lien au terroir - bilan des travaux de recherche, Institut d'Economie Rurale (antenne Romande), Zurich, mai 1998, 27 p.

Barjolle, D. and J.M. Chappuis, 2000. Transaction costs and artisanal food products, ISNIE, Lausanne séminaire 2000, 21 p.

Chevalier. B., 2002. Filière Epoisses - Guide du producteur de lait, Syndicat de l'Epoisses, 1.

Faure, M., 2000. Du produit agricole à l'objet culturel. Le processus de patrimonialisation des production fromagères dans les Alpes du Nord. Université Lumière Lyon II, thèse de doctorat en Sociologie et Anthropologie, 401 p.

Giraud-Heraud, E., L.G. Soler, S. Steinmetz and H. Tanguy. 1998. La régulation interprofessionnelle dans le secteur viti-vinicole est-elle fondée économiquement ? In : Qualité des produits liée à l'origine, actes du séminaire INRA des 10 et 11 décembre 1998 à PARIS, p. 154-179.

Letablier, M.T., 2000. La logique du lieu dans la spécification des produits référés à l'origine. Revue d'économie régionale et urbaine, no. 3, p. 475-488.

Ministère de l'Industrie des postes et télécommunications et du commerce extérieur and MFQ, 1995. Prix Français de la qualité: qualité totale, outil de progrès, 26 p.

Parguel, P., L. Berard, P. Marchenay, N. Jehl, C. Dutertre, T. Gadenne, D. Vallot, A. Audiot, J.B. Coulon, A. Lombardet, F. Roncin, J. Tourmeau, C. Danchin, C. Laithier, L. Catalon, D. Bastien and M. Barnaud, 2002. Pour-quoi cet animal ici ?. Institut de l'Elevage, Réseau ACTA Animal et AOC, CR n° 2023118, 12 p.

Parguel, P., 1997. Adaptation de l'assurance qualité à l'élevage laitier. Travaux et innovations, n°43, p. 94-96.

Renard, D, 2003. La qualification des élevages en filière Comté. CIGC, agenda du Comté, p. 10-11.

Risoud, G. and P. Parguel, 2002. Renforcement des conditions de production dans les AOC dites de plaine, Sfer, Economie rurale, no. 270, p. 50-64.

Sylvander, B. , 1998. Les stratégies d'acteurs dans les Appellations d'Origine Contrôlées - application aux secteurs viti-vinicoles, fromagers et charcutiers. Séminaire AOP-AOC de Paris. Qualité des produits liés à leur origine, 11 p.

Valceschini, E., 2000. La dénomination d'origine comme signal de qualité crédible. Revue d'Economie Rurale et Urbaine, no. 3, p. 489-500.

Qualification of the origin of beef meat in Europe
Analysis of socio-technical determinants based on French practices

F. Casabianca[1], N. Trift[2] and B. Sylvander[3]*
[1]INRA SAD LRDE Corte, France, [2]INRA DARESE Paris France, [3]INRA SAD SICOMORE Toulouse France.
**Correspondence: fca@corte.inra.fr*

Summary

Quality has been the subject of various definitions: we make a distinction between the generic quality of the mass market and the specific quality, which allows divisions on the basis of distinctions. Therefore, the origin already defined and protected is a component of this specific quality. The criteria to be applied are the result of grading operations. As regards beef, very few situations can be used to analyse how these criteria are structured. The obstacles which hinder the transfer of the origin from the live animal down to the cut of meat to be consumed are reviewed and their causes studied.

Taking as a basis meats whose origin is qualified, we analyse two contrasting situations in France: (i) the bull from the Camargue, an animal whose carcass is ill-shaped and which offers a strong referent to coordinate the agents in the chain, and (ii) the Full marbled of Mézenc, which, in contrast, has an average carcass and undergoes a quite special fattening system which forces stockmen and butchers to organize around the live animal.

A model of the ways in which the origin of these two bovine meats is qualified shows recurrent elements which, when put to the test in other situations, offer a pertinent framework for future studies. They explain why so few projects have gone ahead. But bovine meats from specific origin do not seem able to travel well beyond their locality of origin, due to the close interaction between the agents - including the consumer - which they imply.

Keywords: bovine meat, protected origin, structuring of criteria, specific technical devices.

Introduction

The idea of quality has been the object of varied definitions, referring to different approaches. Its diverse and partially contradictory implications originate from this initial non-determination. **Two essential ideas** support the use of the term “quality”:

- The first concerns the listing of the characteristics which make something what it is, as regards the ends for which it will be used; these are the properties expected to be present. The ISO 8042 norm gives a rather broad definition: “quality is the totality of the properties and characteristics of a product, a process or a service that bear on its ability to satisfy implied or stated needs”. Stress is then put on the importance of needs and on the comparative character of their satisfaction.
- The second idea supports the manifestation of a level of excellence, a way of setting something apart as compared to similar products and which justifies it being sought.

However, in both cases, nothing sets out who has to define the content of quality, between the provider and the consumer. Nothing indicates how it is fashioned and how is it ensured. Can we be satisfied by an interaction of proximity between the agents concerned directly, or do we have to appeal to institutions, and even the State itself, to regulate the scope of the needs to be fulfilled? We find here a first theoretical paradox: where is quality defined?

These two broad ideas are often merged within a large array of situations, which refer, for instance, to supposed needs, to expressed expectations, to matters of non-quality (lack of properties) or to disqualifying processes (disappointment due to the absence of excellence). This multiple meaning characteristic does give rise to frequent confusions and it would then be important to share clear definitions, through the organisation of a discussion in the European countries about the definitions of quality, so as to no longer speak of quality as such, but to state precisely what is the meaning retained.

Concerning foodstuffs, the word "quality" is logically the object of the same queries. It has also known different main uses during successive periods. This brings us to consider three approach levels:

1. Historically, quality is first understood as the absence of flaws, frauds and fakes. Public power intervention expressed itself very early concerning this aspect, by the establishment of specific regulations. A *de facto* harmonisation seems to be on the way in the different European States, while a consensus is evolving on the need for a generalised implementation of this approach.
2. More recently, quality rests upon expected properties such as organoleptic and nutritional characteristics and useful value. This creates the need to take into account the legitimate expectations of the users and require from professionals to guarantee this acknowledgement. The function of the State here is to look after the interests of its citizens, including substituting itself to their expression. In this manner, the needs of the consumers remain largely implicit and encompassed in the concept of public interest. This is the case of regulations concerning the sanitary safety of foods and other normative characteristics, specially contributing to nutritional equilibrium or to services.
3. Finally quality denotes sought after characteristics likely to allow an increase in value, for instance the production methods (organic agriculture, environmentally friendly production, and animal well-being), the production zones (territory of origin, mountains) and their inherent traditions. These characteristics must be explicitly stated in the offer of the products so as to pinpoint the necessary interventions, the responsibility of each operator and to bring about the expected increase in value.

These three approach levels do not substitute each other; they are permanently superposed and justify different intervention levels of public authorities, of the operators and of the consumers. In particular, the drawing up of national or even international norms, which bind everyone, is based on very different proceedings from those that sanction voluntary interventions where only certain operators involve themselves.

Concerning foodstuffs in Europe, we can consider that at present the two first levels refer to all the products on the mass markets, and they can be included in the designation of "generic quality" which is expected to be thorough and without ambiguities. On the contrary, the third level entails differentiation strategies for the products and segmentation of the markets. This third level, identified as "specific quality" is the one which implies the affixing official quality signs and entails special arrangements (Sylvander *et al.*, 2000).

To give value to an **origin** seems to be advantageous for the marketing of products bearing the name of a territory. Studies have demonstrated that consumers, especially during a period of crisis, feel reassured by information concerning the locality of production of a foodstuff (Barjolle and Sylvander, 2000). It becomes then essential to protect this name, since it is associated to specific properties: quality linked to the production territories is one of the forms of specific quality

(Casabianca and de Sainte Marie, 1998). We undertake to deal in depth with the terms of quality linked to the origin.

The position which we have adopted consist in considering quality **as a social construct,** where the convergence between the various agents involved is not spontaneous or natural. It arises from a complex process through which partial and provisional compromises are worked out between purposes sometimes difficult to harmonise. Therefore, within the networks, the operators must carry out negotiations concerning the technological qualities of the goods exchanged. Likewise, in the markets, the heterogeneity of the preferences must be taken into account when organising the offer, and compromises are made between the signals sent (price levels, information attached to the products, trademarks) and societal demands (environmental performances, rural or territorial development).

Quality criteria become then the **results of qualification operations** through which the construction of a socio-technical convention is effected, allowing the majority of the agents to reach their aims. To qualify a product is then not reducible to analyses regarding some characteristics chosen on the final products. It is in fact necessary to reach convergent representations of the aims and the means employed (including bio-technical resources and know-how) to develop a concerted project based on quality linked to the origin (Casabianca *et al.*, 1994). Researchers have a particular responsability to face in these qualifications.

In this approach, beef meat is an interesting case, as an apparent paradox can be seen: the stocking terroirs (and specially the pastures) are strongly highlighted to praise the properties of certain meats or of certain breeds. Nevertheless, very few beef meats show quality signs linking their characteristics to the origin. By itself, this paradox is a challenge for researchers and authorises opening a series of enquiries about the qualification of origin of beef meats in the European context.

In the first part, we set out the field of study of our work, detailing the general framework within which the origin is now protected in Europe and the special questions raised by beef meat. Then, in the second part, we will endeavour to understand the obstacles which hinder the transfer of the origin of the animal to its meat, reviewing their deep-seated causes, including the role of researchers themselves. Then in the third part, we bring forth our field work on the study of specific cases where local agents have started (and in some cases carried to completion) the steps for the qualification of origin of beef meats. Finally a fourth part will allow us to set out the formal elements of an analysis protocol for situations that can be observed in Europe.

Field of study

A juridical framework to specify origin and quality

A juridical framework exists, and gives status to the exclusive association between a name and a product and ensures its protection: a geographic indication is understood as intellectual property. In effect the TRIPS agreements[1] held within the scope of the WTO (World Trade Organisation) as well as the provisions of the PDO (protected origin appellation) and PGI (protected geographic indication) in the EU (regulation n° 2081 of 1992), foresee the protection of the "geographic indications": the name of a region, a specific locality or in some exceptional cases of a country, serving to differentiate an agricultural product or foodstuff originating in of that region, that

[1] TRIPS agreements are dedicated to the rights on Intellectual Properties at world level, including the geographical indications.

locality or that country. These, considering that a specific quality, repute or other characteristic can be attributed to that geographic origin and whose production and/or processing and/or elaboration take place within the delimited geographic area.

The definition of the IG in the TRIPS agreements and that of the PGI in European regulation 2081 are identical. From this point of view, the protected origin appellation can be considered as a special instance of the general case of geographic indication : "the name of a region, of a specific locality, or in exceptional cases, of a country, used to designate an agricultural product or a foodstuff originating in that region, of that specific locality or that country and whose quality or characteristics are essentially or exclusively due to the geographic setting, including natural and human factors and whose production, processing and elaboration take place within the delimited geographic area".

The origin is then not a plain source which would be documented by an efficacious traceability. The basis of the legitimacy to allow exclusiveness to certain producers is constituted by qualities, characteristics, a reputation attached to a determined zone. The origin can result:

- From a strong link between the product and the "terroir". This link being expressed by specificities inherent to these localities and elicited by human activities. It is the principle of the "appellations d'origine" in France, first used in viticulture and later on extended to other products, especially cheeses. The product is then presumed to express a "special type" linked to the terroir.
- A strong anchoring to the production territory, which makes the elaboration of the product very difficult to carry out in other localities. In this case, the strength of local solidarities both ensures the product's reputation and justifies preventing the geographic name to be used freely, to avoid its becoming "generic". This set is what confers to the product the legitimacy to bear a protected name.

The provisions arising from regulation 2081 had been preceded in certain countries (mainly those following the "Roman law" tradition) by national laws concerning the "designations of origin". While in the wine and cheeses sectors numerous products enjoyed the protection of their origin, meat in general and especially beef, stayed largely out of these interventions. Therefore, until the beginning of the 90's, there was only one meat in France under the AOC (the Appellation d'origine Controlée, the french version of the PDO): the fowl of Bresse (Sylvander, 1995).

On the other hand in France, quality signs also comprise, besides Organic Agriculture which corresponds to a set of obligatory norms for the means of production (production methods which do not use synthetic chemical products), the Red Label and the certificate of compliance whose definitions are as follows:

- The Red Label: "certification attesting that an agricultural product or foodstuff has a set of characteristics fixed beforehand which guarantee a superior level of quality that distinguishes it from similar standard products. The Red Label is a collective brand belonging to the Ministry of Agriculture and Fisheries".
- The Certification of Compliance of the product: "attests that a foodstuff or an agricultural non-alimentary and non-processed product conforms to specific characteristics or rules established beforehand, concerning its production, processing or packaging. These characteristics must be objective, traceable, measurable and significant for the consumer".

The Red Label has been the object of a concerted action of the meat sector, above all to point out the fitness of breed types of meat (Charolais, Limousin etc.). Thus, it is the quality characteristics of the meat from animals belonging to specialised breeds - whatever the locality of their breeding

- which are guaranteed (de Fontguyon, 1995). It can be said that, until 1992, bovine meat was left outside the proceedings and systems for the qualification of the origin (Béranger *et al.*, 1999).

Beef in an unprecedented crisis

The upheavals which followed the crisis due to the "mad cow" disease revealed to the consumers "the existence of breeding practices (specifically as concerns feeding) which they did not suspect" (Sans et de Fontguyon, 1999a). This recent awakening has led, in a few weeks, to a shift in the "hierarchy of the attributes sought by the buyers of beef" (Sans and de Fontguyon, 1999b). The organoleptic dimensions of the meat (tenderness, taste, etc.) lose importance in favour of food safety, which entails the guarantee of the source of the meat. Exigencies regarding the labelling of the source and transparency requirements within the chain had then to be reinforced, leading to a partial questioning of the ruptures which had taken place before (Sans et de Fontguyon, 1999a).

However, the obligations to inform the consumer deal only essentially with traceability and placing of the animal into broad and scarcely explicit categories (for instance meat breed heifer). From the point of view of regulations (European regulation 1760/2000 and the ISO 8042[2] norm concerning documented source), these obligations translate in terms of identification and registration of the bovines associated to a labelling system for the meat and the products based on beef. From the point of view of the inter-professional organisation, the establishment of two collective brands "Bœuf de Tradition Bouchère" "Traditional Butcher's Beef" (BTB)[3] and "Green Pastures Beef" "Bœuf Verte Prairie" (BVP)[4] which guarantee, by means of their listing of specifications the French origin of the animals, the "meat breed" type, the "traditional" feeding of the animals and the tenderness of the cuts to be braised. Sans and de Fontguyon (1999b) mention collective mass differentiation. This differentiation does not rely on a strong innovation of the product, but rather on an organisational change of the whole meat network. The surfacing of this organisational dimension, together with the hierarchical change of attribution, has deeply disrupted the capacity for initiative of the agents within the chain. In effect, "the differentiation interventions promoted by the processing and mass distribution industrialists, are not in a position to react to this new situation" (Mazé, 2000). The study of relationships within the sector shows that before the crisis, the processing and mass distribution industrial groups did not set out special exigencies in the matter of stocking. Also, collective interventions often initiated upstream (such as BTB or BVP) are today able to assimilate the new regulatory exigencies, since they define stocking practices in a set of specifications controlled by independent third party organisations. We find here the two constraints mentioned by Valceschini (1999) for the elaboration and use of a "quality signal". The first refers to the "relevance of the signal" and the need of the agents to recur to it. Within the context of the consumers' change of attribution, described by Sans and de Fontguyon (1999a), the sign developed by the inter-profession is fully relevant. The second concerns "the credibility of the quality sign". The control via third party organisms permits to guarantee "that the information affixed on the product complies with the stated definition and the commitments made" (Valceschini, 1999).

[2] The ISO 8042 norm defines traceability as the aptitude to find the history, utilisation or localisation of an entity by means of registered identifications.

[3] Intended for retail butcher shops

[4] Intended for mass distribution

Responses to reassure the consumers

During decades, the rules concerning the labelling of bovine meat remained unclear. When introducing the last European regulation 1760/2000 of July 17, 2000[5], the lawmaker mentions the instability of the beef market caused by the ESB crisis and the lack of efficacy of the previous directives. In spite of the existence of a decree, dictated in 1969, regulating the identification of the animals, and their zoo-technical registration[6], and of European directives 90/425[7] and 92/102[8] "experience has shown that the implementation of the directive [at least that of 1992] was not completely satisfactory and must still be improved. It is therefore necessary to adopt a specific regulation for bovine animals so as to reinforce the provisions of that directive" (Regulation 1760/2000).

The decrees of application of the law concerning stocking and the directives at the community level did then exist before the first crisis of 1996, but their shortcoming was their lack of operation.

The second drawback of those first regulations regarding what was not yet called traceability (evidently) is that they support the ruptures between the production systems: on the one hand a decree concerning the identification of the animals and on the other that about labelling. Under those conditions, the traceability of beef was not practicable at all, since none of the laws and regulations covered the chain as a whole.

The crisis due to the mad cow, as it directly involved the consumer, led public authorities to work on the request to label the meat all along its processing and especially the labelling of the final product. Therefore, it was deemed essential to address in the same text the identification of the animals (as described in directive 92/102) and the specific labelling for the beef sector.

The hierarchical change of attribution in benefit of the source observed in 1998 by Sans and de Fontguyon (1999b) was confirmed during that same period by Giroud's consumer studies (1998) stating that an 89% of consumers consider this criterion as determining their choice of beef to purchase.

Faced with pressure from the consumers, following the crisis due to the BSE (Bovine Spongiform Encephalopathy), public authorities have provided a first series of information about the **source** of the animal. However, there seems to be a place for a second type of information regarding the **origin** of the meat.

The establishment of traceability as a mean to document the source of the beef is an essential tool to ensure the origin. Without traceability, and a generalised labelling of the meats, it would be impossible to control their origin.

In fact, one of the major stakes is not only to investigate the source of the cut of meat only, but also to become interested in its origin. In other words, can the situation observed in a particular region be transferred to another region? What is the influence of the terroir (local know-how and

[5] This regulation establishes a system of identification and registration of the bovines and the labelling of beef and beef-based products, and abrogates regulation (CE) n° 820/97 of the Council.

[6] Decree n° 69-422 of May 6, 1969.

[7] Council of June 26 1990, concerning veterinary and zootechnical controls of identification and registration of live animals, so as to facilitate exchanges within the Community.

[8] Council of November 27 1992, concerning the identification and registration of animals.

environment) in the making up of the product? Is it a typical product or is it specific only? This series of queries allows an appraisal of the very different nature of these two ideas.

Analysis of the obstacles to the transfer of the origin

In spite of these stakes, the pressure brought to bear by the consumers and the reinforcement of regulatory provisions, today in France there are only three Controlled Origin Labels (AOC) for meat and two for beef. This low level of activity shows either only moderate interest for this type of differentiation or a true difficulty to qualify the origin of meats and specially beefs. A similar situation is observed in other european countries as Italy and non-EU country as Switzerland.

This low utilisation shows either mild interest only for this type of differentiation or a true difficulty to qualify the origin of meat and specially beef. If we add up in parallel bovine meats labelled as to origin by the IGP system (the french version of the PGI), we find 5 instances[9]. This number of IGPs compared to the two existing AOC shows that the differentiation of the products according to their origin is sought by groups furthering recognition projects. But the notion of origin in these two official quality signs refers to different concepts. One - the IGP - stems from Anglo-Saxon case law, referring to the origin rather in view of historical precedence or the reputation of the product, while the other - the AOC - endeavours to demonstrate the origin as a link to the terroir including both environmental aspects and know-how employed by stockmen as well as butchers. The justification of the link to the origin is often more difficult to demonstrate in this last case, since it is needful to establish a bio-physical and socio-historical link between the special type of the product and its terroir.

Our analysis leads us to identify the different obstacles found between the animal and its meat, to hinder the transfer of the characteristics of origin along the whole chain.

Obstacles linked to the agents present in the chain

The insufficiency in the labelling of the origin of bovine meats is first associated to a lack of communication between the operators of the chain (Quilichini, 1998), maintained by the power of particular professional categories.

The empowering of the pegger/grader trade during the 1970 decade, foretold a new organisation of the chain in which the activity of the stockmen and that of the butchers are conveyed into separate registers. In his analysis of the Belgian bovine meat chain, from the point of view of socio-technical innovation networks, Pierre Stassart (2002) recognises "the extraordinary capacity [of the pegger/grader] to assemble heterogeneous worlds, to hold in a single movement both stockmen-breeders and butchers". He adds that at the same time the pegger/grader "avoids putting into contact the butchers among themselves and the latter rarely go to the abattoirs to compare carcasses". This intermediary has then a fund of knowledge acquired through experience, which allow him on the one hand to link the live animal to its carcass and on the other the carcass to the cuts of meat issuing from it. This double connection around the carcass, which he is the only able to effect, places him at the obligatory point of transit of both the products and information (price, volume) which circulate. This point of transit is linked to the "difficult to access skills format" (Stassart, 2002) that the pegger/grader employs. These skills and data are hardly encoded (no contracts, no classification grid) and are then not detachable from his person.

[9] They all employ a Red label on the national level (calves and heavy bovines together).

This capacity to **set apart** stockmen and butchers within their single activity ranges (the live animal or the meat) creates strong rupture lines in the continuity of the technical functioning of the activities and therefore by itself weakens the link to the origin.

Quilichini (1998) re-examines these disconnections from an economic point of view, through the mechanisms of price formation of a carcass or of a cut of meat along the successive commercial transactions. He explores the nature of "the quality criteria which are in fact endorsed by commercial practice" (*ibid.*). He ascertains that these quality criteria vary whether " the producer, the transformer or consumer, to mention only the three main groups of this sector" is addressed. The consumer still very often associates the quality of meat to its tenderness (Mainsant and de Fontguyon, 1988). In spite of this, the price of meat is not directly linked to its tenderness, quite simply because it is not known how to measure it easily at this time. It is the factors scientifically or empirically recognized concerning the management of the stocking to obtain an optimum tenderness that the butchers point out. "Not being able *to access directly to tenderness [...] it is the factors that make up this tenderness, which are then sold*" (Quilichini, 1998).

Finally, at the end of the chain, the morphological characteristics of the carcass (type, weight, conformation) have only very limited influence on the price of the cuts. The shift - even the rupture - between the criteria concurring into the definition of the quality level of a carcass or of a cut of meat is based, for Quilichini (*ibid.*), on a differential valuation of the various cuts obtained from the carcass. *"The categories approach is concerned above all, with setting, a price to a category of meat (based on the type of cut and its level of elaboration)* " and in consequence attenuates the reference to the carcass. In fact it introduces a discontinuity between what makes up the carcass price and the price of meat. In a practically independent manner from the qualification of the carcass, it is in fact the category of meat through the destination of the cut of meat and the manner of cooking it which will largely establish its final price.

Setting equivalences to the carcasses neutralises the link to their origin

The industrialisation process of the bovine meat network (particularly at the processing industry stages) is not totally finalized, as compared with other agronomical food processing sectors. The beef processing network is composed of a juxtaposition of firms with operating patterns which relate more to that of artisan entities than solely industrial ones. The main reasons for this are the strong heterogeneousness of the carcasses processed, which preclude any mechanisation of the processing chains and requires an ample complement of qualified manpower. Attempts at the mechanization of carving-up chains have nevertheless been experimented with, but the rare actual tests have finally failed. The experiment which went farthest towards success lasted about ten years at the INRA Centre at Theix. It led to the design of a carving-up robot which was finally abandoned some years ago, due to its lack of operational adaptation to the carving-up chains.

This aborted attempt witnesses the generalised movement towards industrialisation of the bovine sector. It translates, up-stream, by a will to **'format'** the animals (in a likeness to the 100 kg swine, or 42 days' pullets) and downstream, by the production of huge quantities of merchandises, ever increasingly normalised, standardised and to put on the market more and more elaborated products. Soufflet (1988) had already observed, 15 years ago, the beginning of this industrialisation trend. Sans and Fontguyon (1999 a) confirmed this trend and noted its intensification in the area of meat processing.

This increasing industrialisation required setting up technical devices with the capability of normalising and standardising products stemming from bovine stocking.

Thus, the different devices for the classification of carcasses which have accompanied this trend, since the France grid to the SEUROP grid, have had as their main target that of setting the equivalence of the products, mainly in order to reduce the costs of transactions. This establishment of equivalences (which makes possible the standardisation of procedures), instead of highlighting the value of the characteristics of carcasses, denies their specificities.
This reasoning suppresses any and all reference to the origin of the meat. Under such conditions, the bovine carcasses and their meat are considered only as an undifferentiated 'ore'. The differentiation of the products is, in such a case, provided by the technological processing innovations downstream[10].

The recent progress in the classification of carcasses, intended to replace the OFIVAL agents at the slaughterhouses, do not change in any aspect the existing SEUROP classification grid.

Thus, the introduction of the Classification Machine (MAC) at the slaughterhouses, approved by the authorities, reinforces the tendency to set equivalences to products stemming from the stocking of bovine stocks. The manufacturers justify its implementation by the 'objectification' of the classification of carcasses which it provides, until then carried out by the subjective and fallible eyes of men. This innovation constitutes a new shift in the distribution of the functions within the bovine network, but does not change in the least the disconnections between the various agents in the network.

Beyond the shifts in functions provoked by the machine, "to delegate thus the handling of the information to the calculation centres is not devoid of risks" (Stassard, 2002). The possibility of disseminating throughout the network the information collected and kept by the MAC remains hypothetical. Consequently, the MAC risks maintaining, after the peggers/graders, the disconnections within the network, by keeping all information related to the carcasses in the hands of a single agent.

However, while the pegger/grader was able to change his reference base according to the clients, the MAC, once set to standards, will evaluate all carcasses according to the same equation and will attribute to them a price level. In fact, depending on the butchers in line with their preference, a given carcass could have different prices which the pegger/grader knew how to negotiate (a carcass lightly covered with fat could satisfy a given one as it lengthened the maturing period and displease the other in view of his customers' rejection). Or even further, a carcass covered with fat could supply the butchers in Bordeaux while a lightly covered one would suit the Butchers at Lille.

On the contrary, the definite set of standards of the MAC will automatically class a carcass in its price category and no longer allow counterbalancing this classification having regard to requests of each butcher or wholesaler, or even recognising consumer specificities of each region.

How, under these conditions, can recognition be given to existing local projects of differentiation on criteria other than solely that of conformation? A set of standards on that single parameter would mean excluding projects based on valorising ill-conformed animals or on feeding characteristics based on local resources and guaranteeing at times imperfectly the expected conformations.

In the first case, the MAC would issue a disqualifying classification (if the carcass is ill-conformed) and in the second case, it would not be capable of fully assessing the value of the carcass (due to lack of visible inscriptions on the carcass of given cattle-stocking practices).

[10] This is noticeably the case of the products under commercial brand "Charal".

A matter of disciplines in closed-compartments

These disconnections as a whole, easily observed in productive realities, have justified from a scientific point of view, the rupture among the fields of disciplines concerned: Animal production science, the technology of meats, or the economy of the organisations as well. In fact, the need to establish a link between the ways of producing, the transformation of the meat products and the relationship between the agents was not perceived.

We can succinctly set apart:

- among specialists in animal production, the matter of qualification of the origin is considered in relation to measurable effects of production factors in trying to maximise the biological response of the animal, mainly centred on muscle tissue growth.
- Equally, among specialists in the transformation of meat products, the analysis focuses on the study of the properties of the product (Physical-chemical, technological, organoleptic). It aims at optimising the transformation of muscles into meat, essentially centred on the maturing processes, and indicators of tenderness.
- Finally, among economists, this question is treated classically in terms of interaction among agents in the midst of the information network. Furthermore, the equilibrium theories of classical economy do not allow addressing emergence phenomena, just as transaction costs cannot totally account for organisational phenomena.

The existing technical devices, as well as those being developed (markedly the classification machine) attempt to summon more and more scientific knowledge and technologies, so as to improve objectivity regarding the assessment of any carcass. In fact they represent an obstacle to the qualification of the origin of bovine meats.

Consequently, most of the technical devices, as they have developed at present, are ineffective towards qualifying the origin of bovine meats. Worse still, they preclude the groups involved in the project from obtaining the necessary means to justify and achieve recognition of the origin of their meat. In all cases, the parties directly involved in the branding of origin project appear to be forced to build case-specific technical devices in order to facilitate the transport of the origin.

Some relevant situations

We have chosen two contrasting situations to analyse how the procedures of qualification of the origin of bovine meats could be carried out. Their main characteristics are presented succinctly in tables 1 and 2. It concerns the 'Taureau de Camargue', Camargue Bull first ever French AOC having achieved recognition and the 'Fin Gras du Mézenc' Full marbled from Mézenc, for which the AOC request is under consideration.

The Camargue Bull

The Camargue Bull AOC is based on the breeding of two stocks (raço di biou and brava) in an area matching that of the Camargue proper, expanded to the East by the flats of Bourg 'plan du Bourg' and to the West by the Gard river "Camargue" i.e. 'Camargue gardoise'. The bull is bred first and foremost to sire animals fit for racing or bullfights and the AOC meat comes from those animals eliminated from bullfighting games. The animals graze out in the open on two areas: one dry and the other humid. The reasoning at work in the management of blood lines, to ensure durably the aptitude for bullfighting, results in the shaping of animals of very specific characteristics. A Group has been constituted with at its core the Tarascon Slaughterhouse and its director.

Table 1. Synthesis of the essential characteristics of the Camargue bull.

Production	• Total volume in 2001: 254 T, that is 1,614 Camargue bulls • Average weight of the carcasses: 165 kg • AOC conformity rate: 83 %
Terroir	• AOC area, identical to the area of origin of the Camargue horse (Decree of March 1978) • Two dovetailing areas: – AOC Geographic area (a triangle whose points are: North Uzès; West Montpellier and East Istres) – Humid zone (corresponding to the Petite Camargue) where the animals graze during six months in winter • AOC area over 2 regions and 3 departments (causing a coordination problem between the actions of elected authorities and Chambers of Agriculture)
Product	• Mainly 3 to 6 year old oxen killed as unsuitable for taurine games • Light conformation of the animals and of the carcasses. (ideally situated in the P2 class) • Breed "Camargue bull" renamed as "raço du biou" to avoid confusion between the name of the appellation and the name of the breed and to differentiate the two breeds in the area: *raço di biou* and *brava*. • Animals with status of wild • Sensory tests conclude it is a special type of beef meat: – A difference of the meat of Camargue bull from other bovine breeds (Charolais) – A difference between Camargue bull meats, according to the stockbreeding and feeding on pastures
Know-how breeding	• The logic employed to manage the breeds is to ensure that an aptitude for taurine games be maintained. This leads to the production of animals with special characteristics. • Changing pasture sites in summer and winter (= humid zone) • Complex selection strategy seeking to obtain traits whose genetic transmittal is not known ("drive" of the animal) • Testing of the selection during races "à la cocarde" and elimination of unsuitable bulls
Core of the project group	• A local businessman is the core of the project group • Good representation. Quite legitimate. Has been the core of the project since the beginning of the AOC certification effort, but periodically questioned. • Strategy of integration of the stockmen in the construction of technical devices

The ideal animal is that which knows how to race and fight. It is clear that this does not provide it with any particular aptitude for the production of meat. This is reflected logically in its carcass which is, on the average, classed P2. Normally, this classification would lead it towards a low-grade use in food for household animals (pet-food).

As it constitutes the core of the qualification of origin system, the carcass is an easily identifiable point of reference due to its unique conformation. Paradoxically, this extreme originality in the manner of producing the meat confers to the carcass of the Camargue Bull its characteristic type,

Table 2. Synthesis of the essential characteristics of the Full Marbled of Mézenc.

Production	• Total volume in 2002: 390 animals were marketed under the Full Marbled of Mézenc brand in 2002 • Average weight of the carcasses: 358 kg • Conformity rate: 73 %
Terroir	• The highest permanent habitat in Europe (up to 1500 m) • The zone proposed for the Full Marbled of Mézenc AOC is defined by its altitude. Above 1,100 m, hay is characterized by the presence of wild fennel. • Zone placed over 2 regions and 2 departments (causing a coordination problem between the actions of elected authorities and Chambers of Agriculture)
Product	• The Fine Marbled is an anachronistic animal in a bovine sector seeking rather lean meats. • Carcasses have a slight fat covering, but the meat is marbled. • Sensory testing evidences a difference with other types of animals (stew preparation method, which reveals the special type Full Marbled stock meat) • Establishment of technical devices to grade the Full Marbled type animal (Full Marbled evaluation grid, Full Marbled commission, grading of Full Marbled during livestock fairs)
Know-how breeding	• The local breed has disappeared (employing pure breeds or crosses : Salers, Aubrac, Limousin, Charolais crossing with Montbéliarde or abundantly with Charolais or Limousin bulls) • Several gradings during the life of the animal, more or less intensive according to the availability of will fennel hay in place in the stables, the number of heifers in a cohort • Special fattening based of hay from the Mézenc. • Classing of the hay, mainly for its floristic characteristics. • Distribution of the hay 3 times a day, and elimination of the rejected fodder. • Concentrated fodder added so as to modulate fat content. • Different strategies based of the utilisation of hay and the number of Full Marbled to be fattened. • Management of the stock (especially fattening and finishing) oriented to the production of marbled meat
Core group of the project	• The core group of the project is made up by stockmen and local elected representatives. • Representative and legitimate for the professionals (stockmen, butchers, restaurant operators) and the public

in the sense that it is not comparable to any other existing beef category. This singular type makes it an ideal referent as a basis for technical devices.

The re-focusing of the technical devices around the carcass itself gives the participating parties a tangible object on which to base the qualification tests. In fact, from its visual defects the carcass is a sturdy indicator of the quality of the meat. **The carcass of the Camargue bull is therefore informative of the type of meat which is found on it**.

Reality tests focused on the carcass allow using the technical devices already existing in the bovine meat network. The SEUROP grid can thus be easily summoned for the evaluation of Camargue bull carcasses. It cannot however be used as it now stands since it would not lead to qualify the carcasses, but to disqualify them in view of their bad shape. The group which is furthering the project has had then to adjust the qualification grid to the type of carcass to be graded.

The stockmen are then but appended to the technical arrangements. They are merely summoned to the work-sites only through some types of carcasses which they deliver to the slaughterhouse. Even when they are not directly present however, **the specified stocking methods are inscribed in the carcasses**: One look at a bull's carcass suffices to notice them. The characteristic type of the carcasses easily witnesses the transfer of the live animal's origin, to its carcass.

In as far as knowing whether the transfer from the origin extends to the meat itself, this is a matter for the butchers. There also, the skills of the butcher (or the de-boning technician in the carving-up workroom) are very important, for the registry of the origin from the bull up to its meat. In the Camargue, he must be capable of carving up whole carcasses, and must as well adapt his cuts to the specific type of the carcasses. In view of the variability of the carcasses (considering the erratic composition of slaughterhouse cattle lots) and their conformation (let us also think about an ideal carcass classed P2), butchers must develop specific cutting patterns.

The fact that the carcass is indicative of the type of meat and that there are specific cutting-up devices (which are not yet specified), allows the marketing of Camargue bulls carcasses and meat in supermarkets. The customers no longer need to find a butcher behind the meat counter, considering that the origin of the meat is visibly inscribed.
In the Camargue case, it is possible to imagine the modalities of transfer of the origin as and when transformations take place: from a live animal to the cut of meat (Figure 1) with the carcass in a pivotal position.

The Full marbled of Mézenc

This concerns a small volume production. The animals (mainly heifers of somewhat less than 3 years, of different breeds) are marketed only during 6 months of the year, from January to June.

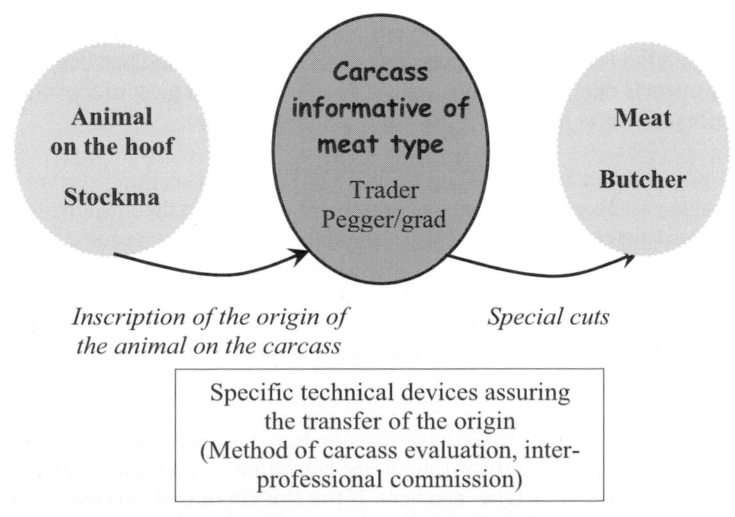

Figure 1. Modes of transfer of the origin in the case of the Camargue bull.

The originality of the Full marbled rests of an *ad hoc* selection of the animals followed by a lengthy Winter feed essentially of hay of the Mézenc Plateau. This altitude hay is special, as it contains 'cistre' (wild fennel). This is sorted out; usually the best fraction is distributed to the Full marbled being fattened, three times a day. These animals are marketed almost exclusively among retail butchers.

The construction of technical devices (even if it is not finalised) must go through the definition of a context constituted by objects identified as pertinent to offer a pattern of action common to stockmen and butchers. From this point of view, the carcass is not the focal point either to the stocking or butchering trades.

On the other hand the carcasses of the Full marbled animals are not very typified, that is to say that they cannot be considered, by themselves, to belong to a specific type. They do not offer then reliable points of reference to decide on the origin of the animal and its meat, as they are difficult to identify in comparison to a more classical production of single breed heifers.

Consequently, the technical devices get organised around a centrifugal movement of the carcass expelled from the qualification of origin devices. **The carcass of the Full marbled from Mézenc is not indicative of the type of meat which can be found on it.** Therefore, there is no testing of the carcass, which is dumb and incapable of refer back to its own origin. There remains then only the live animal and its method of stocking which can bear witness to the origin of the meat.

However, it is impossible to put this link directly to a test since it does not exist as a material object, but only as a virtual thread between two states of the animal fundamentally disconnected. Only the procedure based on an evaluation of the animal can justify that thread, bearing in mind that it is not entirely foreseeable.

The reality tests imposed by the Full marbled commission and validated indirectly by the butchers through prices agreed for the purchase of Full marbled animals confirms the link between the fattening methods and the type of meat. In other words, these reality tests witness the transfer of the origin of the animal to its meat and control the possible drifts of the qualifying system of the origin.
This direct relation of the stockman and the butcher demands that both parties involved should have sufficient common points of reference to talk to each other. The purpose is not so much the sharing all points of reference available since, were it so; the stocking and butchering trades would be confused. The intent is rather to rely on the animals and the graded meat to constitute significant points of reference (animal fit or unfit for fattening, fixing the price).

In such a case both interested parties agree that the specific breeding system permits the registration of origin at the level of the animal, and which is found in the meat thanks to the work of the butcher. Consequently, this type of carcass and its meat must be accompanied by a set of information, without which the thin thread of the origin which links the animal to its meat risks being lost. The sale at a butcher's shop is therefore indispensable, so that the butcher may inform his customers on the stocking method used for the animal and the name of the stockman who fattened it. In a Supermarket, having regard to the fact that the product can hardly speak for itself, the origin would be obliterated.

In these conditions (and for reasons opposed to the situation in Camargue), it is fairly logical that the group which carries the project forward should be constituted around a core of stockmen from the 'Estables' who then enrol some retail butchers. **If the stockmen had carried the project with the support of locally elected officials only, they would not have succeeded to express the**

potential characteristic type of the animal without summoning the assistance of butchers capable of carving up whole carcasses.

Considering that the peggers/graders have been separated from the grading system for the 'Full marbled' (in view of the absence of a device to qualify the carcasses), the butchers, besides purchasing the cattle on the hoof, must cut up entirely and trade the whole carcass. At present, fewer and fewer butchers possess such skills, but here also the work on' ready-to-carve' pieces of meat would not have allowed to transfer the origin to the cut of meat. These skills relate to the aptitude to set values to all pieces including fore quarters, to manage the stock as a function of sales, and to know their customers sufficiently well to trade the meat which is suitable.

It is possible to imagine ways for the transfer of the origin in the case of the Full marbled of Mézenc through setting in direct connection the animal on the hoof and its meat (Figure 2).

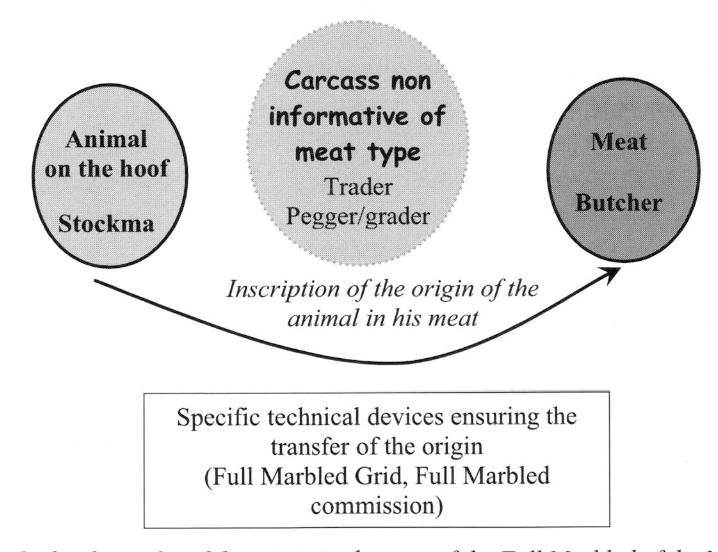

Figure 2. Methods of transfer of the origin in the case of the Full Marbled of the Mézenc.

This type of grading, far from a reversal to an obsolete tradition, is integrated in the modernity of social demands and scientific research. In fact, the identification of the origin of the animal through the specified ways of production is well established in the current concerns of consumers. Furthermore, the interest focused on the meat in the Full marbled project (even if the purchase is done on a live animal, it is first and foremost a specific type of meat which is purchased by the butchers) goes in the sense of a increased awareness of the criteria for evaluation of meat in the establishment of the price of the animal. Finally, establishing a direct relationship of the stocking system (especially cattle feed) and the characteristics of the meat follows the same trend as current research[11].

[11] Particularly the European "Healthy Beef" programme which researches the relationship between the nutritional qualities of meat regarding fatty acids, and feeding modes. This programme is presented in an in-depth article in this book (Scollan et al., 2005).

Proposal for a model

Table 3 presents broadly the possible situations depending on whether the carcass is informative, or not, of its origin. These two archetypes are not meant to represent the overall range of possible situations. They delimit a field of situations by presenting what we analyse as its two extremes.

The technical devices are centred either on the carcass at the slaughterhouse or on the stocking activity. Thus, depending on the archetype, the displacement concerns simultaneously the distance (from stable to slaughterhouse), the objects of classification (from live animal to its carcass) as well as the trades along the transformation chain (from butcher to the pegger-grader).

It is not possible to go any deeper here in the total number of cases with which this model has been confronted. However, in his Thesis for a Doctor's Degree of one of us (Trift, 2003) one can find specifically how three other situations, which have been the object of a detailed research, are positioned:

- the 'Manzu de Corse', whose characteristics clearly place it near the archetype 'Taureau de Camargue',
- the 'Bœuf de Charolles' which would be relatively close to the 'Full marbled of Mézenc',
- and the 'Maine-Anjou' whose carcasses are very informative but whose mode of construction brings them close to the archetype 'Full marbled of Mézenc', achieving through this fact a sort of synthesis.

Table 3. Properties of the technical devices in each typical situation.

Technical devices	Properties sought in the technical devices according to the specific situations	
	Carcass scarcely informative of meat type Full Marbled of Mézenc	Carcass informative of the meat type, Camargue bull
Project Core group	Stockmen	traders/Institutions
Production sequence	Selection and then orientation of the animals	Orientation and then selection of the animals
Slaughterhouse lots	Homogeneous	Heterogeneous
Qualification System	Disregards the carcass	Centered on the carcass
Qualification methods	Evaluation	Prediction
Parties involved in the qualification of the origin	Stockmen and butchers	Traders (and breeders)
Place of qualification	Stables (and cattle fair)	Slaughterhouse (and arena)
Nature of relationships between stockmen and butcher/trader	Shared uncertainty and bilateral negotiation	Uncertainty borne by the stockman and preset prices
Connection of the products	Direct connection: animal on the hoof - meat	Successive connection: animal on the hoof - carcass - meat
Ascertaining of the origin	Process of evaluation of the link live animal/meat (demonstration)	Testing of the carcass (show)
Aim of the work on the carcass	Valuation of the origin	Preservation of the origin
Marketing	Essentially in butcher shops	Possible in supermarkets

This last case shows that granting validity to the origin stimulates operators upstream to exert their capacity to build specific technical devices while closely associating butchers to this effort. This working group has succeeded in leading the necessary apprenticeships and has just obtained (early 2004) its AOC, the second one in France for bovine meat after that of the 'Taureau de Camargue'.

Through informations from some colleagues in other countries, we tried to verify the relevance of such model. A case in Switzerland named "Dry beef meat of the Valais" is informed by Barjolle (personal communication) and another case from Italy "Vitellone bianco dell'Apennino centrale" is studied by Marescotti (Personal communication). In the Table 4, we summarise the main points to be compared. Obviously, the swiss beef meat is on the same trend as 'Taureau de Camargue' because of the essential place of the negociants in the creation of the qualification and carcass is put at the very center of the project. At the contrary, the Italian case is more or less comparable to the case of 'Full marbled of Mezenc' because of the main role of breeders and the central place given to the animals (Marescotti, 2000). These two examples encourage the use of such a framework in an overview all around Europe.

Conditions for grading

Bring the stocking and butchering activities out of their closed compartments: have the stockman establish a dialogue with "his" butcher.

The separation of the activities of stockmen and butchers, through the multiplication of intermediaries between the animal on the hoof and its meat, creates rupture lines among the parties participating in the network. So that the origin gets diluted through every change of hands and transformations of the product, and no longer warrants its transfer to the cut of meat.

On the contrary, to place in direct connection the technical functioning of the stocking and butchering activities - which would not be limited to a simple flux of animals and carcasses - but would also be an occasion for exchanges makes it possible to envisage the transfer of the origin.

Table 4. Properties of the technical devices in two European situations.

Technical devices	Vitellone Bianco dell' Apennino Centrale (I)	Viande séchée du Valais (CH)
Project Core group	Stockmen (Consorzio Carne Bovine Italiane)	Traders/Institutions
Qualification System	Centered on the animals	Centered on the carcass
Parties involved in the qualification of the origin	Stockmen and butchers	Traders (and breeders)
Place of qualification	Stables (and cattle fair)	Slaughterhouse (and arena)
Ascertaining of the origin	Process of evaluation of the link live animal/meat (demonstration)	Testing of the carcass (show)
Aim of the work on the carcass	Valuation of the origin	Preservation of the origin through processing know-how
Marketing	Essentially in butcher shops	Essentially in supermarkets for exportation

Establishing the connection requires re-learning to work together. This requires re-defining the boundaries of everyone's activities. Thus the work of the stockman will no longer stop when the animal leaves the stocking farm, but will extend beyond its slaughter.

Discussions between the stockman and (his) butcher will allow him to apprehend what happens with the carcass of the animal he bred. It is not so much that he will be interested in which butcher shop or supermarket it will be sold, or to what type of client. He will be concerned instead with the **judgement criteria** on the basis of which it will be evaluated, so that he may eventually respond by adjusting the type of animal to be bred.

Today, the butcher's function is reduced to cutting-up work ever more distant from live animals and their carcasses. This to the point where it disappears completely when it is the carving up shop which supplies him with vacuum-packed ready-to-cut pieces (RTC) which he has but to slice. Is it possible, in actual fact, to still call "butcher" a person whose task could be summed-up as just cutting slices of meat? The carving up robot, if it should have proved operational in slaughterhouses, would have pushed this reasoning to its limit by excluding definitely the butcher from his original trade. Establishing a teamwork approach between the stockman and butcher would require that the latter re-assume several facets of his trade:

- Evaluation of the animal on the hoof which he negotiates with the stockman.
- The purchase and cutting-up of the whole carcass which requires both technical skills, but also commercial ones as well so as to be able to sell the whole carcass. This problem does not come up when the "butcher" works with RTC pieces. He orders replenishments when he needs them, but composed exclusively of fine cuts of meat.

In this context, the butcher will be able to talk to his clients about the meant that he sells and reconstruct the link to its origin, while this will be hidden when the butcher works only with RTCs.

However, establishing this connection is not spontaneous, either from the point of view of the objects (live animal and meat) or of the participating parties and their activities. It will require some form of adjustment of judgement criteria among the professionals. A stockman considers preferentially the level of fattening of the animal and will not judge it in the same manner as a butcher, who will mainly be looking for the yield in meat. The participating parties must therefore construct **specific technical devices** which recognise these different judgement criteria, but which maintain them in compatibility. Finally, the team perspective of criteria brought forward by stockmen, and those used by butchers allows then to ensure reciprocal validations of production choices.

To state the origin of a meat: recognize the distinctiveness

The development of industrial strategies imposes setting in equivalence the types of bovine meats in order to obtain objective measures independently of time and space. In such a case all the meats are evaluated on the same basis and reference to the origin is completely left out. However, origin is singular by definition. It is not then reducible to pre-existing categories which, in general, are not suited for the judgement of animals and carcasses which should be atypical as compared with the remainder of bovine production.

By the same token, origin is not transferable. The technical devices specially created to qualify the origin of meats in a specific situation are not transferable elsewhere. This non reproducible nature stems from the fact that they were developed on the basis of specific stocking and butchering know-how and practices. Nevertheless, it is sometimes difficult to develop such specific devices against the current of technical advice being circulated.

The stake is then to construct collectively **a relevant offer**. The choice of the registry on which the character of the meat is based seals a strategic agreement: for the Taureau de Camargue, the bullfighting games mould the type of animal and its carcass. The technical elaboration is naturally not easily coded through the classic criteria used for the description of indifferent stocking concerns: the 'Full marbled of Mézenc' shows a combination of stocking conditions, type of fattening based on mountain hay and of conducts which are not directly legible on the carcass and do not fit the standard zoo-technical precepts (Hoquette *et al.*, 2001). Finally, relevance is evaluated by the intended consumers, through various confidence aspects: the statutory and sanitary basis, the recognised skills of the butcher at the counter, as well as being culturally and gastronomically familiar with the product itself. All this will translate into a willingness to pay representing the recognition of the efforts exerted.

Collective action dynamics

Evidently, all this presupposes that a negotiation is conducted to its conclusion concerning the construction of this offer. The concepts of production costs, seasonal variations and quality management remain central ones. The cases studied show that important apprenticeship efforts must be carried out during the inception and development of the project, with leaders having a personality capable of gathering around them the various interests and trades.

But it is also a matter of showing themselves as collectively capable (Nicolas and Valceschini, 1995) of :

- obtaining legal recognition of the origin constructed,
- ensuring the legitimacy of this construction in the view of the local society,
- working out effectively the jointly defined product,
- showing that it is subjected to credible controls, with sanctions for those who do not respect the general regulations,
- accompanying-assisting the operators through research and development.

It is clear that the public mechanisms set in place following European regulation 2081 authorise such processes, even if it is noted that up to now very few have succeeded. The nature of the obstacles identified, as well as the archetypes of the technical devices to be imagined as we have analysed them, explain this low number.

Conclusion

Among the products which claim a quality linked to their origin, meats remain a category of foods which does not fit easily in the definitions and frames foreseen. The interest of the micro-networks which we have identified does not stem from their capacity to supply huge volumes for mass-markets: it is clear that the constraints of geographical limits and technical choices (breeds, feeds, and ages) severely reduce the productive capacity of the associations constituted. For all that, to qualify the origin of a bovine meat forces the consideration of **criteria which are of interest to all agents**. And that is a fairly original situation: quality cannot be concentrated in an object which can be isolated from operational relationships. The criteria are constructed on technical objects which gather several points of view to be linked. They re-enquire the territorial anchoring through the insertion of local resources at good spots in the technical regulations: those that structure the local grading devices (Trift and Casabianca, 2002). These steps do not seem circumscribed to micro-networks or "niche" markets: it's the whole bovine meat sector which is invited to carry out a reflexive analysis at the hour of globalisation and alignment of European prices to worldwide rates.

Nevertheless, contrary to other quality signs which allow the product to travel, the construction of AOC's on bovine meats does not give much scope for this. Normally, the signs of origin allow the transportation of the product beyond the region in which it has been elaborated, as is the case for cheeses or wines. In the absence of direct contacts between the producer and his customer, the presence of an indication (most frequently in the form of a logo vouching for the observance of procedures attaching to the certification) guarantees to a extra-local third party the characteristics of the product he is buying (in the case at hand its origin if it is a product recognised as AOC or in PGI).

In the case of meat, it seems that the recognition as AOC does not easily permit to have it travel outside of its geographical area of origin. In fact, the high level of interaction between local butchers and their customers implicitly includes the latter in the process of construction of the origin. It is the nature of the meat imperfectly transformed on leaving the butcher's shop, which definitely requires the participation of customers in the completion of the elaboration processes. When cheese exits a maturing shed, a wine the cellar or bread the bakery, they can be consumed without other process but the initiation to the art of tasting them. Beef meat on the contrary, requires a culinary preparation to be edible, which calls for other skills in addition to its tasting. These culinary preparations are still largely present in regional gastronomy traditions. They act strongly for a return to the methods of butcher carving of the carcasses and cuts, since given preparations demand specific cuts of meat (then of specific types of cutting it).

Consequently, the characteristic types of bovine meat (including those with an indication of their origin) seem scarcely transportable outside their production cradle. This is valid for carcasses as well as meats. To make characteristic types of meat travel, means risking that the butchers will not take their originality into account when cutting them up and neutralise thus their specificity potential (fatty carcass, ill shaped) (Ventura and Van der Mulen, 1994).

To have specially typed meats travel, is also to take the risk that the customers ill-accustomed to these types of meat (redder, firmer or more marbled) prepare them in an inappropriate manner and erase their specificity. However, there are specific culinary preparations which bring out the originality of meats (the "gardiane" for the 'Taureau de Camargue', the stew of the Plateau du Mézenc, or the thick marbled Maine Anjou sirloin steak). The development of culinary preparations (especially as ready to serve dishes) is an interesting response to the transport of typical meats, but in turn, hinder the recognition of their origin in a wide area.

References

Barjolle, D. and B. Sylvander, 2000. Some factors of success for origin labelled products in agri-food supply chains in Europe: market, internal ressources and institutions. In: The socio-economics of origin labelled products in agrifood supply chains : spatial, institutional and coordination aspects, B. Sylvander, D. Barjolle, and P. Arfini (editors), Actes et Communications no.17.

Barjolle, D., J.M. Chappuis and B. Sylvander, 1998. From individual competitiveness to collective effectiveness : a study on cheese with Protected Designation of Origin, 59. EAAE Seminar *"Competitiveness : does economic theory contribute to a better understanding of competitiveness ?"* La Hague (NLD), 1998/04/28, 17 p.

Béranger, C., G., Monin and F. Casabianca, 1999. La codification des liens entre le terroir et le produit dans le cas des produits carnés: Analyse de situation et perspectives. In: Lagrange L. (editor), p. 91-103.

Casabianca, F. and Ch. de Sainte Marie,1998. Concevoir des innovations pour les produits typiques. Quelques enseignements des charcuteries sèches corses. In: F. Arfini and C. Mora, Proceedings of the 52nd EAAE Seminar, Parma (Italy), June 18-22 1997: *EU typical products and traditional productions : rural effects and agro-industrial problems,* p. 59-76.

Casabianca, F., Ch. de Sainte Marie, P.-M. Santucci, F. Vallerand and J.-A. Prost, 1994. Maîtrise de la qualité et solidarité des acteurs. La pertinence des innovations dans les filières d'élevage en Corse. Etudes et Recherches sur les Systèmes Agraires et le Développement, Cerf *et al.* éds., 28, p. 343-357.

De Fontguyon, G., 1995. La différenciation par la qualité en viande fraîche de gros bovin. In: ENITA, p.113-117.

Giroud, J., 1998. Marché des viandes: segmentation du marché et poids des viandes identifiées. Bétail et Viandes 32: p. 14-17.

Hocquette, J.F, D. Micol and G.Renand, 2001. Quelle production pour quelle qualité : que maîtrisons-nous ? In: Actes de la session du 14-15/11/2001 "Quelle viande bovine pour demain?", Nouan le Fuzelier, 9 p. http://www.clermont.inra.fr/commission-bovine/textes/8hocquette.pdf

Mainsant P., de Fontguyon G., 1988. Bœuf et veau. La démarcation de qualité. Viandes Prod. Carnés, Vol. 9: p. 267-273.

Marescotti, A., 2000. Marketing channels, quality hallmarks and the theory of conventions. In: The socio-economics of origin labelled products in agrifood supply chains : spatial, institutional and coordination aspects, B. Sylvander, D. Barjolle, and P. Arfini (editors), Actes et Communications no. 17.

Mazé, A., 2000. Le choix des contrats à l'épreuve de la qualité. Une analyse des mécanismes de gouvernance dans le secteur de la viande bovine. Thèse de doctorat, Université Paris 1 Panthéon-Sorbonne, p. 57-217.

Nicolas, F., and E. Valceschini (editors) 1995. Agro-alimentaire: une économie de la qualité, Economica, Paris.

Quilichini, Y., 1998. L'appréciation des carcasses et des viandes par les professionnels. La qualité et les transactions. Revue française génétique et reproduction, 22 (88), p. 8-16.

Sans, P., G. de Fontguyon, 1999a. Choc exogène et évolution des formes organisationnelles hybrides : les effets de la crise dite de la " vache folle " sur la filière bovine. Science de la société 46: p. 173-190.

Sans, P., G.de Fontguyon, 1999b. Différenciation des produits et segmentation de marché : l'exemple de la viande bovine en France, Cahiers d'Economie et Sociologie Rurales 50: p. 55-76.

Scollan, N.D., I. Richardson, S. De Smet, A.P. Moloney, M. Doreau, D. Bauchart and K. Nuernberg, 2005. Enhancing the content of beneficial fatty acids in beef and consequences for meat quality. In: Indicators of milk and beef quality, J.F. Hocquette and S. Gigli (editors), Wageningen Academic Publishers, Wageningen, The Netherlands, this book.

Soufflet, J.-F., 1988. La filière bétail et viande bovine. Fonctionnement et évolution de 1960 à 1985, perspective 1990. Essai sur la dynamique des structures et des comportements. Thèse de doctorat, Université de Montpellier I, 670 p.

Stassart, P., 2002. Produits fermiers : entre qualification et identité. Thèse de doctorat, Fondation Universitaire Luxembourgeoise, 379 p.

Sylvander, B. 1995. Formes de coordination et marché des produits de qualité spécifique. Analyse sur le cas de la filière volaille. In: G. Allaire and R. Boyer, La grande transformation de l'agriculture. INRA-Economica. Paris.

Sylvander, B., D. Barjolle and F. Arfini, (editors), 2000. The socio-economics of Origin Labelled Products in Agri-Food Supply Chains: spatial, institutional and co-ordination aspects, INRA, Serie Actes et Communications, no. 17, Paris.

Trift, N., 2003. Qualification de l'origine des viandes bovines selon les manières de produire Le rôle des savoir-faire professionnels et les enjeux de leur couplage. Thèse de doctorat, INA Paris-Grignon, 354 p. + annexes

Trift, N. and F. Casabianca, 2002. Socio-technical construction of the beef meat origin in France. IFSA European symposium. *Farming and Rural Systems Research and Extension. Local identities and Globalisation*, Florence, p. 390-398.

Valceschini, E., 1999. L'étiquetage des aliments est-il la meilleure solution pour les consommateurs? Paris, INRA.

Ventura, F., H. Van der Mulen, 1994. Transformation and Consumption of High-quality meat: the case of Chianina meat in Umbria, Italy. In: J.D. Van der Ploeg and A. Long, p.128-159.

Short communications: Milk and dairy products

α_{S1}-casein yield and milk composition are associated with a polymorphic regulatory element in the bovine α_{S1}-casein gene

A.W. Kuss, T. Peischl, J. Gogol, H. Bartenschlager and H. Geldermann**
Department of Animal Breeding and Biotechnology, University of Hohenheim, D-70593 Stuttgart, Germany.
**Correspondence: tzunihoh@uni-hohenheim.de*

Summary

In this study the protein synthesised from the bovine α_{S1}-casein gene was to be quantified and investigated in view of associations with DNA-variants in the 5'-flanking region. Blood and milk samples from cows of the breeds German Holstein Friesian and Simmental were analysed. Alkaline Urea-PAGE in combination with densitometry was applied for quantification of milk protein fractions (α-lactalbumin, β-lactoglobulin, α_{S1}-, β- and κ-casein) in at least four milk samples per cow. A DNA variant in the 5'-flanking region of the α_{S1}-casein encoding gene was analysed by PCR-RFLP. The findings reveal associations with the relative proportion as well as the yield of α_{S1}-casein, and could represent an example for a potentially specific influence of variable regulatory elements on gene expression and gene product quantities, therefore being of relevance for further breeding on milk protein yield and composition.

Keywords: AP-1, α_{S1}-casein, cattle, polymorphism, promoter

Introduction

α_{s1}-casein (α_{s1}-Cn) is with a proportion of 35 -45 % the major contributor to the protein fraction of bovine milk. It's encoding gene (*CSN1S1*) is part of the casein cluster, located on BTA 6 and paralogous with the α_{s2}- and β-casein encoding genes (for review see e.g. Rijnkels, 2002).

DNA variants occur in the coding region of the *CSN1S1* (Mercier *et al.*, 1971; Grosclaude *et al.*, 1972; Grosclaude *et al.*, 1970; Prinzenberg *et al.*, 1998) as well as the non-coding areas such as the promoter region (Schild and Geldermann, 1996). Variants of the coding region of bovine *CSN1S1* and their associations with quantitative effects on milk performance parameters have hitherto been investigated abundantly (e.g. McLean *et al.*, 1984; Ng-Kwai-Hang *et al.*, 1987; Ehrmann *et al.*, 1997b).

Also for polymorphisms in the 5'-flanking region of milk protein encoding genes observations showing significant associations with relative content as well as amount of single milk proteins have been made (Ehrmann *et al.*, 1997b, for review see Martin *et al.* 2002). Regarding the bovine β-lactoglobulin encoding gene, we have already been able to find associations of a variable AP-2 element with relative content as well as amount of the gene product, which strongly suggest being due to allelic differences in gene expression (Kuss *et al.*, 2003).

The study presented here was designed as a population analysis for single milk protein traits in association with a polymorphic putative regulatory element in the promoter region of *CSN1S1*. This element comprises an A to G SNP (Schild and Geldermann, 1996) affecting the 5' end of an AP-1 consensus sequence (TGAGTCTA; Vogt and Bos, 1990; Angel and Karin, 1991).

Material and methods

Animals and samples

Samples and data were collected between April 1997 and March 1999 from 80 cows of a German Holstein Friesian (HF) herd and 62 cows of a Simmental (SM) herd, both from experimental stations of the Hohenheim University. For pedigree information see Kuss *et al.* (2003). At least four milk samples per cow were collected in intervals of eight weeks during lactation and defatted samples were stored at -80 °C. Genomic DNA was isolated from blood by means of the NucleoSpin® Blood Quick Pure kit (Macherey Nagel, Düren, Germany) according to the instructions of the manufacturer.

Quantification of milk proteins

Milk proteins were separated by vertical Urea-PAGE (8 M Urea, 8.4 % Acrylamide; see Ehrmann *et al.*, 1997a). The gels were then stained (staining buffer: 8 % w/v Trichloroacetic acid, 26.7 % v/v Methanol, 9.3 % v/v Acetic Acid, 0.033 % w/v Coomassie Brilliant Blue R250) over night with agitation. After standardized destaining (29 % v/v Methanol, 5 % v/v Acetic Acid; 4 h with agitation), milk protein fractions (α_{s1}-casein, β-casein, κ-casein, α-lactalbumin, β-lactoglobulin) were quantified using a densitometer (ATH Elscript 400, Hirschmann, Germany) with software of the manufacturer.

Genotyping of DNA variants and statistical analysis

A 192 bp fragment of the bovine *CSN1S1* 5'-flanking region between -317 bp and -125 bp was amplified by PCR from genomic DNA. The amplicon was digested with *Nmu*Cl (MBI Fermentas, St.Leon-Rot, Germany) and subsequently submitted to agarose gel analysis for the RFLP (restriction fragment length polymorphism) determination of the polymorphic AP-1 site R1.3 (SNP at -175 bp, described by Schild and Geldermann, 1996). Genotype nomenclature refers to the respective nucleotide (A or G) at Position -175 bp. Statistical analysis was carried out as previously described (Kuss *et al.*, 2003).

Results

Homozygotes for the G-allele were not observed, A-homozygotes occurred at a four fold frequency as compared to heterozygotes and the allele frequencies of the G-allele were 0.05 in HF and 0.16 in SM.

The results of an analysis across both breeds are presented in Table 1, showing that the carriers of the G-variant were significantly superior both in yield (g/day) as well as content (% of protein fraction) of α_{s1}-Cn as compared to A-homozygotes.

The content of β-Cn and κ-Cn was lower in the milk of homozygous carriers of the A-allele (α_{S1}-Cn, -175bp) in both breeds (not shown). Concerning the amount of β-Cn and κ-Cn the observed differences did not reach the level of significance. The percentage of total protein was 3.41 ± 0.01 for A-homozygotes as compared to 3.29 ± 0.02 for heterozygotes ($p<0.001$). The whey proteins β-lactoglobulin and α-lactalbumin did not show associations in SM, while in HF an increase ($p< 0.001$) in content and amount of α-lactalbumin appeared to be associated with the G-allele (not shown). This was accompanied by a significantly enhanced total protein yield while the total protein percentage (% of whole milk) in HF carriers of the G-allele was reduced ($p< 0.001$).

Table 1. Associations between content (Percentage of protein fraction) and amount (g/day) of α_{S1}-Cn and genotypes of the variable AP1 binding site (-175bp) in the promoter region of CSN1S1.

Genotype (Frequency)[1]	Number of animals	Percentage of protein fraction LSQ mean ± SE	Amount of protein [g/day] LSQ mean ± SE
AA (0.8)	114	$36.76^a \pm 0.12$	$258.44^a \pm 2.38$
AG (0.2)	28	$40.08^b \pm 0.28$	$294.99^b \pm 5.29$
	P[2]	***	***

[1]Average for both breeds; [2]***: $p < 0.001$ (F-Test); Significant differences ($p<0.01$; t-Test) between genotypes are indicated by different letters;

Discussion

In view of previous findings for β-lactoglobulin (Lum *et al.*, 1997; Kuss *et al.*, 2003), as well as for promoter variants in *CSN1S1* and *CSN1S2* (Martin *et al.*, 2002) the differences we observed between carriers of the AP-1 alleles A and G concerning yield and amount of α_{s1}-Cn indicate differential protein binding properties of the variable promoter element. This was confirmed by EMSA results (not shown) that revealed reduced protein binding of oligos containing the G-variant which represents only about 87% of the AP-1 consensus sequence. Hence our findings suggest that the involved AP-1 factor functions as a repressor for the expression of α_{s1}-Cn and that therefore an impaired consensus sequence can be conducive to enhanced α_{s1}-Cn expression. This interpretation is corroborated by the observation that c-jun, the major component of AP-1 dimers (for review see e.g. Karin *et al.*, 1997), plays a role in cell proliferation and survival also as a repressor rather than an enhancer (reviewed by Shaulian and Karin, 2001). The presence of AP-1 factors in mammary epithelial cells and their involvement in mammary gene regulation have been established e. g. by Olazabal *et al.*, (2000).

Our findings as to differences in total protein content between A-homozygotes and heterozygotes in HF are in agreement with the observations of Prinzenberg *et al.* (2003) who, by analysis of variance using a granddaughter design, found associations between SSCP genotypes of the *CSN1S1* promoter and protein percentage in Holstein cattle. In Finnish Ayrshire cows, QTLs for both total protein percentage and yield have been assigned to the location of *CSN1S1* (MARC97: 82.6 cM; Maki-Tanila *et al.*, 1998 Ikonen *et al.*, 1999).

Interestingly, the increased total protein yield we observed in HF coincided with decreased total protein percentage in milk from carriers of the G-allele. This can be explained by the approximately 35% higher amount (g/day) of α-lactalbumin (not shown) in the milk from these animals: Since α-lactalbumin enhances the concentration of lactose which is the major osmole in milk (e.g. Ramakrishnan *et al.*, 2001) its elevated production is likely to cause a higher water ratio and thus a reduced protein percentage. As these observations were not made in SM, we assume that they represent coincidental effects of breeding on high milk performance in HF.

Associations between *CSN1S1* exon variants and α_{s1}-Cn traits which were previously observed in the same breeds e.g. by Cardak *et al.*, (2003) might be due to the very close linkage of these loci with the AP-1 variant and therefore can, in accordance with suggestions by Koczan *et al.*, (1993) and Ehrmann *et al.* (1997a) imply the presence of "intragenic haplotypes". The same assumption can be made with regard to the impact of specific milk proteins on the quality of milk such as the implication of *CSN1S1* exon variants in superior protein and casein content (reviewed by Buchberger and Dovc, 2000). For cheese making ability traits however, polymorphisms affecting protein characteristics might have direct effects, independent of promoter variants.

Still, the observed close association between differential binding capacity of a potentially functional polymorphic promoter element in *CSN1S1* and quantitative characteristics of the gene product requires further supporting analyses, e.g. reporter gene studies, in order to reveal additional evidence as to the true causal coherences.

References

Angel, P. and M. Karin, 1991. The role of Jun, Fos and the AP-1 complex in cell-proliferation and transformation. Biochim. Biophys. Acta 1072: p. 129-157.

Buchberger, J. and P. Dovc, 2000. Lactoprotein Genetic Variants in Cattle and Cheese Making Ability. Food technol. biotechnol. 38: p. 91-98.

Cardak, A. D., H. Bartenschlager, and H. Geldermann. 2003. Effects of polymorph milk proteins on the individual milk protein content of Holstein-Friesian and Simmental cows. Milchwissenschaft 58: p. 235-238.

Ehrmann, S., H. Bartenschlager and H. Geldermann,1997a. Polymorphism in the 5' flanking region of the bovine lactoglobulin encoding gene and its association with beta-lactoglobulin in the milk. J. Anim. Breed. Genet. 114: p. 49-53.

Ehrmann, S., H. Bartenschlager and H. Geldermann,1997b. Quantification of gene effects on single milk proteins in selected groups of dairy cows. J. Anim. Breed. Genet. 114: p. 121-132.

Grosclaude, F., M. F. Mahe, J. C. Mercier and B. Ribadeau-Dumas, 1970. The A variant of bovine alpha(s1) casein is devoid of the segment of 13 amino acid residues which occupies the 14th to 26th position from the NH(2)-terminal in the polypeptide chain (198 residues) of the B and C variants. FEBS Lett. 11: p. 09-112.

Grosclaude, F., M. F. Mahe, J. C. Mercier and B. Ribadeau-Dumas, 1972. Characterization of genetic variants of a S1 and bovine caseins. Eur. J. Biochem. 26: p. 328-337.

Ikonen, T., M. Ojala, and O. Ruottinen, 1999. Associations between milk protein polymorphism and first lactation milk production traits in Finnish Ayrshire cows. J. Dairy Sci. 82: p. 1026-1033.

Karin, M., Z. Liu and E. Zandi, 1997. AP-1 function and regulation. Curr. Opin. Cell Biol. 9: p. 240-246.

Koczan, D., G. Hobom and H. M. Seyfert, 1993. Characterization of the bovine alpha S1-casein gene C-allele, based on a MaeIII polymorphism. Anim Genet. 24: p. 74.

Kuss, A. W., J. Gogol and H. Geldermann, 2003. Associations of a polymorphic AP-2 binding site in the 5'-flanking region of the bovine beta-lactoglobulin gene with milk proteins. J. Dairy Sci. 86: p. 2213-2218.

Lum, L. S., P. Dovc and J. F. Medrano, 1997. Polymorphisms of bovine beta-lactoglobulin promoter and differences in the binding affinity of activator protein-2 transcription factor. J. Dairy Sci. 80: p. 1389-1397.

Maki-Tanila, A., D. J. de Koning, K. Elo, S. Moisio and R. Velmala, 1998. Mapping of multiple quantitative trait loci by regression in half sib designs. Proceedings of the 6th World Congress on Genetics Applied to Livestock Production, Armidale, Australia. 25: p. 269-272.

Martin, P., M. Szymanowska, L. Zwierzchowski and C. Leroux, 2002. The impact of genetic polymorphisms on the protein composition of ruminant milks. Reprod. Nutr. Dev. 42: p. 433-459.

McLean, D. M., E. R. Graham, R. W. Ponzoni and H. A. McKenzie, 1984. Effects of milk protein genetic variants on milk yield and composition. J. Dairy Res. 51: p. 531-546.

Mercier, J. C., F. Grosclaude, and B. Ribadeau-Dumas, 1971. Primary structure of bovine s1 casein. Complete sequence. Eur. J. Biochem. 23: p. 41-51.

Ng-Kwai-Hang, K. F., J. F. Hayes, J. E. Moxley and H. G. Monardes, 1987. Variation in milk protein concentrations associated with genetic polymorphism and environmental factors. J. Dairy Sci. 70: p. 563-570.

Olazabal, I., J. Munoz, S. Ogueta, E. Obregon and J. P. Garcia-Ruiz, 2000. Prolactin (PRL)-PRL receptor system increases cell proliferation involving JNK (c-Jun amino terminal kinase) and AP-1 activation: inhibition by glucocorticoids. Mol. Endocrinol. 14: p. 564-575.

Prinzenberg, E. M., P. Anglade, B. Ribadeau-Dumas and G. Erhardt, 1998. Biochemical characterization of bovine alpha s1-casein F and genotyping with sequence-specific primers. J. Dairy Res. 65: p. 223-231.

Prinzenberg, E. M., C. Weimann, H. Brandt, J. Bennewitz, E. Kalm, M. Schwerin and G. Erhardt, 2003. Polymorphism of the bovine CSN1S1 promoter: linkage mapping, intragenic haplotypes, and effects on milk production traits. J. Dairy Sci. 86: p. 2696-2705.

Ramakrishnan, B., P. S. Shah and P. K. Qasba, 2001. alpha-Lactalbumin (LA) stimulates milk beta-1,4-galactosyltransferase I (beta 4Gal-T1) to transfer glucose from UDP-glucose to N-acetylglucosamine. Crystal structure of beta 4Gal-T1 x LA complex with UDP-Glc. J. Biol. Chem. 276: p. 37665-37671.

Rijnkels, M., 2002. Multispecies comparison of the casein gene loci and evolution of casein gene family. J. Mammary. Gland. Biol. Neoplasia. 7: p. 327-345.

Shaulian, E. and M. Karin, 2001. AP-1 in cell proliferation and survival. Oncogene 20: p. 2390-2400.

Schild, T. A. and H. Geldermann, 1996. Variants within the 5'-flanking regions of bovine milk-protein-encoding genes. III. Genes encoding the Ca-sensitive Caseins alpha-s1, alpha-s2 and beta. Theor. Appl. Genet. 93: p. 887-893.

Vogt, P. K. and T. J. Bos, 1990. jun: oncogene and transcription factor. Adv. Cancer Res. 55: p. 1-35.

Genetic improvement of milk quality traits for cheese production

*M. Ojala**, *A.-M. Tyrisevä and T. Ikonen*
University of Helsinki, Department of Animal Science, PO Box 28, 00014 Helsinki, Finland.
**Correspondence: matti.ojala@helsinki.fi*

Summary

The general goals in present dairy cattle breeding are not in conflict with the goal to improve milk coagulation properties. The suitability of raw milk for cheese production may be improved through indirect or direct selection of breeding animals, or through a combination of them. One option for indirect improvement of milk coagulation traits in the Finnish Ayrshire population is to select against the κ-CN E allele. Estimates of heritability for milk coagulation traits are moderately high, which provides good basis for direct selection. This option may, however, suffer from laborious and costly measurement of the milk coagulation traits. If major genes affecting noncoagulation of milk can be detected, these could be used to improve milk coagulation properties, particularly in dairy cattle populations where the phenomenon of noncoagulation of milk is relatively common.

Keywords: milk, coagulation, cheese, genetic, dairy cattle

Introduction

Milk with good coagulation properties is expected to give more cheese with desirable composition than milk with poor coagulation properties. Also a high casein concentration is beneficial in cheese production. Some milk protein genotypes, especially κ-casein genotypes including the B allele, show favourable associations with the above characteristics (e.g., Ikonen, 2000).

In order to improve cheese production properties of milk through selection of dairy cattle, efficient and cost-effective options have to be sought. An obvious option would be to select breeding animals for milk coagulation traits directly. In this case, reliable estimates of genetic parameters for milk coagulation, quality and yield traits are needed. Milk protein genotypes of animals could be used as an indirect selection criterion. However, before using milk protein genotypes in selection of breeding animals, their effects on milk production traits and other economically important traits have to be established.

The milk of some cows did not coagulate in the applied standard test, resulting in an extreme value of zero for curd firmness (e.g., Ikonen *et al.*, 1999a; Tyrisevä *et al.*, 2003; Ikonen *et al.*, 2004; Tyrisevä *et al.*, 2004). A relatively high frequency of cows in some family groups produced noncoagulating milk. This led to the hypothesis that major genes may explain, in part, the phenomenon of noncoagulation of milk. If genetic markers can be detected, they could be used in selection against the noncoagulation of milk.

The objectives of this report were to summarize our results regarding the effect of milk protein polymorphism on milk coagulation traits (milk coagulation time or curd firmness), other milk quality traits (milk composition, protein composition, somatic cell score or pH) and milk yield traits (daily or 305-d milk, fat or protein yield) and the estimates of genetic parameters (heritability and genetic correlations) for the above traits in the Finnish dairy cattle population, and to discuss various options to improve cheese production through dairy cattle breeding.

Materials and methods

Milk coagulation traits were analyzed together with other milk quality traits and milk yield traits from several data sets of varying size: data from two experimental herds with 114 Finnish Ayrshire (FAy) and Finnish Friesian (FFr) cows with 329 records (Ikonen *et al.*, 1997); data on 875 FAy and FFr cows with one record per cow (Ikonen *et al.*, 1999a); data on 94 FAy cows with 1005 records (Tyrisevä *et al.*, 2003); data on 4664 FAy cows with one record each (Ikonen *et al.*, 2004), and data on 1408 cows of FAy, Holstein-Friesian (Hol) and crossbreds (Tyrisevä *et al.*, 2004). The effect of composite β- and κ-casein (β-κ-CN) genotypes and β-lactoglobulin (β-LG) genotypes on milk coagulation traits was estimated in the studies by Ikonen *et al.*, (1997); Ikonen *et al.*, (1999a); Tyrisevä *et al.*, (2003).

The effect of milk coagulation properties on yield and composition of Emmental cheese was studied using herd bulk milks from two herd groups with extreme milk coagulation properties. Each group included four herds and 67 and 45 cows, respectively (Ikonen *et al.*, 1999c).

The effect of β-κ-CN and β-LG genotypes on 305-d milk, fat and protein yield, and fat and protein content was analyzed in a data set of about 18 700 first lactation FAy cows (Ikonen *et al.*, 1999b). The effect of β-κ-CN haplotypes on the same traits was analyzed in data of about 17 000 cows (Ikonen *et al.*, 2001). The effect of milk protein polymorphism on milk production traits and somatic cell score (SCS) was estimated also from a data set of about 45 900 records from the first three lactations of the above cows (Ojala *et al.*, 2004a,b).

The data sets were edited and preanalyzed based on various fixed effects models using the WSYS-L program package (Vilva, 2004). Variance components for the random effects and estimates of the genetic parameters with their SE were computed using linear mixed animal models and the REML VCE4 program package (Groeneveld, 1997). The effect and the statistical significance of the fixed factors were computed and tested using linear mixed animal models and the PEST program package (Groeneveld, 1990).

Results

Effect of lactation stage, herd and breed on milk coagulation, quality and yield traits

Milk coagulation properties were best at the beginning of lactation and poorest during mid lactation, and improved again at the end of lactation (Ikonen *et al.*, 1999a; Tyrisevä *et al.*, 2003; Ikonen *et al.*, 2004; Tyrisevä *et al.*, 2004). The changes in fat, protein and casein content, and SCS were parallel to those in milk coagulation traits, whereas the change in pH showed an inverse trend.

Herd effects had a markedly lower impact on the variation in milk coagulation traits than in milk yield traits (Ikonen *et al.*, 2004; Tyrisevä *et al.*, 2004). Consequently, the differences in feeding and management practices between herds had only a small effect on the variation in milk coagulation traits. Frequent feeding of concentrates was associated with a slight improvement in milk coagulation traits as well as in milk, fat and protein yield (Tyrisevä *et al.*, 2004).

Milk coagulation properties were on average better for the FFr and Hol cows than for the FAy cows (Ikonen *et al.*, 1999a; Tyrisevä *et al.*, 2004). The inferiority of FAy to FFr and Hol in milk coagulation traits was explained, in part, by a higher pH of milk, and by a higher proportion of cows which produced noncoagulating milk.

Effect of milk protein polymorphism on milk coagulation, quality and yield traits

The composite β-κ-CN genotypes including the κ-CN B allele and β-CN A_1 allele were associated with the best milk coagulation properties and the highest κ-casein content, whereas the β-κ-CN genotypes including κ-CN A or E alleles and β-CN A_2 or A_1 alleles were associated with the poorest milk coagulation properties (Ikonen *et al.*, 1997, 1999a).

Milk yield was higher for the β-κ-CN genotypes and haplotypes with the β-CN A_2 allele than the A_1 allele. Both κ-CN B and E alleles were associated with increased fat content (Ikonen *et al.*, 1999b; Ikonen *et al.*, 2001; Ojala *et al.*, 2004a,b). The κ-CN B allele was associated with increased protein content and decreased SCS, whereas the E allele was associated with decreased protein content and increased SCS. The relatively rare β-κ-CN haplotype A_1B was associated with the lowest milk yield, highest fat and protein content, and lowest SCS. The haplotype A_1E was associated with low milk yield, high fat content, the lowest protein content and highest SCS. The most common β-κ-CN haplotype, A_2A, was associated with the highest milk yield, lowest fat content, moderate protein content and slightly decreased SCS (Ojala *et al.*, 2004b).

The extreme β-κ-CN haplotypes differed about 15% of σ_P in milk yield, about 20% in SCS, close to 30% in fat content, and over 30% in protein yield and content. The β-LG AA genotype was associated with increased milk yield, protein yield and protein content, and decreased fat content, as compared to the AB and BB genotypes (Ikonen *et al.*, 1999b; Ojala *et al.*, 2004a,b).

Effect of milk coagulation traits on cheese production

In a study by Ikonen *et al.* (1999c), herds were selected for extreme milk coagulation properties. The herd bulk milks of group A coagulated moderately, whereas bulk milks of group B did not coagulate. The proportion of FFr cows, the frequency of κ-CN B allele and β-CN A_1 allele, and the content of κ-casein, α-lactalbumin and β-lactoglobulin were higher in group A than in group B. Two Emmental cheeses from both herd groups were manufactured, each cheese made from 650 l of milk. For group A, moisture in non-fat substance was lower and ash content higher in fresh cheeses, and fat content in whey lower, compared to group B. Calcium and phosphorus content in both fresh and ripened cheeses were higher in group A than B (Ikonen *et al.*, 1999c).

Estimates of heritability and genetic correlations

The estimates of heritability (h^2) for milk coagulation traits were moderate or moderately high (Ikonen *et al.*, 1999a; Ikonen *et al.*, 2004; Tyrisevä *et al.*, 2004). The estimates of h^2 from the largest data set were 0.28 for milk coagulation time, 0.39 for curd firmness as a continuous trait and 0.26 as a binary trait (noncoagulated or coagulated milk samples), and 0.29 for protein content, 0.35 for casein content, 0.06 for SCS and 0.38 for pH (Ikonen *et al.*, 2004).

In general, genetic correlations (r_g) between the milk coagulation traits and the other milk quality and milk yield traits were low or relatively low (Ikonen *et al.*, 1999a; Ikonen *et al.*, 2004). Low values of SCS and pH were associated with good curd firmness (r_g = -0.45 and -0.32). Thus, selection to improve curd firmness would not be in conflict with the general breeding goals of Finnish dairy cattle.

Discussion

Selection based on milk protein polymorphism

The β-κ-CN polymorphism had a statistically significant effect on the milk coagulation, quality and yield traits. The milk coagulation properties could be improved indirectly by selection to increase the frequency of the κ-CN B allele and/or decrease the frequency of the κ-CN E allele. In the FAy population, the latter option would give a faster response because of the low frequency of the κ-CN B allele. Selection against the κ-CN E allele would improve both the milk coagulation traits and the other milk quality and milk yield traits. Effects of the milk protein genotypes on fertility traits (Ruottinen *et al.*, 2004) or on body weight (Ojala *et al.*, 2004a) were negligible in the FAy. Thus, the overall effect of selection against the β-κ-CN haplotype A_1E would be advantageous.

Selection based on milk coagulation traits

The estimates of heritability for the milk coagulation traits were moderately high which provide good possibilities for direct selection on these traits. In addition, the genetic correlations between the milk coagulation traits and the milk quality and yield traits imply that selection for improved coagulation properties would not be in conflict with the goals in the dairy cattle breeding. A practical limitation is, however, that milk coagulation traits are laborious and costly to measure with the available techniques in a large scale in the milk recording framework.

Selection against noncoagulation of milk

A relatively high frequency of cows in some family groups produced noncoagulating milk. This led to the hypothesis that noncoagulation of milk is, in part, a genetically determined phenomenon in FAy (Ikonen *et al.*, 1999a). The hypothesis is further supported by the occurrence of cows with repeated noncoagulating milk samples during the lactation (Tyrisevä *et al.*, 2003), and by a relatively high heritability estimate for curd firmness as a binary trait (Ikonen *et al.*, 2004). Noncoagulation of milk was also observed in Hol (Tyrisevä *et al.*, 2004).

The association of milk coagulation traits with SCS was relatively strong, with noncoagulating milk being associated with high SCS (Ikonen *et al.*, 2004). Thus, the present breeding goal with the aim to reduce SCS is favorable for the improvement of milk coagulation traits. The estimates of genetic correlation between curd firmness, and protein and casein content were close to zero. This may, in part, be due to a nonlinear association because both noncoagulating and well coagulating milk were associated with high protein and casein content. Thus, the present breeding goal for high protein content tends to improve curd firmness, but may simultaneously slightly increase the occurrence of noncoagulating milk.

Variation in the frequency of daughters with noncoagulating milk within large paternal half-sib groups led to the hypothesis that one or a few major genes might explain the phenomenon of noncoagulation of milk. Research is in progress to detect genetic markers, and possibly the causative genes, for the noncoagulation of milk. The gene information would be an efficient tool to select against noncoagulation of milk.

Conclusions

The properties of raw milk should be improved to increase the efficiency in cheese production. One or more of the various options (selection of breeding animals based on milk protein polymorphism, or selection for milk coagulation traits or against noncoagulation of milk) provide good basis for improving the milk coagulation properties.

References

Groeneveld, E. 1990. PEST User s Manual. Inst. Anim. Husbandry Anim. Behav., Fed. Agric. Res. Ctr., Neustadt, Germany.

Groeneveld, E. 1997. VCE4 User's Guide and Reference Manual. Inst. Anim. Husbandry Anim. Behav., Fed. Agric. Res. Ctr., Neustadt, Germany.

Ikonen, T. 2000. Possibilities of genetic imrovement of milk coagulation properties of dairy cows. Academic Diss., University of Helsinki, Department of Animal Science, Publications no. 49, http://ethesis.helsinki.fi/julkaisu/kotie/vk/ikonen/

Ikonen, T., K. Ahlfors, R. Kempe, M. Ojala, and O. Ruottinen. 1999a. Genetic parameters for the milk coagulation properties and prevalence of noncoaculating milk in Finnish dairy cows. J. Dairy Sci. 82: p. 205-214.

Ikonen, T., H. Bovenhuis, M. Ojala, O. Ruottinen, and M. Georges. 2001. Associations between casein haplotypes and first lactation milk production traits in Finnish Ayrshire cows. J. Dairy Sci. 84: p. 507-514.

Ikonen, T., S. Morri, A.-M. Tyrisevä, O. Ruottinen, and M. Ojala. 2004. Genetic and phenotypic correlations between milk coagulation properties, milk production traits, somatic cell count, casein content, and pH of milk. J. Dairy Sci. 87: p. 458-467.

Ikonen, T., M. Ojala, and O. Ruottinen. 1999b. Associations between milk protein polymorphism and first lactation milk production traits in Finnish Ayrshire cows. J. Dairy Sci. 82: p. 1026-1033.

Ikonen, T., M. Ojala, and E.-L. Syväoja. 1997. Effects of composite casein and β-lactoglobulin genotypes on renneting properties and composition of bovine milk by assuming an animal model. Agric. and Food Sci. in Finland 6: p. 283-294.

Ikonen, T., O. Ruottinen, E.-L. Syväoja, K. Saarinen, E. Pahkala, and M. Ojala. 1999c. Effect of milk coagulation properties of herd bulk milks on yield and composition of Emmental cheese. Agric. and Food Sci. in Finland 8: p. 411-422.

Ojala, M., T. Seppänen, A.-M. Tyrisevä, and T. Ikonen. 2004a. Effects of milk protein genotypes on body weight and milk production traits in Finnish Ayrshire cows. X Baltic Anim. Breed. Conf. Tartu, Estonia, 13-14 May 2004. Proc. Animal Breeding in the Baltics: p. 68-73.

Ojala, M., T. Seppänen, A.-M. Tyrisevä, and T. Ikonen. 2004b. Associations of milk protein polymorphism and milk production traits and udder health traits in Finnish Ayrshire cows. The 55th Annual Meet. of EAAP. Bled, Slovenia, 5-9 September 2004. Book of Abstracts No. 10: 4, Theatre GPh1.7.

Ruottinen, O., T. Ikonen, and M. Ojala. 2004. Associations between milk protein genotypes and fertility traits in Finnish Ayrshire heifers and first lactation cows. Livest. Prod. Sci. 85: p. 27-34.

Tyrisevä, A.-M., T. Ikonen, and M. Ojala. 2003. Repeatability estimates for milk coagulation traits and non-coagulation of milk in Finnish Ayrshire cows. J. Dairy Res. 70: p. 91-98.

Tyrisevä, A.-M., T. Vahlsten, O. Ruottinen, and M. Ojala. 2004. Nongoagulation of milk in Finnish Ayrshire and Holstein-Friesian cows and effect of herds on milk coagulation ability. J. Dairy Sci. 87: p. 3958-3966.

Vilva, V. 2004. WSYS-L-program package. University of Helsinki, Department of Animal Science. http://www.animal.helsinki.fi/wsys/

Breed effect on milk coagulation and cheese quality parameters of dairy cattle

M. Cassandro[1], M. Povinelli[1], D. Marcomin[1], L. Gallo[1], P. Carnier[1], R. Dal Zotto[2], C. Valorz[2] and G. Bittante[1]*
[1]University of Padova, Department of Animal Science, Viale dell'Università, 16, Agripolis, 35020, Legnaro, Padova, Italy, [2]Superbrown consortium of Bolzano-Bozen and Trento, Via Lavisotto, 125, 38100 Trento, Italy.
**Correspondence: martino.cassandro@unipd.it*

Summary

This study aimed to estimate the effect on milk coagulation and cheese quality traits of Holstein Friesian (HF) and Brown Swiss (BS) cattle reared in the Trento province. Data from 153 dairy cows (BS: 82 and HF: 71), herded in 9 mixed-breed dairy farms, were used for the estimation of the breed effect on the following milk coagulation traits: casein yield (CY), casein content (CC), acidity (SH50), calcium (Ca) and phosphorus (P). The milk produced by a random sample of 41 dairy cows (BS: 22 and HF: 19), herded in 3 mixed-breed dairy farms, was processed to "*Casolet*" cheese, a local 60 d aged product of Trentino region, using either milk by BS or HF or mixed milk by BS+HF. Cheese quality traits considered were: acidity (pH), yellow index for cheese paste (b*) and tenderness (Te). Results on milk coagulation traits did not show any statistical difference between BS and HF for CY, SH50, Ca and P, whereas for CC, BS performed better than HF ($P<0.05$). Concerning cheese quality traits, the breed effect was negligible for pH, Te, and content of ϖ-3 polyunsaturated fatty acid, whereas cheese of BS showed higher b* value and lower content of ϖ-6 polyunsaturated fatty acid respect to cheese by HF ($P<0.001$). Brown Swiss showed an higher cheese yield respect to HF breed over all time maturing and an absence of additive effect on cheese yield was found for milk of BS+HS in comparison with milk of BS and HF.

Keywords: dairy cattle, milk, cheese, coagulation properties, quality traits

Introduction

The economic importance of Italian dairy sector can be estimated as the incidence of Italian dairy sector on agriculture gross domestic product that is about the 10% with an overall milk production of 11.5 million of tons by 91.1% cattle, 6.6% sheep, 1.2% buffalo and 1.1% goat (Cassandro, 2003). The cow milk production in Italy is 10.5 million of tons (AIA *et al.*, 2003) and about 73% of the milk yield is used for cheese production (Osservatorio del Latte, 2003).

In a modern livestock production system the food safety, the traceability and the origin identification of the animal products are new requirements for consumers and social community. In the Trento province a recent study (Nomisma, 2003) showed that quota of consumers which associated food products with the most high quality level was 70% for cheese products, 8% for mushroom, 5% for meat, 4% for water, 3% for fruit, 2% for wine 8% others. Therefore, cheese products seem to be in the mind of the consumers the most guaranteed product with an high quality level. However, the recent negative impact of the food scandals involving cow milk (e.g. aflatoxin) might progressive decline the good perception and opinion of the consumer on milk products. Moreover, Cassandro and Marusi (2001) showed, in the last 10 years, an evident unfavorable phenotypic trend in the milk rennet-coagulation ability considering 108,065 herd samples from 1,317 Italian dairy farms. Hence, more studies for knowing the effects on milk coagulation and cheese quality parameters of dairy cattle should be planed. Breed effect is one of the most studied effects on milk yield and cheese products (Custer, 1979; Mariani and Battistotti,1999) but very few studies compared cheese yield by mixed and separated milk of different breeds. Moreover, few studies planed researches where breed effects was followed along the whole dairy chain production, considering milk quality and coagulation properties and cheese yield and quality traits.

Aims of this study were to estimate the breed effect on milk coagulation and cheese quality traits by mixed and separated milk of Holstein Friesian (HF) and Brown Swiss (BS) cattle reared in the Trento province.

Material and methods

Milk samples were recorded in mixed-breed dairy herds, by dairy cows enrolled in the national herd-books of Brown Swiss and Holstein Friesian breeds, located in the Trento province, a mountain area of the north part of Italy. Therefore, the environment, diet and management practices are assumed to be the same for the two breeds in comparison.

Milk samples data

A single test day milk of 153 random dairy cows (BS: 82 and HF: 71) herded in 9 mixed-breed dairy farms were used. Cows were milked twice a day and for milk coagulation properties were used fresh milk without conservation additive.

For estimating breed effect the following milk coagulation traits were analysed: casein yield in kg/d (CY), casein content in % (CC), titratable acidity in °SH/50 ml(SH50), total calcium content in mg/100 ml (Ca) and total phosphorus content in mg/100 ml (P). Data were recorded in May (herds 1, 2 and 3), in July (herds 4, 5 and 6) and in November (herds 7, 8 and 9) 2003. Milk coagulation parameters and lactodynamographic chart per each milk sample, were determined by Concast laboratory of Trento (Italy). A milk with a normal coagulation properties (lactodynamographic type "A") was defined as a milk sample with a coagulation time between 11 and 18 minutes, a curd-firming time between 10 and 13 minutes and a curd firmness of 20-25 millimetres. Analyses for Ca, P and chloride contents were determined by laboratory of the Department of Animal Science of Padova (Italy) using frozen milk. Sources of variation as parity, days in milk, k-casein genotype, somatic cells count, milk yield and contents were extracted by official milk recording database of Provincial Breeder Federation of Trento.

Cheese samples data

The single test day milk yield, recorded in four days of May 2003, by a random sample of 41 dairy cows (BS: 21 and HF: 19), herded in 3 mixed-breed dairy farms, was processed to *Casolet* cheese of "Val di Sole", a local 60 d aged product of Trentino region, using either mixed milk by BS cows or HF cows or mixed milk by both BS and HF cows (BS+HF). The *Casolet* is a crude soft paste cheese using crude whole milk. The weight of *Casolet* cheese is around 2-3 kg after 2 months of maturing time.

In the four consecutive days were produced *Casolet* cheeses, using for the first and third day, separated milk of BS and HF cows, whereas, for second and fourth day a mixed milk of BS+HF processed in the same tank (Table 1).

A total of 125 *Casolet* cheeses were yielded (37 of BS, 33 of HF and 55 of BS+HF). Cheese yield and quality were evaluated on 30 cheeses (10 of BS; 10 of HS and 10 of BS+HF) random chosen, 5 per breed/day. The cheese yield was recorded 24-h, 10-d, 30-d and 60-d storage. The quality

Table 1. Distribution of number of cheeses yielded per breed within milk processing day.

Milk processing day	Brown Swiss (BS)	Holstein Friesian (HF)	BS+HF
1	18	16	-
2	-	-	28
3	19	17	-
4	-	-	27

parameters at 60-d of cheese maturing, were: acidity (pH), yellow index for cheese paste (b*), tenderness (Te) and polyunsaturated fatty acids (ϖ-3 and ϖ-6 fatty acids).

Statistical analyses

Analyses of variance were performed using GLM procedure of SAS (1999) package. For milk samples the following independent variables were used: breed (BS and HF), herd-test day (9 levels), lactodynamographic milk class (2 levels: normal coagulation properties or type "A" and abnormal coagulation properties or type "no-A"), k-casein genotype class (2 levels: BB or AB+AA), parity (4 levels), stage of lactation in class (3 levels: 8-100 d, 101-200 d and >200 d), class of milk yield within class of stage of lactation (high, medium and low level for the 3 classes of stage of lactation were for high level: >37 kg/d, >34 kg/d and >32 kg/d, respectively; for medium level were: 27-37 kg/d, 24-34 kg/d and 22-32 kg/d, respectively and for low level were <27 kg/d, <24 kg/d and <22 kg/d, respectively), class of somatic cells count (3 levels: >200.000 cells/ml; 50.000-200.000 cells/ml and <50.000 cells/ml), class of chloride (3 levels: <81 mg Cl-/100 g, 81-110 mg Cl-/100 g and >110 mg Cl-/100 g) class of total calcium (3 levels: <116 mg/100 ml, 116-135 mg/100 ml and >135 mg/100 ml), class of total phosphorus (3 levels: <110 mg/100 ml, 111-120 mg/100 ml and >120 mg/100 ml) and class of titratable acidity (3 levels: <3.21 °SH/50ml, 3.21-4.00 °SH/50ml and > 4.00 °SH/50ml). In the linear models of SH50, Ca and P traits, were excluded, as independent variable, the corresponding traits. The effect of breed was tested on breed nested on herd-test day effect, whereas all the other effects were tested on error line.

For cheese yield the following independent variable were used: breed (BS, HF and BS+HF), maturing time (24-h, 10-d, 30-d and 60-d) and date of processing (2 levels). For cheese quality traits, determined at 60-d of maturing, the following independent variables were used: breed (BS, HF and BS+HF), date of cheese yielding (4 levels) and cheese weight (as covariate). The effect of breed was tested on breed nested on date of cheese yielding effect, whereas all the other effects were tested on error line.

Results and discussion

Brown Swiss has an higher genotype frequency of k-casein-BB and a lower genotype frequency of k-casein-AA respect to Holstein Friesian (37 vs 13 % and 17 vs 46 % respectively). Descriptions for milk yield and components, lactation traits, casein yield and content and milk coagulation properties per breed are reported in Table 2.

As expected, Holstein Friesian cows showed a higher milk yield respect to Brown Swiss that reported, at the contrary, an higher fat and protein percentage. Similar values between the two breeds were found for somatic cell count, days in milk and lactation numbers. Brown Swiss breed showed a better casein content, lactodynamographic profile, express as normal coagulation

Table 2. Statistical descriptions for milk yield and composition, lactation traits, casein yield and content, and milk coagulation properties per breed.

Trait	Brown Swiss		Holstein Friesian	
	Mean	Std	Mean	Std
Milk yield and composition:				
Milk yield, kg/d	25.6	6.6	29.3	7.47
Fat, %	3.72	0.67	3.43	0.89
Protein, %	3.52	0.34	3.22	0.34
Somatic cells count, n/ml x1000	285	580	325	1042
Lactation traits:				
Days in milk	164	78	161	96
Lactation numbers	2.60	1.57	2.56	1.65
Casein yield and content:				
Casein, kg/d	0.67	0.16	0.69	0.16
Casein, %	2.60	1.57	2.37	0.23
Milk coagulation properties:				
Normal coagulation samples[1],%	75.6	0.43	62.0	0.49
Titratable acidy, °SH/50 ml	3.64	0.33	3.47	0.35
Calcium, mg/100 ml	130.5	12.7	120.2	17.3
Phosphorus, mg/100 ml	113.1	13.9	105.3	14.4
Chloride, mg Cl-/100 g	92.2	23.5	97.5	22.9

[1] lactodynamographic type 'A'.

samples, titratable acidity, and minerals content respect to Holstein Friesian breed. Similar values were reported by Mariani and Battistotti (1999) and Summer *et al.* (2004) using the same breeds reared in others Italian provinces. These results suggesting that milk samples analysed in this research are representative of the two breeds considered.

Breed effect on casein and milk coagulation properties

Analyses of variance performed on test-day milk samples showed a statistically significant effect of breed only for CC, whereas any statistical difference between BS and HF was found for CY, SH50, Ca and P. Similar results were also found including in the linear model the protein percentage, as covariate. The effect of breed showed a +9% of CC on milk of BS respect to HF milk (Table 3).

The k-casein BB genotype was statistically significant for CC and SH50 whereas lactodynamographic profile was high related with titratable acidity of milk ($P < 0.001$). Total contents of calcium and phosphorus were highly influenced one to each other and, in particular, by herd test-day effect. Moreover, phosphorus content showed and high relation with titratable acidity as reported by Mariani and Battistotti (1999).

Breed effect on cheese yield and quality

Breed effect on cheese yield at 24-h, 10-d, 30-d and 60-d of maturing time is shown in Figure 1.

Table 3. Least square means and contrasts of milk coagulation properties for Brown Swiss (BS) and Holstein Friesian (HF) breed.

Trait	Least square mean		Contrast
	BS	HF	BS vs HF
Casein, kg/d	0.71	0.72	ns
Casein, %	2.66	2.44	*
Titratable acidy, °SH/50 ml	3.48	3.40	ns
Calcium, mg/100 ml	126.6	123.7	ns
Phosphorus, mg/100 ml	111.4	109.9	ns

* P < 0.001; ** P < 0.05; *** P < 0.001; ns: not significant

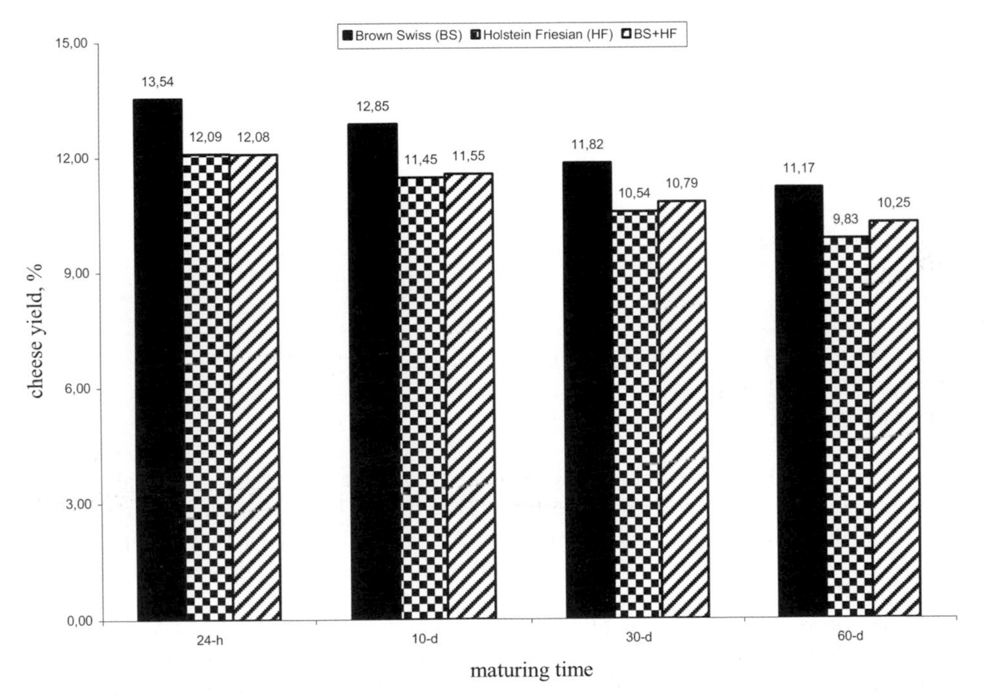

Figure 1. Cheese yield (%) at different maturing time using milk by Brown Swiss (BS) or Holstein Friesian (HF) or mixed milk (BS+HF).

The *Casolet* cheese yield, at the end of maturing time (60-d), was 10.42% of the total milk processed. The BS showed a +14% and +9% of cheese yield, after 2 months, respect to HF and BS+HF, respectively. The cheese yielded with mixed milk by BS+HF showed values closer to HF than an intermediate value, as expected. This result suggesting the absence of additive effect on cheese yield when is processed mixed milk by two breeds. The lower milk coagulation properties of HF seem to influence more cheese yield and quality than the better coagulation properties of BS. Possible reasons of the superiority of cheese yield by milk of BS respect to HF milk might be the high milk composition (3,72% vs 3.43% for fat and 3,52% vs 3.22% for protein, respectively), the high genotype frequency of k-casein-BB (37% vs 13%, respectively) the high casein content (2.66% vs 2.44%, respectively) and the percentage of normal coagulation milk samples (75.6% vs 62.0%, respectively).

Concerning cheese quality traits (Table 4), the breed effect was negligible for pH, Te and ϖ-3, whereas, cheese yielded by BS showed higher b* and content of ϖ-6 with respect to those by HF and BS+HF milk ($P<0,05$). Cheese yielded by BS reported higher yellow value on cheese paste and lower ϖ-6 content than HF. This low content of ϖ-6 seem to be interesting in order to reduce the level of ϖ-6/ϖ-3 ratio which should be 2:1 for guarantee a good nutritional effect in human health because of reduce clots and constricts arteries that increase risk of heart attacks (Mirkin, 2001).

Table 4. Least square means and contrasts of cheeses yield and cheese quality traits for Brown Swiss (BS), Holstein Friesian (HF) and BS+HF.

Trait	Least square means			Contrasts		
	BS	HF	BS+HF	BS vs HF	BS vs BS+HF	HF vs BS+HF
Cheese yield, %	12.34	10.98	11.17	***	***	ns
Yellow index, b*	10.55	9.32	9.90	***	ns	ns
Acidity, pH	5.01	5.04	5.05	ns	ns	ns
Tenderness, Nw/cm^3	6.12	5.79	6.86	ns	ns	ns
Polyunsaturated fatty acids:						
ϖ-3	0.54	0.60	0.56	ns	ns	ns
ϖ-6	2.64	3.09	2.88	***	***	***

P < 0.001; ** P < 0.05; *** P < 0.001; ns: not significant

Conclusions

In the present research the effect of breed was studied along the dairy chain of a local cheese of Trento province, called *Casolet,* used a case study. The effect of breed was related on coagulation and cheese quality traits of milk yielded by Holstein Friesian (HF), Brown Swiss (BS) and mixed milk of BS+HF cattle. The BS milk showed a better casein content (+9%) than HF milk, corrected for k-casein BB genotype, lactodynamographic profile and mineral contents. The cheese yield of BS milk confirmed to be superior, in term of milk rennet-coagulation ability, of +14% than HF milk, moreover, the cheese by BS showed a higher yellow index of paste and a lower ϖ-6 content. Interesting was the results that showed the absence of additive effect on cheese yield when mixed milk of BS+HF is processed. In conclusion, the lower milk coagulation properties of HF seem to influence more cheese yield and quality than the better coagulation properties of BS.

Acknowledgements

The authors thank Concast laboratory and Provincial Breeder Federation of Trento for laboratory facilities and data provided, respectively, and the provincial council of Trento for providing funds for the research.

References

Aia, Ismea and Osservatorio del Latte, 2002. Il mercato dei lattiero caseari nel 2002. Quaderni di ricerca sul mercato dei prodotti zootecnici, Marzo, n.4.

Cassandro, M., 2003. Status of milk production and market in Italy. Agriculturae Conspectus Scientificus, Vol. 68, No. 2: p. 65-69.

Cassandro, M. and M. Marusi, 2001. La caseificabilità del latte nelle frisone italiane. bianconero, XL, 9: p. 43-46.

Custer, E.W., 1979. The effect of milk composition on the yield and quality of cheese. II. The effects of breeds. J. Dairy Sci., 62: Suppl. 1, p. 48-49.

Mariani, P. and B. Battistotti, 1999. Milk quality for cheesemaking. In: Proceedings of the A.S.P.A. XIII Congress, June 21-24, p. 499-516.

Mirkin, G., 2001. Who's afraid of ϖ-6 polyunsaturated fatty acids? Methodological considerations for assessing whether they are harmful. Nutrition Metabolism and Cardiovascular Diseases, Vol 11, Iss. 3, pp 181-188. EM Berry. Hebrew Univ Jerusalem, Hadassah Med Sch, Fac Med, Dept Human Nutr and Metab, IL-91010 Jerusalem, ISRAEL.

Nomisma, 2003. IX Rapporto Nomisma sull'Agricoltura Italiana. Il Sole 24 ORE S.p.A. (editor), first edition, June.

Osservatorio del Latte, 2003. Annuario del latte 2003. Ed. Franco Angeli, Italy.

SAS®, 1999. User's guide: Basic, Version 8.00, Editor. SAS Inst., Inc., Cary, NC.

Summer, A., M. Pecorari, E. Fossa, M. Malacarne, P. Formaggioni, P. Franceschi and P. Mariani, 2004. Frazioni proteiche, caratteristiche di coagulazione presamica e resain formaggio parmigiano-reggiano del latte delle vacche di razza Bruna italiana. In: Proceedings of the 7th World Conference of the Brown Swiss cattle breeders. Verona 3-7 march.

The impact of total somatic cells and polymorphonuclear leucocyte cells on the quality of milk and on the chemical composition of cheese

B. O'Brien[1], B. Gallagher[1, 2], P. Joyce[2], W.J. Meaney[1] and A. Kelly[3]*
[1]Teagasc, Moorepark Research Centre, Fermoy, Co. Cork, Ireland; [2]Zoology Department, University College Dublin, Ireland; [3]Food Science and Technology Department, University College, Cork, Ireland.
**Correspondence: bobrien@moorepark.teagasc.ie*

Summary

This study investigated the effect of somatic cell count (SCC) and polymorphonuclear leucocytes (PMN) on the processability of milks from individual udder quarters. Four milks of SCC $6x10^3$/ml and $920x10^3$/ml and PMN $<1x10^3$/ml and $700x10^3$/ml were selected. The low and high SCC and PMN milks were mixed in various proportions to give four artificial mix milks of SCC and of PMN $200x10^3$/ml and $400x10^3$/ml approximately. All milks were analysed for gross composition, renneting properties and N-fractions. Miniature cheeses were manufactured and analysed for composition after 30 d and for proteolysis after 30 d and 90 d of ripening. Fat and protein contents of the milks generally decreased with increasing SCC and PMN and there was no effect on N-fractions in the milks or on total solids, moisture, NaCl or pH in the miniature cheeses. Elevated SCC and PMN had a negative effect on rennet coagulation of milk. The urea-PAGE electophoretogram indicated that the patterns of proteolysis differed quantitatively and qualitatively with increasing milk SCC and PMN during cheese ripening. High SCC and PMN milk resulted in reduced levels of residual intact α_{s1}- and β-caseins, indicating cell-associated proteinase activity. In conclusion, high SCC and PMN milks had inferior composition and processing characteristics.

Keywords: milk, PMN, processability, SCC

Introduction

The primary characteristic change that occurs in milk during mastitis inflammation is a significant influx of white blood cells (somatic cells) into milk. Milk contains three main somatic cell types, leucocytes (PMN cells), macrophages and lymphocytes. PMN cells account for approximately 10% of total cells in normal milk, but account for greater than 90% of cells in mastitic milk. The total number of cells in milk is commonly used (somatic cell count) as an indicator of mastitis, but little attention is given to the type of cells and their particular function (Concha *et al.*, 1978). Since the increase in the number of cells in mastitic milk has been shown to be predominantly PMN, measurement of these cells may allow the early diagnosis of subclinical mastitis and thus, provide a more specific indicator of mastitis and inflammation than measuring total somatic cells in milk. O'Sullivan *et al.* (1992) developed a rapid ELISA test to measure PMN antigens using horse radish peroxidase conjugated rabbit polyclonal anti-PMN antisera and a mononclonal antibody specific for PMN cells.

Elevated bulk milk SCC has been shown to have significant implications for the processing properties of such milk, and in particular, reduced yield and quality of Cheddar cheese (Auldist *et al.*, 1996a). Recently, high SCC has also been linked to early gelation of UHT milk (Auldist *et al.*, 1996b). However, little information is available on the contribution of different somatic cells in milk to cheese-making properties of the milk. Additionally, bulk milk SCC is used by co-operatives and dairies to identify milk unsuitable for processing. However, a bulk milk SCC reflects the SCC of different udder quarter milks of a number of cows contributing to that milk pool. Thus the objective of this study was to investigate the effect of SCC and PMN levels on the processing characteristics of quarter milks and on the quality and ripening characteristics of miniature cheeses manufactured from such milks.

Material and methods

One-litre volume milk samples were collected from a single udder quarter in each of four cows. Milk from two udder quarters had SCC of $6x10^3$/ml and $920x10^3$/ml, respectively, and milks from the remaining two quarters had PMN levels of <$1x10^3$/ml and $700x10^3$/ml, respectively. The milks of SCC $6x10^3$/ml and $920x10^3$/ml were mixed in various proportions to give two artificial mix milks of SCC $200x10^3$/ml (Mix 1 SCC) and $400x10^3$/ml (Mix 2 SCC), approximately. The remaining milks of PMN <$1x10^3$/ml and $700x10^3$/ml were also mixed in various proportions to give two artificial mix milks of PMN $200x10^3$/ml (Mix 1 PMN) and $400x10^3$/ml (Mix 2 PMN) approximately.

A Bentley Somacount 300 somatic cell counter (Agri York 400 Ltd, York YO4 2QW,UK) was used to measure SCC in milk. PMN levels were determined using the ELISA method of O'Sullivan *et al.* (1992). The four original milks, the two SCC artificial mixes and the two PMN artificial mixes were analysed for gross composition (Milkoscan 605; Foss Electric, DK-3400 Hillerød, Denmark), rennet coagulation characteristics (McMahon and Brown, 1982), N-fractions of total protein (International Dairy Federation, 1993) and casein (International Dairy Federation, 1964). The renneting characteristics of the milks determined were: rennet coagulation time in min (RCT), rate of curd aggregation in min (K20) and curd firmness at 60 min in mm of amplitude (A60).

Miniature cheeses were manufactured from each of the eight milks according to the method of Shakeel-Ur-Rehman *et al.*, 1998. Cheeses were analysed for compositional parameters 30 d after manufacture. Proteolysis of miniature cheeses was measured by urea-polyacrylamide gel electrophoresis after 30 d and 90 d of ripening.

Results

The effect of SCC and PMN level on gross composition, rennet coagulation properties and N fractions of high and low SCC and PMN milks together with the artificial mixes of approximately $200x10^3$/ml and $400x10^3$/ml SCC and PMN are shown in Table 1. The composition of miniature cheeses manufactured from these milks are also shown in Table 1. As SCC and PMN increased the fat and protein contents generally decreased ($p<0.05$). Analysis of variance indicated that RCT increased as both SCC and PMN increased ($p<0.01$) (correlation, R^2= 98% and R^2= 98%, respectively). A60 decreased as both SCC and PMN increased ($p<0.05$) (correlation, R^2= 93% and R^2= 93%, respectively). SCC and PMN had no significant effect on the K20 value and on the N-fractions of the original milks or on the artificial mixes made from the original milks. Analysis of variance indicated that the SCC and PMN levels investigated had no significant effect ($p>0.05$) on NaCl content and pH of miniature cheeses.

The Urea-PAGE electophoretogram (Figure 1) shows the effect of SCC and PMN level on individual casein fraction levels in miniature cheeses manufactured from low and high SCC and PMN milks, together with the artificial mixes, of approximate SCC $200x10^3$/ml and $400x10^3$/ml at 30 d (lanes 1-4 and lanes 9-12 [repeat]) and at 90 d (lanes 5-8) of ripening. There was a clear effect of milk SCC on proteolysis during ripening of the Cheddar type cheese; the patterns of proteolysis differed quantitatively and qualitatively with SCC. Comparing lanes 1-4, increasing

Table 1. Effect of SCC and PMN level on gross composition, rennet coagulation properties and N fractions of high and low SCC and PMN milks together with the artificial mixes of approximately 200x10³/ml and 400x10³/ml SCC and PMN and on composition of miniature cheeses manufactured from these milks.

	Low SCC	Mix 1 SCC	Mix 2 SCC	High SCC	Sig.	Low PMN	Mix 1 PMN	Mix 2 PMN	High PMN	Sig.
SCC (x10³ cells/ml)	6	236	488	920		4	380	545	850	
PMN (x10³ cells/ml)	<1	130	350	830		<1	205	430	700	
Gross composition										
Fat (g/100 g)	4.20	4.50	4.40	3.50	*	4.40	4.20	4.10	3.80	**
Protein (g/100 g)	3.80	3.60	3.50	3.20	*	3.40	3.40	3.40	3.30	*
Renneting properties										
RCT (min)	20.7	30.7	35.2	51.5	**	21.0	28.7	32.7	41.5	**
K20 (min)	6.5	10.2	12.5	17.5	ns	6.7	10.7	15.5	18.2	ns
A60 (mm)	48.0	40.0	34.0	10.0	*	46.0	44.5	32.5	17.0	*
N fractions										
Total protein (g/100 g)	2.00	3.66	2.75	3.30	ns	3.53	3.52	3.44	3.40	ns
Casein (g/100 g)	1.13	2.80	1.89	2.46	ns	2.70	2.72	2.56	2.48	ns
Composition of miniature cheeses										
NaCl (g/100 g)	1.51	1.31	1.66	1.48	ns	1.19	1.41	1.38	1.69	ns
pH	5.2	5.2	5.1	5.0	ns	5.2	5.1	5.0	5.1	ns

** = p<0.01; * = p<0.05; ns = p>0.05

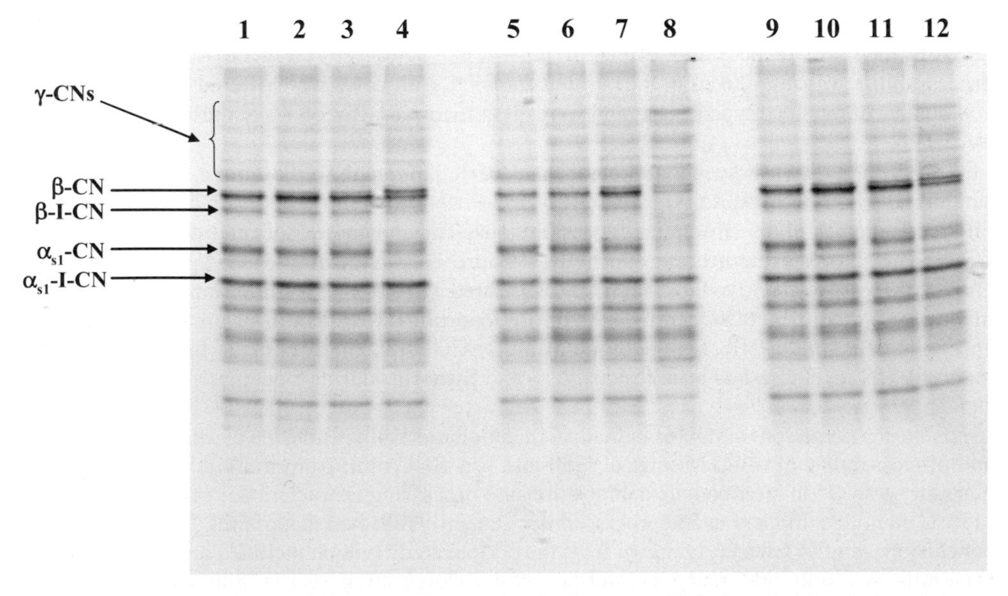

Figure 1. Effect of SCC and PMN level on individual casein (CN) fraction levels in miniature cheeses manufactured from low SCC, Mix 1 SCC, Mix 2 SCC and high SCC milks at 30 days (lanes 1-4, respectively) (lanes 9-12 [repeat]) and at 90 days (lanes 5-8, respectively) of ripening.

numbers of bands of slow electrophoretic mobility were apparent towards the top of the gel with progressively increasing SCC. These bands could be either products associated with high SCC milk which became entrapped in the curd during manufacture, or the products of proteolytic enzymes associated with the somatic cells acting during ripening. The same general trends could be seen in replicate trials (lanes 9-12). Comparison of the same samples after 90 d of ripening (lanes 5-8) supports the latter interpretation, as these bands had increased in level of intensity. In addition, after 90 d of ripening the cheese made from the milk of highest SCC (lane 8) had much less residual intact α_{s1}- and β-caseins, again indicating cell-associated proteinase activity. Hence, it appears that enzymes associated with the cells were retained in the cheese curd and contributed to the proteolysis during ripening, and even the addition of small volumes of high SCC milk had an obvious impact (lanes 10 and 11).

Discussion

Reports as to the effect of elevated milk SCC on protein content are conflicting and varied. The decline in milkfat content associated with elevated milk SCC is logical, given the reduced synthetic and secretory ability of the mammary gland (Auldist, 2000). Meanwhile Verdi *et al.* (1987) indicated associated extended rennet clotting time in cheese making. The results of the current study are in general agreement with these authors. It is difficult to associate any particular SCC level with the onset of specific defects in dairy products. Some researchers have reported that SCC begins to affect product as it increases above 100×10^3/ml (Barbano *et al.* 1991), while others have suggested that the threshold is closer to 500×10^3/ml (Politis and Ng-Kwai-Hang, 1988). Whether elevated bulk milk SCC is due to milk from a small number of cows with extremely high SCC being included with milk from a predominantly healthy herd, or to large numbers of cows with low-level sub-clinical infections, probably contributes to variation in the effects of SCC on dairy products. Additionally, some pathogens affect milk composition in different ways, irrespective of SCC level. It is also likely that nutritional status, stage of lactation or somatic cell type could affect the magnitude of the effects of mastitis on milk composition and dairy products. The potential effect of nutritional status and stage of lactation on milk processability was minimized in the current study since all cows were offered a similar quantity and quality of one feed type (grass) and were all at the mid-lactation stage during the trial.

Le Roux *et al.* (1995) showed that proteolysis occurred in quarter milk samples with SCC as low as 250×10^3/ml. The impact of elevated SCC and PMN milks on proteolysis during ripening of miniature cheeses in the current study was obvious. The enzymes associated with the cells were retained in the cheese curd and contributed to proteolysis during ripening. The patterns of proteolysis were different to those normally associated with cheese ripening, and even the addition of low amounts of high SCC milk had an obvious impact. Addition of high PMN milk had approximately similar effects to addition of high SCC milk. Cheeses made from milk with the highest PMN level also had least residual α_{s1}- and β-caseins during ripening.

In conclusion, variation in SCC of individual udder quarter milk results in variation in composition and processability of milk. Mixing of high and low SCC milk from individual udder quarters generally resulted in intermediate composition and processing characteristics of milk. The impact of SCC on milk composition and processability generally reflected that of milk PMN. While these conclusions may be tentatively drawn from the current study (which included one replicate within each milk SCC and PMN category), further work incorporating a larger number of replicates is required in order to strengthen the statistical power of the data and hence confidence in the resultant trends. An SCC standard of 400×10^3/ml for bulk milk is being adopted in milk quality schemes generally as a result of the European Union requirements which came into force in January 1998. This level of SCC should minimise the effects of mastitis on product quality, although evidence for potential deleterious effects on the quality of dairy products has been suggested by the results in this study.

References

Auldist, M.J., S. Coats, B. J. Sutherland, J.J. Mayes, G.H. McDowell and G.L. Rogers, 1996a. Effects of somatic cell count and stage of lactation on raw milk composition and the yield and quality of Cheddar cheese. J. Dairy Sci. 63: p. 269-280.

Auldist, M.J., S. Coats, B.J. Sutherland, J.J. Mayes, G.H. McDowell and G.L. Rogers, 1996b. Effects of somatic cell count and stage of lactation on the quality and storage life of ultra high temperature milk. J. Dairy Res. 63: p. 377-386.

Auldist, M.J. 2000. Effects of mastitis on raw milk and dairy products. In: Proceedings of Pacific congress on milk quality and mastitis control. 13-16 November, 2000. Nagano, Japan. p. 191-205.

Barbano, D.M., R.R. Rasmussen and J.M. Lynch, 1991. Influence of milk somatic cell count and milk age on cheese yield. J. Dairy Sci. 74: p. 369-388.

Concha, C., O. Holmberg and B. Morein, 1978. Proportion of β- and T-lymphocytes in normal bovine milk. J. Dairy Res. 45: p. 287-290.

International Dairy Federation 1993. Milk: Determination of Nitrogen Content. Brussels: IDF (FIL-IDF Standard no. 20B).

International Dairy Federation. 1964. Determination of the casein content of milk. IDF (FIL-IDF Standard no. 29).

Le Roux, Y., O. Colin and F. Laurent, 1995. Proteolysis in samples of quarter milk with varying somatic cell counts. (1) Comparison of some indicators of endogenous proteolysis in milk. J. Dairy Sci. 78: p. 1289-1297.

McMahon, D. J. and R.J. Brown, 1982. Evaluation of Formagraph for comparing rennet solutions. J. Dairy Sci. 65: p. 1639-1642.

O'Sullivan, C.A., P.J. Joyce, T. Sloan and A.G. Shattock, 1992. Capture Immunoassay for the diagnosis of bovine mastitis using a monoclonal antibody to polymorphonuclear granulocytes. J. Dairy Res. 59: p. 123-131.

Politis, I. and K.F. Ng-Kwai-Hang, 1988. Effect of somatic cell counts and milk composition on the coagulating properties of milk. J. Dairy Res. 71: p. 1740-1746.

Shakeel-Ur-Rehman, P.L.H. McSweeney and P.F Fox, 1998. Protocol for the manufacture of miniature cheeses. Lait, 78: p. 607-620.

Verdi, R.J., D.M. Barbano, M.E., Dellavalle and G.F. Senyk, 1987. Variability in true protein, casein, non-protein nitrogen and proteolysis in high and low somatic cell milks. J. Dairy Sci. 70: p. 230-242.

Relationship between soluble carbohydrate/nitrogen ratio in grass and milk fatty acid composition in cows: role of ruminal metabolism

M. Doreau[1], J.L. Peyraud[2] and D. Rearte[2,3]*
[1]INRA-Unité de Recherches sur les Herbivores, Theix, 63122 Saint-Genès-Champanelle, France,[2]INRA Unité Mixte de Recherches sur la Production Laitière, 35590 Saint-Gilles, France, and [3]EEA INTA 7620 Balcarce, Argentina.
**Correspondence: doreau@clermont.inra.fr*

Summary

The effect of grass composition on milk fatty acid (FA) composition has been studied. Perennial ryegrass was offered to 4 dairy cows with duodenal cannulae. Grass differed by the ratio between soluble carbohydrate (S) and N, high ratio (S+N-) or low ratio (S-N+); differences were due to level of N fertilisation, days of regrowth and hour of cutting. The S+N- grass contained less FA, of which less linolenic acid, than S-N+ (1.82 *vs.* 2.49%, and 60.9 *vs.* 65.6%, respectively). Milk FA was lower in linolenic acid, vaccenic acid and rumenic acid for S+N- than for S-N+ (0.67 *vs.* 0.91%, 3.27 *vs.* 4.92%, 1.58 *vs.* 2.10% for these 3 FA, respectively). Differences in grass FA intake resulted in differences in duodenal FA flow, but only in moderate differences in duodenal FA composition (on average 1.51%, 11.31% and 0.05% for linolenic, vaccenic and rumenic acids, respectively). The percentages of these 3 FA in milk fat was related to their percentage at the duodenum (r = 0.47 to 0.79).

Keywords: milk, polyunsaturated fatty acids, CLA, grass, rumen

Introduction

Improving the nutritional quality of milk fatty acids (FA) is now a frequent demand of consumers. Milk and dairy products are rich in saturated FA, of which some are considered as atherogenic. Increasing the concentration of FA beneficial for human health may restore consumers' confidence in milk. Polyunsaturated FA (PUFA) of the *n-3* series (ω3) are recognised as reducing the prevalence of heart diseases, and conjugated linoleic acids (CLA), especially the *cis*-9, *trans*-11 isomer (rumenic acid), may prevent carcinogenesis (Williams, 2000). Milk fat generally contains less than 1% ω3 FA, and less than 1% CLA. A way to increase these FA is to increase PUFA content in cow diets. Forages, especially grass, are rich in linolenic acid (18:3 *cis*-9, *cis*-12, *cis*-15). Despite extensive hydrogenation in the rumen, these feeds may contribute to increase ω3 and CLA intake in humans (Chilliard *et al.*, 2001).

Grass feeding increases the concentration of ω3 and CLA in milk especially at early stage, probably due to a higher PUFA content in young grass (Dewhurst *et al.*, 2001). However, the direct effects of chemical composition have not been studied. Grazing management widely modifies both nitrogen (N) and soluble carbohydrates (S) in grass. The present experiment aimed to analyse the consequences of variations of S / N ratio in perennial ryegrass. This ratio may increase by decreasing N fertilisation, by increasing the age of regrowths, and is higher when grass is cut in the evening rather than in the morning (Delagarde *et al.*, 2000). These 3 factors, which are expected to be additive, have been associated in this experiment. The consequences on grass and milk FA composition, and the role of the rumen in FA biohydrogenation have been studied.

Material and methods

Experimental design

The experiment was carried out at the INRA Research Centre of Rennes (France) during 3 successive periods, with 2 groups of 2 mid-lactation (180 days in milk) Holstein cows fitted with ruminal and duodenal cannulae, fed perennial ryegrass indoors without concentrate. The two grass managements were 1) N fertilisation of 80 kg/ha, cutting at 21 days of regrowth, at 09.00 h (S-N+), and 2) N fertilisation of 40 kg/ha, cutting at 28 days of regrowth, at 16.30 h (S+N-). The first group of cows was allocated to S+N- in periods 1 and 3, and to S-N+ in period 2, the second group was allocated to S-N+ in periods 1 and 3, and to S+N- in period 2. Each period lasted 10 days: 5 for adaptation and 5 for measurements. An unique sward was divided into 3 plots, one for each period, and each plot was divided into two subplots, one per management. Pasture was managed so that regrowths of 21 and 28 days corresponded to the middle of experimental periods. Animals were fed at 90% of ad libitum intake, measured in a preliminary period. Grass was given 3 times per day, at 07.30 h, 13.30 h and 18.30 h, in the proportions of 40, 20 and 40% of daily offer, respectively. Grass distribution to animals was managed so that for both groups the maximum duration of storage did not exceed 22 h, and that the mean time between harvesting and offering grass to cows was on average 12 h.

Measurements and analyses

Duodenal flows of dry matter have been determined by the double marker method as specified by Rearte *et al.* (2003). Fatty acids in forages and duodenal digesta were methylated according to Sukhija and Palmquist (1988) with modifications. Analysis was performed by gas liquid chromatography using a 100-m CPSil88 capillary column. Details about the different steps of FA extraction and analysis are given in Loor *et al.* (2004). Fatty acids in milk have been extracted then transesterified with butanol/HCl for butyl esters and with methanol/NaOH and methanol/BF3 for methyl esters. Analyses were made by gas liquid chromatography on butyl esters for all FA with a 25-m OV-1 capillary column and on methyl esters for a more efficient separation of 18-carbon FA with a 100-m SP2560 capillary column.

Results of chemical composition of grass were analysed by a one-way analysis of variance, considering periods as replicates. After a preliminary analysis which showed the absence of interaction between type of grass and period, all other data were analysed by a three-way analysis of variance, with type of grass (1 df), period (2 df) and cow (3 df) as main factors. The GLM procedure of SAS was used.

Results and discussion

Grass fatty acid composition and intake

As expected, the differences in crude protein (CP) and water soluble carbohydrates (WSC) were large and significant (Table 1). The S+N- grass contained less FA than the S-N+ grass, contrary to results of Scollan *et al.* (2003) who compared two varieties of perennial ryegrass, with 16.1 and 23.4% S, but with the same N content, and found a higher FA content with the grass rich in S. On another hand, Miller *et al.* (2001) found no difference in FA content between two perennial ryegrass species varying in S. In the present experiment, FA content may depend on the number of days of regrowth. This effect has been confirmed by Dewhurst *et al.* (2001). The S+N- grass contained less linolenic acid than S-N+ grass (60.9 vs. 65.6%), and more saturated FA from 14 to 24 carbons, the percentage of other FA being unchanged. As DM intake did not vary between

Table 1. Grass dry matter (DM) content and composition.

	DM (%)	CP (% DM)	WSC (% DM)	FA (% DM)	NDF (% DM)
S-N+	17.4	20.7	13.7	2.49	47.4
S+N-	22.8	12.5	24.6	1.82	46.4

managements (15.7 kg on average), the same differences were observed for FA intake: 287 and 388 g/d for S+N- and S-N+, respectively.

Milk fatty acids

Milk yield averaged 22.2 and 22.5 kg/d, and milk fat was reduced (37.4 *vs.* 38.8 g/d, $P < 0.05$) for S+N- compared to S-N+, respectively. Increasing S resulted in an increase in short- and medium-chain FA, including palmitic acid, and in a decrease in stearic, vaccenic and linolenic acid, and in a trend ($P=0.054$) to a decrease in rumenic acid (Table 2). Fatty acids expected to be beneficial to human health are thus decreased. On average, linolenic acid concentration was low compared to other data obtained at pasture (Chilliard *et al.*, 2001). On the contrary, the CLA concentration was higher than in most experiments at pasture. However, similar concentrations have been found by Lawless *et al.* (1998). These latter authors and Rego *et al.* (2004) even observed CLA concentrations reaching 2.5% of FA in individual cows at pasture.

Table 2. Fatty acid composition of milk fat (% of total FA).

	S+N-	S-N+	SEM	*P*
Short chain fatty acids, (<14)	13.48	11.57	0.140	**
C14:0 (myristic)	12.29	11.17	0.129	**
C16:0 (palmitic)	30.98	27.50	0.617	**
C18:0 (stearic)	8.83	11.27	0.349	**
C18:1 *cis*-9 (oleic)	18.28	20.64	0.731	NS
C18:1 *trans*-11 (vaccenic)	3.27	4.92	0.22	**
C18-2 *cis*-9, *cis*-12 (linoleic)	1.09	1.18	0.078	NS
CLA *cis*-9, *trans*-11 (rumenic)	1.58	2.10	0.144	*
C18-3 *cis*-9, *cis*-12, *cis*-15 (linolenic)	0.67	0.91	0.033	**

NS: non significant ($P > 0.05$)

Fatty acid metabolism in the rumen

Fatty acid duodenal flow was lower ($P<0.05$) for S+N- than for S-N+ (391 *vs.* 513 g/d), and exceeded FA intake by 37 and 31%, respectively. This net synthesis corresponds to the general trend (Doreau and Ferlay, 1994) but is not always observed in grass diets, which sometimes result in FA net losses. In particular, Scollan *et al.* (2003) found a net synthesis with a normal ryegrass and a net loss with a ryegrass rich in S. Duodenal FA composition (Table 3) showed an extensive ruminal hydrogenation (50.33% stearic acid in duodenal FA *vs.* 1.51% in grass), a high proportion of C18:1 *trans*, and a very low proportion of rumenic acid. Although rumenic acid is the main CLA isomer in milk, it represents only 14 % of peaks present in the CLA area in duodenal digesta, where CLA *trans*-11, *trans*-13 is the predominant isomer. These results confirm recent data obtained with forage diets (Doreau *et al.*, 2003, Loor *et al.*, 2004). Ruminal hydrogenation of

Table 3. Fatty acid composition of duodenal contents (% of total FA).

	S+N-	S-N+	SEM	*P*
C14:0 (myristic)	0.74	0.69	0.07	**
C16:0 (palmitic)	13.10	13.27	0.16	NS
C18:0 (stearic)	49.88	50.77	0.76	NS
C18 :1 *cis*-9 (oleic)	2.30	2.24	0.09	NS
C18:1 *trans*-11 (vaccenic)	10.65	11.97	0.38	NS
C18:2 *cis*-9, *cis*-12 (linoleic)	1.40	1.27	0.02	**
CLA *cis*-9, *trans*-11 (rumenic)	0.05	0.06	0.01	NS
C18:3 *cis*-9,*cis*-12, *cis*-15 (linolenic)	1.58	1.73	0.08	NS

NS: non significant ($P > 0.05$)

PUFA, defined as the percentage of disappearance of these FA between mouth and duodenum, was 85.6 and 86.4% for linoleic acid, and 96.4% and 96.6% for linolenic acid, for S+N- and S-N+, respectively (P>0.05). These values are in the range of data obtained in forages (Doreau *et al.*, 2003), although lower values have been found for linoleic acid of fresh perennial ryegrass (80%, Scollan *et al.*, 2003) or linolenic acid of red clover silage (86%, Lee *et al.*, 2003).

Relationship between grass, duodenal and milk fatty acids

The differences in linolenic acid concentration of pastures (11.1 vs. 16.3 g/kg in DM for S+N- and S-N+) resulted in differences of the same extent in milk linolenic acid, but also in milk CLA and 18:1 *trans*, which are intermediates of PUFA hydrogenation.

When individual data are considered (4 cows, 3 periods) a close relationship was shown between milk and duodenal percentages of total FA for linolenic acid, but not for linoleic acid (Table 4). Vaccenic acid in milk can be better predicted than rumenic acid by duodenal precursor FA. The ratio between rumenic acid and vaccenic acid is extremely low at the duodenum, confirming that almost all rumenic acid synthesis is post-ruminal. It can be calculated that about 30% of vaccenic acid taken up by the mammary gland is desaturated into rumenic acid.

Conclusions

In this experiment, increasing S/N ratio decreased grass FA and especially linolenic acid. This resulted in a decrease in linolenic, vaccenic and rumenic acids in milk. Their percentage in milk can be explained both by grass FA and by duodenal FA composition. Despite the extensive ruminal FA metabolism, grass characteristics can influence milk FA composition. Other studies on grass are necessary to predict or to manipulate the amount of beneficial FA in milk.

Table 4. Correlations between duodenal FA and milk FA composition (% of total FA).

Milk fatty acids (%)	Duodenal fatty acids (%)	r
linolenic	linolenic	0.79
linoleic	linoleic	0.00
vaccenic	vaccenic	0.66
vaccenic + rumenic	vaccenic + rumenic	0.62
rumenic	vaccenic + rumenic	0.48

Acknowledgements

This experiment has been partly granted by the EU (project QLK1-CT-2000-01423 "HealthyBeef").

References

Chilliard, Y., A. Ferlay and M. Doreau, 2001. Effect of different types of forages, animal fat or marine oils in cow's diet on milk fat secretion and composition, especially conjugated linoleic acid (CLA) and polyunsaturated fatty acids. Livest. Prod. Sci. 70, p. 31-48.

Delagarde, R., J.L. Peyraud, L. Delaby and P. Faverdin, 2000. Vertical distribution of biomass, chemical composition and pepsin-cellulase digestibility in a perennial ryegrass sward: interaction with month of year, regrowth age and time of day. Anim. Feed Sci. Technol. 84: p. 49-68.

Dewhurst, R.J., N.D. Scollan, S.J. Youell, J.K.S. Tweed and M.O. Humphreys, 2001. Influence of species, cutting date and cutting interval on the fatty acid composition of grasses. Grass For. Sci. 56: p. 68-74.

Doreau, M. and A. Ferlay, 1994. Digestion and utilisation of fatty acids by ruminants. Anim. Feed Sci. Technol. 45: p. 379-396.

Doreau, M., K. Ueda and C. Poncet, 2003. Fatty acid ruminal metabolism and intestinal digestibility in sheep fed ryegrass silage and hay. Trop. Subtrop. Agroec. 3: p. 289-293.

Lawless, F., J.J. Murphy, D. Harrington, R. Devery and C. Stanton, 1998. Elevation of conjugated cis-9, trans-11-octadecadienoic acid in bovine milk because of dietary supplementation. J. Dairy Sci. 81: p. 3259-3267.

Lee, M.R.F., L.J. Harris, R.J. Dewhurst, R.J. Merry and N.D. Scollan, 2003. The effect of clover silages on long chain fatty acid rumen transformations and digestion in beef steers. Anim. Sci. 76: p. 491-501.

Loor, J.J., K. Ueda, A. Ferlay, Y. Chilliard and M. Doreau, 2004. Biohydrogenation, duodenal flow, and intestinal digestibility of *trans* fatty acids and conjugated linoleic acids in response to dietary forage:concentrate ratio and linseed oil in dairy cows. J. Dairy Sci. 87: p. 2472-2485.

Miller, L.A., J.M. Moorby, D.R. Davis, M.O. Humphreys, N.D. Scollan, J.C. MacRae and M.K. Theodorou, 2001. Increased concentration of water-soluble carbohydrate in perennial ryegrass (Lolium perenne L.): milk production from late-lactation dairy cows. Grass For. Sci. 56: p. 383-394.

Rearte, D.H., J.L. Peyraud and C. Poncet, 2003. Increasing the water soluble carbohydrate / protein ratio of temperate pasture affects the ruminal digestion of energy and protein in dairy cows. Trop. Subtrop. Agroec. 3: p. 251-254.

Rego, O.A., P.V. Portugal, M.B. Sousa, H.J.D. Rosa, C.M. Vouzela, A.E.S. Borba and R.J.B. Bessa, 2004. Effect of diet on the fatty acid pattern of milk from dairy cows. Anim. Res. 53: p. 213-220.

Scollan, N.D., M.R.F. Lee and M. Enser, 2003. Biohydrogenation and digestion of long-chain fatty acids in steers fed on Lolium perenne bred for elevated levels of water-soluble carbohydrate. Anim. Res. 52: p. 501-511.

Sukhija, P.S. and D.L. Palmquist, 1988. Rapid method for determination of total fatty acid content and composition of feedstuffs and feces. J. Agric. Food Chem. 36: p. 1202-1206.

Williams, C.M., 2000. Dietary fatty acids and human health. Ann. Zootech. 49: p. 165-180.

CLA content and n-3/n-6 ratio in dairy milk as affected by farm size and management

L. Bailoni, G. Prevedello, S. Schiavon, R. Mantovani and G. Bittante*
Department of Animal Science, University of Padova, Agripolis, Viale dell'Università 16, 35020 Legnaro (PD), Italy.
**Correspondence: lucia.bailoni@unipd.it*

Summary

Conjugated linoleic acid (CLA) and fatty acids from the n-3 series have been related to several beneficial effects on human and animal health. The main objectives of this study were to determine fatty acid profile, and particularly CLA content and the n-3/n-6 ratio, in milk samples collected in about 250 dairy farms located in Veneto region (north-eastern Italy) and to identify possible relationships between production levels and/or hygienic conditions of dairy farms and CLA and n-3/n-6 ratio in milk. The farms belonged to 21 classes according to 7 milk production levels (PL: <20, 20-40, 40-60, 60-100, 100-150, 150-300, and >300 t/year) and to 3 hygienic conditions of milk (HC: <20%, 20-50%, and >50% of test day milk records with somatic cell count higher than 400.000/ml and/or with total bacterial count higher than 100.000/ml). Fatty acids of milk were analysed by gas chromatographic method as the methyl ester derivatives. Mean and standard deviation for CLA and n-3/n-6 ratio were: 0.467 ± 0.123 % on the total fatty acids and 0.336 ± 0.156, respectively. Both parameters were affected significantly (P<0.001) by milk production level. CLA content and n-3/n-6 ratio decreased linearly (P<0.001) from the lowest-producing farms (on average 0.552 % on the total fatty acids and 0.591, respectively) to highest-producing farms (on average 0.415 % on the total fatty acids and 0.194, respectively). Effects of hygienic condition of milk and PL x HC interaction were not statistically significant (P>0.10).

Keywords: milk, fatty acid profile, n-3/n-6 ratio, CLA

Introduction

The interest of nutritionists for the polyunsaturated n-3 and n-6 fatty acids (FA) and conjugated linoleic acid (CLA) has recently increased due to their putative effect on human health and well-being (Mc Guire and Mc Guire, 1999; Simopoulos, 2002).

Particularly, FA from n-3 and n-6 series are precursors of eicosanoids with opposite effects on human health status. Many chronic diseases (cardiovascular diseases, several cases of diabetes, cancers and obesity, rheumatoid arthritis, autoimmune diseases, asthma and depression) seems associated to a greater amount of FA from the n-6 series and a decrease of those from the n-3 series in the diet (Simopoulos, 2002). Due to the ruminal biohydrogenation the level of polyunsaturated FA (both n-6 and n-3 series) in milk and dairy products is very low but some variation of n-3/n-6 ratio are likely by dietary changes (Cant *et al.*, 1997). Optimum n-3 and n-6 ratio varies from 1:1 to 4:1 according to specific diseases.

CLA is a group of positional and geometric isomers of linoleic acid (Parodi, 1994). The most active isomer in nature is *cis*-9 *trans*-11 (Pariza *et al.*, 2000; Chouinard *et al.*, 1999). Several in vivo and in vitro studies (McGuire and Mc Guire, 1999; Lock and Garnsworthy, 2002) put in evidence, besides anti-carcinogenic activity of CLA, also anti-atherogenic, anti-diabetes, and immune-stimulating properties. It is therefore necessary to upgrade the fatty acid profile of foods of animal origin and, in particular, of milk and dairy products. This objective may be achieved largely through a nutritional approach. For example, CLA content could be increased in milk obtained by cows consuming pasture or receiving diets supplemented with full fat seeds, vegetable oils, calcium soaps etc. (Chilliard *et al.*, 2002; Antongiovanni *et al.*, 2003).

Few investigation are performed to evaluate the factors affecting milk fatty acid profile in practical conditions. The objective of this research was to identify possible relationships between production levels (annual milk production) and/or hygienic conditions (somatic cell and bacterial counts) in about 250 dairy farms located in Veneto region (north-eastern Italy) and characteristics of milk fatty acid profile, in particular n-3/n-6 ratio and CLA content.

Material and methods

In this study 242 dairy farms were selected from a database of 3654 farms located in Veneto associated to A.Pro.La.V (Regional Association of Milk Producers). Farms have been identified to realize an homogeneous distribution according to 7 milk production levels (PL: <20, 20-40, 40-60, 60-100, 100-150, 150-300, and >300 t/year) and to 3 hygienic conditions of milk (HQ: high quality; MQ: medium quality; LQ: low quality). Data of quantity and quality parameters are biweekly collected by associations of milk breeders and dairy industries. To this regard dairy farms were classified into three classes: <20%, 20-50%, and >50% of test day milk records with somatic cell count (SCC) higher than 400.000/ml and/or with total bacterial count (TBC) higher than 100.000/ml. Relationship between these two parameters (SCC and TBC) defines the three quality classes as shown in Table 1.

Table 1. Classes of milk hygienic condition based on the % of test day milk records over the defined threshold for both somatic cell count (SCC) and total bacterial count (TBC).

	SCC (> 400.000)		
	< 20 %	20-50 %	> 50 %
TBC (>100.000):			
- < 20 %	1	2	3
- 20-50 %:	2	2	3
- > 50%	3	3	3

Table 2 shows the distribution of dairy farms with different production levels and different quality classes. Milk samples were collected during the first six months of the year from dairy farms' milk tank (morning and previous evening milking). FA were extracted (Chouinard *et al.*, 1999), trans-esterificated and finally identified by gas-chromatographic method as the methyl ester derivatives by using gas-chromatography (GC 8000, Top Series, Termoquest Italia, Milano, Italy). FA methyl esters were separated through a capillary column (SP-2560, 100 m x 0.25 mm, Supelco, Bellefonte, PA, USA). Helium was used as the carrier gas with flow of 1.40 ml/min. The oven temperature was programmed from 140 to 240°C at 4°C/min. Injector and detector were kept at 250°C and 260°C respectively.

All data were analysed by ANOVA according to SAS/STAT (1990) and considering milk production level (PL), milk hygienic condition (HC) and interaction PL x HC effects. Linear and quadratic components were also tested by using the mean squares of PL effect.

Table 2. Distribution of dairy farms sampled

	Milk hygienic condition			
	HQ	MQ	LQ	Total
Milk production levels (ton/year):				
<20	13	8	9	30
20-40	16	17	12	45
40-60	14	8	8	30
60-100	13	13	7	33
100-150	11	12	4	27
150-300	20	12	3	35
>300	27	14	1	42
Total	114	84	44	242

Information on animal characteristics, feeding strategies and management system of each dairy farm were also recorded using suitable assessment forms. These data are now subjected to statistic analysis and then will be discussed in a following report.

Results

Table 3 shows results of ANOVA for milk FA profile. Almost all variables analysed were affected significantly (P<0.001) by milk production level, with a significant linear component for all variables, indicating a strong relationship between production level and milk contents of polyunsaturated FA (PUFA), total n-3 FA, total n-6 FA, n-3/n-6 ratio and CLA). Effects of hygienic condition of milk and PL x HC interaction were not statistically significant (P>0.10).

Table 3. Results of ANOVA for milk FA profile.

	Milk production levels (PL)			F value of milk quality classes (HC)	F value of PL x HC interaction	Error mean squares
	F value	Linear . comp	Quadratic comp			
SFA	1.25	**	n.s.	0.95	0.19	9.09
MUFA	0.69	*	n.s.	0.54	0.29	6.70
PUFA	6.30***	***	***	0.71	1.11	0.27
Saturated/un-saturated FA	1.70	**	n.s.	0.91	0.26	0.12
Total n-3	13.27***	***	***	0.88	1.57	0.03
Total n-6	14.67***	***	n.s.	0.48	0.90	0.20
n-3/n-6	20.76***	***	**	2.19	1.34	0.02
CLA	3.93***	***	n.s.	0.64	0.59	0.01

SFA: saturated FA; MUFA: mono-unsaturated FA; PUFA: poly-unsaturated FA
n.s.: not significant; *: P<0.05; ** P<0.01; *** P<0.001

With the increase of production level (PL), saturated FA decreased significantly (linear component: P<0.01) from 71.7 to 69.4 g/100 g of the total FA (Tables 3 and 4). The polyunsaturated FA content in milk was affected significantly (P<0.001) by production level of dairy farm with a tendency to increase (from 3.1 to 3.9 g/100 g of the total FA, quadratic component: P<0.001) particularly for the highest producing classes (150-300 and >300 ton/year).

In this study the mean and standard deviation for n-3/n-6 ratio in dairy milk was 0.336 ± 0.156. The values are in agreement with those of our previous experiments (Bailoni *et al.*, 2004) and in general with literature (Hepburn *et al.*, 1986). Figure 1 shows that n-3/n-6 ratio decreased significantly (linear component: P<0.001) from 0.591 of the lowest-producing (<20 ton/year) to 0.194 of the highest-producing farms (more than 300 ton/year).

The mean and standard deviation for CLA content in milk fat from the analysed dairy farms was 0.467 ± 0.123 g/100 g of the total FA, according to data reported in literature (Chilliard *et al.*, 2002, Antongiovanni *et al.*, 2003, Bailoni *et al.*, 2004). Also in this case the effect of milk production level was highly significant (P<0.001) with a linear decrease (P<0.001) from the lowest to the highest-producing levels (0.552 and 0.415 g/100 g on the total FA respectively).

Table 4. Least square means of fatty acids (g/100 g of the total fatty acid) according to different production level.

	Saturated FA	Mono-unsaturated FA	Poly-unsaturated FA	Saturated/unsaturated FA
Milk production levels (ton/year):				
<20	71,72	25,15	3,12	2,54
20-40	71,42	25,51	3,06	2,50
40-60	71,35	25,56	3,10	2,49
60-100	71,30	25,54	3,16	2,48
100-150	71,01	25,81	3,17	2,45
150-300	70,48	26,01	3,51	2,39
>300	69,38	26,71	3,91	2,27

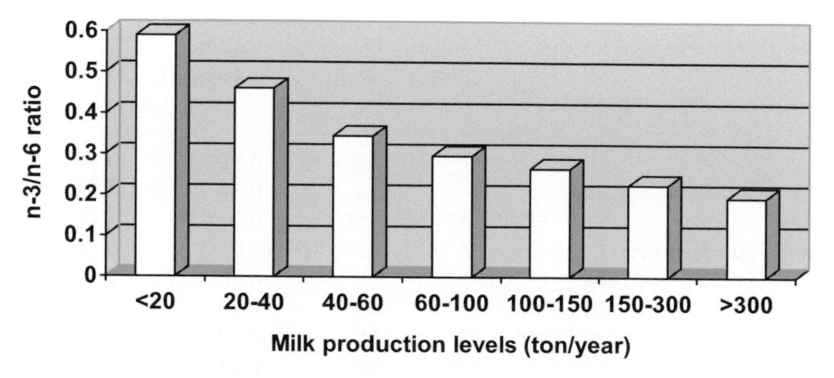

Figure 1. Effect of milk production level on milk n-3/n-6 ratio.

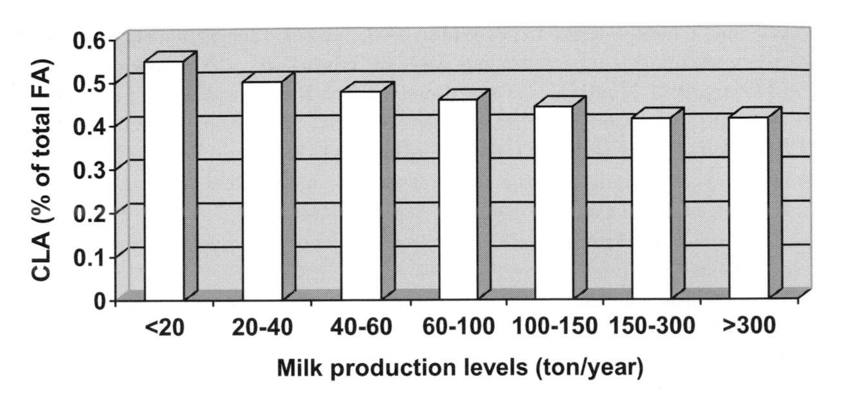

Figure 2. Effect of milk production level on CLA content (% on the total FA).

Discussion

Although it is well know that feeding strategies can greatly change the CLA content in and n-3/n-6 FA ratio (Chilliard *et al.*, 2000), It would be interesting to study the effects of different feeding strategies (use of pasture, forage:concentrate ratio, separate distribution of forages and concentrate, total mixed ration, fat and oil supplementation etc.) adopted in practical conditions by dairy farms on the FA profile of milk. On the basis of our survey among farms located in the Veneto region we have found that different milk production levels (PL) are associated with changes in farm feeding and management systems. Indeed, the small dairy farms are usually located in the mountain areas and cows are fed pasture or mixed hay with the manual distribution of low amounts of concentrates. On the other hand large farms in the flat land are suitable for intensive dairy system based on complete mixed diets and large use of maize silage and concentrates (www.aia.it).

In conclusion, preliminary results of this research showed significant relationships between milk production levels and milk n-3/n-6 ratio as well as conjugated linoleic acid (CLA) content. The better FA profile of the lowest producing dairy farms should be considered in order to increase the value of milk obtained from the human health point of view. On the contrary, FA profile seems not related to somatic cell count and total bacterial count of milk, and, more generally, to hygienic condition in the farm.

Acknowledgements

This research was supported by "Regione Veneto" and "Associazione Produttori Latte del Veneto".

References

Antongiovanni, M., A. Buccioni, F. Petacchi, P. Secchiari, M. Mele and A. Serra, 2003. Upgrading the lipid fraction of foods of animal origin by dietary means: rumen activity and presence of trans fatty acids and CLA in milk and meat. Ital. J. Anim. Sci. 2: p. 3-28.

Bailoni, L., A. Bortolozzo, R. Mantovani, A. Simonetto, S. Schiavon and G. Bittante, 2004. Feeding dairy cows with full fat extruded or toasted soybean seeds as replacement of soybean meal and effects on milk yield, fatty acid profile and CLA content. Ital. J. Anim. Sci. 3, p. 243-258.

Cant, J.P., A.H. Fredeen, T. MacIntyre, J. Gunn and N. Crowe, 1997. Effect of fish oil and monensin on milk composition in dairy cows. Can. J. Anim. Sci. 77: p. 125-131.

Chilliard, Y., A. Ferlay, J. Loor, J. Rouel and B. Martin, 2002. Trans and conjugated fatty acids in milk from cows and goats consuming pasture or receiving vegetable oils or seeds. Ital. J. Anim. Sci. 1: p. 243-254.

Chilliard, Y., A. Ferlay, R.M. Mansbridge and M. Doreau, 2000. Ruminant milk fat plasticity: nutritional control of saturated, polyunsaturated, trans and conjugated fatty acids. Ann. Zootech. 49(3): p. 181-205.

Chouinard, P.Y., L. Corneu, D.M. Barbano, L.E. Metzeger and D.E. Bauman, 1999. Conjugated linoleic acid alter milk fatty acid composition and inhibit milk fat secretion in dairy cows. J. Nutr. 129: p. 1579-1584.

http://www.aia.it (Italian Official Control System for Animal Products)

Hepburn, F.N., J. Exler and J.L. Weihrauch, 1986. Provisional tables on the content of omega-3 fatty acids and other fat components of selected foods. J. Am. Diet. Ass. 86: p. 788-793.

Lock, A.L. and P.C. Garnsworthy, 2002. Independent effects of dietary linoleic and linolenic fatty acids on the conjugated linoleic acid content of cows' milk. Anim. Sci. 74: p. 163-176.

Mc Guire, M.A. and M.K. Mc Guire, 1999. Conjugated linoleic acid (CLA): a ruminant fatty acid with beneficial effects on human health. J. Anim. Sci. 77(suppl. 1): p. 118-125.

Pariza, M.W., Y.H. Park and M.E. Cook, 2000. Mechanisms of action of conjugated linoleic acid: evidence and speculation. Proceedings of the Society of Experimental Biology and Medicine. 223 (1): p. 8-13.

Parodi, P.W., 1994. Conjugated linoleic acid: an anticarcinogenic fatty acid present in milk fat. Australian J. Dairy Tech. 49(2): p. 93-97.

Simopoulos, A.P,. 2002. The importance of the ratio of omega-6/omega-3 essential fatty acids. Biomedicine and Pharmacotherapy. 56(8): p. 365-379.

SAS-STAT, 1990. SAS User's guide. Statistical Analysis Institute, inc. Cary, N.C.

Fatty acid composition and CLA content of milk fat from Italian Buffalo

P. Secchiari[1], G. Campanile[2], M. Mele[1], F. Zicarelli[2], A. Serra[1], M. Del Viva and L. Amante[2]*
[1]DAGA, Sezione Scienze Zootecniche, Università di Pisa, via del Borghetto, 80, Pisa. Italy,
[2]Dipartimento di Scienze Zootecniche e Ispezione degli Alimenti, Università di Napoli, via Delpino 1, Napoli, Italy.
**Correspondence: psecchia@agr.unipi.it*

Summary

Aim of the present work was to characterize the fatty acid composition and CLA content of milk from buffalo fed hay or fresh forage. Milk samples from 88 lactating buffalo belonging to seven herd located in South Italy were collected. Two samplings have been performed: the first when animals fed a ration included green forage and the second when fresh forage was substituted by hay in a total mixed ration (TMR) diet. Milk samples were analysed for fatty acid composition. Results showed that when buffalo cows are fed fresh forage percentages of medium-chain fatty acids and saturated fatty acids significantly decreased in milk fat, while that of monounsaturated fatty acids and polyunsaturated fatty acids increased. The conjugated linoleic acid average content (CLA) of milk was enhanced by the inclusion of fresh forage in the TMR diet (0.66 vs. 0.46), in a similar way to what reported for dairy cattle. A wide individual variation of milk CLA content may be observed, when animals were submitted to a similar dietary regimen. The slope of the relationship between rumenic acid (RA, *cis*9, *trans*11 CLA) and vaccenic acid (VA, *trans*11, 18:1) was comparable to that reported for dairy cattle (RA = 0.12 + 0.23 (VA+RA); $p< 0.01$; $R^2 = 0.89$).

Keywords: buffalo, milk, CLA, fatty acids

Introduction

In Italy, buffalo milk is transformed in Mozzarella cheese, a PDO (Protected Designation of Origin) product, which represents a very important voice in the economy of two regions (Campania and Lazio) located in the south Italy. In the last years, the consumer has shown more attention to traits related to food safety, health and nutritional value.

Conjugated linoleic acid (CLA) occurs naturally in dairy products and other foods derived from ruminant animals (Chin, S. F. *et al.*, 1992). Moreover CLA remains stable in processed dairy products (Lin *et al.*, 1995), and the content in such products is a function of the concentration of CLA in raw milk.

Pasture intake can increase the levels of CLA in cow's milk (Kelly *et al.*, 1998; Dhiman *et al.*, 1999). Bergamo *et al.* (2003), reported that buffalo milk fat from organic farms contained significantly higher CLA and vaccenic acid (VA) amounts than did milk from conventional farm where pasture is not included in the diet. Similar results are reported also for dairy ewes and goats fed on pasture (Nudda *et al.*, 2003).

The aim of this trial was to evaluate the effect of different forage in a TMR diet (hay vs fresh grass) on fatty acid composition and CLA content in the milk from Italian Mediterranean buffalo cows.

Table 3. Fatty acid composition of milk from buffalo fed diets with fresh forage or hay (g/100 g fat).

	Fresh forage	Hay	SE	P
$C_{4:0}$	5.08	5.22	0.06	NS
$C_{6:0}$	2.52	2.58	0.04	NS
$C_{8:0}$	1.38	1.37	0.04	NS
$C_{10:0}$	1.52	1.56	0.05	NS
$C_{12:0}$	2.05	2.12	0.062	NS
$C_{14:0}$	9.1	10.16	0.206	**
$C_{14:0}$ iso	0.16	0.18	0.004	*
$C_{14:1}$ *cis* 9	0.54	0.61	0.021	*
$C_{15:0}$	1.08	1.21	0.022	**
$C_{15:0}$ anteiso	0.54	0.56	0.012	NS
$C_{16:0}$	26.77	29.58	0.522	**
$C_{16:0}$ iso	0.33	0.39	0.008	**
$C_{16:1}$ *cis* 9	1.67	1.93	0.049	**
$C_{17:0}$	0.49	0.51	0.011	NS
$C_{17:0}$ anteiso	0.33	0.39	0.013	*
$C_{18:0}$	10.7	9.99	0.349	NS
$C_{18:1}$ *trans* 6 - *trans* 8	0.14	0.08	0.009	**
$C_{18:1}$ *trans* 9	0.27	0.21	0.012	**
$C_{18:1}$ *trans* 10	0.32	0.18	0.021	**
$C_{18:1}$ *trans* 11	1.84	1.08	0.098	**
$C_{18:1}$ *trans* 12	0.24	0.13	0.017	**
$C_{18:1}$ *cis* 7	0.21	0.14	0.013	**
$C_{18:1}$ *cis* 9	17.81	17.44	0.332	NS
$C_{18:1}$ *cis* 11	0.35	0.35	0.008	NS
$C_{18:1}$ *cis* 12	0.27	0.16	0.012	**
$C_{18:1}$ *cis* 13	0.04	0.03	0.001	NS
$C_{18:1}$ *cis* 14	0.29	0.19	0.012	**
$C_{18:1}$ *cis* 15	0.04	0.03	0.001	NS
$C_{18:2}$ *cis* 9, *trans* 11	0.66	0.46	0.003	**
$C_{18:2}$ *cis* 9, *cis* 12	1.62	1.34	0.048	**
$C_{18:3}$ *cis* 9, *cis* 12, *cis* 15	0.26	0.21	0.007	**
$C_{20:0}$	0.19	0.19	0.005	NS

*$P< 0.05$; ** $P < 0.01$. NS: not significant

PUFA content of the cut grass. During the interval between cutting and feed supply, in fact, the plant lipoxygenases may oxidize an aliquot of the linoleic and linolenic acids, that represent the most common substrates of these enzymes (Porta and Rocha-Sosa, 2002).

In this study, a wide individual variation of milk CLA content may be observed when buffalo cows were submitted to a similar dietary regimen (min 0.27, max 0.84 for H diet and min. 0.37, max 1.58 for F diet), as previously reported for dairy cows (Peterson *et al.*, 2002; Kelsey *et al.*, 2003). A strictly linear relationship between VA and RA was found (RA = 0.12 + 0.23 (VA+RA); $p< 0.01$; $R^2 = 0.89$) and the value of the regression coefficient is within the range of values observed in other studies with dairy cows (Peterson *et al.*, 2002).

Conclusions

The substitution of hay with fresh forage in TMR diets allowed to modify slightly the fatty acid composition of milk from lactating Italian Mediterranean buffalo cows toward a more desirable characteristics for human health. The levels of MCFA and SFA in milk fat, in fact, decreased, while those of PUFA and LCFA increased. Also CLA and VA increase, but with a minor extent than did when dairy cows are fed totally or partially on pasture.

References

Bergamo, P., E. Fedele, L. Iannibelli and G. Marzillo, 2003. Fat-soluble vitamin contents and fatty acid composition in organic and conventional Italian dairy products. Food Chem. 82: p. 625-631.

Chin, S.F., W. Liu, J.M. Storkson, Y.L. Ha and M.W. Pariza, 1992. Dietary sources of conjugated dienoic isomers of linoleic acid, a newly recognized class of anticarcinogens. J. Food Comp. Anal. 5: p. 185-197.

Christie, W.W., 1982. A simple procedure of rapid transmethylation of glycerolipids and cholesteryl esters. J. Lipid Res. 23: p. 1072-1075.

Dhiman, T.R., G.R. Anand, L.D. Setter and M.W. Pariza, 1999. Conjugated Linoleic Acid content of milk from cows fed different diets. J. Dairy Sci. 82: p. 2146-2156.

Griinari, J.M., B.A. Corl, S.H. Lacy, P.Y. Chouinard, K.V.V. Nurmela and D.E. Bauman, 2000. Conjugated linoleic acid is synthesized endogenously in lactating dairy cows by Δ^9 desaturase. J. Nutr. 130: p. 2285-2291.

Harfoot, C.G. and G.P. Hazelwood, 1988. Lipid metabolism in the rumen. In: The rumen microbial ecosystem, P.N. Hobson, (editor). Elsevier Science Publishing, New York. USA, p. 285-322.

Kelly, M.L., E.S. Kolver, D.E. Bauman, M.E. Vanamburgh and L.D. Muller, 1998. Effect of intake of pasture on concentrations of conjugated linoleic acid in milk of lactating cows. J. Dairy Sci. 81: p. 1630-1636.

Kelsey, J.A., B.A. Corl, R.J. Collier and D.E. Bauman, 2003. The effect of breed, parity and stage of lactation on conjugated linoleic acid (CLA) in milk fat from dairy cows. J. Dairy Sci. 86: p. 2588-2597.

Lin, H, T.D. Boylston, M.J. Chang, L.O. Luedecke and T.D. Shultz, 1995. Survey of the conjugated linoleic acid contents of dairy products. J. Dairy Sci. 78: p. 2358-2365.

Molkentin, J. and D. Precht, 2000. Validation of a gas-chromatographic method for the determination of milk fat contents in mixed fats by butyric acid analysis. Eur. J. Lipid Sci. Technol. 102: p. 194-201.

Nudda, A., M. Mele, G. Battacone, M.G. Usai and N.P.P. Macciotta, 2003. Comparison of conjugated linoleic acid (CLA) content in milk of ewes and goats with the same dietary regimen. Ital. J. Anim. Sci. 2 (suppl. 1): p. 515-517.

Peterson, D.G., J.A. Kelsey and D.E. Bauman, 2002. Analysis of variation in cis-9, trans-11 conjugated linoleic acid (CLA) in milk fat of dairy cows. J. Dairy Sci. 85: p. 2164-2172.

Porta, H. and M. Rocha-Sosa, 2002. Plant lipoxygenases. Physiological and molecular features. Plant Physiology 130, p. 15-21.

Schroeder, G.F., G.A. Gagliostro, F. Bargo, J.E. Delahoy and L.D. Muller, 2004. Effects of fat supplementation on milk production and composition by dairy cows on pasture: a review. Livest. Prod. Sci. 86: p. 1-18.

Secchiari, P., M. Antongiovanni, M. Mele, A. Serra, A. Buccioni, G. Ferruzzi, F. Poletti and F. Petacchi, 2003. Effect of kind of dietary fat on the quality of milk fat from Italian Friesian cows. Livest. Prod. Sci. 83: p. 43-52.

SAS User's Guide, 1999. Statistics, Version 8.0 Edition, SAS Inst. Inc., Cary, NC.

Wolff, R.L. and C.C. Bayard, 1995. Improvement in the resolution of individual *trans*-18:1 isomers by capillary gas liquid chromatography: use of a 100 m CP-Sil 88 column. J. AOCS. 72: p. 1197-1201.

Vermorel, M., J.B. Coulon and M. Journet, 1987. Révision du système des unités fourragères (UF). Bull. Techn. C.R.Z.V. Theix, I.N.R.A. 70: p. 9-18.

Zicarelli, L., 1999. Nutrition in dairy buffaloes. In: Perspectives of buffalo husbandry in Brazil and Latin America. H., Tonhati, V.H., Barnabe and P.S., Baruselli (editors). FUNEP, Jabuticabal. Brazil. p. 157-178.

Flavonoids and other phenolics in milk as a putative tool for traceability of dairy production systems

J.M. Besle[1], J.L. Lamaison[2], B. Dujol[1], P. Pradel[3], D. Fraisse[2], D. Viala[1] and B. Martin[1]*
[1]INRA, URH, Clermont Ferrand-Theix, 63122 St Genès Champanelle, France, [2]U.F.R. Pharmacie, pl. H. Dunant, 63000 Clermont Ferrand, France, [3]Domaine de la Borie, INRA, 15190 Marcenat, France.
**Correspondence: besle@clermont.inra.fr*

Summary

Forages are rich in polyphenolic compounds. The biodisposable fraction contains flavonoids and other hydrolysable phenolics (a part of tannins and cell wall phenolics). This paper presents results obtained in recent experiments in which flavonoids and other simple, ethanol-water-soluble phenolic compounds (FPCs) were analysed in forages fed to cows and in their milk.

Six different diets were given to 6 groups of cows: a diet rich in concentrates (CR), maize silage, rye grass silage, rye grass hay, native mountain hay and native mountain pasture. The FPC content of these forages, determined by HPLC, varied from 0.8 to 8 g kg^{-1} DM, with natural pasture being by far the richest. Several FPCs were specific to each forage.

About 54 FPCs compounds were detected in the milks by HPLC. About half were found in all of the milks whatever the diet, with five of them being predominant. The other compounds were more specific to the diet. It is suggested that the first FPC group present in all milks (about 98 % of the peak areas after HPLC separation) was degradation products, mostly derived from the bioavailable cell wall phenolics, whereas the second FPC group originated from the flavonoid pool or other cellular phenolics after transformations in the rumen. The estimated total FPC content in milk varied from 4.0 to 10.8 mg L^{-1} milk for concentrate rich and natural pasture diets, respectively.

The combination of relative amounts in common compounds and in specific compounds may be used to establish a chromatographic fingerprint for the traceability of the diet. In addition, FPCs might contribute to flavour. Lastly, specific molecules of FPC might have a positive effect on human health.

Keywords: milk, flavonoids, phenolics, traceability

Introduction

Flavonoids and several other cellular phenolic compounds (FPCs) are widely present in fruit and vegetables (Hollman and Arts, 2000) and are important source of micronutrients that have a beneficial effect on human health. However, few studies have been made on these compounds in forages and animal products such as milk. Studies on FPCs have mainly involved specific fractions of forages for pharmaceutical purposes (Harborne, 1999) and only rarely whole grassland plants (Jeangros *et al.*, 1999). Hence, we recently analysed this fraction in single species forages and in natural pasture (Poulet *et al.*, 2002). The latter was by far the richest but only 9 compounds were identified out of about 100 peaks separated by HPLC.

Milk contains several micronutrients which have recently been studied: β carotene, vitamin E, polyunsaturated fatty acids, (Martin *et al.*, 2002) and terpenes (Cornu *et al.*, 2002). They have a nutritional role or can be used as liposoluble tracers of the diet, or both. FPCs have also been evidenced in milk but again our current knowledge is scant. A few compounds were observed in small quantities by Brewington *et al.* (1974) and by Lopez and Lindsay (1993) but the effect of the diet was not studied. Among flavonoids, isoflavones were analysed in the milk of cows receiving red clover (King *et al.*, 1998, Sakakibara *et al.*, 2003). This indicates that different compounds may be recovered according to diet.

The aim of this study was to characterise FPCs in the milk of cows receiving the same single species diets or natural pasture, as described by Martin *et al.* (2002) for the study of other micronutrients.

Material and methods

Animals

Six groups of five cows at mid lactation stage were used in the present experiment.

Four groups were reared during winter and were fed the following diets: a concentrate rich diet (CR, barley and soybean meal), maize silage (MS), perennial rye grass silage (RGS) or rye grass hay (RGH). In CR, the concentrate contained 95% barley and 5% soybean meal and accounted for 68% of the dry matter ingested by the animals. In the other diets, the concentrate contained more soya (25 %) and accounted for 10 to 14 % of the intake. In addition, all the groups received the same proportion of mineral and vitamins (MV) supplement with the concentrate.

The two other groups of cows were reared during summer. The first one grazed fresh natural pasture (NPF) with only MV supplement, while the other received natural pasture hay (NPH) with the same supplement as in the winter diet.

The intake and refused fractions were measured for all the preserved diets. An aliquot of the weekly milk production of each cow was kept at -15°C.

Analysis

The FPCs of the forages and concentrates were analysed according to Poulet *et al.* (2002).

The milks of the five cows per diet were analysed as followed. FPCs were extracted, purified and the conjugated forms were hydrolysed according to King *et al.* (1998), using acetonitrile to extract the phenolics instead of methanol. They were then separated by HPLC equipped with a diode array detector according to Petitjean-Freytet *et al.* (1991). In parallel, a control milk containing 10 additional pure compounds of different types of phenolic molecules (acids, simple phenols, flavones, flavonols, isoflavones) was used to calculate a survival factor for each type of compound. The compounds from the experimental samples were classified into different families of phenolic molecules according to their retention time and their spectrum (Markham, 1982). The amount of each compound was assessed at 265 nm in comparison to standard molecules representing each type of compound: syringic acid (for simple phenols and C6-C1 acids), ferulic acid (guaiacyl), luteolin (flavones), quercetin (flavonols), and genistein (isoflavones).

Results

The FCPs contents of the diets CR, MS, RGS, RGH, NPH and NPF were 2.1, 0.8, 3.0, 4.1, 2.5 and 8.2 g kg^{-1} DM, respectively. For the diets CR, MS, RGS, RGH, NPH and NPF, the amount of FCPs ingested by the cows was thus estimated at 35, 15, 129, 111, 39, 148 g/day, respectively. For all diets, about 54 peaks having a typical phenolic compound spectrum were separated in milk. They were all different (according to the retention time and the UV spectrum) to the FCPs in the diets. Half of them (27 peaks) were common to at least two diets (Table 1). The identification of these compounds is currently in progress. Preliminary analysis showed that most of these compounds (13 peaks) were simple phenols or phenolic acids (SP), whereas the others were probably flavonols, flavones and isoflavones (seven, five and one peaks, respectively). Five compounds (peaks 4, 7, 22, 23, 27) accounted for 90% of the area. The other compounds seemed to be specific to a particular diet, they were present generally in small amounts, often as traces.

Milks from diets MS and RGH contained six specific compounds (one flavone and five SP) while RGS gave only two SP and one flavone, NPH gave 2 SP and CR two flavonols and one flavone. As NPF is a mixture of several plant species, some of the "specific" compounds are also found in single species forages which form part of the NPF. This is the case for two SP which were also observed with RGH. In total, NPF gave the greatest number of "specific" compounds; of the 9 peaks observed, four gave a signal sufficiently high to be recorded (Figure 1) and five were present as traces but with a clear UV spectrum. It is noteworthy that one peak (n° 8, a flavone) seemed to be characteristic of hays and that another was only seen in the two rye grass diets. The total amount of phenolic compounds in milk was highest for NPF (10.8 mg L^{-1}), and lowest for CR (4 mg L^{-1}) with intermediate values for the other diets

Table 1. Estimated content[1] (µg L^{-1}) in fractions of milk FCPs according to diet.

Diet[2]	Common compounds [3]								Specific Compounds[4]	Total
	A	4 SP	7, SP or flavone	8 flavone	22 SP	23 SP	27 flavonol	B		
CR	77 ^{a}	202 ^{a}	211 ^{a}	0 ^{a}	73 ^{a}	322 ^{a}	2750 ^{a}	277 ^{a}	53	3964 ^{a}
MS	222 ^{b}	951 ^{bc}	590 ^{b}	0 ^{a}	174 ^{a}	638 ^{a}	6030 ^{b}	131 ^{a}	486	9222 ^{bc}
RGS	217 ^{abc}	380 ^{ab}	300 ^{a}	0 ^{a}	870 ^{b}	1110 ^{b}	5482 ^{b}	244 ^{a}	20	8621 ^{b}
RGH	249 ^{b}	1465 ^{c}	0 ^{c}	28 ^{b}	38 ^{a}	808 ^{ab}	5085 ^{b}	169 ^{a}	82	7923 ^{b}
NPH	219 ^{bc}	817 ^{abd}	91 ^{ac}	17 ^{b}	68 ^{a}	537 ^{ab}	3162 ^{a}	247 ^{a}	tr	5157 ^{a}
NPF	73 ^{ac}	1377 ^{cd}	84 ^{ac}	0 ^{a}	0 ^{a}	2908 ^{c}	5852 ^{b}	208 ^{a}	298	10799 ^{c}
RSD	*128*	*693*	*152*	*12*	*299*	*494*	*1203*	*272*	*186*	*1631*

[1] Data with different superscripts are significantly different ($P<0.05$), tr: traces, RSD: residual standard deviation; [2] See abbreviations in text; [3] Within the 54 separated peaks corresponding to phenolic compounds in milk, the common compounds were: peaks 4, 7, 8, 22, 23 and 27 = major peaks, A = sum of 10 minor peaks eluted in the first half of the chromatogram, B = sum of 11 minor peaks eluted in the the second half of the chromatogram; [4] Specific compounds according to the diets: CR, MS, RGS, RGH, NPH and NPF gave 3, 6, 3, 6, 2 and 9 peaks respectively.

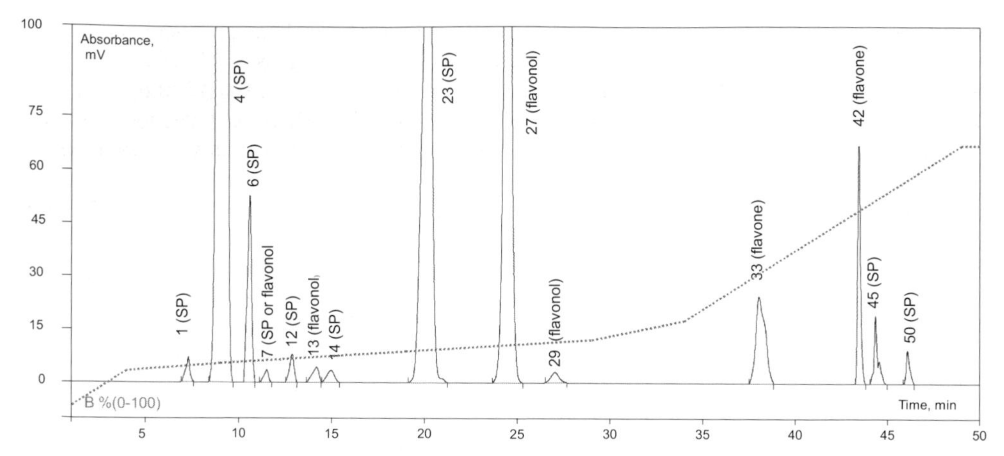

Figure 1. HPLC separation of milk FPCs for NPF diet (absorbance at λ max). Peaks n° 12, 33, 45, and 50 are specific for the NPF diet ; SP = simple phenol or phenolic acid, ... eluent gradient (B = acetonitrile 70%).

Discussion

Identification of phenolic compounds in milk

In a previous analysis of cow's milk, Brewington *et al.* (1974) and Lopez and Lindsay (1993) detected about 15 aromatic compounds, of which 5 or 6 were more abundant than the others. The major compounds detected in both studies were simple phenols (phenol, cresol, ethylphenol, propylphenol), diphenols (catechol group) and C6-C2 compounds (phenylacetic acid or acetovanillone). According to the authors, they result essentially from the detoxification process or from the degradation of cell wall phenolics by the animal catabolic activity.

In another study with a clover rich diet, the cows received known isoflavonoids. It has been calculated that 0.2 p 1000 of the ingested compounds were secreted in milk (Sakakibara *et al.*, 2003). In the present assay, with the NPF diet, the amount of ingested FCPs was similar to that for clover diet but the secreted FCPs were 5-fold higher. Thus, most of the secreted phenolics are probably not the result of a transfer from the ingested FCP fraction to the milk. They probably derive from the catabolism or from the bio-disposability of cell wall phenolics (lignin, hydroxycinnamic acids), which are present in at least 20 fold higher amounts than FCPs in the forage. This would also explain the discrepancies between the relative amounts of FCPs in the diet and in the milk. For example, the ingested FCPs were 10-fold lower for MS diet than for NPF whereas in milk the FCPs were present at a closely similar level with MS and NPF diets. The phenolics in milk probably came from the biodisposable hydroxycinnamic acids of the silage (about 25 $g.kg^{-1}$ DM) rather than from the flavonoids (0.8 $g.kg^{-1}$ DM).

However, it has been shown in other studies that specific compounds of the diet, such as isoflavones or polychlorinated aromatics (O'Connell and Fox, 2001), are partially recovered in milk. It is therefore not surprising that several specific compounds are found in milk.

Interest for animal diet traceability or human nutrition

The specific compounds observed for each diet and the relative ratios between common compounds constitute a chromatographic fingerprint that can be used to characterise the diet. They can be very specific, especially in the case of grazed natural pasture whose botanical composition depends on altitude, exposure to sunlight, climate and season (Jeangros *et al.*, 1999). FCPs are currently the only water-soluble tracer, and provide additional information to other fat-soluble tracers such as terpenes (Cornu *et al.*, 2002) or polyunsaturated fatty acids. In this approach, it is not necessary to identify each compound, but the detection method must be highly sensitive.

The estimated FCP content in milk in our study was greater than in that of Lopez and Lindsay (1993). Although not negligible, this amount is lower than that in several plant sources (*ie* 30 and 2800 mg kg^{-1} fresh weight in apples and in oranges, respectively, according to Remesy *et al.*, 1996). Therefore, the nutritional effect of FCPs provided by milk is likely to be lower compared to that of fruits. However, several compounds can have a nutritional role at a low amount. The effect of some molecules may be beneficial, as in the case of isoflavones (Sakakibara *et al.*, 2003), or detrimental, as for polychlorinated phenolics (Baars *et al.*, 2004), or they may contribute to the organoleptic properties of the dairy products. This could in part explain the variations in milk flavour between the studied animal diets observed by a sensory panel (Dubroeucq *et al.*, 2002).

Conclusion

Several phenolic compounds were found in milk, some of which had a spectrum close to that of flavonoids. Some molecules were common to all diets, and they likely derived from the cell walls of plants or originate from animal metabolism while others were found to be specific to a particular diet. This is the basis to obtain a chromatographic fingerprint of a water-soluble tracer characteristic of a diet. A mathematical model is currently being developed and further work is needed to confirm these results.

Acknowledgements

This work was supported by the French Research Ministry (Food-Quality-Safety programme)

References

Baars, A.J., M.I. Bakker, R.A. Baumann, P.E. Boon, J.I. Freijer, L.A.P. Hoogenboon, R. Hoogerbrugge, J. D. Van Klaveren, A. K. D. Liem, W. A. Traag and J. De Vries, 2004. Dioxins, Dioxin-Like Pcbs and Non-Dioxin-Like Pcbs in Foodstuffs: Occurrence and Dietary Intake in the Netherlands. Toxicology Letter 151: p. 51-61.

Brewington, C.R., O.W. Parks and D.P. Schwartz, 1974. Conjugated compounds in cow's milk. II. J. Agr. Food Chem. 22: p. 293-294.

Cornu, A., N. Kondjoyan, B. Martin, A. Ferlay, P. Pradel, J.B. Coulon and J.L. Berdagué, 2002, Vers une reconnaissance des principaux régimes alimentaires des vaches à l'aide des profils terpéniques du lait, Renc. Rech. Ruminants 9: p. 370.

Dubroeucq, H., B. Martin, A. Ferlay, P. Pradel, I. Verdier-Metz, Y. Chilliard, J. Agabriel and J.B. Coulon, 2002. L'alimentation des vaches laitières est susceptible de modifier les caractéristiques sensorielles des laits. Renc. Rech. Ruminants 9: p. 351-354.

Harborne, J.B. (editor), 1999. The handbook of natural flavonoids. Whiley and Sons, New York, vol 1: 889 p., vol 2: 879 p.

Hollman, P.C.H. and I.C.W Arts, 2000. Flavonols, Flavones and Flavanols - Nature, Occurrence and Dietary Burden. J. Sci. Food Agric. 80: p. 1081-1093.

Jeangros, B., J. Scehovic, J. Troxler, H.J. Bachmann and J.O. Bosset, 1999. Comparison of the botanical and chemical characteristics of grazed pastures, in lowlands and in the mountains. Fourrages 159: p. 277-292.

King, R.A., M.M. Mano, and R.J. Head, 1998. Assessment of Isoflavonoid Concentrations in Australian Bovine Milk Samples. J. Dairy Res. 65: p. 479-489.

Lopez, V. and R.C. Lindsay, 1993. Metabolic conjugates as precursors for characterizing flavor compounds in ruminant milks. J. Agric. Food Chem. 41: p. 446-454.

Markham, K. R. (editor), 1982. Ultraviolet-visible absorption spectroscopy. In: Techniques of flavonoid identification. Academic Press, New York, 82: p. 1-51.

Martin, B., A. Ferlay, Ph. Pradel, E. Rock, P. Grolier, D. Dupont, D. Gruffat, J.M. Besle, Y.Chilliard and J.B. Coulon, 2002. Variabilité de la teneur des laits en constituants d'intérêt nutritionnel selon la nature des fourrages consommés par les vaches laitières. Renc. Rech. Ruminants 9: p. 347-350.

Petitjean-Freytet, C., A. Carnat and J.L. Lamaison, 1991. Teneurs en flavonoïdes et en dérivés hydroxycinnamiques de la fleur de *Sambucus nigra* L. J. Pharm. Belg. 46: p. 241-246.

O'connell, J., and P. Fox, 2001. Significance and applications of phenolic compounds in the production and quality of milk and dairy Products: a review. Int. Dairy J. 11: p. 103-120.

Poulet, J.L., D. Fraisse, D. Viala, A. Carnat, P. Pradel, B. Martin, J.L. Lamaison and J.M. Besle, 2002. Flavonoids in forages: composition and possible effects on milk quality. In: Multi-Function grasslands. Quality forages, animal products and landscapes, J.J. Durand., J.C. Emile, C. Huyghe and C. Lemaire (editors), 19th EGF Congress, La Rochelle, 27-30 May 2002.Vol. 7, p. 590-591.

Remesy, C., C. Manach, C. Demigne, O. Texier and F. Regerat, 1996. Nutritional interest of flavonoids. Med. Nutr. 32: p. 17-27.

Sakakibara, H., D., Viala, M. Doreau and J.M. Besle 2003. Clover isoflavones move to cow's milk. 1st International Conference on Polyphenols and Health, 18-21 November 2003, Vichy, France: p. 296.

Milk quality and automatic milking: effects of free fatty acids

B.A. Slaghuis, K. Bos, O. de Jong and K. de Koning*
Applied Research of the Animal Sciences Group of Wageningen UR, PO Box 2176, 8203AD Lelystad, the Netherlands.
**Correspondence: Betsie.Slaghuis@wur.nl*

Summary

With the introduction of automatic milking (AM) systems, increased levels of free fatty acids (FFA) in milk were observed, which might result in off-flavours in milk and dairy products. The aim of this study was to investigate the factors contributing to elevated FFA levels: influence of the milking frequency, technical parameters of the milking system, and finally, farm management aspects. Milking frequency was studied in a Latin square design with milking intervals of 4, 8 and 12 hours and showed increased FFA -levels for the shorter intervals. Technical factors were studied in a laboratory study using milking machine components of AM-systems and conventional systems. With susceptible milk, milking machine components of AM systems showed more increased FFA levels. In some herds, FFA problems remained after solving technical and milking frequency problems. Indications were found that feed composition and feeding regime might influence susceptibility of milk for FFA-formation. These farm management aspects are subject of ongoing research.

Keywords: milk quality, automatic milking, free fatty acids

Introduction

At the introduction of milk pipe lines and farm milk tanks in the seventies, problems with increased levels of free fatty acids (FFA) were observed, resulting in off-flavours in milk and dairy products. Therefore, measurement of FFA was integrated in the quality control system in the Netherlands (determination twice a year, in spring and autumn). The problems of increased FFA levels were tackled practically by avoiding blind pumping, reducing air leakage and improving maintenance of the milking equipment. Especially the introduction of a national milking machine maintenance program contributed largely to overcome problems with FFA.

When milking three or more times a day instead of two times is applied in conventional milking parlours, increased milk yields, decreased fat and protein contents and increased free fatty acid levels have been reported (Ipema and Schuiling, 1992; Jellema, 1986, Klei *et al.*, 1997, Svennersten-Sjauna *et al.*, 2002).

With the introduction of AM systems, the problem of increased FFA levels (also known from milking 3 times per day in conventional farms) occurred again (Klungel *et al.*, 2000; Vorst and Koning, 2002). Increased FFA levels seemed to be related with the higher average milking frequency with AM systems and to technical features of this system compared to conventional milking systems. However, some other management factors also may play a role, as Chazal *et al.* (1987) and Ferlay *et al.* (2002) found influences of feeding on lipolysis and FFA levels.

As increased FFA levels are a result of lipolysis, the possible mechanism of lipolysis is explained below.

Possible mechanism of lipolysis

Lipolysis is enzymatic hydrolysis of milk fat by milk lipoprotein lipase (mLPL), causing accumulation of FFA. The milk fat present in milk fat globules is protected against the action of mLPL by the milk fat globule membrane. If this membrane is disrupted by e.g. agitation, fatty acids can be split from the glycerol part by mLPL.

But in untreated milk, lipolysis also occurs due to spontaneous lipolysis. Jellema (1986) defined susceptibility as FFA level in untreated milk after 24 hours storage at 4°C.

Late lactation milk and milk from cows milked three times per day or more is more susceptible to spontaneous lipolysis (Jellema, 1986). The milk fat globule is not disrupted, but some factors (activators) present in milk may favour the interaction of mLPL with milk fat resulting in higher degree of FFA. Activators are related to blood serum components, as addition of blood serum to raw milk increases FFA levels (Jellema, 1975, 1986). Inhibitors, that inhibit contact between mLPL and milk fat globule membrane, have been determined as proteose pepton component 3 fractions (Cartier *et al.*, 1990).

Several treatments, referred to as activation treatments, enhance lipolysis (induced lipolysis) and are often used to study lipolysis of milk fat by endogenous mLPL. Treatments which cause activation include agitation, homogenisation, temperature changes and the addition of blood serum or heparin to milk. Milk from individual cows differs in susceptibility to these treatments. Correlations between spontaneous and induced lipolysis have been reported (Jellema, 1986; Chazal *et al.*, 1987; Cartier and Chilliard, 1989).

The mechanisms that promote lipolysis are not fully understood. The aim of this study was to investigate the factors contributing to elevated FFA levels in AM systems: influence of the milking frequency, technical parameters of the milking system, and finally farm management aspects.

Material and methods

Milking frequency

In the first study spontaneous lipolysis (Cartier and Chilliard, 1990) was studied in a Latin square design experiment with 12 cows, three periods of four days, three milking intervals (4, 8 and 12 hours corresponding to 6, 3 and 2 times milking a day) and 4 cows per group (Table 1). Cows were selected to form equal groups (parity and lactation stage). Cows were milked conventionally according to this scheme for 4 days and on day 4 all cows and milk was sampled.

Table 1. Latin square design of spontaneous lipolysis study, with groups of cows defined as A, B and C.

Period	1	2	3
Milking interval			
2x per day	A	B	C
3x per day	B	C	A
6x per day	C	A	B

Milk samples were divided into two sub samples: one was inactivated by hydrogen peroxide immediately (0,02%, Jellema, 1979) to stop lipolysis, the other one was stored at 5°C ±1°C for 24 hours and then inactivated with hydrogen peroxide. Difference between the sub samples is defined as spontaneous lipolysis (Cartier and Chilliard, 1990). Samples were analysed according to the BDI-method (IDF,1991).

Technical parameters of the AM-system

Fresh milk, susceptible for lipolysis (from 4-5 cows at the end of lactation), was divided in two parts. One part was passed through a reconstructed AM-system (two brands: teat cups and receiver part) and milk was sampled before and after passing the system (air inlet at the teat cups was 24 l/min at 42 kPa vacuum) . The other portion of milk was passed through a conventional system and sampled before and after passing the system (air inlet at the milking cluster was 8 l/min at 42 kPa vacuum). The milk was mixed and used again to repeat the experiment.

Farm management

Variation in farm systems

In a study on variation of FFA levels in farm tank milk of 12 farms (4 conventional (milking twice a day), 4 conventional milking (three times a day) , 4 AM; Table 2), FFA levels were determined monthly and in two periods of 14 days from every bulk tank. Groups of farms were selected as high (>0.80 mmol/100 g fat during the last year) or low (<0.70 mmol/100 g fat), based on the milk quality research of the previous year.

Table 2. Number of farms in design of study on variation of FFA levels

FFA level	high	low
Type of farm		
Conventional 2x per day	2	2
Conventional 3x per day	2	2
AM	2	2

Farm characteristics

The effect of farm management aspects was studied at 8 farms using the same brand of AM-system and the same cooling system and 6 farms milking three times a day (Table 3). Half of the farms were classified as high (FFA >0.75 mmol/ 100 g fat during at least the last year) and the other half as low (<0.70 mmol/100 g fat).This selection was based on the results of milk quality research of the previous year.

Table 3. Number of farms in design of study on farm management.

FFA level	high	low	remarks
Type of farm			
Conventional 3x per day	3	3	different brands, same cooling system
AM	4	4	same brand, same cooling system

Bulk tank milk was sampled and analysed for FFA. Detailed farm (including milk frequencies), robotic and management information regarding feeding, housing conditions and animal health was obtained by a questionnaire during a visit at the farm and correlations among these factors with FFA levels were calculated.

Results and discussion

Milking frequency

Analysis of variance of log transformed FFA levels showed slight differences between intervals after 0 hour storage and significant differences after 24 hours of storage (Table 4). The increase after 24 hours storage, defined as spontaneous lipolysis (Cartier and Chilliard, 1990) was significant for the different milking intervals. Initial levels of FFA were rather low compared to other studies.

As all bulk milk samples contain milk of 3 days old and the increase in FFA is the highest in the first 24 hours (data not published), no hydrogen peroxide was used anymore during the rest of the experiments and no sub samples were taken anymore.

Table 4. Effect of milking interval on FFA contents (mmol/100 g fat) and standard deviations (between brackets) in raw milk after 0 hours and 24 hours storage after sampling.

Interval (times/day)	6	3	2
0 hours at 5°C	0.20a (0.04)	0.19ab (0.07)	0.15b (0.07)
24 hours at 5°C	1.23a (0.98)	0.71b (0.47)	0.42c (0.31)
increase in 24hours	0.97a (0.98)	0.49b (0.45)	0.25c (0.26)

a,b,cstatistically significant difference on the same row $P<0.05$)

Technical parameters of the AM-system

The milk used in test 1 on day 2 with brand 1 was not susceptible enough, because initial FFA level was low (Table 5). Selection of susceptible milk was not properly. The milk used for test 2 with brand 2 was susceptible enough to show difference in FFA increase, but different initial FFA

Table 5. Mean FFA level before and after passing through a conventional or an AM system.

Milking system	n	mean FFA before (mmol/100 g fat)	mean FFA after (mmol/100 g fat)	increase FFA (mmol/100 g fat)
Test 1				
Conventional	12	0.43a	0.50a	0.07a
AM brand 1	12	0.42a	0.58b	0.16b
Test 2				
Conventional	8	0.58a	0.65a	0.07a
AM brand 2	8	0.45b	0.65a	0.21b

Different superscripts mean significant difference ($P<0.05$), n = number of tests, 4 tests per day, figures with superscripts are least square means.

levels in milk passing the conventional and in milk passing the reconstructed AM system made results difficult to interpret. In AM systems more air is used to transport the same amount of milk in comparison with conventional milking, so FFA increase may be expected (O'Brien *et al.*, 1998). The maximal air inlet advised by producers of AM systems varies from 24-48 l/min (Wemmenhove, 2001), whereas ISO- norms for conventional milking systems give a maximum air inlet of 12 l/min (ISO, 1996a,b and c).

Farm management

Variation in farm systems

Some characteristics of the studied farms are given in Table 6. Results until April 2004 showed more variation for high than for low FFA levels on different farms (Figure 1). Also on farms milking three times a day (3x), selection based on low levels in the previous year, was not adequate. Seasonality and/or differences in feeding regimes or other factors seemed to influence the FFA levels and the variation on these farms. For farms milking two times a day (2x) the selection was appropriate: farms with high FFA levels had high milk lines and tied stalls, where it is more difficult to achieve low FFA levels. Construction of milking equipment is important to avoid high FFA levels. One of the farms milking 3x also had a tied stall with a high milk line (farm 6). On two of the four farms milking 3x (farm 5 and 8) FFA levels increased when cows were pastured. Ferlay *et al.* (2002) found a higher lipolysis for milk of pasturing cows than for milk from cows feeding a diet rich in concentrate or a corn silage-based diet. The cows grazing on pasture were slightly underfed. Increased lipolysis due to underfeeding in cow milk were observed previously (Jellema, 1975). So for two farms there might be a feeding problem, especially in summer. Apparently variation in FFA level on farms milking 3x was due to other factors than distinction between high and low (Figure 1.). Farms milking 3x had higher production levels than the other farms. Maybe these high production levels, combined with frequent milking might indicate higher demands from individual cows (e.g. energy metabolism on higher levels). Also late lactation cows are milked three times a day, resulting in extra risk for high FFA levels. But not all farms milking 3x had high FFA levels. For farms milking 3x BMSCC were lower than for the other type of farms. Maybe the high production levels on these farms resulted in these BMSCC's.

Table 6. Characteristics of 12 farms with regularly sampling for FFA.

Farm	Type	Level	Cows	Milk yield (kg)	Housing	Grazing (month)	Mastitis* cases (% per year)	BMSCC ($x10^3$cells/ml)	TBC ($x10^3$/ml)
1	2x	high	44	6973	Tied stall	0	0-10	480	10
2	2x	high	50	-	Tied stall	6	0-10	±300	-
3	2x	low	187	8900	Loose	2	0-10	187	5
4	2x	low	60	8000	Loose	6	10-20	165	9
5	3x	high	55	11000	Loose	6	0-10	104	5
6	3x	high	45	8500	Tied stall	6	0-10	202	11
7	3x	low	135	9000	Loose	0	10-20	-	-
8	3x	low	38	9500	Loose	5	0-10	±200	±7
9	AM	high	100	8100	Loose	0	10-20	±300	-
10	AM	high	55	7500	Loose	0	20-30	400	9
11	AM	low	60	7000	Loose	6	0-10	250	10
12	AM	low	53	9500	Loose	0	-	194	12

= not known, * = reported by the farmer, ± = estimated, based on what farmer mentioned

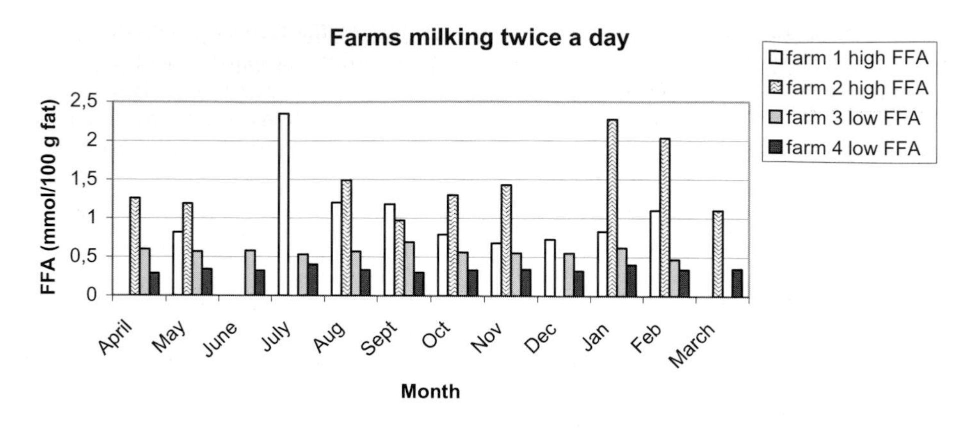

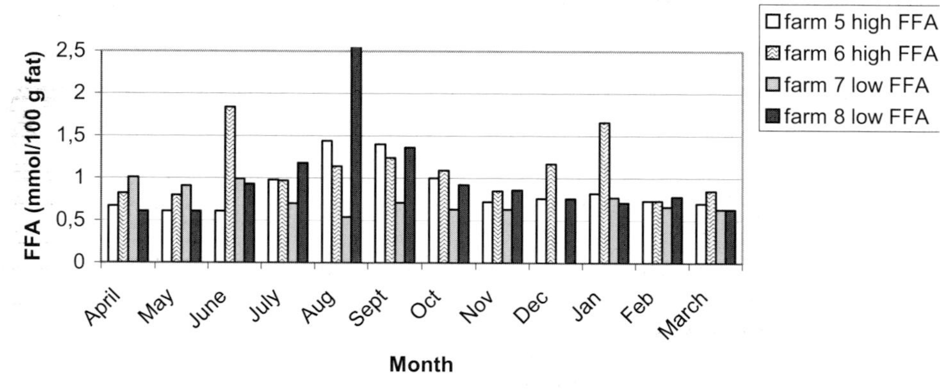

Farms with AM system

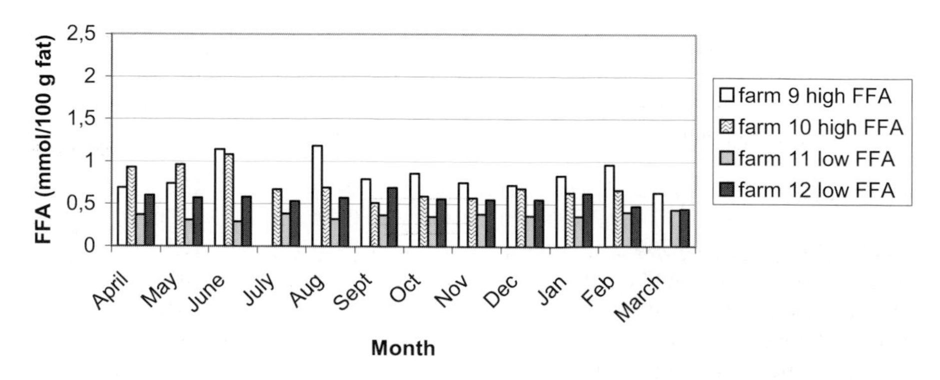

Figure 1. FFA levels on twelve farms based on monthly sampling.

For automatic milking (AM) differences between high and low were not extreme, but still more variation in farms with higher levels could be concluded. However, levels for AM farms were relative lower than on the other type of farms. Farm 10 adjusted the number of failed milkings in June and the FFA level decreased from that time and remained relative low. This was an example of excessive mixing of milk and air. For farm 9 such an easy explanation could not be concluded.

Apparently some management factors not easily defined were due to this variable FFA levels. Vorst *et al.* (2002) found indications that farms with higher milk production might give lower risk for high FFA levels with AM systems.

Farm characteristics

Selection of farms, based on FFA levels in the previous year, in the study on farm management aspects seemed to be adequate for 3x milking per day but not for AM-systems (Table 7).

On farms milking 3x with high FFA levels quota and number of cows on the farm were significant lower than on farms with low FFA levels. No differences between high and low FFA level on farms milking 3x were found for average milk production per cow and fat percentage of the milk.

For AM farms the selection was made on the same brand and cooling systems, all quite recently installed, but FFA results for at least one year were available. After solving some initial problems most of the FFA levels decreased and although significant difference was found between AM farms with high and low FFA levels, the levels were not as high as in the previous year as the selection was based on > 0,75 mmol/100 g fat. On the AM farm with the highest FFA levels, the milking frequency was not adequately adjusted. Cows producing less (7 kg per milking) were milked too often.

On AM farms no differences in quota, number of cows, milk production and fat percentage of the milk were found for high and low FFA level.

Table 7. Mean FFA and standard deviation (between brackets) of 6 farms milking three times a day and 9 farms milking with the same AM system. All farms are sampled twice within six weeks in autumn (oct-nov 2003).

Farm type	FFA (mmol/100 g fat)					
	n	High FFA		n	Low FFA	
3x	3	0.87a	(0.17)	3	0.48b	(0.05)
AM	4	0.69a	(0.13)	4	0.53b	(0.12)

a,b Different superscripts within one row means significant difference

Conclusions

Increased FFA levels were due to increased milking frequencies (both for AM and conventional: 3x milking per day) and probably due to milking machine components of AM systems. Milking frequencies seemed to be of more importance than technical parameters of the AM system, because FFA levels for farms milking three times per day and AM systems were comparable. However, technical aspects cannot be excluded. Compared to conventional milking, the air/milk ratio is threefold higher, probably resulting in more disruption of milk fat globule membranes.

Apart from milking frequencies and technical parameters, management aspects probably play a role. Feeding might be a factor of importance and should be studied in more detail. More fundamental research is needed regarding the susceptibility of cows and the basic mechanism of increased FFA levels.

References

Cartier, P., and Y. Chilliard, 1989. Lipase redistribution in cows' milk during induced lipolysis. I. Activation by agitation, temperature change, blood serum and heparin. J. Dairy Res. 56: p. 699-709.

Cartier, P. and Y. Chilliard, 1990. Spontaneous Lipolysis in Bovine Milk: Combined Effects of Nine Characteristics in Native Milk. J. Dairy Sci 73: p. 1178-1189.

Cartier, P., Y. Chilliard and D. Paquet, 1990. Inhibiting and activating effects of skim milks and proteose-peptone fractions on spontaneous lipolysis and purified lipoprotein lipase activity in bovine milk. J. Dairy Res. 73: p. 1173-1177.

Chazal, M.P., Y. Chilliard and J.B. Coulon, 1987. Effect of nature of forage on spontaneous lipolysis in milk from cows in late lactation. J. Dairy Res. 54: p. 13-18.

Ferlay, A., B. Martin, Ph. Pradel and Y. Chilliard, 2002. Effect of the nature of forages on lipolytic system in cow milk. In: Proceedings Congrilait 2002, 26th IDF World Dairy Congress, 24-27 September 2002, Paris, France. poster B4-38.

IDF, 1991. Determination of free fatty acids in milk and milk products. IDF Bulletin No. 265. Brussels, International Dairy Federation.

Ipema, A.H., Schuiling,E., 1992. Free fatty acids; influence of milking frequency. In: Ipema, A.H., A.C. Lippus, J.H.M. Metz and W. Rossing (editors) Proceedings of the International Symposium on Prospects for Automatic Milking, Wageningen, EAAP series 65: p. 491-496.

ISO 3918, 1996a. Milking machine installations - Terms and definitions.

ISO 5707, 1996b. Milking machine installations - Construction and performance.

ISO 6690, 1996c. Milking machine installations - Mechanical tests.

Jellema, A., 1975. Susceptibility of bovine milk to lipolysis. Neth. Milk andDairy J. 29: p. 145-152.

Jellema, A., 1979. Behandeling van monsters i.v.m. vetsplitsingsonderzoek. Zuivelzicht 71: p. 24.

Jellema, A., 1986. Some factors affecting the susceptibility of raw cow milk to lipolysis. Milchwissenschaft 41: p. 553-558.

Klei, L.R., J.M. Lynch, D.M. Barbano, P.A. Oltenacu, J. Lednor and D.K. Bandler, 1997. Influence of milking three times a day on milk quality. J. Dairy Sci. 80: p. 427- 436.

Klungel, G.H., B.A. Slaghuis and H. Hogeveen, 2000. The effect of the introduction of automatic milking systems on milk quality. J. Dairy Sci. 83: p. 1998-2003

O'Brien, B., E. O'Callaghan and P. Dillon, 1998. Effect of various milking machine systems and components on free fatty acid levels in milk. J. Dairy Res. 65: p. 335-339

Svennersten-Sjauna, K., S. Persson and H. Wiktorsson, 2002. The effect of milking interval on milk yield, milk composition an raw milk quality. In: Proceedings of the first North American Conference on Robotic Milking, J. McClean, M. Sinclair and B.West (editors) Wageningen Pers, Wageningen, The Netherlands: p. V43 -V48.

Vorst, Y. and K. de Koning, 2002. Automatic milking systems and milk quality in three European countries. In: Proceedings of the first North American Conference on Robotic Milking, J. McClean, M. Sinclair and B.West (editors) Wageningen Pers, Wageningen, The Netherlands: p. V1-V12.

Wemmenhove, H., 2001. Quality control of automatic milking systems. Construction and performance. Report nr. 217 Research Institute for Animal Husbandry, Lelystad, the Netherlands, ISSN 0169-3689.

Milk quality and automatic milking: effects of teat and system cleaning

B.A. Slaghuis, J.A.M Verstappen, R.T. Ferwerda, C.H. Bos and H.J. Schuiling*
Applied research of the Animal Sciences Group of Wageningen UR, PO Box 2176, 8203AD Lelystad, the Netherlands.
**Correspondence: Betsie.Slaghuis@wur.nl*

Summary

With the introduction of automatic milking (AM) systems, some increases in total bacterial count (TBC) and of free fatty acids (FFA) in milk were observed. Contamination of milk, resulting in elevated TBC, originates from four main sources: inside of the udder, outside of the udder, the milking machine and bulk tank. The aim of this study was to investigate the efficacy of teat cleaning devices and to study the effect on milk quality of two and three system cleanings per day. AM systems have special teat cleaning devices. The efficacy of these teat cleaning devices was studied by contaminating the teats first, followed by swabbing the teats before and after teat cleaning. Per brand of AM system on two farms were taken. The effect of system cleaning frequency was studied on 13 farms by performing two or three system cleanings per day. Milk quality was determined on bulk tank samples. All teat cleaning systems showed positive effects. Differences were found between brands of AM systems. The level of housing hygiene influenced the level of teat contamination. Three system cleanings per day resulted in a significant lower TBC (10.000 vs. 13.000 cfu/ml), number of coliforms, thermodurics and psychrotrophs compared with two system cleanings per day. In both cases the levels found for TBC were far within the penalty limits.

Keywords: milk quality, automatic milking, total bacterial count, teat cleaning, system cleaning

Introduction

Teat cleaning

Besides mastitis pathogens and bacteria from milk contact surfaces, bacteria from udder and teat surfaces belong to the three main causes of microbial contamination of raw milk (Slaghuis, 1996, Sumner, 1996). Clean udder and teats before milking are demanded by Directive 89/362/EEC on General Conditions of Hygiene in Milk Production Holdings to avoid negative influences on milk quality.

Different systems of teat cleaning are applied by Automatic Milking (AM) systems: with use of brushes or in the teat cup itself and with specific cleaning teat cups with air and water. Common to all systems is that no control of the teat cleaning effect on individual cows is performed. Limited information is available on teat cleaning efficiency by AM systems. Schuiling (1992) compared udder cleaning with brushes to no cleaning and found removal of 69 % of manure due to cleaning after artificial contamination of teats with lithium. Melin *et al.* (2002) used spores for artificial contamination and found a better reduction of carry over of spores into milk by an automated teat cleaning procedure compared to manual cleaning (98.0 versus 66.5 %). In contrast, TenHag and Leslie (2002) could not determine significant differences between effects of automated and manual teat cleaning when a swab method with a simplified determination (please explain was is a simplified determination of BC) of bacterial counts was applied. Only one system each was included in these studies with different approaches to determine the teat cleaning effect, so making comparisons between studies is difficult.

Therefore an investigation including all AM systems currently on the market was performed. The systems were evaluated by a set of different methods (Knappstein *et al.*, 2004). Artificial contamination with poppy seed is only reported here.

System cleaning

Automatic milking is a continuous process throughout the day. Every hold-up in this process should be avoided, in order not to disrupt cow traffic and to reduce throughput (Ipema, 1997). On the other hand, the hygienic quality of the milk should not be decreased.

Starting with a clean and disinfected system and milking the first cows, it will take some time for bacteria to adapt to the environment. Depending on the type of bacteria, the temperature of the milk and other factors adaptation time will vary. It is important to clean and disinfect the machine before bacteria have adapted. Earlier research, with simulated AM-systems and with first generation AM-systems has shown that after eight hours bacterial growth is increasing (Verhey, 1992; Ordolff and Bolling, 1992; Frost *et al.*, 1999). Based on these results it became a demand in several European countries that milk producers with AM-systems must perform a system cleaning 3 times per day.

Bacterial growth however is not only depending on cleaning frequency, but also on how well milk residues in the system are flushed away during the milking of the next cow. In this regard the construction in general, and in particular the inner surface, the slope towards the receiver and the absence of 'dead ends' are important, as is the remaining volume in the receiver and in the delivery line (Schuiling, 1995; IDF 1996; Verstappen *et al.*, 1999). A sanitary construction will also increase the efficiency of cleaning (Vorst *et al.*, 2002). It is known that some farmers do clean the AM-system only twice a day, often with good results for milk quality.

Materials and Methods

Teat cleaning efficiency

Teat cleaning efficiencies of the following brands of AM systems were determined: DeLaval, Insentec, Lely Industries, Fullwood, Prolion/Gascoigne Melotte and Westfalia Landtechnik GmbH.. For AM systems codes were used for competition reasons.

This study was based on artificial contamination of teats with a mixture of poppy seed and manure (20 % w/w) and determination of carry over of poppy seed (Papaver somniferum) into milk. This approach was applied on 12 farms working with AM systems of 6 different brands (2 farms per brand) and two farms with conventional milking systems. Per farm teats of ten cows were contaminated, 5 cows were milked without cleaning and 5 cows after teat cleaning. The number of poppy seeds in the milk of each cow was determined by filtering the composite milk through a no woven filter and counting the number of poppy seeds.

The levels of number of poppy seeds in the milk were analysed for the fixed effects of brand and teat cleaning efficiency using a mixed logistic model. In the model the relationship between the probability p ($0<p<1$) of observing the poppy seeds and the explanatory variables was described using the following logit-link function:

$$\mathrm{Logit}(p_{ijk}) = \ln(p_{ijk} / 1 - p_{ijk}) = c + fm_i + br_j + tcleffect_k + br_j * tcleffect_k \qquad (1)$$

where c represents the constant term and is the mean for the combination with all factors at the lowest level, fm_i is the random effect of farm i, considered to be normally distributed with mean 0 and variance equal to σ^2_{farm}, br_j is the fixed effect of brand j and $tcleffect_k$ is the fixed effect of teat cleaning effect k.

The model assumed that the variance of the observed counts Y can be adequately described by variance $(Y|p) = np(1-p)$. Estimates of model parameters and components of variance were obtained using the Generalized Linear Mixed Model (GLMM) Genstat procedure (2002). Fixed effects were assessed using chi-squares for the Wald statistics.

System cleaning

The effect of the cleaning frequency, 2 versus 3 system cleanings per day, was studied on commercial farms. The farms were selected on their milk quality in the past six month (average TBC<= 15 ($x10^3$/ml) without large fluctuations) and on the willingness to participate in the project under the project conditions. During 9 weeks the system was cleaned 2 times/day and during 9 weeks 3 times per day. The order of the cleaning frequency was set at random. During both periods of 9 weeks the bulk milk quality was analysed: total bacterial count (TBC) each collection of bulk milk, coliform count (CC), thermoduric count (TC) and psychrotrophic count (PC) once per two weeks. Also the freezing point of the milk was measured once every two weeks. The frequency of bulk milk collection was every 3 days or 3 times/week (2, 2 and 3 days), depending on regional systems.

Samples for bacteriological quality were taken by the lorry driver who is collecting the milk and is approved for taking and handling samples. The samples were put on ice, transported to the MCS Nederland (Milk Control Station the Netherlands) and analysed within 24 hours after sampling. Milk collected on Saturdays was not sampled, due to the impossibility to analyse the samples within 24 hours.

TBC was analysed using the Bactoscan 8000, CC conform NEN 6874, TC conform NEN 6807 and PC conform ISO 6730.

On each farm the system cleaning was analysed and the hygienic condition of the AM-system was checked at the start of the project, at change-over of the frequency and at the end of the project. These check consisted of measuring the volumes, concentration and temperatures of the cleaning fluids, the efficiency of the post rinse during one cleaning. For this test a procedure was developed.

The hygienic condition of the installation was tested by visual checking of known weak points in the installation (f.i. liners) and ATP-measurements of the housing of the inline milk filter. The choice for ATP-measurements in the housing of the inline filter is based on the fact that this place is easily assessed in all AM-systems and is one of the weak points in the system due to minimal turbulence and low flow rate during cleaning. To collect a sample, the swab is brought inside the tube 10 cm deep and a sample is collected by making a spiral motion upwards, using 10 spins and covering 5 cm of tube length.

The collected data was evaluated by an analysis of variance using GenStat. In the fixed part of the model brand, treatment, period and interactions between these were tested. No significant interactions whereas obtained. In the random part of the model were farm within brand and residuals. The random components were assumed to be normal distributed. For TBC, TC and PC bacteria log transformation was used a log transformation to meet the demands.

Results and discussion

Teat cleaning

Results of the statistical analysis of the most important results in milk are given in Table 1. Because an significant interaction of the effect of teat cleaning and brand was found, the results per brand with and without teat cleaning are presented. In all cases an effect of teat cleaning was found, but differences between brands could be concluded. Teat cleaning of brand 4 was comparable with conventional teat cleaning. Brand 2, 3 and 6 seemed to be less effective in teat cleaning (50-70% reduction).

Differences between brands may be due to the different way of teat cleaning, but brands with less effective teat cleaning were not all with cup cleaning or with brushes. So cleaning with cups being better than cleaning with brushes could not be concluded.

The level of reduction of the manual teat cleaning was rather high, compared to for instance the removal of bacterial spores where a maximal reduction of 96% (moist washable towel + dry paper (10 + 10s)) compared to no cleaning was found (Magnusson *et al.*, 2002). Melin *et al.* (2002) found better removal of spores with a VMS™ teat cleaning procedure (98% reduction) than with conventional manual cleaning (66,5% reduction). The manual cleaning was performed according to normal preparation procedure in the milking parlour. In our study the manual teat cleaning was also a normal preparation procedure (about 10 sec per cow with a moist washable towel on one farm and on the other farm with dry paper during about 10 sec per cow). Our reduction of 99% was better than for spores, probably because of better removal properties of poppy seed or less adhesion of poppy seed to the teats. Also less poppy seed than spores could be applied to the teats because of different diameters of the seed and spores. Vries and Stadhouders (1977) found a maximal reduction of butyric acid spores of 90% with intensive udder preparation and not too high levels of spores.

Obvious is also the difference between brands in level of poppy seed in milk from cows without teat cleaning. These differences may be partly caused by some practical problems occurring during the experiments.

Table 1. Back-transformed mean percentages of probability of poppy seeds in milk from cows with and without teat cleaning. Reduction of effect of teat cleaning is calculated from these probabilities.

Brand	Teat cleaning		
	No (%)	Yes (%)	Reduction (%)
Conventional	8.3	0.1	99
1	10.5	1.3	88
2	15.0	6.6	56
3	23.2	8.4	65
4	18.1	0.2	99
5	24.6	2.5	90
6	11.0	5.3	51

System cleaning

Thirteen farms participated in the project. On these farms the system cleaning was checked before the start of the experiment: in six cases the system cleaning had to be adjusted to improve the concentration of the cleaning fluid (3 x), to raise the water quantity for the after rinse (2 x) or for the pre rinse (1 x). Problems with the water quantity do often occur when the quantity is based on filling time and the pressure in the water supply is below normal values. At the change-over of frequency and at the end of the experiment these checks of the system cleaning and of the hygienic condition of the installation were repeated. No defects or problems were found then.

In case a raise in TBC occurred (TBC > 50(x10^3)) the farmer was informed, questioned about the cause of the problem and in case no explanation could be found, the farm was visited and the installation was checked.

In seven cases the TBC in the bulk milk showed raised values. The cause of the problems was variable; a broken valve in the cleaning system, insufficient hot water (2 x), faulty attachment of teat cups to the jetter blockage in the cleaning system, an alarm which was not handled properly and one unknown cause. In some cases the cause was hard to find, so elevated TBC was found in consecutive samples.

For the analysis the bacterial counts are log 10 transformed, in order to create a normal distribution. The results in Table 1 are transformed again to numbers of cfu/ml.

For all bacterial groups there is a significant difference between two and three cleanings per day. The difference for TBC is however small and in average for both frequencies far below the penalty limit (100*10^3 cfu/ml). 8 results on 4 farms showed TBC over 100; the cases were equally divided over both cleaning frequencies.

As expected, the freezing point of the milk was slightly higher when more system cleanings are performed.

On five farms, using four types of AM-systems, there was no difference in the average log TBC between 2 and 3 times cleaning per day (Table 2), indicating that also with a lower cleaning frequency it is possible to have the same milk quality. The influence of the farm and type of AM-system is small for coliform bacteria compared to TBC and to TC and PC. As can be seen from the causes of hygienic problems, it is important to have a proper installation check, a good maintenance of the AM-system and a keen operator.

Table 2. Effect of cleaning frequency on milk quality.

Quality parameter	Frequency		
	2 times/day	3 times/day	significance
TBC (10^3 cfu/ml)	13	10	<0.001
Coliform bacteria (cfu/ml)	173	13	<0.001
Thermoduric bacteria (cfu/ml)	877	320	<0.001
Psychrotrofic bacteria (cfu/ml)	1047	522	<0.001
Freezing point (°C)	-0.520	-0.519	0.003

Conclusions

Use of artificial teat contamination with poppy seed showed that automatic teat cleaning removes most of the contamination, with differences in efficiency per brand.
Cleaning AMS three time a day resulted in lower TBC, coliforms, thermoduric and psychrotrophic counts that cleaning the same AMS twice a day.

References

Frost, A.R., T.T. Mottram, C.J. Allen and R.P. White, 1999. Influence of milking interval on the total bacterial count in a simulated automatic milking system. J. Dairy Res 66: p. 125-129.

IDF (`1996) General recommendations for the Hygienic Design of Dairy Equipment. Bulletin of IDF 310. 5 p.

Ipema, A.H. 1997. Integration of robotic milking in dairy housing systems. Review of cow traffic and milking capacity aspects. Comput. Electron. Agric. 17: p. 79-94.

Knappstein, K., J. Reichmuth. and G. Suhren. 2002. Influence on bacteriological quality of milk in herds using automatic milking systems and experiences from selected German farms. Proc. First North American Conference on Robotic Milking, March 20-22, 2002, Toronto, Ontario, Canada. p. V13-24.

Knappstein, K., N. Roth, H.G. Walte, J. Reichmuth, B.A. Slaghuis, R.T. Ferwerda- van Zonneveld and A. Mooiweer, 2004. Report on the effectiveness of cleaning procedures applied in different automatic milking systems. Report D14, EU project Automatic Milking. www.automaticmilking.nl

Magnusson, M., A. Christiansson, B. Svensson and C. Kolstrup, 2002. Effect of different manual cleaning methods on spores in milk. In: The First North American Conference on Robotic Milking- March 20-22, 2002 J. McClean, M. Sinclair and B.West (editors) Wageningen Pers, Wageningen, The Netherlands, p. IV-67-70.

Melin, M., H. Wiktorsson and A. Christiansson, 2002. Teat cleaning efficiency before milking in DeLaval VMS versus conventional manual cleaning, using Clostridium tyrobutyricum spores as marker. In: The First North American Conference on Robotic Milking- March 20-22, 2002, J. McClean, M. Sinclair and B.West (editors) Wageningen Pers, Wageningen, The Netherlands, p. II-60-63.

Ordolff, D. and D. Bölling, 1992. Effects of milking intervals on the demand for cleaning the milking system in robotized stations. In: Proceedings for Automatic Milking A.H. Ipema, A.C. Lippus, J.H.M. Metz and W. Rossing (editors), (EAAP Publication No. 65), Wageningen Pers, Wageningen, the Netherlands, p. 169-174.

Schuiling, E. 1995. Eisen aan de reiniging bij automatisch melken. In: Reiniging en afvalwater rond de melkwinning, Praktijkonderzoek Rundvee, Schapen en Paarden, p. 29-31.

Schuiling, E. 1992. Teat cleaning and stimulation. In: Prospects for Automatic Milking. Pudoc, A.H. Ipema (editor), Wageningen Pers, Wageningen, Netherlands, p. 164-168.

Slaghuis, B. 1996. Sources and significance of contaminants on different levels of raw milk production. In: Symposium on bacteriological Quality of raw milk proceedings. IDF, Austria, p. 19-27.

Ten Hag, J., and K.E Leslie 2002. Preliminary investigation of teat cleaning procedures in a robotic milking system. Proc. of The First North American Conference on Robotic Milking, March 20-22, 2002, Toronto, Canada, p. V 55-58.

Verheij, J.G.P. 1992. Cleaning frequency of automatic milking equipment. In: Proceedings for Automatic Milking, H. Ipema, A.C. Lippus, J.H.M. Metz and W. Rossing (editors), (EAAP Publication No. 65), Wageningen Pers, Wageningen, the Netherlands, p. 175-178.

Verstappen, J., G. Klungel, G. Wolters and H. Hogeveen, 1999. Lange persleiding bij automatisch melken geen probleem, Praktijkonderzoek Rundvee, Schapen en Paarden 12 (1), p. 18-19.

Vorst, Y. van der, K. Knappstein and M.D. Rasmussen 2002. Effect of automatic milking on the quality of produced milk. Report D8, EU project Automatic Milking. www.automaticmilking.nl

Vries, Tj. De and J. Stadhouders, 1977. Boterzuurbacteriën in melk. Bedrijfsontwikkeling 8: p. 123-127.

Short communications: Beef

An association of leptin gene polymorphism with carcass traits in cattle

J. Oprządek, E. Dymnicki, U. Charytonik and L. Zwierzchowski*
Institute of Genetics and Animal Breeding, Polish Academy of Sciences. Jastrzębiec 05-552 Wólka Kosowska, Poland.
**Corresponding author: j.oprzadek@ighz.pl*

Summary

The objective of this study was to characterize genetic variation at two LEP loci and to evaluate their allelic effects on carcass traits in cattle. Genotypes of leptin (LEP) were analysed using the PCR-RFLP technique with *Hph*I and *Kpn*2I restriction endonucleases. The allele frequencies at the respective loci were 0.77/0.23 for LEP A/B variants and 0.60/0.40 for LEP C/T variants. The weight of valuable cuts in carcass was highest in LEP/ *Hph*I BB genotype bulls - 50.4 kg as compared to 49,4 in AA and 49,2 in AB animals. The BB genotype had simultaneously significant higher lean weight and higher fat weight of valuables cuts. In this study the effect of LEP/ *Kpn*2I genotype was observed only for dressing percentage. Both homozygotes had a highest dressing percentage as compared to heterozygotes CT animals.

Keywords: leptin, polymorphism, cattle, carcass traits

Introduction

In 1994, the mouse *ob* gene was shown to encode a secreted protein of 16 kDa. The protein was named leptin (from Greek *leptos*=thin). In *ob*$^-$/*ob*$^-$ mice the gene contains a missense mutation resulting in the absence of leptin and overt obesity (Zhang *et al.*, 1994). Subsequently, the leptin cDNA was cloned in sheep (Dyer *et al.*, 1997), cattle (Ji *et al.*, 1998), pigs (Bidwell *et al.*, 1997) and chicken (Taouis *et al.*, 1998). Mutations in the leptin or leptin receptor genes results in morbid obesity, infertility and insulin resistance (Houseknecht and Portocarrero, 1998). The leptin protein is highly conserved across all of these species. Leptin is a hormone secreted mainly by adipose tissue. One of its essential roles is to inform the organism about the level of fat reserves.

The leptin gene is expressed in bovine and ovine adipose tissues. Recent results on variations in plasma leptin and/or levels of leptin mRNA in adipose tissues showed positive effects of body fatness and feeding level. Progress in knowledge about leptin will allow to better understanding and controlling the adaptations of energy metabolism and reproductive activity of ruminants to seasonal variations in day length and food supply, as well as variations in carcass fatness of growing ruminants (Houseknecht and Portocarrero, 1998). Molecular probes were developed using polymerase chain reaction (PCR) technology to evaluate leptin expression in adipose depots and to evaluate the tissue-dependent nature of expression reported in other species (Ji *et al.*, 1998). *In vitro* studies suggest that leptin can directly modulate energy metabolism in peripheral tissues and may antagonize insulin activities in adipose and muscle. These physiological properties support leptin as strong candidate gene for evaluation of genetic polymorphisms that could affect carcass fat and muscle in cattle.

Several polymorphisms within the leptin gene in cattle have been described but only few studies have been performed of the effect of leptin gene polymorphism on performance traits in cattle. Buchanan *et al.* (2002) have reported that the T allele was associated with fatter carcasses and C allele with leaner carcasses.

The aim of this study was to characterise the genetic variation at LEP *loci* in young Polish Friesian bulls and to evaluate their allelic effects on carcass traits.

Material and methods

The study included data from 115 Polish Friesian bulls, originating from 20 Holstein sires. The animals were born and maintained at the local station and were housed in tie-stalls and fed *ad libitum* on silage, hay and concentrate, according to age, under a standardised feeding regimen (Polish Norms, 1993). Approximately 10 ml blood was withdrawn from each animal to test tubes containing K_2 EDTA. The bulls were slaughtered at the age of 11 months, at the local abattoir (0 to 2 h after arrival). The carcasses were split sagittaly and chilled at 4°C. After a period of 24 h, the right side of the carcass was divided into valuable cuts and rest of joints. The valuable cuts included fore and best ribs (1-13 thoracic vertebrae), sirloin (1-6 lumbar vertebrae), shoulder and round of beef. The weights of chilled sides, valuable cuts and all dissection products were recorded. Only the valuable cuts were dissected into cleaned bones, meat and fat.

Genotypes of leptin (LEP) were analysed using the PCR-RFLP technique. Crude DNA was prepared from blood samples according to Kanai *et al.* (1994).

The LEP/*Hph*I genetic variants were identified according to Haegeman *et al.* (2000). This polymorphism is situated in exon 3 of the leptin gene and results in amino acid change from alanine to valine. The sequences of primers were: 5'GGGAAGGGCAGAAAGATAG3' and 3' AGGCAGACTGT*TGAGGATC5'. Thermal cycling conditions included an initial denaturation at 94ºC for 1 min, followed by 35 cycles of 94°C for 30 s, 55°C for 1 min and 72° C for 30 s, followed by a final extension for 15 min at 72°C. The length of digestion fragments were 331 bp and 311 bp + 20 bp.

The LEP/*Kpn*2I genotypes were identified according to Buchanan *et al.* (2002). The sequences of primers were: 5'ATGCGCTGTGG-ACCCCTGTATC3' and 5' TGGTGTCATCCTGGACCTCC3'. The amplification program consisted of an initial denaturation at f 94°C - 2 min, followed by a cycle that ran 35 times: 94ºC - 45 s; 52ºC - 45 s, and 72ºC for 55 s. A final extension of 72°C was maintained for 3 min. The SNP in exon 2 was distinguished by digestion with the restriction enzyme *Kpn*2I. The C allele was cleaved into two fragments of 75 and 19 bp, while T allele remained uncut at 94 bp. The digested DNA fragments were then separated by electrophoresis in 2% agarose (Gibco, BRL, England) in 1 x TBE buffer (0.09 M Tris-boric acid, 0.002 M EDTA) with 0.5 mg/ml ethidium bromide (Et-Br) added to the gels, visualised under UV light, and scanned in FX Molecular Imager apparatus (Bio-Rad).

The effects of individual genotypes on the traits under study were analysed by the least-squares method as applied in the general linear model (GLM) procedure of SAS according to the following statistical model:

$$Y_{ijkl} = \mu + G_i + R_j + S_{\cdot k} + \beta (x_{ijkl} - x) + e_{ijkl}$$

where: Y_{ijkl} = studied traits;
μ = overall mean;
G_i = the fixed effect of LEP genotype;
R_j = the fixed effect of year at start of fattening;
$S_{\cdot k}$ = the fixed effect of season at start of fattening;
$\beta (x_{ijkl} - x)$ = regression on the cold carcass weight;
e_{ijkl} = the random residual effect.

The differences were tested by Duncan's test.

Results

The genotype frequencies are given in Table 1. The allele frequencies at the loci studied were 0.77/0.23 for LEP (*Hph*I) A/B variants and 0.60/0.40 for LEP (*Kpn*2I) C/T variants. Significant differences were found between LEP/*Hph*I genotypes (Table 2). The weight of valuable cuts in carcass was highest in BB genotype bulls - 50.4 kg as compared to 49.2 in AA and 49,4 in AB animals. The weight of lean and fat in valuable cuts was highest in BB homozygotes.

In this study the effect of LEP/*Kpn*2I genotype was observed only for dressing percentage (Table 2). The CT heterozygotes had a lowest dressing percentage as compared to CC and TT animals. No other important relationships between leptin genotypes and carcass traits were found.

Discussion

Only a few studies have been performed on the effect of leptin gene polymorphism on performance traits in cattle.

Jiang and Gibson (1999) found a possible association between the C/T substitution at position 3469 in LEP gene and the fatness in pigs, but the evidence was not conclusive. In beef cattle an

Table 1. Observed numbers of leptin (LEP) genotypes and corresponding allele frequencies.

Genotype		n	%	frequency
LEP/*Hph*I	AA	67	58.3	A=0.77
	AB	44	38.3	B=0.23
	BB	4	3.4	
LEP/*Kpn*2I	CC	36	31.3	C=0.60
	CT	65	56.5	T=0.40
	TT	14	12.2	

Table 2. Overall least squares means (LSM) and standard errors (SE) of carcass traits across the LEP/HphI genotypes and LEP/Kpn2I genotypes.

LEP/*Hph*I	AA		AB		BB	
Traits	LSM	SE	LSM	SE	LSM	SE
Dressing percentage (%)	49.7 [a]	0.22	50.3 [b]	0.29	50.6 [b]	0.70
Valuable cuts of carcass-side (kg)	49.4 [a]	0.10	49.2 [a]	0.10	50.4 [b]	0.40
Lean weight of valuable cuts (kg)	34.3 [a]	0.20	34.2 [a]	0.20	35.2 [b]	0.70
Bone weight of valuable cuts (kg)	10.3 [A]	0.06	9.9 [B]	0.07	10.3 [AB]	0.23
Fat weight of valuable cuts (kg)	5.1 [a]	0.12	5.4 [ab]	0.14	6.2 [b]	0.50
Valuable cuts share in carcass-side (kg)	60.8 [a]	0.16	60.9 [a]	0.20	62.1 [b]	0.50
LEP/*Kpn*2I	**CC**		**CT**		**TT**	
Trait	LSM	SE	LSM	SE	LSM	SE
Dressing percentage (%)	50.6 [a]	0.31	49.5 [b]	0.23	50.5 [a]	0.39

[aA] within traits means bearing different superscripts differ significantly: small letters - P ≤0.05; capitals - ≤0.01

association was found between carcass traits and polymorphism of BM 1500 microsatellite located closely to the leptin gene (Fitzsimmons *et al.*, 1998). Bierman *et al.* (2003) showed that genotype was not significantly associated with carcass weight or rib eye area. Associations of genotype with external fat thickness and overall cutability were small and generally not statistically significant. Similar differences between genotypes in marbling scores were observed when adjusting to a constant carcass weight or external fat thickness.

Our previous results (Zwierzchowski *et al.*, 2001, Oprządek *et al.*, 2003), showed that LEP/*Sau3*AI AB heterozygotes gave better results as regards weight of lean in selected carcass parts. The weight of valuable cuts in carcass was highest in BB genotype bulls - 50.4 kg as compared to 49.2 in AA and 49.4 in AB animals. The genotype was significantly associated with higher fat content in carcass. LEP/Sau3AI polymorphism affected also certain feed intake and feed conversion traits. In general bulls of AA LEP genotype consumed more feed and used more feed for gain of body weight than AB heterozygotes.

Thus, the leptin may prove to be one of several molecular genetic markers which could enhance genetic selection for improved livestock characteristics.

References

Bidwell, C.A., S. Ji, G.R. Frank, S.G. Cornelius, G.M. Willis and M.E. Spurlock, 1997. Cloning and expression of the porcine obese gene. Anim. Biotech. 8: p. 191-206.

Bierman, C.D., D.M. Marschall, E. Campbell and N.H. Granholm, 2003. Associations of leptin gene polymorphism with beef carcass traits. 2003 South Dakota Beef Report, South Dakota State Univ. p. 22-25.

Buchanan, F.C., C.J. Fitzsimmons, A.G. Van Kessel, T.D. Thue, D.C. Winkelman-Sim and S.M. Schmutz, 2002. Association of a missense mutation in the bovine leptin gene with carcass fat content and Leptin mRNA levels. Genet. Sel. Evol. 34: p. 105-116.

Dyer, C.J. J.M. Simmons, R.L. Matteri and D.H. Keisler, 1997. cDNA cloning and tissue-specific gene expression of ovine leptin, NPY-Y1 receptor, and NPY-Y2 receptor. Domest. Anim. Endocrinol. 14: p. 295-303.

Fitzsimmons, C.J., S.M. Schmutz, R.D. Bergen and J.J. Mckinnon, 1998. A potential association between the BM 1500 microsatellite and fat deposition in beef cattle. Mamm. Genome 9: p. 432-434.

Haegeman, A.A. Van Zeveren and L.J. Peelman, 2000. New mutation in exon 2 of the bovine leptin gene. Anim. Genet. 31: p. 79.

Houseknecht, K.L. and C.P. Portocarrero, 1998. Leptin and its receptors: regulators of whole-body energy homeostasis. Domest. Anim. Endocrinol. 15: p. 457-475.

Ji, S., G.M. Willis, R.R. Scott and M.E. Spurlock, 1998. Partial cloning and expression of the bovine leptin gene. Anim. Biotech. 9: p. 1-14.

Jiang, Z-H. and J.P. Gibson, 1999. Genetic polymorphism in the leptin gene and their association with fatness in four pig breeds. Mamm. Genome.10: p. 191-193.

Kanai, N., T. Fujii, K. Saito and T. Yokoyama, 1994. Rapid and simple method for preparation of genomic DNA from easily obtained clotted blood. J. Clinic.Path. 47: p. 1043-1044.

Oprządek, J., K. Flisikowski, L. Zwierzchowski and E. Dymnicki, 2003. Polymorphism at loci of leptin, Pit-1, and STAT5 and their association with growth, feed conversion and carcass quality in Black-and-White bulls. Anim. Sci. Pap. Rep. 21: p. 135-146.

Taouis, M., J-W. Chen, C. Davi, J. Dupont J, M. Derouet and J. Simon, 1998. Cloning the chicken leptin gene. Gene 208: p. 239-242.

Zhang, Y., R. Proenca, M. Maffei, M. Barone, L. Leopold and J.M. Friedman, 1994. Positional cloning of the mouse obese gene and its human homologue. Nature 372: p. 425-432.

Zwierzchowski, L., J. Oprządek, E. Dymnicki and P. Dzierzbicki, 2001. An association of growth hormone, κ-casein, β-lactoglobulin, leptin and Pit-1 loci polymorphism with growth rate and carcass traits in beef cattle. Anim. Sci. Pap. Rep. 19: p. 65-77.

Molecular and biochemical muscle characteristics of Charolais bulls divergently selected for muscle growth

I. Cassar-Malek[1], Y. Ueda[1], C. Bernard[1], C. Jurie[1], K. Sudre[1], A. Listrat[1], I. Barnola[1], G. Gentès[1], C. Leroux[1], G. Renand[2], P. Martin[2] and J.-F. Hocquette[1]*
[1]INRA, 63122 St-Genès-Champanelle, [2]INRA, 78352 Jouy-en-Josas, France
**Correspondence: cassar@clermont.inra.fr*

Summary

Up to now, genetic selection of beef cattle has been directed in favour of muscle growth. Consumers today seek products of high and reliable sensory quality (tenderness, flavour, etc), so we examined the consequences of growth selection on muscle biochemical and molecular characteristics putatively associated with meat quality. Charolais young bulls were divergently selected on the basis of their muscle growth capacity. Biochemical studies and transcriptome analysis using bovine muscle cDNA macro-arrays were performed with *Rectus abdominis* (RA, oxidative) and *Semitendinosus* (ST, glycolytic) muscle samples which were collected at 15 months of age from two groups of 3 extreme bulls of high (H) and low (L) muscle growth capacity. The two groups were characterised by differences in muscle mass and fat proportion in the carcass. Citrate synthase (a mitochondrial enzyme) activity was lower in RA muscle of H compared to L bulls. Transcriptome studies on pooled samples enabled the identification of 11 out of 400 genes (mostly of the contractile apparatus) that were differentially expressed in muscles between L and H bulls. In conclusion, both biochemical and transcriptomic data indicate that selection on muscle growth potential is associated with higher fast-glycolytic muscle characteristics, especially in RA muscle. Further studies are required to understand the physiological importance of the genes whose expression is regulated by selection.

Keywords: cattle, muscle growth, selection, gene expression profiling

Introduction

In France, genetic selection of specialized beef cattle has been directed in favour of muscle growth in order to produce lean carcasses with the ultimate objective of increasing production of meat at the expense of fat. Selective breeding in cattle has been indeed successful in increasing growth rates in beef cattle. In addition, consumers are looking for bovine meat of high and consistent quality. These demands are forcing farmers and processors to rethink their production systems. Meat eating quality traits are affected both by the *post-mortem* technological treatments applied to the carcass, and by the anatomical, physical and chemical characteristics of the muscles themselves. The latter depend upon the genetic potential of the animals and the production systems.

One key question is to clarify the consequences on muscle characteristics and on meat quality traits of a selection on muscle growth capacity. So, we have analysed the muscles of divergently selected animals using two complementary strategies: (i) the biochemical studies of muscles, (ii) the identification of differentially expressed genes in muscles using DNA array technology. This innovative technology is indeed an appropriate method to understand the biological mechanisms which determine eating quality traits of meat and to identify the underlying key genes. To achieve this aim, we have constructed a cDNA library and derived from it a specific set of relevant muscle gene probes devoted to muscle growth and meat quality traits (Sudre *et al.*, in press). We have thus performed gene expression profiling in two bovine muscles of bulls divergently selected on their muscle growth capacity.

Material and methods

Animals, samples and biochemical studies

The study was conducted with 64 young Charolais bulls. They were all born from pure breed Charolais cows from an INRA experimental herd, weaned at 32 weeks and then kept in an open shed. They were fed a complete pelleted diet distributed *ad libitum* with a limited amount of straw until slaughter. Animals were slaughtered at 15 months of age. The warm carcass and the internal fat deposit weights were recorded. The next day the 6th rib was dissected and the carcass composition (muscle and fat contents) was estimated using the Robelin and Geay (1975) prediction equation.

Bull calves were progeny of 25 Charolais sires divergently selected on their muscle growth capacity among 80 progeny-tested sires. This progeny testing was previously conducted in this herd with 793 slaughtered bull calves. The sires were ranked on a synthetic index combining their breeding value for a high muscle weight and a low carcass fat percentage. The sires used for procreating the current generation of experimental animals were chosen from each extreme of this selection index. The breeding value of the experimental animals was estimated in an animal model using all available information. They were ranked on a similar synthetic index as their sires and three bull calves from each extremity of its distribution were used in the present study. They were characterised either by a high (H) or a low (L) muscle growth genetic potential.

Two muscles of the carcass differing in their contractile and metabolic characteristics were excised from each animal: a slow oxidative muscle (*Rectus abdominis*, RA) and a fast glycolytic muscle (*Semitendinosus*, ST). The samples were immediately frozen in liquid nitrogen and stored at -80°C until analysed. Enzyme activities characteristics were determined as described (Piot *et al.*, 1998).

Transcriptome analysis

Total RNA was extracted from the RA and ST muscles of the three H and the three L bulls respectively. RNA samples were pooled to get one sample for each muscle and each genetic type. Poly(A)+ RNA were then isolated from the 4 pooled samples (RA-H, RA- L, ST-H, ST-L) and 500 ng were labelled during reverse transcription by incorporating [α-^{33}P]dCTP.

High density filters were constructed from a collection derived from a non-normalised directed bovine muscle cDNA library (Sudre *et al.*, in press). A set of 530 cDNAs signing 400 distinct muscle genes was spotted together with controls (water, buffer) in duplicate onto 8x12 cm Nylon membranes (672 x 2 spots). Each cDNA spot of the array was a PCR-amplified cDNA clone ranging between 200 bp and 500 bp. The bovine muscle gene set (Figure 1) comprised cDNAs representative of the contractile apparatus, metabolism, mitochondrial genes, proteolysis, replication, transcription, translation. Most of these sequences are registered in the EMBL Nucleotide sequence Database with accession numbers CR382441 to CR383573. In order to control technical variability, four filters were incubated with each radiolabelled cDNA target as previously described (Sudre *et al.*, in press). After washing, they were exposed 48 hours to phosphor screens before scanning on the PhosphorImager imaging plate system for quantitative analysis of signal intensity.

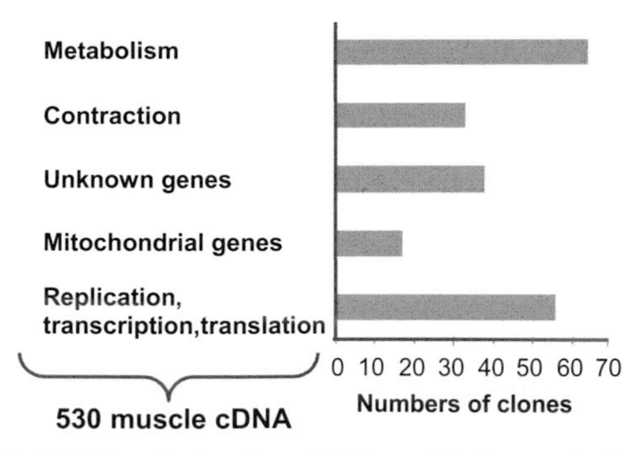

Figure 1. Categories of cDNA in the bovine cDNA set (Sudre et al., in press).

After acquisition, the scanned images were analysed using the GenePix Pro V4.1 software (Axon instrument, Inc). Data were analysed using a combination of the Statistical Analysis of Microarrays (SAM, Tusher *et al.*, 2001) and of the fold change method. Genes were declared as differentially expressed for a ratio ≥ 1.5 between two samples and a false discovery rate (FDR) < 5%.

Results

Bulls with a high muscle growth capacity were characterised by a muscle mass in the carcass 32% higher and by a proportion of fat in the carcass 26% lower than bulls with a low capacity ($P < 0.02$).

The activity of LDH was higher (+24%, $P < 0.05$) and that of ICDH lower (-24%, $P = 0.15$) in ST than in RA. These results are in agreement with the oxidative or glycolytic status of RA or ST muscles respectively.

Whereas ICDH and LDH activities did not differ between the two groups of animals, the mitochondrial citrate synthase activity was 1.8-fold higher in RA muscle of L bulls ($P < 0.05$) with a similar tendency in ST (x 1.5, $P < 0.20$) (Figure 2). This indicates that a high growth potential is associated with a less oxidative metabolism, especially in RA.

Analysis of hybridisation results revealed 171 to 185 reliable signals over 530 muscle clones (Table 1) which corresponds to a proportion of 32 to 35 %. Most of the genes (85 to 99 %) were

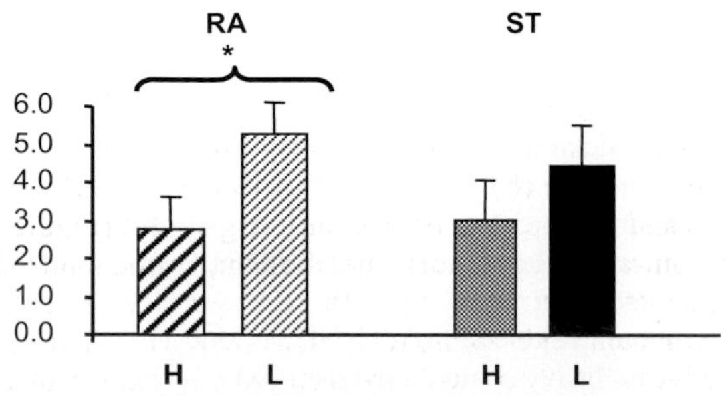

*Figure 2. Activity of citrate synthase in the Semitendinosus (ST) and Rectus abdominis (RA) muscles of bulls divergently selected for high (H) or low (L) muscle growth. One unit of enzyme is defined as the amount which, under assay conditions, catalyses the liberation of 1 µmol of coenzyme A per min and per g tissue fresh weight. ** P< 0.003.*

Table 1. Numbers of hybridised positive clones.

	H versus L		*RA versus ST*	
	in RA	in ST	in H	in L
Reliable clones	171	183	185	169
Significantly differentially[1] expressed clones	9	2	14	25

[1]Differential expression was based on SAM analysis with a minimal fold change of 1.5 (FDR < 5%).

similarly expressed in the four samples. This is in good agreement with data showing 20 to 40% of reliable signals and up to 10% of differential expression with macro-array technology (Hampson and Hughes, 2001).

Table 2. List of transcripts differentially[1] expressed between RA and ST muscles.

Function	Identity	Ratio RA/ST in H bulls	Ratio RA/ST in L bulls
Contraction	Troponin T cardiac (TNNT2)	3.21	
Cell regulation	Magic roundabout	2.41	2.14
Contraction	Myosin regulatory light chain 2 (CMLC2)	2.36	1.76
Myofibril assembly	Clone weakly similar to murine nebulin-related anchoring protein (N-RAP)	2.27	1.84
Transposon	Mariner1 transposase	2.08	1.65
Contraction	Troponin C slow (TNNC1)	1.94	
Heat shock protein	Alpha B-crystallin (CRYAB)	1.69	1.74
Ribosomal protein	Ribosomal protein S2 (RPS2)	0.65	0.47
Contraction	Myosin heavy chain MHC 67	0.64	0.47
Contraction	Myosin heavy chain 2a (MyHC-2a)	0.62	0.22
Contraction	Troponin T fast skeletal (TNNT3)	0.56	0.25
Cell regulation	Fetuin	0.55	0.32
Contraction	Myosin heavy chain 2x (MyHC-2x) (clone 1)	0.55	0.41
Contraction	Myosin heavy chain 2x (MyHC-2x) (clone 2)	0.49	0.29
Contraction	Myosin heavy chain 2a (MyHC-2a)		0.65
Contraction	Alpha actin, cardiac muscle		0.65
Contraction	Tropomyosin 1 (alpha, TPM1)		0.65
Metabolism	Aldolase A		0.65
Contraction	Alpha actin		0.62
Metabolism	Glyceraldehyde 3-phosphate dehydrogenase		0.61
Contraction	Myosin light chain 1, cardiac (CMLC1)		0.59
Contraction	Myosin light chain 1 (MLC1)		0.58
Contraction	Myosin regulatory light chain 2 (MRLC2)		0.58
Contraction	Gamma actin		0.56
Contraction	Troponin I skeletal fast (TNNI2) (clone 1)		0.49
Contraction	Myosin heavy chain 2x (MyHC-2x)		0.31
Contraction	Troponin I skeletal fast (TNNI2) (clone 2)		0.30

[1]Differential expression was based on SAM analysis with a minimal fold change of 1.5 (FDR < 5%).

The number of genes significantly differentially expressed between RA *vs* ST muscles was higher than between H and L genetic types (39 *vs* 11, Table 1). The identities of these two groups of genes are presented below in Tables 2 and 3 respectively.

Genes encoding proteins involved in fast contraction (Myosin heavy chain fast isoforms 2a and 2x, troponin T fast isoforms) and in glycolytic metabolism (Glyceraldehyde-3-phosphate dehydrogenase, aldolase A) appeared to be under-expressed in RA compared to ST muscle. This finding is consistent with histochemical and biochemical data indicating that RA is more oxidative than ST (Talmant *et al.*, 1986). Independent methods (RT-PCR, Northern-Blot) were used to validate the macro-array data (Figure 3A).

Interestingly, genetic selection on muscle growth capacity was associated with an increased expression of several transcripts involved in muscle contraction (isoforms of Myosin Heavy and Light Chains, tropomyosins, and troponins) or metabolism (lactate dehydrogenase A) as shown in Table 3. Some differentially expressed genes were confirmed by Northern blot analysis (Figure 3B). Thus, the genetic-type related muscle gene expression was in favour of a fast-glycolytic twitch orientation of muscles, mostly in RA.

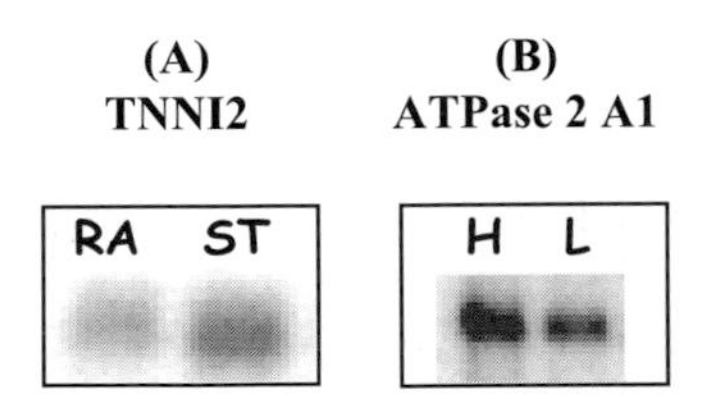

Figure 3. Examples of validation of differential expression between RA and ST muscles or H and L genetic types by Northern Blot analyses of poly(A+) RNA (500ng) from muscles using cDNA probes. (A) Troponin I2 (TNNI2); (B) ATPase Ca2+ transporting cardiac, fast twitch 1 (ATP 2A1).

Table 3. Transcripts up-regulated[1] in the muscles of high growth bulls potential (H) versus low growth potential bulls (L).

Function	Denomination	muscle	ratio H/L
Metabolism	Lactate dehydrogenase A (LDH-A)	RA	2.07
Contraction	Tropomyosin beta (TPM2)	RA	1.90
Contraction	Gamma actin (ACTG)	RA	1.74
Contraction	ATPase Ca2+ transporting cardiac, fast twitch 1 (ATP 2A1)	RA	1.64
Contraction	Myosin heavy chain 2x (MyHC 2x)	RA	1.58
Contraction	Troponin T, cardiac (TNNT2)	RA	1.57
Contraction	Tropomyosin alpha (TPM1)	RA	1.56
Contraction	Troponin I skeletal muscle (TNNI2)	RA	1.53
Contraction	Myosin light chain 1 (MLC1)	RA	1.51
Calcium/calmodulin signalling	Calcineurin A	ST	1.89
Proteolysis	Cullin	ST	1.67

[1]Differential expression was based on SAM analysis with a minimal fold change of 1.5 (FDR < 5%).

Discussion

The *in vivo* determinants of meat eating quality are not yet fully understood, but it is clear that the intrinsic characteristics of the muscle play an important role. They include proteases, collagen content, fat and fatty acid content and muscle fibre type of the animal (Geay *et al.*, 2001). However, muscle fibre, collagen and lipid characteristics only explained one fourth to one third of the variability of 2 day mechanical strength and 15 day tenderness or flavour scores (Renand *et al.*, 2001). Many studies indicate that a number of meat quality traits are heritable and variable between genotypes depending on many factors including growth potential (Burrow *et al.*, 2001). Thus, a better understanding of this genetic variability may allow the development of discriminatory tests for eating quality. It is therefore required to know more about the molecular, genetic and biochemical differences between animals with different ability to growth or which produce meat of different eating qualities.

In this study, the two groups of divergently selected bulls were characterised by important differences in carcass composition. The number of animals we have analysed is low (3 per group) but the differences in their carcass composition was high enough to detect significant differences in muscles which are associated with muscle growth capacity.

As expected, most of the genes differentially expressed between the two muscle types are involved in contractile and metabolic properties, which made us confident of our analyses. Interestingly, many of the genes differentially expressed between genetic types are genes involved in contractile muscle traits. The selection on muscle growth potential is indeed associated both with gene signing fast contractile apparatus and, to a lesser extent, glycolytic muscle metabolism, confirming biochemical data (a low citrate synthase activity) and other results from the literature (for review, see Hocquette *et al.*, 1998).

Lastly, transcriptomic analysis has allowed the identification of new putative markers of the genetic selection (e.g. cullin, a scaffolding protein involved in ubiquitin-mediated proteolysis) or of the muscle type (e.g. Magic roundabout, encoding a protein similar to axon guidance protein roundabout; fetuin, a protein interfering with signalling of TGF-β, HGF and insulin, putatively involved in cell differentiation and plasticity). Further studies are required to understand the putative implications of those genes in muscle biology.

Conclusions

The differences in muscle characteristics were higher between muscle types than between genetic types by both enzymatic and molecular studies. The selection in favour of muscle growth induces an orientation towards the fast glycolytic type, especially in the oxidative RA muscle. The specific library of bovine muscle genes has allowed the identification of 11 genes differentially expressed between the two genetic types; these genes are mostly involved in contractile traits but also in metabolism. These genes may be new indicators of muscle characteristics related to muscle growth and, by extension, to meat quality.

Acknowledgements

This research has been supported by a grant of INRA within the frame of the AGENAE program related to structural and functional genomics applied to Animal Science. The authors thank the French National Center of Biological Resources for Livestock Species Genomics (INRA, Jouy-en-Josas) for the preparation of the arrays and the SIGENAE team for the analysis of the cDNA sequences.

References

Burrow, H.M., S.S. Moore, D.J. Johnson, W. Barendse and B.M. Bindon, 2001. Quantitative and molecular genetic influences on properties of beef: a review. Aust. J. Exp. Agr. 41: p. 893-919.

Geay, Y., D. Bauchart, J.F. Hocquette and J. Culioli, 2001. Effect of nutritional factors on biochemical, structural and metabolic characteristics of muscles in ruminants; consequences on dietetic value and sensorial qualities of meat. Reprod. Nutr. Dev. 41: 1-26, Erratum, 41, p. 377.

Hampson, R., and S.M. Hughes, 2001. Muscular expressions: profiling genes in complex tissues. Genome Biol. 2: p. 1.033.1-10033.3.

Hocquette, J.F., I. Ortigues-Marty, D.W. Pethick, P. Herpin and X. Fernandez 1998. Nutritional and hormonal regulation of energy metabolism in skeletal muscles of meat-producing animals. Livest. Prod. Sci. 56: p. 115-143.

Piot, C., J.H. Veerkamp, D. Bauchart and J.F. Hocquette 1998. Contribution of mitochondria and peroxisomes to palmitate oxidation in rat and bovine tissues. Comp. Biochem. Physiol. B 121: p. 69-78.

Renand, G., B. Picard, C. Touraille, P. Berge and J. Lepetit, 2001. Joint variability of meat quality traits and muscle characteristics of young Charolais beef cattle. Meat Sci. 59: p. 49-60.

Robelin, J. and Y. Geay 1975. Estimation de la composition des carcasses de jeunes bovins à partir de la composition d'un morceau monocostal prélevé au niveau de la 11ème côte. Ann. Zootech. 24: p. 391-402.

Sudre, K., C. Leroux, I. Cassar-Malek, J.-F. Hocquette and P. Martin, 2005. A collection of bovine cDNA probes for gene expression profiling in muscle. Mol. Cell. Probes *In press.*

Talmant, A., G. Monin, M. Briand, L. Dadet and Y. Briand, 1986. Activities of metabolic and contractile enzymes in 18 bovine muscles. Meat Sci. 18: p. 23-40.

Tusher, V.G., R.Tibshirani and G. Chu, 2001. Significance analysis of micro-arrays applied to the ionizing radiation response. Proc. Natl. Acad. Sci. USA 98: p. 5116-5121.

Proteomics applied to the analysis of bovine muscle hypertrophy

B. Picard[1], J. Bouley[1], I. Cassar-Malek[1], C. Bernard[1], G. Renand[2] and J.F. Hocquette[1]*
[1]INRA, Unité de Recherches sur les Herbivores, Theix, 63122 Saint-Genès-Champanelle, France,
[2]INRA, Station de Génétique Quantitative et Appliquée 78352 Jouy-en-Josas, France.
**Correspondence: picard@clermont.inra.fr*

Summary

Muscle hypertrophy is of particular interest in beef meat production as it has a strong economic importance. High muscle development is accompanied by particular characteristics in favour of improvement of meat tenderness but not of its flavour. This study was therefore conducted to identify some markers of muscle hypertrophy. We have studied two models of cattle with different origins of muscle hypertrophy: monogenic (Belgium Blue bulls double-muscled: DM) or polygenic (divergent lineages of Charolais bulls: with high (H) or low (L) muscle growth rate). Differential proteomic analysis of *Semitendinosus* muscle (ST, mixed fast glycolytic) was performed using two-dimensional gel electrophoresis (4-7 pH gradient in the first dimension and 11% SDS-PAGE in the second) followed by mass spectrometric analysis of interesting spots. Among the proteins differentially expressed between high and low muscled cattle, results revealed 17 proteins common of the two models DM and H. Eight proteins were over-expressed in high-muscled cattle. They corresponded mainly to Myosin Binding Protein H, isoforms of fast Troponin T, Myosin Regulatory light chain 2 and 3, Phosphoglucomutase, etc. Nine other proteins such as slow Troponin T isoforms, slow isoforms of Myosin light chain 1, p20 were under-expressed in high muscled cattle. These proteins variably expressed in the different models could be good markers for muscle hypertrophy. Our results provide further evidence for a reinforced fast glycolytic phenotype in *Semitendinosus* muscle of high-muscled bulls. These properties have been confirmed by biochemical studies and a transcriptomic analysis of ST and *Rectus abdominis* (RA, slow oxidative) muscles of the same animals belonging to the divergent lineage of Charolais bulls.

Keywords: cattle, muscle hypertrophy, proteomics

Introduction

One of the aims of beef production has been to increase muscle mass. This could be obtained by using specific genotypes with high muscle growth potential or by selection. Different studies (Picard *et al.*, 2002) have clearly shown a higher proliferation of myoblasts in muscles from double-muscled (DM) cattle inducing a higher total number of fibres in adult muscles. The results of Deveaux *et al.*, (2001) showed that this higher proliferation was specific for the second generation of myoblasts giving fast IIX fibres. This suggests a negative control of proliferation of this second generation of muscle cells by the myostatin growth factor which is mutated in DM cattle. The work of Duris and Picard, (1999) in cell cultures of Charolais muscles selected for high muscle growth (H), revealed a higher proliferation of myoblasts as observed in DM muscle.

The aim of the present study was to identify some markers of muscle hypertrophy by the functional genomics approach in order to better understand the mechanism controlling muscle hypertrophy and to provide some markers which could be included into the selection schema.

Materials and Methods

Animals and muscle samples

The *Semitendinosus* muscle (ST, mixed fast glycolytic) was collected just after slaughter of animals, immediately frozen in liquid nitrogen and stored at - 80°C until analysis. Two groups of young bulls 15-month-old on average with different origins of muscle hypertrophy were studied. One group with monogenic origin of muscle hypertrophy was comprised of Belgium Blue young bulls: homozygote double-muscled (DM) compared to heterozygote double-muscled (HDM) and homozygote non double-muscled (NDM). All animals were genotyped according to the protocol of Grobet *et al.*, (1998). The other group, with polygenic origin of muscle hypertrophy, was comprised of Charolais young bulls divergently selected on their muscle growth capacity: high-muscled (H) compared to low-muscled (L).

Proteomic analysis

Sample preparation and two dimensional gel electrophoresis (2-DE) were performed according to Bouley *et al.*, (2004). Briefly, frozen muscle tissue was homogenized in a lysis buffer containing 8.3 M urea, 2 M thiourea, 1 % DTT, 2 % CHAPS and 2 % IPG buffer pH 3-10 and centrifuged at 10000 g for 30 min. Three muscle samples of each genotype were analyzed (with three gels of each sample). In the first dimension 800 μg of proteins were subjected to isoelectric focusing (IEF) in an immobilized pH gradient (IPG strips pH 4-7, 18 cm) (73.5 kVh) in a Multiphor II gel apparatus at a temperature of 20.5°C. In the second dimension, proteins were separated on 11%T, 2.6% C separating polyacrylamide gels (SDS-PAGE) using Hoefer DALTsix system. 2-D gels were stained using a colloidal CBB G-250 procedure (Neuhoff *et al.*, 1988). Spot detection and quantification were performed with ImageMaster 2D Elite software (Bouley *et al.*, 2004). A statistical analysis derived from the standard *t* test used for analysis of microarray data, the SAM method (for Significance Analysis of Microarrays) (Tusher *et al.*, 2001) and adapted for proteomic analysis (Meunier *et al.*, 2004) was used to detect proteins differentially expressed between two groups. Proteins of interest were identified by reference to the protein map established by Bouley *et al.* (2004) or by mass spectrometry. Protein polypeptides underwent trypsin digestion (Montage In-Gel Digestion Kit, Millipore, Bedford, MA, USA) followed by a MALDI-MS analysis using a MALDI-Tof Voyager-DE Pro (Perspective Biosystems, Inc, Farminghan, MA, USA). The masses were compared with known trypsin digest databases (NCBI non redundant and SWISS-PROT) using Mascot (www.matrixscience.com) and Profound (http://prowl.rockefeller.edu/cgi-bin/ProFound) search engines.

Results and Discussion

The analysis of muscle fibre proportions in Belgium Blue young bulls showed a lower proportion of slow type I fibres and a higher proportion of type IIX fibres (Figure 1) in DM muscle than in NDM in accordance with the data of the literature (Picard *et al.*, 2002). The muscle of HDM presented intermediate properties. The same trend was observed in ST muscle of H Charolais comparatively with L, but the differences were not significant (Bouley *et al.*, 2005).

Among the proteins differentially expressed in DM muscle, eight were related to contractile apparatus including myosin-binding protein H (MyBP-H, SSP 23, 26 and 132), several myosin light chains including MLC2 (SSP 135), MLC3 (SSP 113), MLC2s (SSP 133, 134 and 114), MLC1sa (SSP 102) and MLC1sb (SSP 94) and troponin T including slow TnT (sTnT, SSP 67, 68, 79 and 80) and fast TnT (fTnT, SSP 72, 74, 81, 83 and 85). The spot 68 (sTnT) was the only contractile apparatus protein that was significantly altered in HDM muscle. Two proteins were

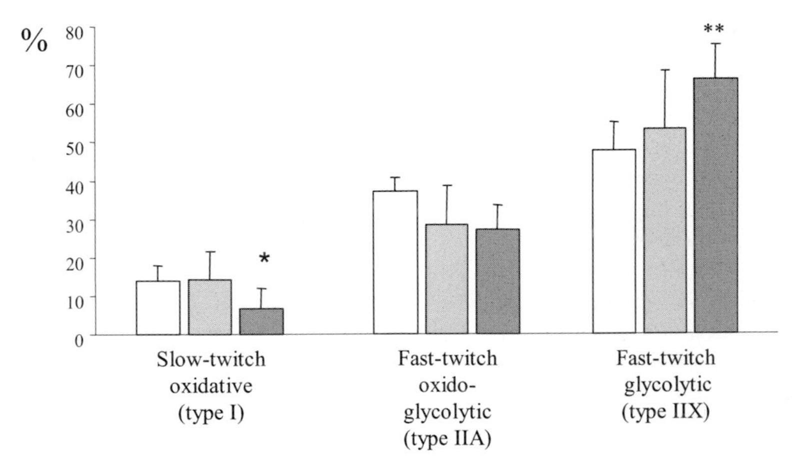

Figure 1. Fibre type composition in bovine Semitendinosus muscle.
*NDM (homozygote non double-muscled, in white), HDM (heterozygote double-muscled, in grey) and DM (homozygote double-muscled, in dark grey) (n=6, for each genetic type). Each bar represents the mean percentage (± SEM) of each type of fibre calculated on muscle cross sections. * P< 0.05 vs. control NDM. ** P< 0.01 vs. control NDM.*

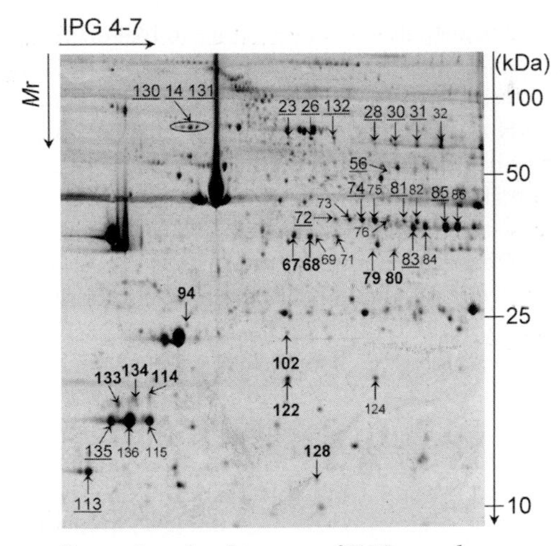

Figure 2. Representative two dimensional gel image of DM muscle.
Proteins (800 µg) were separated in the first dimension using a pH range of 4-7 and visualized by a colloidal blue staining. Spot numbers refer to Table 1. Protein spots significantly up-regulated (underlined) and down-regulated (bold type) in muscle from homozygote (DM) animals for the myostatin deletion (FC≥ 2-fold, FDR< 0.01 vs. control NDM).

involved in metabolic pathways, including phosphoglucomutase (PGM, SSP 28, 30, 31) and heart fatty acid-binding protein (H-FABP, SSP 128). Finally, three proteins were significantly altered including sarcosin (SSP 130, 14 and 131), sarcoplasmic reticulum 53kDa glycoprotein (SR53G, SSP 56) and p20 (SSP 122). Interestingly, an intermediate level of protein expression was generally found in HDM muscle (Figure 3)

According to these data it seems that contractile properties are more modified than metabolic ones in DM muscles. However, most of the proteins implied in energetic metabolism have a basic pH.

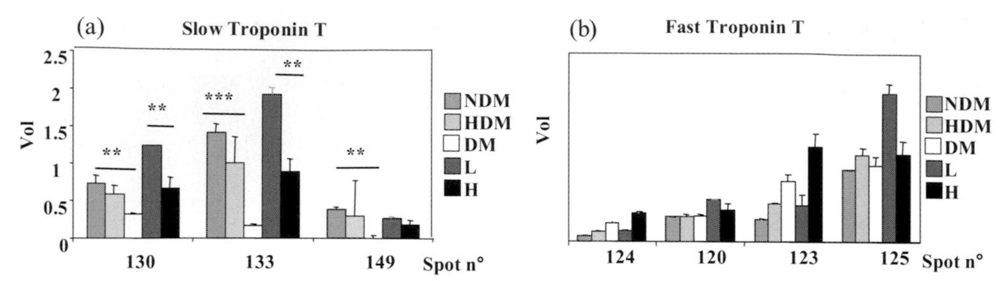

Figure 3. Differential expression of some slow (a) and fast (b) Troponin T isoforms, in Semitendinosus muscle of young bulls with different muscle growth rate. Data are expressed in volumes of the protein spots on the gels (defined as the product of spot area and spot integrated optical density in arbitrary units). DM: Belgium blue homozygote double-muscled; HDM: Belgium blue heterozygote double-muscled; NDM: Belgium blue homozygote non double-muscled; H: Charolais bulls with high muscle growth rate; L: Charolais bulls with low muscle growth rate.

So, it would be interesting to complete this study by a 2DE in a basic pH gradient in the first dimension to better study proteins of the metabolism. However, most of the modifications of protein expression are in favour of fast glycolytic properties of DM muscle. This is in agreement with the data of the literature showing more of a fast glycolytic pattern of hypertrophied muscles.

Interestingly, a similar analysis conducted on ST muscle from H and L young bulls revealed that the proteins differentially expressed in DM muscle were also differentially expressed in H muscle (Table 1). Only the expression of H-FABP which was decreased in DM muscle was not significantly modified in H muscle. The expression of SR53G (over-expressed in DM) was not modified in H muscle. All other proteins differentially expressed in DM muscle were also differentially expressed in H muscle.

Examples of differentially expressed proteins are given in Figure 3. Spots of slow Troponin T (sTnT) were under-expressed in DM and H muscles, and had an intermediate expression in HDM (Figure 3a). On the contrary, fast isoforms (spots 124 and 123) (Figure 3b) were over-expressed in DM and H muscles. Interestingly, we observed that spots 120 and 125 corresponding also to fast troponin T isoforms (fTnT), were not different between the different genotypes. Eleven spots of fast Troponin T were analysed more precisely. The expression levels of five of them were significantly increased in DM and H muscles. The levels of expression of the other six ones were unchanged or decreased in DM and H muscles. These fTnT isoforms differ in terms of the presence of the alternative splicing region corresponding to mutually exclusive exons 16 and 17 (Muroya *et al.*, 2003; Bouley *et al.*, 2004).

All these data demonstrated that the troponin T fast isoforms over-expressed in the muscle of high muscle growth potential (DM and H) contained the exon 16 and the others contain the exon 17. This could constitute a good protein marker of muscle hypertrophy. This higher expression of fTnT with exon 16 in muscles of high growth potential showing an over expression of fast contractile proteins is in agreement with the literature. For example, Bucher *et al.*, (1989) showed a correlation between the expression of fTnT exon 16 and fast contractile status of muscle. Also, Jin *et al.* (1998) found a higher proportion of fTnT cDNA encoding the mutually exclusive fTnT exon 16 in a fast glycolytic muscle (*Pectoralis*).

The orientation toward a fast glycolytic pattern of hypertrophied muscle is in accordance with biochemical data (Picard *et al.*, 2002). For instance, citrate synthase activity (a mitochondrial enzyme of oxidative metabolism) is higher in muscles from the H Charolais young bulls compared

Table 1. Proteins differentially expressed in Semitendinosus muscle of double-muscled (DM) young bulls and of Charolais young bulls selected on high muscle growth rate (H). The numbers correspond to the access in databases SWISS-PROT or NCBI. Statistical analysis was performed using SAM as described in methods (ratio ≥ 2-fold, FDR (false discovery rate) < 0.01 vs. normal heterozygote double-muscled (HDM) or low muscled Charolais young bulls (L) for DM and H animals respectively -, proteins not changed).

Spots	Proteins	Number of accession	Muscle DM (Ratio)	Muscle H (Ratio)
Contractile apparatus				
23, 26, 132	Myosin-binding protein H (MyBP-H)	Q13203	Increased 3.8, 20, 5.2	Increased 15, 3.5, 3
67, 68	Slow troponin T (sTnT high M_r)	gi|21039008	Decreased 4.2, 5.1	Decreased 2.4, 2
69, 71	Slow troponin T (sTnT high M_r)	gi|21039008	Decreased 1.5, 1.4	Decreased 3, 2
79, 80	Slow troponin T (sTnT low M_r)	gi|21039010	Decreased 2.2, 3.6	Decreased 2.7, 1.5
72, 74	Fast troponin T (fTnT with exon 16)	gi|21038992	Increased 2.6, 3.7	Increased 2, 2.2
73	Fast troponin T (fTnT with exon 17)	gi|21038994	Decreased 1.7	Decreased 1.3
75, 76	Fast troponin T (fTnT with exon 17)	gi|21038994	-	Decreased 1.4, 1.4
81, 83, 85	Fast troponin T (fTnT with exon 16)	gi|21038996	Increased 4.9, 2.2, 2.1	Increased 2.5, 2.2, 3
82	Fast troponin T (fTnT with exon 17)	gi|21038998	-	decreased 1.2
84, 86	Fast troponin T (fTnT with exon 17)	gi|21038998	Decreased 1.4, 1.5	Decreased 1.2, 1.4
102	Myosin light chain 1, slow-twitch muscle A isoform (MLC1sa)	P14649	Decreased 3.3	Decreased 2.1
94	Myosin light chain 1, slow-twitch muscle B isoform (MLC1sb)	P16409	Decreased 3.0	Decreased 2.6
135	Myosin regulatory light chain 2, skeletal muscle isoform (MLC2)	P24732	Increased 2.0	Increased 2.3
136, 115	Myosin regulatory light chain 2, skeletal muscle isoform (MLC2)	P24732	-	-
133, 134, 114	Myosin regulatory light chain 2, cardiac muscle isoform (MLC2s)	P10916	Decreased 3.1, 2.7, 3.3	Decreased 1.4, 1.8, 1.6
113	Myosin light chain 3, skeletal muscle isoform (MLC3)	P02603	Increased 2.7	Increased 1.6
Metabolism				
28, 30, 31	Phosphoglucomutase (PGM)	P38652	Increased 2.0, 2.6, 3.0	-
32	Phosphoglucomutase (PGM)	P38652	Increased 1.4	Increased 1.3
128	Fatty acid-binding protein, heart (H-FABP)	P10790	Decreased 2.7	-
Others				
130, 14, 131	Sarcosin	O60662	Increased 3.4, 2.8, 3.1	Decreased 4, 2.6, 2.6
110	Peptide methionine sulfoxide reductase	P54149	Increased 1.4	Increased 2
117	Parvalbumin	P80050	Increased 1.9	Increased 4
56	Sarcoplasmic reticulum 53kDa glycoprotein (SR53G)	P13666-2	Increased 2.9	-
122	p20	O14558	Decreased 2.0	Decreased 1.3
124	p20	O14558	-	Increased 1.8

to the L group (Cassar-Malek *et al.*, 2005). The same tendency was confirmed by transcriptomic analysis which was conducted for Charolais young bulls H and L only. The results showed an increased expression of several transcripts involved in muscle contraction (Myosin Heavy Chain and Light Chain isoforms, Tropomyosins and Troponins) or in metabolism (Lactate dehydrogenase A). The differences between H and L muscles were more important in RA muscle (slow oxidative) than in ST (Cassar-Malek *et al.*, 2005).

Conclusion

These data indicate that the high muscle growth potential of monogenic origin (Double-muscled) or obtained by divergent selection, induces an over-expression of fast glycolytic genes. Genes controlling contractile properties seem to be more affected than that of metabolism. Among the differential expression, the fast troponin T isoforms containing the exon 16 appear to be good markers of muscle hypertrophy. Further studies at the protein and RNA levels are in progress to confirm and complete these preliminary data.

References

Bouley, J., C. Chambon and B. Picard, 2004. Mapping of bovine skeletal muscle proteins using two-dimensional gel electrophoresis and mass spectrometry. Proteomics 4: p. 1811-1824.

Bouley, J., B. Meunier, C. Chambon, S. DeSmet, J.F. Hocquette and B. Picard, 2005. Proteomic analysis of bovine skeletal muscle hypertrophy. Proteomics, In press.

Bucher, E.A., F.C. de la Brousse and C.P. Emerson, 1989. Developmental and muscle-specific regulation of avian fast skeletal troponin T isoform expression by mRNA splicing. J. Biol. Chem. 25: p. 12482-12491.

Cassar-Malek, I., Y. Ueda, C. Bernard, C. Jurie, K. Sudre, A. Listrat, I. Barnola, G. Gentès, C. Leroux, G. Renand, P. Martin and J.F. Hocquette, 2005. Molecular and biochemical muscle characteristics of Charolais bulls divergently selected for muscle group. In: Indicators of milk and beef quality, J.F. Hocquette and S. Gigli (editors), Wageningen Academic Publishers, The Netherlands.

Deveaux, V., I. Cassar-Malek and Picard P., 2001. Comparison of contractile characteristics of muscle from Holstein and Double-Muscled Belgian Blue foetuses. Comp. Biochem. Physiol. 131: p. 21-29.

Duris, M.P. and Picard B., 1999. Genetic variability of fœtal bovine myoblasts in primary culture. Histochem. J. 31: p. 753-760.

Grobet, L., D. Poncelet, L.J. Royo, B. Brouwers, D. Pirottin, C. Michaux, F. Ménissier, M. Zanotti, S. Dunner and M. Georges, 1998. Molecular definition of an allelic series of mutations disrupting the myostatin function and causing double-muscling in cattle. Mamm. Genome 9: p. 210-213.

Jin, J.P., J. Wang and O. Ogut, 1998. Developmentally regulated muscle type-specific alternative splicing of the COOH-terminal variable region of fast skeletal muscle troponin T and an aberrant splicing pathway to encode a mutant COOH-terminus. Biochem. Biophys. Res. Commun. 26: p. 540-544.

Meunier, B., J. Bouley, I. Piec, C. Bernard, B. Picard, J.F. Hocquette, 2004. Data analysis methods for accurate detection of differential protein expression in two-dimensional gel electrophoresis. Proceedings of the French-Polish Symposium, September 2004, Paris, France.

Muroya, S., I. Nakajima and K. Chikuni, 2003. Amino acid sequences of multiple fast and slow troponin T isoforms expressed in adult bovine skeletal muscles. J. Anim. Sci. 81: p. 1185-1192.

Neuhoff V., Arold N., Taube D., Ehrhadt W., 1988. Improved staining of proteins in polyacrylamide gels including isoelectric focusing gels with clear background at nanogram sensitivity using Coomassie Brilliant Blue G-250 and R-250. Electrophoresis 9, p. 255-262.

Picard, B., L. Lefaucheur, C. Berri and M.J. Duclos, 2002. Muscle fibre ontogenesis in farm animal species, Reprod. Nutr. Dev. 42: p. 415-431.

Tusher, V.G., R. Tibshirani and G. Chu, 2001. Significance analysis of micro-arrays applied to the ionizing radiation response. Proc. Natl. Acad. Sci. USA 98: p. 5116-5121.

Pasture-based beef production systems may influence muscle characteristics and gene expression

I. Cassar-Malek[1], C. Bernard[1], C. Jurie[1], I. Barnola[1], G. Gentès[1], D. Dozias[2], D. Micol[1] and J.F. Hocquette[1]*
INRA, [1]Herbivore Research Unit, Theix, France, [2]Domaine expérimental du Pin au Haras, Domaine de Borculo, 61310 Le Pin-au-Haras, France.
**Correspondence: cassar@clermont.inra.fr*

Summary

Extensive beef production systems on pasture are promoted to improve animal welfare and beef quality. This study aimed to compare the influence on muscle characteristics of two management approaches representative of intensive and extensive production systems. One group of 6 Charolais steers was fed maize-silage indoors and another group of 6 Charolais steers was fed on pasture. Activities of glycolytic (Lactate dehydrogenase [LDH], phosphofructokinase) and oxidative (hydroxyacyl-CoA dehydrogenase [HAD], Isocitrate dehydrogenase [ICDH], citrate synthase [CS] and cytochrome-c oxydase [COX]) muscle enzymes were assessed in *Rectus abdominis* (RA) and *Semitendinosus* (ST) muscles. Activities of oxidative enzymes ICDH, CS and HAD were higher ($P = 0.08$; $P < 0.01$ and $P < 0.01$ respectively) in muscles from pasture-fed animals. Transcriptome studies using a multi-tissue bovine cDNA repertoire were performed to compare gene expression profiling in RA and ST muscles between both production groups. Variance analysis showed that the muscle type has an important effect on the expression level of genes. Interestingly, the effect of the production system was less marked. A list of the 30 most variable genes was established, of which 15 muscle genes were considered. Amongst them, the Selenoprotein W, which was found to be under-expressed on pasture, could be considered as an indicator of pasture-based system. In conclusion, enzyme-specific adaptations and gene expression modifications were observed in response to the production system.

Keywords: cattle, pasture, muscle characteristics, gene expression profiling, Selenoprotein W

Introduction

Genetic, environmental and production factors influence muscle characteristics, and hence meat quality. The influence of production factors (age, sex, feeding plan) on muscle characteristics has been studied extensively (for review, see Geay *et al.*, 2001). However, there is less knowledge on the possible impact of production factors on muscle gene expression. The advent of high-throughput techniques for the study of gene expression makes it possible, not only to identify new predictors of meat quality, but also to monitor beef quality through production systems.

This study aimed to compare the influence of two rearing methods representative of two French production systems (intensive and extensive) on the muscle characteristics of 30-month-old Charolais steers and on muscle gene expression using cDNA macro-arrays.

Material and methods

Animals, samples and biochemical studies

The study was conducted with 12 Charolais steers. The animals were offspring of pure-bred Charolais cows and bulls of an INRA experimental herd, weaned at 32 weeks and then housed in open sheds. At 8 months of age, they were allotted to 2 groups and were either fed indoors with a maize silage diet or grazed on pasture. The animals were slaughtered at 30 months of age. Two muscles of the carcass were excised from each animal within less than ten minutes after slaughter: *Rectus abdominis* (red and oxidative muscle, RA) and *Semitendinosus* (white and glycolytic muscle, ST). The samples were immediately frozen in liquid nitrogen and stored at -80°C until analysed.

The maximal activity levels of the following enzymes, reflecting the potential of β-oxidation (β-hydroxyacyl-CoA dehydrogenase, HAD), mitochondrial density (citrate synthase, CS; isocitrate dehydrogenase, ICDH), oxidative phosphorylation (cytochrome-c oxidase, COX), or glycolytic metabolic pathway (phosphofructokinase, PFK; lactate dehydrogenase, LDH) were determined spectrophotometrically, as described elsewhere (Gondret *et al.*, 2004; Cassar-Malek *et al.*, 2004). One unit of enzyme was defined as the amount which catalyses the disappearance of 1 μmol of NADH per min for HAD, PFK and LDH, the appearance of 1 μmol per min of NADPH for ICDH, the liberation of 1 μmol of coenzyme A per min for CS, and the oxidation of 1 μmol per min of cytochrome-c for COX. The results were expressed as unit per g tissue fresh weight. Variance of muscle enzymatic activities was analysed using the General Linear Models (GLM) procedure of SAS (SAS, 1996) with three effects: production system (P), animal nested within production system and muscle (M) as previously described (Cassar-Malek *et al.*, 2004).

Transcriptome analysis

RNA extraction and radioactive labelling was performed as previously described. Briefly, total RNA was extracted from the RA and ST muscles of the six steers per group. RNA samples were pooled to obtain one sample for each muscle and each group (RA/maize silage, RA/pasture, ST/maize silage, ST/pasture). Poly(A)$^+$ RNA was then isolated from the pools and 500 ng of poly (A)+ was labelled during reverse transcription by incorporating [α-^{33}P]dATP (Sudre *et al.*, 2003).

High density filters were constructed from a collection derived from 3 cDNA libraries: a non-normalised directed bovine muscle library (Sudre *et al.*, in press), a non-normalised 14-day-old embryo library (Degrelle *et al.*, in preparation) and a lactating mammary library (Le Provost *et al.*, 1996). A set of 2304 spots corresponding to 637 muscle, 882 embryo and 377 mammary gland amplified cDNA fragments was printed together with controls (*Arabidopsis thaliana* C554 clone, cDNA from the AMERSHAM *Lucidea* kit, and water) onto 8x12 cm Nylon membranes at the National Biological Resources Centre for Animal Genomics (Jouy-en-Josas). Sequences of the cDNAs are available on the http://sigena.jouy.inra.fr website.

Hybridisations were performed in duplicate (at the probe labelling and hybridisation levels). For each experiment, 4 filters were incubated with each radiolabelled cDNA target as previously described (Sudre *et al.*, 2003). After washing, they were exposed 48 hours to phosphor screens before scanning on the PhosphorImager (STORM 840, Molecular Dynamics) imaging plate system for quantitative analysis of signal intensity. After acquisition, the scanned images were analysed using the GenePix Pro V4.1 software (Axon instrument, Inc). Data were analysed using a standard analysis of variance (ANOVA) with three fixed effects (experiment, production system, muscle) using a program called GeneANOVA (Didier *et al.*, 2002). Data of gene expression were previously log transformed before variance analysis. To confirm differential expression, data were also analysed using a combination of the fold change method and the Statistical Analysis of Microarrays (SAM, Tusher *et al.*, 2001). Genes were declared as differentially expressed for a ratio ≥ 1.5 and a false discovery rate (FDR) $< 5\%$. Validation of differential expression of some genes was performed by Northern Blot hybridisations, as previously described (Sudre *et al.*, 2003).

Results

Muscle type had a significant effect on the activities of metabolic enzymes (Table 1). For either muscle type, the activities of glycolytic enzymes did not differ between the two production systems (Table 1). The activity of oxidative enzymes responded to the production system, except for COX activity. The activity of ICDH ($P = 0.08$), CS and HAD ($P < 0.01$) increased on pasture (Table 1). For ICDH, an interaction ($P < 0.01$) was found between muscle and production system. A muscle-specific increase of ICDH activity was recorded in the RA muscle on pasture ($P < 0.0001$). Overall, these results are consistent with an oxidative switch of muscle metabolism when animals were fed on pasture.

Table 1. Influence of the production system on metabolic activities in muscles.

	Maize silage	Pasture	SEM	Effect
ICDH	1.05	1.26	0.034	M***; P^t; P x M**
CS	4.39	5.16	0.087	M***; P**
HAD	1.53	1.92	0.046	M**; P**
COX	9.99	11.11	0.655	M**
LDH	841	891	14.09	M***
PFK	23.69	22.26	1.532	M***

Values are LSMeans (µmol/min per g fresh tissue) of 6 animals per production group for both muscles. M: effect of muscle; P: effect of production system; P x M: production system x muscle interaction. Level of significance: ^{t}: $P = 0.08$; *: $P < 0.05$; **: $P < 0.01$; ***: $P < 0.001$.

Variance analysis of the transcriptome analysis data showed an effect of the hybridisation experiment (technical variability), the muscle, and the production system on gene expression ($F = 16.38$, $F= 16.26$ and $F=1.33$ respectively; $P< 0.0001$). The 20 most variable gene probes according to muscle type were annotated according to library of origin and gene ontology (Table 2). This analysis revealed that they all belonged to the subset of cDNA probes derived from the muscle library and were mainly related to metabolic enzyme and contractile protein pathways.

The production system (maize silage *vs* pasture) had less effect on gene expression than muscle type. Amongst the 30 most variable genes, 17 belonged to the muscle cDNA library subset whereas 8 and 5 belonged to the embryo and mammary gland cDNA library subsets respectively. Table 3 shows a list of the 15 most variable genes from the muscle cDNA library subset in response to the production system. They correspond mostly to genes encoding metabolic enzymes, contractile and ribosomal proteins. Three of the muscle-derived cDNA probes could not be identified using BLASTN and BLASTX searches.

Interestingly, a muscle clone corresponding to Selenoprotein W was found to be under-expressed in the RA of pasture-fed compared to maize-silage fed steers using either ANOVA (Table 3) or a combination of the fold-change method and SAM analysis (with a differential ratio ≥ 1.5 and a FDR of 5%, Figure 1A). Northern Blot analysis confirmed the under-expression of a 0.9 kb transcript using a cDNA probe prepared from the clone used for spotting (Figure 1B).

Discussion

We have conducted this study in order to better understand the global gene expression characteristics of different skeletal muscles in the bovine and the muscle gene expression response to the production system. As previously observed (Ortigues-Marty *et al.*, 2002), the muscles of

Table 2. Examples of macroarray gene probes from the muscle cDNA library, which were differentially expressed in the two muscle types studied.

Functional category	Identity	F	P-value
Calcium pump	sarcoplasmic/endoplasmic reticulum calcium ATPase 2 (serca2)	415	<0,0001
Contraction	myosin heavy chain IIx/d (myhc-IIx/d) clone 1	345	<0,0001
Contraction	myosin heavy chain IIx/d (myhc-IIx/d) clone 2	345	<0,0001
Metabolism	carbonic anhydrase III	327	<0,0001
Metabolism	NADH-ubiquinone oxidoreductase chain 6	255	<0,0001
Metabolism	glyceraldehyde 3-phosphate dehydrogenase (GAPDH)	244	<0,0001
Metabolism	phospholipase C, DKFZP564M182 (chimeric clone)	232	<0,0001
Metabolism	aldolase A	205	<0,0001
Calcium pump	sarcoplasmic/endoplasmic reticulum calcium ATPase 1 (serca1)	204	<0,0001
Cell regulation	ubiquitin-like 3	187	<0,0001
Contraction	tropomyosin beta chain, skeletal muscle (TPM2)	179	<0,0001
Metabolism	fructose-bisphosphate aldolase a	161	<0,0001
Metabolism	glyceraldehyde 3-phosphate dehydrogenase (GAPDH)	161	<0,0001
*	unknown (sigenae accession number: bcaj0009a.c.08)	158	<0,0001
Cell regulation	heat shock protein 27 kDa (HSPB 27)	156	<0,0001
Contraction	troponin I, fast skeletal muscle	154	<0,0001
Contraction	myosin light chain 1, skeletal muscle isoform (mlc1f)	152	<0,0001
Metabolism	NADH-ubiquinone oxidoreductase chain 1	145	<0,0001
Contraction	myosin regulatory light chain 2, cardiac muscle isoform (mlc-2)	143	<0,0001
Ribosomal protein	40s ribosomal protein S2	139	<0,0001

* sequences corresponding to unknown cDNA are available on the http://sigena.jouy.inra.fr website.

steers fed on pasture have more oxidative characteristics. This may be related to a better vascularisation (Vestergaard *et al.*, 1999) and an increased lipid utilisation in grazing animals as observed for moderate exercise (Schmitt *et al.*, 2003).

Most of the genes differentially expressed between muscle types are from the muscle cDNA library probe subset and, as already shown they are involved in contractile and metabolic properties. The effect of the production system appears to be associated mainly with genes belonging to the muscle and embryo cDNA library subsets. Most of the genes from the embryo cDNA library subset could not be identified (using BLASTN or BLASTX searches). Most of the differentially expressed genes from the muscle cDNA library subset were associated with muscle contraction, metabolism, and protein synthesis.

An interesting finding was the differential gene expression of selenoprotein W (SeWP), which was under-expressed in steers grazing on pasture. The encoded protein is a selenocysteine-containing protein. Although the metabolic function of Selenoprotein W is not yet known, it is speculated that it is involved in muscle and cardiac metabolism (Whanger, 2000) and may play a role in oxidant defence (Jeong *et al.*, 2002). As the abundance of this protein in skeletal muscle and some other tissues is regulated by dietary selenium (Wendeland *et al.*, 1995), the differential expression of SeWP may be related to the selenium content or bioavailability in the diet, but this remains to be studied. Lastly, as differential expression of SePW was observed in muscle atrophy

Table 3. Examples of macroarray gene probes from the muscle cDNA library, which were differentially expressed in the two production systems studied.

Function	Identity	F	P-value
Anti-oxidant	selenoprotein W (SePW)	25	<0,0001
*	unknown (sigenae accession number: bcaj0006a.d.04)	18	0,0002
Metabolism	creatine kinase M chain (m-ck)	17	0,0007
*	unknown (sigenae accession number: bcas0001a.e.12)	14	0,0007
Contraction	troponin T, slow skeletal muscle isoforms	14	0,0008
Contraction	myosin heavy chain IIx/d (myhc-IIx/d)	14	0,0008
Ribosomal protein	60s acidic ribosomal protein p0	14	0,00010
*	unknown (sigenae accession number: bcaj0005a.c.11)	13	0,0010
Metabolism	retinoic acid-binding protein II, cellular (CRABP-II)	13	0,0010
Differentiation	skeletal muscle lim-protein 1 (slim 1)	13	0,0014
Metabolism	carbonic anhydrase III	12	0,0017
Cell regulation	similar to mouse exportin 7	12	0,0019
Contraction	actin, alpha skeletal muscle	12	0,002
Contraction	myosin light chain 1, embryonic muscle/atrial isoform	11	0.0023
Ribosomal protein	60s ribosomal protein pl4	11	0.0026

* sequences corresponding to unknown cDNA are available on the http://sigena.jouy.inra.fr website.

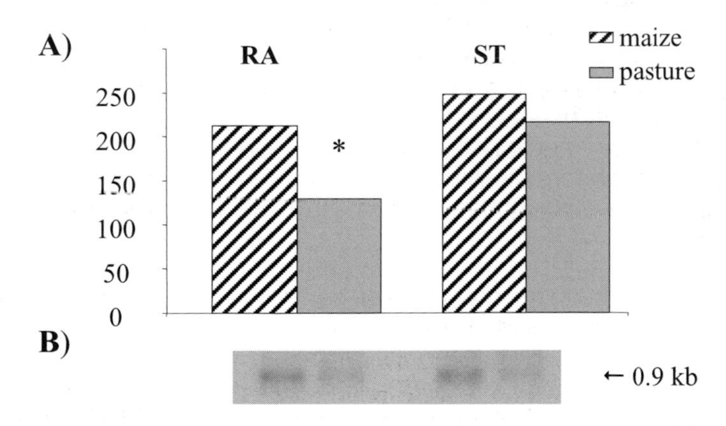

*Figure 1. Expression of Selenoprotein W. A) Hybridisation signal intensity (arbitrary densitometry). * SAM analysis detected a differential ratio ≥1.5 and a FDR of 5%. B) Northern-Blot analysis performed on poly(A)$^+$ RNA pools*

(St-Amand *et al.*, 2001) it may also be regulated by muscle activity. Thus, it can be speculated that SePW expression may be influenced by the mobility of grazing steers.

Conclusions

Using transcriptome studies, we detected genes differentially expressed between two muscle types and two production systems. This allowed us to identify the expression levels of the gene encoding Selenoprotein W as a putative correlate of the production system. However, whether variation of Selenoprotein W gene expression is linked to the feeding regime (grass *vs* maize-silage), to the selenium status of the diet or to the mobility is still unclear.

Acknowledgements

This research has been supported by a grant of INRA within the frame of the AGENAE program related to structural and functional genomics applied to Animal Science and by a grant of the Herbivore Research unit within the frame of the Traceability program. The authors thank Drs C. Leroux (INRA, Theix), I. Hue and P. Martin (INRA, Jouy-en-Josas) for the preparation of the cDNA libraries. The authors also thank the French National Center of Biological Resources for Livestock Species Genomics (INRA, Jouy-en-Josas) for the preparation of the arrays and the SIGENAE bioinformatics team for the analysis of the cDNA sequences.

References

Cassar-Malek, I., J.-F. Hocquette, C. Jurie, A. Listrat, R. Jailler, D. Bauchart, Y. Briand and B. Picard. 2004. Muscle-specific metabolic, histochemical and biochemical responses to a nutritionally induced discontinuous growth path. Anim. Sci. 79: p. 49-59.

Didier, G., P. Brezellec, E. Remy and A. Henaut. 2002. GeneANOVA- gene expression analysis of variance. Bioinformatics 18: p. 490-491.

Geay, Y., D. Bauchart, J.F. Hocquette and J. Culioli. 2001. Effect of nutritional factors on biochemical, structural and metabolic characteristics of muscles in ruminants, consequences on dietetic value and sensorial qualities of meat. Reprod. Nutr. Dev. 41: 1-26, Erratum. 41, p. 377.

Gondret, F., J-F. Hocquette and P. Herpin. 2004. Age-related relationships between fat content and metabolic traits in growing rabbits. Reprod. Nutr. Dev. 44: p. 1-16.

Jeong, D.W., T.S. Kim, Y.W. Chung, B.J. Lee and I.Y. Kim. 2002. Selenoprotein W is a glutathione-dependent antioxidant in vivo. FEBS Lett. 517: p. 225-228.

Le Provost, F., A. Lepingle, and P. Martin. 1996. A survey of the goat genome transcribed in the lactating mammary gland. Mamm. Genome.7: p. 657-666.

Ortigues-Marty, I., C. Jurie, J-F. Hocquette, B. Picard, I. Cassar-Malek, A. Listrat, R. Jailler, D. Bauchart, D. Dozias and D. Micol. 2002. The use of principal component analysis (PCA) to characterize beef steers. In: Multi-Function grasslands. Quality forages, animal products and landscapes, J.J. Durand., J.C. Emile, C. Huyghe and C. Lemaire (editors), EGF, Volume 7, pages 584-585, Grassland Science in Europe

SAS (Statistical Analysis Systems Institute), 1996. SAS/STAT guide for personal computers. SAS Institute Inc., Cary, NC.

Schmitt, B., M. Fluck, J. Decombaz, R. Kreis, C. Boesch, M. Wittwer, F. Graber, M. Vogt, H. Howald and H. Hoppeler. 2003. Transcriptional adaptations of lipid metabolism in tibialis anterior muscle of endurance-trained athletes. Physiol. Genomics. 15: p. 148-157.

St-Amand, J., K. Okamura, K. Matsumoto, S. Shimizu and Y. Sogawa. 2001. Characterization of control and immobilized skeletal muscle: an overview from genetic engineering. FASEB J. 15: p. 684-692.

Sudre, K., C. Leroux, G. Piétu, I. Cassar-Malek, E. Petit, A. Listrat, C. Auffray, B. Picard, P. Martin and J.F. Hocquette. 2003. Transcriptome analysis of two bovine muscles during ontogenesis. J. Biochem. 133: p. 745-756.

Sudre, K., C. Leroux, I. Cassar-Malek, J.-F. Hocquette and P. Martin. 2005. A collection of bovine cDNA probes for gene expression profiling in muscle. Mol. Cell. Probes *in press*

Tusher, V.G., R.Tibshirani and G. Chu, 2001. Significance analysis of micro-arrays applied to the ionizing radiation response. Proc. Natl. Acad. Sci. USA 98: p. 5116-5121.

Vendeland, S.C., M.A. Beilstein, J.Y. Yeh, W. Ream, and P.D. Whanger. 1995. Rat skeletal muscle selenoprotein W: cDNA clone and mRNA modulation by dietary selenium. Proc. Natl. Acad. Sci. U S A. 92: p. 8749-8753.

Vestergaard, M., N. Oksbjerg and P. Henckel. 2000. Influence of feeding intensity, grazing and finishing feeding on muscle fibre characteristics and meat colour of *semitendinosus*, *longissimus dorsi* and *supraspinatus* muscles of young bulls. Meat Sci. 54: p. 177-185.

Whanger, P.D. Selenoprotein W: a review. 2000. Cell Mol Life Sci. 57: p. 1846-1852.

Meat toughness as affected by muscle type

A. Ouali, M.A. Sentandreu, L. Aubry, A. Boudjellal, C. Tassy, G.H. Geesink and G. Farias-Maffet*
INRA-Theix, Meat Research Station, 63122 Saint Genès Champanelle, France.
**Correspondence: ouali@clermont.inra.fr*

Summary

For years, muscle type has been suggested to strongly affect ultimate toughness of beef and other types of meat. However, the conclusions of previous studies are controversial and the exact relationship between muscle type and meat toughness is still unclear. The present study addresses this question by comparing the ultimate toughness of six bovine muscles stored for 8 days and classified as slow oxidative (*M. Diaphragma pedialis* and *Supraspinatus*), fast-glycolytic (*M. Longissimus* and *Semimembranosus*) and intermediate (*M. Triceps brachii caput longum* and *Rectus abdominis*). One important finding of the present work is that no linear relationship was found between ultimate muscle toughness and biochemical variables related to their oxidative/glycolytic capacities and their contraction speed. For all of them, muscles were indeed distributed within triangles delineated by intermediate, slow-oxidative, or fast-glycolytic muscle types. It was concluded that the present variables and those generally used for muscle characterization did not enable a good classification of muscle. The major finding is however that slow-oxidative muscles tenderized as much as fast-glycolytic muscles whilst intermediate muscles were significantly tougher.

Keywords: tenderness/toughness, muscle type, beef, meat

Introduction

According to different surveys carried out in US and in Europe in the beginning of the 1990s, tenderness is the primary quality attribute for consumer acceptance of meat. This is particularly the case for beef and, to a lesser extent, pork. For almost all meat animal species, ultimate tenderness of meat is highly variable and appears to be dependent on *ante-mortem* (genotype, sex, growth rate, feeding conditions, .etc) and *post-mortem* factors (for review, see Monin and Ouali, 1990). In this respect, .the relationship between muscle fibre composition (red slow oxidative, red fast oxido-glycolytic, white fast-glycolytic) and meat tenderness or eating quality has been studied for many years (for review, see Maltin *et al.*, 2003).

A review of the corresponding literature clearly shows that the relationship between meat tenderness and muscle type is complex and highly controversial. Whether some reports suggested no relationship, some others supported a positive relationship between tenderness and red fibres content (fast and/or slow oxidative fibres) or with muscle contraction speed. Few of them stressed a positive relationship between tenderness and fast glycolytic fibres (for review see Maltin *et al.*, 2003 and Monin and Ouali, 1990). Taken together these studies clearly demonstrated that muscle type might greatly influence meat tenderness but the exact nature of this relationship is still unclear.

The present work attempted therefore to clarify this relationship by comparing the ultimate tenderness of six different bovine muscles assessed after 8 days of storage, a storage length corresponding to the most common standard in France and in Europe in general although the tenderization process is not completed at this stage. In addition, this shorter period of storage allow us to maintain a relatively high variability in ultimate toughness between muscles. On the basis of their metabolic characteristics, these muscles were classified as red, slow and oxidative

(*Diaphragma pedialis* and *Supraspinatus M.*), intermediate (*Triceps brachii caput longum* and *Rectus abdominis M.*) and fast and glycolytic muscles (*Semimembranosus and Longissimus M.*).

From our data, it was concluded that slow oxidative muscles are as tender as fast-glycolytic muscles after eight days of storage while intermediate muscles were found significantly tougher. We further confirmed the complex relationship between muscle type and meat toughness.

Materials and Methods

Materials

Six 5-6 year-old Friesian cull cows were purchased from local producers and slaughtered at the INRA Research Centre Abattoir. Muscles *Diaphragma pedialis* (Dp), *Supraspinatus* (SS), *Triceps brachii caput Iongum* (TB), *Rectus abdominis* (RA), *Semimembranosus* (SM) and *Longissimus* (L) were excised within 1 h *post-mortem*, sliced into six 5 cm-thick cuts which were vacuum packed and stored at 12°C up to 24 h *post-mortem* and then transferred to 4°C for 8 days. A sample was immediately treated for biochemical analysis.

Sarcomere length measurement and Measurement of biochemical parameters

Post rigor sarcomere length was determined 24 h *post-mortem* by laser diffraction on fixed samples as described by Cross *et al.* (1980-81).

Lactate dehydrogenase (LDH) and citrate synthase (CS) activities were measured as described previously (Zamora *et al.*, 1996) and expressed as μmoles of NAD/sec per g wet muscle or as μmoles of NTB (Nïtro-Thio Benzoic acid) released /sec per g wet muscle, respectively.

Myoglobin concentration in the above muscles was determined using the Immunochemical technique of Mancini *et al.*, (1965) and expressed as mg/g wet muscle.

Heam iron content was determined according to the method of Hornsey (1956) and expressed as mg/g wet muscle. The protein concentration was assayed according to either the Bradford (1976) or the biuret (Gornall *et al.*, 1949) methods.

Acto-myosin ATPase activity was assayed according to the method previously described (Zamora *et al.*, 1996) and expressed as micromoles KOH/min/mg of myofibrillar proteins.

Rheological assessment of meat toughness

Muscle toughness was assessed on raw meat as previously described (Lepetit *et al.*,1986).

Statistical analysis

Mean values comparison and regression analysis were carried out using the STATITCF software (ITCF, 1991).

Results and discussion

Sarcomere length and physiological characteristics of the muscles

The six muscles were selected on the basis of previous data obtained in our laboratory (Ouali, 1981; Lepetit *et al.*, 1986) dealing with their tenderizing profile and their physiological properties, i.e. metabolic and contractile characteristics. All muscles selected, i.e. *Diaphragma pedialis M.* (Dp), *Supraspinatus M* (SS), *Triceps brachii caput longum M.* (TB) *Recrus abdominis M.* (RA), *Longissimus M.* (L) and *Semimembranosus M.* (SM) are consumed as steak or roast. Furthermore, although variable between muscles, they exhibited a relatively low collagen content, a major muscle component affecting tenderness. Therefore, we expected to assess more accurately the contribution of muscle fibres to variability in tenderness.

The sarcomere length was measured after rigor completion to ensure that no muscle undergoes cold shortening, a factor which might have induced tough meat (Table 1b). Muscles thus exhibited very close sarcomere length (1.83 to 2.04 μm) stressing that no cold shortening took place in the present experimental conditions.

Assessment of muscle type was performed through measurement of a set of five biochemical variables generally measured for this purpose (Table 1a). From the ATPase activity, Dp and SS were classified as slow and L and SM were classified as fast. From Fe content, Dp and SS were classified as oxidative and L and SM as glycolytic. According to the LDH/CS ratio, Dp and SS can be classified as slow-oxidative while L and SM exhibited values characteristic of fast-glycolytic muscles From the analysis of the whole set of variables, RA and TB muscles can be included neither in slow oxidative nor in fast-glycolytic groups. Hence, with regard to these muscles, the high discrepancies between the different quantitative variables forced us to classify them as intermediate (Table 1a, 1b).

Table 1a. Metabolic and contractile properties of the muscles. CS: citrate synthase activity; Fe: heam iron concentration; Mb: myoglobin concentration; LDH: lactate dehydogenase activity; ATPase: acto-myosin ATPase activity. Muscles names were those depicted in the legend of Table 1b.

	Dp	SS	RA	TB	L	SM
CS	4.75±0.05	1.37±0.05	0.83±0.05	1.89±0.05	1.03±0.05	1.16±0.04
Fe	164±16	137±24	112±16	111±22	107±19	105±36
Mb	3.93±0.28	3.47±0.14	2.88±0.12	3.68±0.14	3.11±0.22	3.37±0.14
LDH	3.59±0.22	3.98±0.35	10.22±0.24	10.46±3.40	16.71±2.45	13.15±1.05
ATPase	0.107±0.022	0.140±0.051	0.229±0.058	0.268±0.040	0.254±0.048	0.271±0.086

Table 1b. LDH/CS ratio and Sarcomere length. Muscles are Diaphragma pedialis (Dp), Supraspinatus (SS), Rectus abdominis (RA), Triceps brachii caput longum (TB), Longissimus (L) and Semimembranosus (SM).

	Dp	SS	RA	TB	L	SM
LHD/CS	0.76±0.08	2.91±0.41	12.31±0.98	5.53±0.79	16.22±1.10	11.34±0.93
Sarco.length(μm)	1.85±0.12	1.95±0.09	1.95±0.12	2.04±0.15	1.92±0.09	1.83±0.05

Ultimate toughness and muscle type

After 8 days of storage in similar conditions, RA and TB muscles were significantly tougher than the other muscles followed by SS and Dp, the most tender being the L and SM muscles (Figure 1). It is however worthy to note that no significant differences were found between SS and Dp on the one hand and L and SM on the other hand. This agrees with the positive relationship reported between meat toughness and fibre diameter and area (Maltin *et al.*, 2003).

To clarify the relationship between the ultimate toughness of these muscles and their biological characteristics, meat toughness was plotted versus the mean values of the main quantitative variables including LDH, heam iron content and ATPase activity (Figure 2).

Interestingly, as shown in Figure 2, no linear relationship was observed between muscle toughness at 8 days *post-mortem* and any of the quantitative biological variables measured. For LDH, heam iron content and acto-myosin ATPase activity, muscles were thus distributed within triangles, the apexes of which are more or less well defined. For LDH and heam iron content (Figure 2a and b), the triangles are delineated by slow-oxidative (Dp and SS), fast-glycolytic (L) and intermediate (RA) muscles. For the ATPase activity (Figure 2c), two apexes of the triangle are well identified by the Dp muscle on the one hand and the SM and L muscles on the other hand. The third apex of the triangle is somewhere between TB and RA muscles. Similar findings were obtained for the other quantitative variables measured. Taken together, these findings stress that, irrespective of

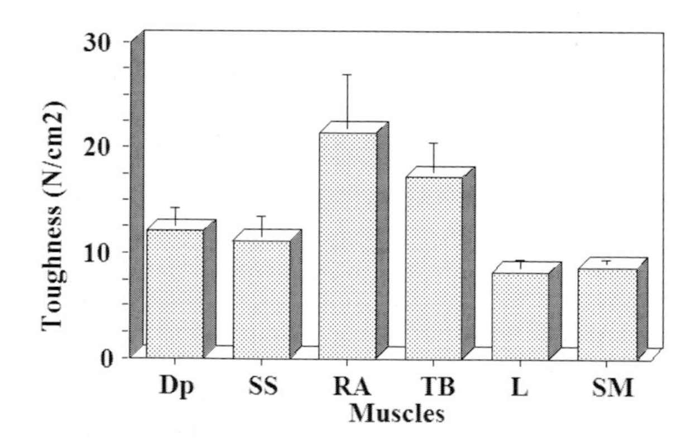

Figure 1. Rheological assessment of muscle toughness investigated after 8 days of storage. Abbreviations used for the identification of each muscle are those mentioned in the legend of Table 1.

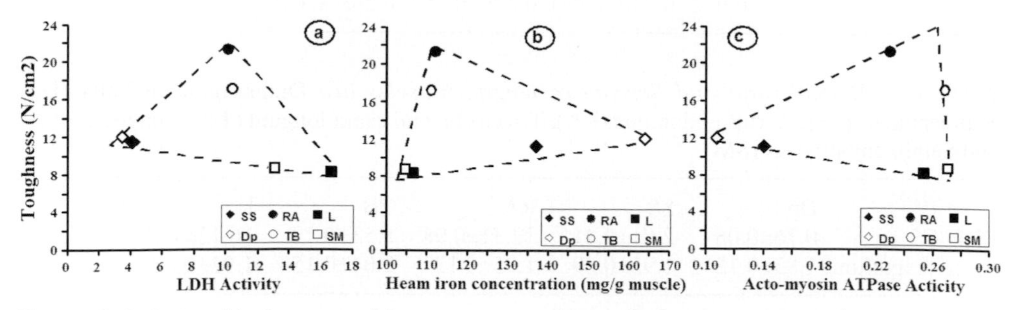

Figure 2. Relationship between ultimate meat toughness (8 days post-mortem) and the major metabolic / contractile characteristics of muscles. Abbreviations used for the identification of each muscle are those mentioned in the legend of Table 1.

the biological characteristic considered, no linear relationship could be expected between meat toughness, and fibre type.

Conclusion

In conclusion, we provide here, for the first time, clear evidence stressing that slow-oxidative muscles tenderized to a similar extent than fast-glycolytic muscles. Furthermore and as expected, the relationship between muscle biochemical or histochemical characteristics and the ultimate toughness of meat was found to be highly complex. These findings led us to conclude that the use of the profile pattern of any of the polymorphic myofibrillar proteins would be very likely a more accurate tool for muscle type characterisation. In this context heavy or light chains, troponin isoforms can be mentioned, the relevance of them being underlined by functional genomic approaches as described in other papers within this EAAP publication.

References

Bradford, M.M., 1976. A rapid and sensitive method for the quantification of microgram quantities of protein utilizing the principle of protein-dye binding. Anal. Biochem. 72: p. 248-254.

Gornall, A.G., E.J. Bardawill and M.M. David, 1949. Determination of serum proteins by means of the biuret method. J. Biol. Chem. 177: p. 751-766.

Hornsey, H.C., 1956. The coulour of cooked cured pork. 1. Estimation of the nitric oxide heam pigments. J. Sci. Food Agric. 7: p. 534-539.

I.T.C.F., 1991. User's guide STATITCF, ITCF, Paris, 80 p.

Lepetit, J., P. Salé and A. Ouali, 1986. *Post-mortem* evolution of rheological properties of the myofibrillar structure. Meat Sci. 16: p. 161-174.

Mancini, G., A.O. Carbonara and J.F. Hermans, 1965. Immunochemical quantification of antigens by single radial immunodiffusion. Immunochemistry. 2: p. 235-254.

Maltin, C., D. Balcerzak, R. Tilley and M. Delday, 2003. Determinants of meat quality: tenderness. Proc. Nutr. Soc. 62: p. 337-347.

Monin, G. and A. Ouali, 1990. Muscle differentiation and meat quality. Dev. Meat Sci. 5: p. 89-157.

Ouali, A., 1981. Variation between muscles of the effect of ageing on myofibrillar ATPase activity. Sci. Alim. 1: p. 1-6.

Zamora, F., E. Debiton, J. Lepetit, A. Lebert, E. Dransfield and A. Ouali, 1996. Predicting variability of ageing and toughness in beef M. *Longissimus.* Meat Sci. 43: p. 321-333.

Blood serum metabolites levels and some meat characteristics in double-muscled young bulls

C. Lazzaroni
Department of Animal Science, University of Torino, Via Leonardo da Vinci 44, 10095 Grugliasco, Italy.
Correspondence: carla.lazzaroni@unito.it

Summary

The double-muscling character is known to induce differences in body composition and a lower intramuscular fat content. Between factors influencing such characteristics there is the animal physiology, so the link between the level of some haematochimic parameters and meat characteristics was studied on two homogeneous groups of double-muscled animals. The trial was carried on 24 Piemontese (P) and 24 Belgian Blue (B) young bulls, maintained in the same environmental conditions until the same age and fattening state, and slaughtered at the same weight (555.7±47.95 *vs.* 557.7±27.91 kg, in P and B respectively). The week before the starting of slaughters a blood sample was withdrawn from each animal to determine the levels of Na, K, glucose, cholesterol, NEFA, urea, total protein, creatinine, ALT, AST, LDH and CK. After slaughter and 7 days of ageing in the same conditions, a sample of *Longissimus thoracis et lumborum* m. was taken from the right side of each animal, on which the meat composition and WHC, as water content, drip losses (DL) and cooking losses (CL), were determined. Even if positive and significant correlations were found between the studied parameters, the Principal Component Analysis method (PCA) based on a matrix correlation between all parameters indicated that it was impossible to discriminate the two breeds, despite lower ether extract, drip and cooking losses in P bulls. In conclusion, the double-muscling character induces similar physiological and metabolic changes in both B and P young bulls.

Keywords: haematochimic parameters, cooking losses, cattle, Piemontese breed, Belgian Blue breed

Introduction

The double-muscling character is known to induce differences in body composition (a muscle hypertrophy and a reduced body fat in the carcass). In addition, beef meat from double-muscled cattle contains much less fat. In the Belgian Blue breed, this was associated to differences in blood metabolite profile compared to normal beef (Hocquette *et al.*, 1999) or dairy breeds (Istasse *et al.*, 1990). However, it is not known if the double-muscling character induces the same physiological and metabolic changes in other breeds. Therefore, the aim of this experiment was to compare the metabolic profile of double-muscled cattle from two different breeds as well as basic meat characteristics highly dependent on animal metabolism and physiology (namely pH, drip and cooking losses, intramuscular fat content).

Particularly the water content, and the water holding capacity (WHC), are two of the most important characteristics of meat, at chilling in the slaughterhouse, in butcher store, at cooking and at consumer level, as they influence carcass weight, meat shelf life, aspect of the cooked meat and its juiciness. The WHC of muscle and meat is influenced by several factors such as species, age, muscular function, protein and energy metabolism. It is indeed well known that in muscles, and in meat, water is mainly linked to proteins, owing to their chemical and physical binding properties: the water content is influenced by the final pH, both as absolute value and as duration of the decrease. A low pH, for example, especially if reached in a short time, causes protein denaturation and a low water retention (Lawrie, 1979).

The haematochimic parameters are studied to understand and explain animal physiology, particularly related with nutrition. There is only little detailed information on metabolites concentrations in beef cattle (Hocquette *et al.*, 1999; Lobley, 1998), particularly in animals with hypertrophied muscles (Istasse *et al.*, 1990; Lazzaroni *et al.*, 1995, 1998; Hocquette *et al.*, 1999), while more data have been published on dairy cows (Bertoni and Piccioli Cappelli, 2000-2001). It is, however, quite clear that muscle metabolism differ between normal and double-muscled cattle (for review, see Hocquette *et al.*, 1998).

To better understand the role of serum metabolites on meat quality, the blood levels of different metabolites were correlated to meat composition and WHC parameters from fattened double muscled bulls of two different breeds.

Material and methods

The trial was performed with 48 double muscled animals, 24 Piemontese (P) and 24 Belgian Blue (B), maintained for an average of 279 d in the same tied stall. The diet, supplied twice a day, consisted in hay and concentrate in increasing amounts, varied each month, to meet at least the INRA requirements of 1.2 kg daily gain for late maturing beef cattle (Jarrige, 1988).

When animals reached the slaughter weight of 550 kg (555.7 ± 47.95 *vs.* 557.7 ± 27.91 kg, in P and B respectively, at about 14-16 month of age) and a specific and uniform fattening state, blood samples (about 10 ml) were withdrawn from the external jugular vein directly into a silicone-coated test tube. After about 3 h, to allow the blood to clot, the samples were centrifuged (2500 *g*) for about 20 min and the serum obtained was frozen until assayed.

The following week, after a transportation time of less than 30 min to a commercial abattoir, the animals were stunned by a captive bolt and slaughtered according to standard procedures. The carcasses were split and the sides were stored at 2 °C in a chilling room. pH was measured at the 13th thoracic vertebra level within 30 min (pH_i) and after 24 hours (pH_u) from slaughter by a pH meter, with a spear electrode and automatic temperature compensator.

Seven days after slaughter a portion of *Longissimus thoracis et lumborum* muscle (between the 9th thoracic and 1st lumbar vertebra) was taken from the right hand side of each animal and brought to the laboratory to perform the meat analysis.

In the laboratory, blood serum concentrations of some metabolites were determined by different standard methods (Lazzaroni *et al.*, 1998): Na and K, by ISE indirect; glucose, cholesterol, and NEFA, by Trinder's enzymatic; urea, by urease-GLDH; protein, by biuret; creatinine, by kinetic picric-acid; ALT and AST, by IFCC 37 °C; LDH, by SCE 37 °C; CK, by DGKCH 37 °C.

On the meat the following analyses were performed (Destefanis *et al.*, 1996): protein (GP), ether extract (EE) and water content (H_2O), by AOAC; drip losses (DL), on a steak weighing about 80 g and 1.5 cm thick and kept for 48 h in a plastic container with a double bottom (Lundströ and Malmfors, 1985); cooking losses (CL), as water boiling losses on a 4 cm thick steak, sealed in a polyethylene bag and heated in a water bath to an internal temperature of 70 °C for 30 min.

The data were analysed by the GLM procedure (SPSS, 2001) and by Principal Component Analysis (PCA; Statistica, 2001). The PCA method aims to have an overview of relationships between all the studied parameters. This approach is based on a matrix correlation between all parameters. The main steps of the calculations are: (i) the transformation of data into normalised values, (ii) the calculation of the correlation matrix of the variables, (iii) the calculation of independent principal components defined as linear combinations of the original variables which explain the variance of the data, (iv) the projection of the variables and then of the individuals on a 2D plan, the two axis being the first two principal components. In this graphic representation, the farther away is the variable from the two axis origin, the better it is represented on the considered plane. Two individuals and variables are positively correlated if they are close together far from the origin of the axis. Two individuals and variables are negatively correlated if they are symmetric from the origin of the axis. When individuals and variables are in orthogonal positions, they are independent (i.e. not correlated).

Results

The average daily feed consumption is reported as dry matter (DM), energy (UFV) and protein (PI) intake in Table 1.

No differences were found between the two breeds for DM and PI intakes, but a significant difference was found in UFV intake which was higher in P than in B (P=.035).

The mean values of the parameters obtained for blood serum and meat samples were in agreement with the normal values reported for cattle and beef meat (Table 2).

Moreover, it could be pointed out that no differences between breeds were found in blood parameters, while in meat there were only small differences in EE (P=.046) and in DL (P=.013), both parameters being higher in B than in P.

The matrix correlation (data not shown), as well as the graphical representation by first two axes of the results from the PCA (Figure 1a), showed that the muscle parameters (EE, H_2O, DL and CL) were positively correlated between them and negatively correlated with feed intakes (UFV, PI and DM), so as to the meat protein content.

Significant and positive correlation coefficients have been found also between all blood serum studied parameters, excepted for NEFA. By contrast the pH values were not significantly correlated to any of the studied characteristics of meat quality and blood metabolism, including glucose and LDH content, and furthermore, they were not correlated between themselves.

The blood parameters were correlated neither to UFV or PI intakes nor to meat characteristics.

The PCA also showed that, despite the little differences in feed intake and in meat composition, it was impossible to discriminate the animals of the two breeds (Figure 1b).

Table 1. Daily feed consumption during the fattening period in Piemontese and Belgian Blue bulls (mean ± std. dev.).

Feed intake	P	B	P<F
Dry matter (kg)	7.0 ± .62	6.7 ± .59	0.110
Energy (UFV)	6.6 ± .46	6.3 ± .49	0.035
Protein (kg)	1.1 ± .08	1.0 ± .08	0.051

Table 2. Blood serum and meat parameters in Piemontese and Belgian Blue bulls (mean ± std. dev.).

		P	B	P<F
Blood serum	Sodium (mmol/l)	144.0 ± 39.76	143.9 ± 28.99	0.990
	Potassium (mmol/l)	5.5 ± 1.72	5.5 ± 1.18	0.968
	Glucose (mg/dl)	117.4 ± 59.19	140.7 ± 94.23	0.322
	Cholesterol (mg/dl)	148.0 ± 48.48	133.3 ± 36.93	0.251
	NEFA (mg/dl)	173.9 ± 74.75	134.1 ± 76.98	0.082
	Urea (mg/dl)	32.7 ± 11.58	32.9 ± 8.17	0.953
	Total protein (g/l)	78.1 ± 25.38	78.4 ± 16.92	0.967
	Creatinine (mg/dl)	2.5 ± .77	2.4 ± .53	0.625
	ALT (mU/ml)	26.8 ± 12.50	26.4 ± 7.74	0.910
	AST (mU/ml)	128.2 ± 70.64	132.8 ± 37.49	0.784
	LDH (mU/ml)	3417 ± 1991	3883 ± 1630	0.391
	CK (mU/ml)	1534 ± 1597	985 ± 580	0.128
Meat	pH_i	6.6 ± .27	6.5 ± .23	0.064
	pH_u	5.6 ± .22	5.6 ± .15	0.944
	Protein (%)	22.6 ± .63	22.5 ± .71	0.621
	Ether extract (%)	.34 ± .12	.45 ± .21	0.046
	H_2O (%)	75.3 ± .59	75.3 ± .61	0.732
	Drip losses (%)	2.3 ± .73	3.0 ± .90	0.013
	Cooking losses (%)	28.4 ± 1.32	29.3 ± 1.95	0.055

Discussion and conclusion

In animals, Na and K are considered as the most important minerals in the control of osmotic and acid-base balance, membrane activity and neuro-muscular stimulation, even if both are kept constant by homeostasis. Similarly cholesterol and protein levels are linked to cell membrane activity; so they influence also the meat pH. Blood serum metabolites associated to energy metabolism (glucose, cholesterol and fatty acids) and some enzymes indexes of cell status are also implicated with effects on pH and on meat water content.

In young bulls between 6 and 16 months of age Toscano Pagano *et al.* (1999) reported that levels of some serum metabolites were different in animals according to muscular hypertrophy. The levels were related to breeds and feeding levels, the metabolites and hormones concentrations depending also on age and on changes in nutrient supply (Beeby *et al.*, 1988; Toscano Pagano and Biagini, 2003).

The Piemontese and Belgian Blue young bulls used in the trial, even if the second ones showed a lower UFV intake, were quite similar in blood serum level for all the studied parameters, while little differences were found in meat quality parameters (EE and DL, higher in B), even if the two breeds differed in shape, more uniform in B than in P animals (Lazzaroni *et al.*, 2001).

The PCA, able to explain more than the 50% of variation, suggested that none of the studied blood parameters was correlated to the meat characteristics within our population of double-muscled cattle, even if some of them are surely involved in the control of fat and water content, both directly (Na, K, glucose, cholesterol, etc.) and indirectly (protein denaturation, cell damages, etc.). It is possible, especially during cooking, that some associated effects occur, with changes in the

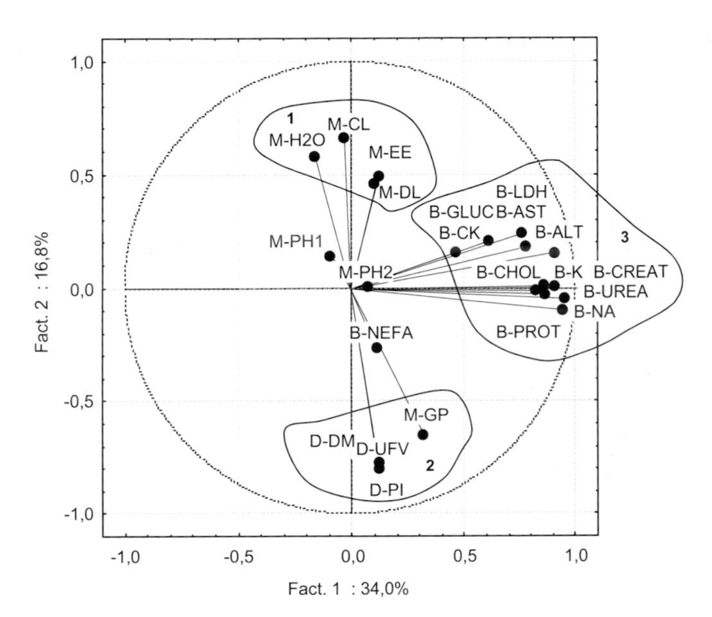

Figure 1a - Principal Component Analysis (PCA) showing relationships between the studied parameters of the diet (D-UFV: energy feed intake (UFV); D-DM: dry matter feed intake; D-PI: protein feed intake) of the blood (B-NA: sodium level; B-K: potassium level; B-GLUC: glucose level; B-CHOL: cholesterol level; B-NEFA: NEFA level; B-UREA: urea level; B-PROT: total protein level; B-CREAT: creatinine level; B-ALT: ALT level; B-AST: AST level; B-LDH: LDH level; B-CK: CK level) and of the meat (M-PH1: pH_i; M-PH2: pH_u; M-GP: protein content; M-EE: ether extract content; M-H2O: water content; M-DL: drip losses; M-CL: cooking losses). Parameters circled together are positively correlated. Parameters of circle 1 are negatively correlated to parameters of circle 2. Parameters of circle 3 (blood characteristics) are in an orthogonal position compared to parameters of circles 1 and 2, they are therefore independent from diet and muscle parameters.

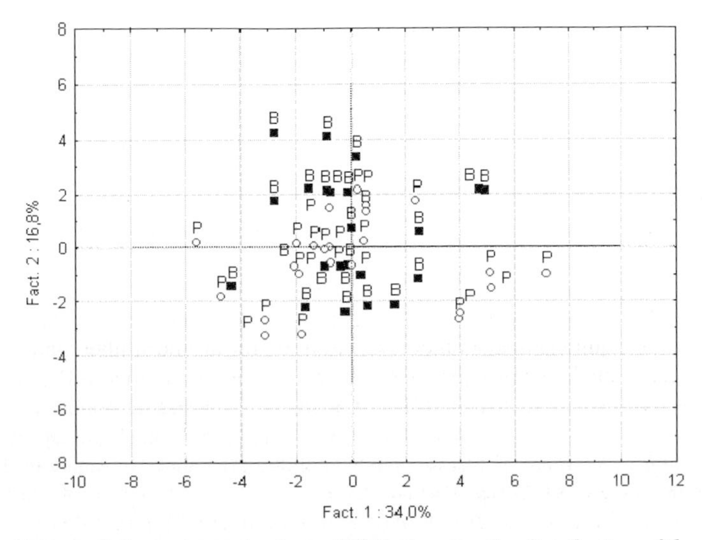

Figure 1b - Principal Component Analysis (PCA) showing the distribution of the animals using the studied parameters (○ P: Piemontese bulls; ■ B: Belgian Blue bulls).

bindings between protein and water inside muscles, so that any interactions between all blood serum parameters could influence particularly CL.

Moreover, the little difference in feed intakes found between breeds did not influence the blood parameters levels, but perhaps only the meat protein content.

It could be interesting to improve the knowledge of such relationships, not only in double muscled animals but also in normal ones, with the aim to be able to predict the WHC of meat from blood parameters.

Acknowledgements

The author wishes to thank: Prof. G. Destefanis, Prof. M.T. Barge and Dr. A. Brugiapaglia (Dept. Animal Science, University of Torino, Italy) who provided the data on meat analysis; Prof. P.G. Pagano and Dr. A. Arnelli (SS. Annunziata Hospital, ASL 17 Piemonte, Italy) for the kind collaboration in performing the blood serum analyses; Dr. J.F. Hocquette (Herbivore Research Unit, Muscle Growth and Metabolism Group, INRA Theix Research Center, France) for the suggestion in data analysis.

References

Beeby J.M., W. Haresign and H. Swan, 1988. Endogenous hormone and metabolite concentrations in different breeds of beef steer on two systems of production. Anim. Prod. 47: p. 231-244.

Bertoni G. and F. Piccioli Cappelli (editors) 2000-2001. Guida all'interpretazione dei profili metabolici in zootecnia. Progress in Nutrition 2 (1-2-4), 3 (1): 49-76, 59-76, 57-88, p. 74-93.

Destefanis G., M.T. Barge and A. Brugiapaglia, 1996. Meat quality in four muscles of hypertrophied Piemontese and Belgian Blue and White young bulls. In: Meat for the consumer, Proceedings of the 42nd International Congress of Meat Science and Technology. K.I. Hildrum (editor), Matforsk, Norwey, ISBN 8290394586, p. 298-299.

Hocquette J.F., I. Ortigues-Marty, D. Pethick, P. Herpin and X. Fernandez, 1998. Nutritional and hormonal regulation of energy metabolism in skeletal muscles of meat-producing animals. Livest. Prod. Sci., 56: p. 115-143.

Hocquette J.F., P. Bass, D. Bauchart, M. Vermorel and Y. Geay, 1999. Fat partitioning and biochemical characteristics of fatty tissues in relation to plasma metabolites and hormones in normal and double-muscled young growing bulls. Comparative Biochem. Physiol., 122 (part A): 127-138.

Istasse L., C. Van Eenaeme, P. Evrard, A. Gabriel, P. Baldwin, G. Maghuin Rogister and J.M. Bienfait, 1990. Animal performance and metabolites in Holstein and Belgian Blue growing-fattening bulls. J. Anim. Sci. 68: p. 2666-2673.

Jarrige R. (editor), 1988. Alimentation des bovins, ovins et caprins. INRA, Paris, France, ISBN 2738000215, 476 p.

Lawrie R.A., 1979. Meat Science. Pergamon Press, Oxford, UK, ISBN 008023173X, 451 p.

Lazzaroni C., G. Toscano Pagano and A. Arnelli, 1995. Study on double muscled cattle: plasma metabolites. In: Proceedings of the 2nd Dummerstorf Muscle Workshop on Muscle growth and meat quality. K. Ender (editor), FBN, Rostock, Germany, ISSN 09461981, p. 226.

Lazzaroni C., G. Toscano Pagano, D. Biagini and A. Arnelli, 1998. Plasma metabolites trend in double muscled cattle. In: Proceeding of the Symposium on Growth in ruminants: basic aspects, theory and practice for the future. J.W. Blum, T. Elsasser and P. Guilloteau (editors), University of Berne, Berne, Switzerland, ISBN 3952106704, P13, p. 299.

Lazzaroni C., G. Toscano Pagano, A. Andrione and D. Biagini, 2001. Body and carcass measurements in Piemontese and Belgian Blue young bulls. "Recent Progress in Animal Production Science. 2. Proceedings of the Associazione Scientifica di Produzione Animale XIV Congress", Firenze, Italy, p. 287-289.

Lobley G.E., 1998. Nutritional and hormonal control of muscle and peripheral tissue metabolism in farm species. Livest. Prod. Sci., 56: p. 91-114.

Lundströ K. and G. Malmfors, 1985. Variation in light scattering and water holding capacity along the porcine longissimus dorsi muscle. Meat Sci., 15: p. 203-215.

Statistica, 2001. Statistica 6. StatSoft, Inc., Tulsa, OK, USA.

SPSS, 2001. SPSS 10.1.3. SPSS Inc., Chicago, IL, USA.

Toscano Pagano G. and D. Biagini, 2003. Serum metabolites levels in double muscled Piemontese fattening young bulls. Ital. J. Anim. Sci. 2 (suppl. 1): p. 349-351.

Toscano Pagano G., C. Lazzaroni, D. Biagini and A. Arnelli, 1999. Na and K serum levels in fattening young bulls. In: Recent Progress in Animal Production Science. 1. Proceedings of the Associazione Scientifica di Produzione Animale XIII Congress. G. Piva, G. Bertoni, F. Masoero, P. Bani and L. Calamari (editors), Ed. F. Angeli, Milano, Italy, ISBN 8846415353, p. 584-586.

Understanding the effect of gender and age on the pattern of fat deposition in cattle

A.K. Pugh[1], B. McIntyre[2], G. Tudor[3] and D.W. Pethick[1]*
[1]Division of Veterinary and Biomedical Sciences, Murdoch University, Murdoch, 6150, Western Australia, Australia, [2]Department of Agriculture, Western Australia, South Perth, 6151, Australia, [3]Department of Agriculture, Western Australia, Bunbury, 6231, Australia.
**Correspondence: apugh@central.murdoch.edu.au*

Summary

This experiment investigated the allometric development of intramuscular (imf), subcutaneous, and intermuscular fat in Australian Angus heifers and steers a. Steers (90) heifers (75) were fed a grain based ration and randomly allocated to a slaughter weight in the range 200kg to 450kg carcass weight. Body composition was estimated from a 6 rib dissection (5-10) and imf content was estimated on of *m. longissimus thoracis et lumborum* (LTL). Analysis of the results using general linear modelling (SAS) showed that at any given level of estimated total body fatness, steers had a significantly higher level of intramuscular fat ($P<0.0001$) when compared to heifers, though at the same carcass weight there was no significant difference ($P >0.05$). There was a significantly linear increase ($P <0.0001$) in imf of the LTL as proportion of total rib fat. As carcass weight increased there was a significant increase in the ratio of gms fat in the LTL/gms of total body fat ($P<0.0001$), although the r^2 was low ($r^2=0.121$), indicating that intramuscular fat tended to develop more strongly as the carcass weight increased. It can be concluded that steers are more efficient than heifers with respect to accumulation of imf, as they have a higher level of imf at the same total carcass fatness. In addition intramuscular fat accumulation occurred primarily in parallel with total body fat synthesis.

Keywords: intramuscular fat, total body fat, steers, heifers

Introduction

Within the total fat depot there are many different depots. The main fat depots are subcutaneous, intermuscular, channel and kidney and intramuscular fat (imf) and a commonly held view is that imf is the last depot to develop (Vernon, 1981). However this conclusion was not supported by (Johnson *et al.*, 1972) who found that as a proportion of total carcass, the absolute amount of intramuscular fat develops at the same rate as intermuscular, subcutaneous, channel and kidney fat. Other work has also shown that imf, over a wide range in fatness levels, is highly correlated with total body fatness (Jones *et al.*, 1990). These results suggest that within a genotype imf and other fat depots increase at the same rate as animals fatten and therefore in this study it was hypothesized that imf will develop at a constant rate, as a proportion of the total fat depot.

There have been many studies on the expression of imf in heifers and steers and in the past the trend is for higher levels of imf in heifers at a given carcass weight (Hardt *et al.*, 1995; Jones *et al.*, 1990; Kazala *et al.*, 1999). Few studies have examined the interaction between imf, sex, and total body fatness in a serial slaughter experiment.

The primary objective of this experiment was to gain a better understanding of the allometric growth of the fat depots in steers and heifers over a wide range of carcass weights. It was hypothesized that heifers will show increased imf (%) at lower carcass weights than steers, but when corrected for total body fatness heifers and steers will show the same level of imf (%) at the same level of total body fatness. In addition intramuscular fat accumulation will accumulate at the same rate as the sum of subcutaneous and intermuscular fat depots.

Material and methods

Animals and diet

Australian Angus steers and heifers were purchased from one property and placed into intensive feeding pens (3 pen replications per gender) at an initial liveweight of 426kg and 415kg for steers and heifers respectively. The steers were castrated at 3-4 months of age by elastration. There were 25 heifers and 30 steers per replicate pen. Cattle were allocated at random to final hot carcass weight endpoints (15 of each gender per endpoint) and were slaughtered when the mean estimated carcass weight was reached for each group. The hot carcass weight endpoints were 220, 260, 300, 340, 380 and 450kg for heifers, and 230, 280, 330, 380, 430 and 480kg and for the steers.

The drafting process also included weighing and scanning (P8and rib fat) and finer detailing on the number, age and sex of the species.The diet consisted of ground hay (15%), rolled barley (67.4%) and lupin grain (15%) and a mineral pre-mix with an estimated metabolisable energy and crude protein of 11.5 MJ/kg and 14% (dry matter basis) respectively. Total time on the grain based ration for the last slaughter group was 385 or 354 days for the steers and heifers respectively.

Measurements

Liveweight was measured every 2 weeks and feed intake was measured by weighing the residues weekly. The animals were slaughtered when the cattle reached the appropriate estimated carcass weights. An estimate of carcass composition was obtained via a 6 rib set dissection (rib 5-10) (Johnson and Charles, 1981). The joint was dissected into subcutaneous and intermuscular fat (total rib fat), muscle, bone and connective tissue (ligamentous nuchae only). Intramuscular fat (% fresh weight) was estimated by Soxhlet extraction (Tume, 1984) in *m. longissimus thoracis et lumborum* (LTL) at the level of the 10th rib.

Statistical analysis

All analyses were performed using the software package SAS, using generalised linear models, with sex as a fixed effect, (SAS, 1997). For all of the models non-significant interactions were sequentially removed from the model until the most significant model ($P<0.05$) was obtained (e.g. linear, quadratic, cubic).
The dependant variable percentage total rib fat was tested using a model which comprised terms for sex, hot carcass weight and percentage imf of the LTL (linear and curve linear) and all first order interactions. The dependant variable percentage imf of LTL was tested using a model, which comprised terms for sex, hot carcass weight and percentage total rib fat and all first order interactions. The dependant variable imf of the LTL/total rib fat, was tested using a model which comprised of terms for sex and hot carcass weight and all first order interactions.

Results

The heifers grew at 1.5kg/day on 0-30days on feed which decreased down to 0.9kg/day in 260-300 days on feed, and the steers grew 2.0kg/day on days 0-37 which also decreased down to 0.92kg/day on days 280-343 on feed.

Carcass weight was a significant predictor of imf (%), total rib fat (%) and gms imf in the LTL/gms total rib fat (Figures 1*(i), (ii) and (iii))*. However, there was no significant difference between imf (%) content of the LTL and gms imf in the LTL/total rib fat in heifers and steers (P>0.05) (Figures 1*(i) and (ii)*). There was a positive linear relationship between imf of the LTL and carcass weight (Figures 1 *(ii)*).

Heifers were significantly fatter at all carcass weights when compared to steers (P<0.0001), but there was no significant difference in the rate of fattening as they grew to maturity (Figure 1*(i)*). There was a weak (r^2=0.12) but significant linear relationship (P<0.0001) between carcass weight (kg) and gms of imf in the LTL/ gms total rib fat but there was no significant difference between heifers and steers (P>0.05) (Figure 1*(iii)*). The percentage of total rib fat was a significant predictor of imf (%) with steers (Figure 1*(iv)*).

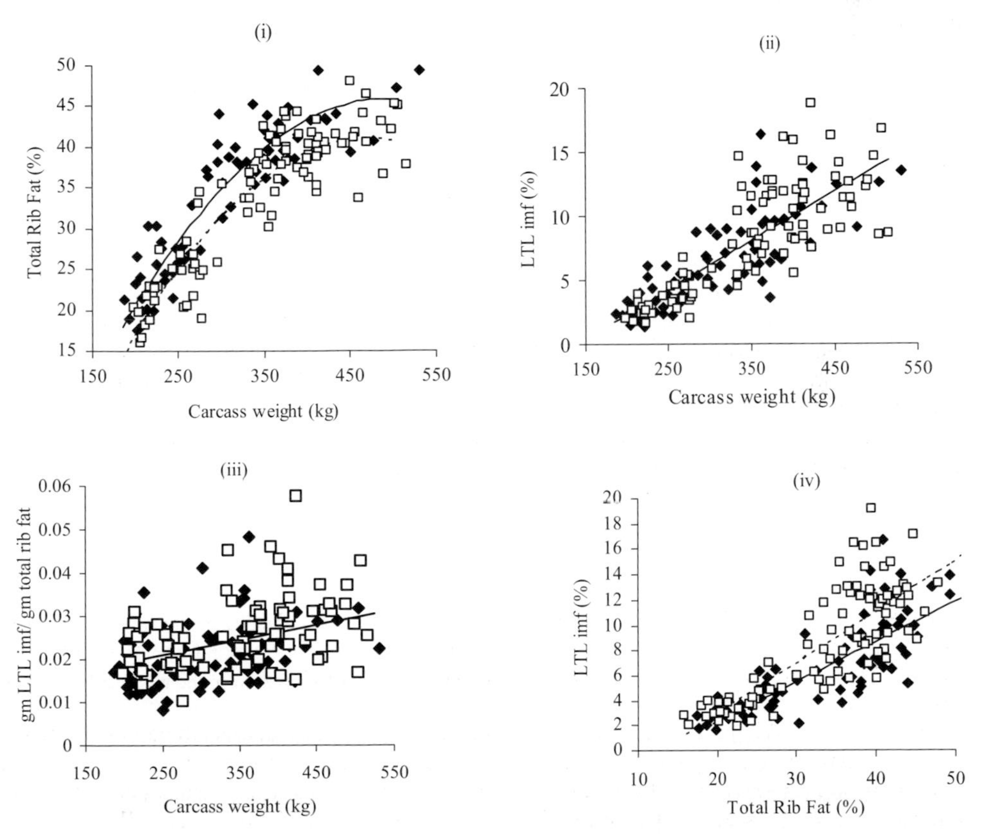

Figure 1. The relationship between carcass weights and (i) percentage total rib fat (heifers $y=-24.81+0.284x - 0.000287x^2$; steers $y=-28.51+0.284x - 0.000287x^2$ $R^2= 0.837$, P<0.0001) (ii) percentage chemical imf of LTL (LTL imf%, $y= -4.99+0.038x$ $R^2= 0.67$, P<0.0001) (iii)gms of imf in the LTL/ gms total rib fat ($y=0.012 + 0.000035x$, $R^2= 0.121$, P<0.0001) and the (iv) relationship between total percentage total rib fat and percentage chemical imf of LTL, (LTL imf%, Heifers $y = -4.07+ 0.32x$; Steers $y = -5.25 + 0.40x$, $R^2 = 0.658$, P< 0.0001) (□---□ Steers ◆—◆ Heifers).

Discussion

We conclude that there is a significant increase at the imf site compared to subcutaneous and intermuscular fat. This is not consistent with findings by (Johnson, 1975) who found that LTL imf does increase largely with other depots i.e. that LTL imf is not late maturing. However the relationship is relatively weak (slope = 0.000035, R^2= 0.121), indicating the trend for late maturing is not strong. Given this data we would suggest that in general, fat development is relatively similar across the 3 depots studies (intramuscular versus the sum of subcutaneous and intermuscular).

Heifers were fatter than steers at any given carcass weight. This is consistent with the findings of (Jones *et al.*, 1990), however heifers did not express more imf at this same carcass weight (The fitted model was linear (Figure 1 (ii)) which suggests that imf (%) would have continued to increase with further increases in carcass weight. This finding is consistent with findings for British breed cattle (Duckett *et al.*, 1993) and Japanese Black x Holstein cattle (Aoki *et al.*, 2001).

When plotted against total rib fat heifers show a lower imf (%) than steers (Figure 1*(iv)*). Given the high energetic demands associated with fat synthesis (NRC), steers will represent a more efficient production system for the development of marbled beef.

Conclusion

We can conclude that the development of imf, within a gender, is driven primarily by total body fatness of the animal, There is a gender effect for imf (%) development when total carcass fatness is accounted for such that steers have significantly higher levels of imf (%) at the same percentage total rib fat (body fatness).

References

Aoki, Y., N. Nakanishi, T. Yamada, N. Harashima and T. Yamazaki, 2001. Subsequent performance and carcass characteristics of japanese black and holstein crossbred steers reared on pasture from weaning at three months of age. Bull. Natl. Grass. Res. Inst. 60: p. 40-55.

Duckett, S.K., D G. Wagner, L.D. Yates, H.G. Dolezal and S.G. May, 1993. The effects of time on feed on beef nutrient composition. J. Anim. Sci. 71: p. 2079-2088.

Hardt, P.F., L.W. Greene and D.K. Lunt. 1995. Alterations in metacarpal characteristics in steers and heifers sequentially implanted with synovex from 45 days of birth. J. Anim. Sci. 73: p. 55-62.

Johnson, E.R., 1975. Relationships between the intramuscular fat and dissected fat in the beef carcass. Aust. J. Agr. Res. 26: p. 777-782.

Johnson, E.R., R.M. Butterfield and W.J. Pryor, 1972. Studies of fat distribution in the bovine carcass. I. The partition of fatty tissues between depots. Aust. J. Agr. Res. 23: p. 381.

Johnson, E.R., and D.D. Charles, 1981. The use of carcass cuts to predict beef carcass composition: A research technique. Aust. J. Agr. Res. 32: p. 987-997.

Jones, D.K., J.W. Savell and H.R. Cross, 1990. The influence of sex-class, usda yield grade and usda quality grade on seam fat trim from primals of beef carcasses. J. Anim. Sci. 68: p. 1987-1991.

Kazala, C.E, F.J Lozeman, P.S Mir, A. Laroche, D.R.C Bailey, and J.R. Weselake, 1999. Relationship of fatty acid composition to intramuscular fat content in beef from crossbred wagyu cattle. J. Anim. Sci. 77: p. 1717-1725

SAS. 1997. Sas applied statistics and the sas programming language (4th ed.).Cary, nc.

Tume, R.K.,1984. Soxhlet extraction method Adac official methods of analysis 24.005 fat (crude) or ether extract in meat. No. 431. Food Science Australia, Canon Hall, Queensland.

Vernon, R.G. 1981. Lipid metabolism in the adipose tissue of ruminants. In: Lipid metabolism in Ruminant Animals, W.W. Christie (editor), 1st edition, Pergamon Press, Oxford, New York, p. 350 *et seq.*

Influence of breed, diet and muscle on the fatty acid content in meat from young finished bulls

C. Cuvelier, O. Dotreppe, J.F. Cabaraux, I. Dufrasne, L. Istasse and J.L. Hornick*
Nutrition Unit, Veterinary Faculty, Liege University, Bld de Colonster, 20, Bât. B43, 4000 Liège, Belgium.
**Correspondence: ccuvelier@ulg.ac.be*

Summary

α-Linolenic acid, C18:3n-3, is an essential fatty acid present in animal products such as meat or milk. Beef meat could be thus considered as a source of C18:3n-3 for the consumer. The aim of the present work was to compare the C18:3n-3 content in different beef meat samples. Three groups of young growing fattening bulls (6 Belgian Blue double-muscled (BB), 6 Limousin (LIM) and 6 Aberdeen Angus (AA)) were offered a concentrate diet based either on barley or on sugar beet pulp. At slaughter, samples from the *Longissimus thoracis* (LT), the *Rectus abdominis* (RA) and the *Semitendinosus* (ST) were obtained for fatty acid analysis.

The fat content of meat was significantly affected by breed (2.19, 4.50 and 6.00 % DM for BB, LIM and AA respectively; $P<0.001$) and by muscle location (6.19, 4.10 and 2.40 % DM for LT, RA and ST respectively; $P<0.001$). The C18:3n-3 content in the fresh meat followed the similar ranking in terms of breed (9.14; 12.14 and 16.00 mg/100 g; $P<0.001$) and in terms of muscle location (16.36, 10.50 and 10.42 mg/100 g; $P<0.001$). There were no significant dietary effects on either fat or C18:3n-3 contents, $P<0.123$ and 0.096). It is concluded that due to changes in fat content, both breed and muscle location influence the C18:3n-3 content of bovine muscle.

Keywords: linolenic acid, fatty acid, breed, diet, muscle

Introduction

Nowadays, consumers attach an increasingly large importance to the sensory and nutritional qualities of beef meat. Until now, beef cattle have been selected essentially on their conformation and growth potential. But meat from Belgian Blue (BB) is yet known to be tender (Clinquart *et al.*, 1998). Moreover, its dietetic reputation is good owing to a low fat content with a high proportion of polyunsaturated fatty acids (PUFA). Thus, meat from BB satisfies current European nutritional advice which is to reduce the ingestion of fats, especially the intake of saturated fatty acids (SFA), and to increase the intake of PUFA. But the low intramuscular fat content in the BB proves to be a disadvantage in terms of flavour.

Modifying rumen fermentation patterns by mean of starch, fibre or fat provision in diet can change the fatty acid composition of the tissues in cattle. Diet can also influence the deposition of intramuscular fats. Cereal-rich diets are known to increase the production of propionic acid in the rumen, while sugar beet pulp-based diets lead to a strong production of acetic acid (Van Eenaeme *et al.*, 1990). Propionic acid is a powerful stimulus for insulin production in ruminants while acetic acid is not (Bhattcharya and Alulu, 1975; Istasse *et al.*, 1987). Furthermore, it has been shown that glucose provides 50-75 % of the acetyl units to fatty acids in the intramuscular fat while acetic acid provides only 10-25 % (Smith and Crouse, 1984). Therefore, cereal finishing diets might be used to support the deposition of intramuscular fats.

α-Linolenic acid is the trivial name given to a PUFA with 18 carbon atoms and 3 double bonds with *cis* configuration. It is an essential fatty acid which plays an important role in structural membrane lipids, particularly in the nerve tissue and in the retina. α-Linolenic acid is a precursor of eicosanoids. The elongation and desaturation products of α-linolenic acid are eicosapentaenoic acid (C20:5n-3) and docosahexaenoic acid (C22:6n-3). Growing evidence suggests that the n-3 PUFA have many beneficial effects for health (Simopoulos, 2002). Current nutritional advice is to increase the intake of PUFA. In addition, the intake of the fatty acids of the n-6 series should be decreased while the intake of the PUFA of the n-3 series should be increased. Indeed, the current western diet is very high in n-6 fatty acids, the n-6:n-3 PUFA value being now 15-16.7:1, while it was 1-4:1 150 years ago (Simopoulos, 2002). This ratio is of metabolic interest as the fatty acids of the both series enter in competition for the enzymes which carry out their elongation and desaturation to form eicosanoids. The eicosanoids from the n-6 series are biologically active in very small quantities and if they are produced in large amounts, they lead to a prothrombotic, proaggregatory and proconstrictive state, they contribute to the occurrence of allergic and inflammatory disorders and they increase the proliferation of cells. Moreover, eicosanoids derived from the n-6 series of fatty acids are found in increased amounts in many chronic states, such diabetes, obesity, asthma, cancer, autoimmune diseases, depression, cardiovascular diseases and rheumatoid arthritis (Simopoulos, 1999, 2002). On the other hand, eicosanoids from the n-3 series are less pro-inflammatory than the eicosanoids from the n-6 series and have opposite properties (Lee *et al.*, 1985).

Finally, PUFA content in beef meat is expected to be influenced by several factors such as breed, type of muscle and diet. The aim of this study was thus to assess the influence of these factors on the fatty acids content in beef meat, and more specifically on the α-linolenic acid content.

Materials and methods

Animals and management

A total of 36 young bulls from 3 different breeds were used in this experiment: 12 BB chosen on their double-muscled phenotype, 12 Limousin (LIM) and 12 Aberdeen Angus (AA). The animals were fattened at the experimental station of the University of Liège. At 14-15 months of age, the 12 animals of each breed were randomly allocated to 2 groups of 6 animals. The first group received a concentrate diet based on cereals (diet 1) and the second group a concentrate diet based on sugar beet pulp (diet 2). The composition and characteristics of the diets are given in Table 1.

Bulls were slaughtered at 18-20 months of age. Within 30 minutes after slaughter, samples of 3 muscle types were taken: *Rectus abdominis* (RA), *Longissimus thoracis* (LT) and *Semitendinosus* (ST). Samples were immediately frozen in liquid nitrogen and stored until fatty acids analysis.

The Animal care and Use Council of our institute approved the use and treatment of animals in this study.

Measurements

The chemical composition (organic matter, ether extract, crude protein) of meat was assessed on freeze-dried samples of the muscles according to official procedures (Association of Official Analytical Chemists, 1975).

The determination of the fatty acid profile in meat samples was performed using gas chromatography (GC) after extraction and trans-esterification of fatty acids according to the method of Sukhija and Palmquist (1988) adapted to freeze-dried meat.

Table 1. Composition and characteristics of the diets.

	Cereal-rich diet	Pulp-based diet
Ingredients, %		
Spelt	10	10
Rolled barley	25	9.5
Crushed maize	25	9.5
Sugar beet pulp	18.3	50
Soy bean meal	8	8.5
Linseed meal	8	8.5
Molasses	3	3
Mineral mixture (16/5)	1	1
Chalk	0.5	
Bicarbonate	1.2	
Chemical composition, g/kg DM		
Crude protein	159	161
Ether extract	29	22
Acid detergent fibre	92	146
Ca	7.4	8.9
P	4.4	3.7
Na	4.9	1.4
Mg	2.0	2.3

A combined one-step extraction and esterification method was carried on using a mixture of solvents containing methanol, benzene and acetyl chloride, to produce the different fatty acid methyl esters. The internal standard was margaric acid (C17:0). A 1 µl aliquot was injected into a Chrompack CP 9001 chromatograph (Middelburg, The Netherlands) fitted with a CP-9010 automatic liquid sampler, a split-splitless injector and a 901A flame ionization detector (Chrompack, Middelburg, The Netherlands).

The GC system was fitted with an Omegawax 320 fused silica capillary column (30 m x 0.32 mm i.d.) with a stationary polyethylene glycol phase (Supelco, Bellefonte, United States of America) coated with a 0.25 µm film thickness. Hydrogen was used as carrier gas at a pressure on the top of the column of 50 kPa.

The column temperature was programmed from 120 to 240 °C at a rate of 5 °C/min. The temperatures of the injection port and detector were 250 °C and 260 °C, respectively. The injection was performed in the split mode with a split ratio of 1:25.

The software Alltech Allchrom Plus Chromatography Data System Version 1.4.2.1. (Alltech Associates Inc., Lokeren, Belgium) was used for data processing. Fatty acids were identified by comparison of their retention times with that of the corresponding standard mix.

Statistical analysis and mathematical modelling

The data were analysed using the general linear models (GLM) procedure of Statistical Analysis Systems with the fixed effects of diet, breed, muscle, animal nested within diet and breed and the interactions diet/breed and breed/muscle (SAS Institute Inc., 1989). Diet and breed effects were tested against animal within diet and breed. Muscle effect was tested using the residual error of the model. Results are presented as least-square means with standard error of the means.

Results and discussion

The fat content of the meat and the fatty acid composition expressed as g/100 g of fatty acids are given in Table 2. The fatty acid content expressed as mg/100 g of fresh meat is provided in Table 3.

Table 2. Dry matter (DM %), fat (% DM) and fatty acid content (g/100 g fatty acids) of muscle as influenced by the diet, the breed (B) and the muscle location (M).

	Diet		Breed			Muscle		
	Pulp	Barley	BB	LIM	AA	LT	RA	ST
% DM	24.37	24.26	23.86^{a}	24.34^{ab}	24.77^{b}	25.08^{a}	23.59^{b}	24.29^{c}
% Fat	4.48	3.98	2.19^{a}	4.50^{b}	6.00^{c}	6.19^{a}	4.10^{b}	2.40^{c}
C14:0	1.46	1.43	0.80^{a}	1.64^{b}	1.90^{c}	1.70^{a}	1.69^{a}	0.95^{b}
C14:1	0.50	0.49	0.33^{a}	0.52^{b}	0.63^{c}	0.25^{a}	0.69^{b}	0.54^{c}
C16:0	21.99	21.42	18.07^{a}	22.83^{b}	24.21^{b}	21.97^{a}	22.72^{a}	20.42^{b}
C16:1	1.92	1.92	1.09^{a}	2.29^{b}	2.38^{b}	2.02^{a}	2.08^{a}	1.66^{b}
C18:0	17.34	17.51	17.88	17.19	17.21	20.12^{a}	17.21^{b}	14.94^{c}
C18:1n9/7	31.34	30.30	21.81^{a}	34.01^{b}	36.64^{c}	35.65^{a}	30.20^{b}	26.61^{c}
C18:2n6	14.58	15.95	24.17^{a}	12.34^{b}	9.29^{c}	11.44^{a}	15.45^{b}	18.90^{c}
C18:3n3	1.57	1.48	2.00^{a}	1.30^{b}	1.29^{b}	1.25^{a}	1.33^{a}	2.01^{b}
C20:0	0.06	0.05	0.02^{a}	0.06^{b}	0.09^{c}	0.10^{a}	0.05^{b}	0.01^{c}
C20:1n9	0.08	0.08	0.00^{a}	0.10^{b}	0.13^{c}	0.13^{a}	0.07^{b}	0.03^{c}
C20:2n6	0.16	0.15	0.16	0.15	0.14	0.15	0.13	0.18
C20:3n6	1.12	1.17	1.82^{a}	0.92^{b}	0.69^{b}	0.72^{a}	1.09^{b}	1.62^{c}
C20:4n6	4.39	4.56	7.01^{a}	3.76^{b}	2.65^{c}	2.56^{a}	4.41^{b}	6.46^{c}
C20:5n3	0.90	0.94	1.17^{a}	0.75^{b}	0.83^{b}	0.47^{a}	0.61^{a}	1.67^{b}
C22:0	0.09	0.10	0.11	0.08	0.09	0.13^{a}	0.05^{b}	0.10^{a}
C22:4n6	0.44	0.44	0.69^{a}	0.38^{b}	0.26^{c}	0.30^{a}	0.44^{b}	0.58^{c}
C22:5n3	1.89	1.84	2.64^{a}	1.56^{b}	1.40^{b}	0.95^{a}	1.65^{b}	3.00^{c}
C22:6n3	0.17	0.19	0.24^{a}	0.13^{b}	0.17^{ab}	0.10^{a}	0.14^{a}	0.31^{b}
SFA	40.94	40.51	36.88^{a}	41.80^{b}	43.49^{b}	44.02^{a}	41.73^{b}	36.43^{c}
UFA	59.06	59.49	63.12^{a}	58.20^{b}	56.51^{b}	55.98^{a}	58.27^{b}	63.57^{c}
MUFA	33.84	32.78	23.23^{a}	36.91^{b}	39.79^{c}	38.05^{a}	33.04^{b}	28.84^{c}
PUFA	25.22	26.71	39.89^{a}	21.29^{b}	16.72^{c}	17.93^{a}	25.23^{b}	34.74^{c}
SFA+MUFA	74.78	73.29	60.11^{a}	78.71^{b}	83.28^{c}	82.07^{a}	74.77^{b}	65.26^{c}
PUFA/SFA	0.67	0.71	1.12^{a}	0.54^{b}	0.41^{b}	0.43^{a}	0.65^{b}	0.99^{c}
n-3	4.53	4.45	6.04^{a}	3.74^{b}	3.69^{b}	2.77^{a}	3.72^{b}	6.98^{c}
n-6	20.69	22.26	33.85^{a}	17.54^{b}	13.03^{c}	15.16^{a}	21.51^{b}	27.75^{c}
n-6/n-3	4.59	5.07	5.91^{a}	4.89^{b}	3.70^{c}	5.12^{a}	5.50^{b}	3.88^{c}

BB = Belgian Blue; LIM = Limousin; AA = Aberdeen Angus; LT = *Longissimus thoracis*; RA = *Rectus abdominis*; ST = *Semitendinosus*; NS = Not Significant (P>0.1); +: P<0.1; *: P<0.05; **: P<0.01; ***: P<0.001; ^{a,b,c}: data within a row and within a group with different superscripts are significantly different (P<0.05).

There were no significant effects of diet on the parameters measured except for the C18:2n-6 and the n-6 content and for the n-6/n-3 ratio (Table 3). Sugar beet pulp, barley and maize are commonly used in growing fattening diets. The 3 ingredients were used in each diet of the present experiment but at different inclusion rates. In diet 1, the cereal based diet, there were rather high and equal proportions of barley and maize (250g/kg) for starch provision. Starch from maize is degraded in the rumen to a lesser extent than that of barley (Mayombo *et al.*, 1997). Barley will therefore induce the production of more propionic acid while more glucose will be absorbed from maize in the small intestine. By contrast, diet 2 with 50% sugar beet pulp is a source of a highly

BB			LIM			AA			Level of significance				
LT	RA	ST	LT	RA	ST	LT	RA	ST	Diet	B	M	B*M	SEM
24.19	23.23	24.15	25.12	23.62	24.28	25.95	23.92	24.44	NS	**	***	NS	0.11
2.69	2.15	1.74	6.51	4.41	2.58	9.37	5.75	2.89	NS	***	***	***	0.25
1.09	0.91	0.40	1.88	1.86	1.17	2.14	2.29	1.27	NS	***	***	NS	0.07
0.08	0.51	0.39	0.30	0.70	0.56	0.38	0.86	0.66	NS	***	***	NS	0.03
18.25	18.12	17.84	23.28	24.09	21.12	24.37	25.96	22.31	NS	***	***	NS	0.35
1.35	1.08	0.84	2.33	2.50	2.04	2.38	2.66	2.10	NS	***	***	NS	0.08
20.55	17.98	15.10	20.03	16.74	14.80	19.77	16.92	14.93	NS	NS	***	NS	0.26
26.92	20.88	17.64	38.49	33.44	30.09	41.55	36.27	32.10	NS	***	***	NS	0.82
20.26	25.15	27.09	8.43	12.45	16.13	5.62	8.74	13.50	NS	***	***	NS	0.78
1.69	1.68	2.63	1.06	1.16	1.67	1.00	1.14	1.73	NS	***	***	NS	0.05
0.05	0.02	0.00	0.12	0.05	0.01	0.14	0.09	0.03	NS	***	***	+	0.01
0.01	0.00	0.00	0.18	0.08	0.03	0.19	0.14	0.07	NS	***	***	***	0.01
0.16	0.13	0.19	0.16	0.12	0.18	0.12	0.13	0.17	NS	NS	NS	NS	0.01
1.36	1.81	2.29	0.51	0.86	1.38	0.30	0.59	1.19	NS	***	***	NS	0.07
4.95	7.43	8.65	1.80	3.58	5.91	0.92	2.21	4.83	NS	***	***	NS	0.27
0.76	0.92	1.83	0.34	0.45	1.47	0.32	0.45	1.71	NS	***	***	NS	0.06
0.17	0.07	0.11	0.12	0.04	0.09	0.11	0.05	0.11	NS	NS	**	NS	0.01
0.53	0.70	0.84	0.23	0.40	0.50	0.13	0.23	0.42	NS	***	***	NS	0.03
1.70	2.47	3.75	0.67	1.38	2.63	0.47	1.10	2.63	NS	***	***	NS	0.11
0.15	0.15	0.42	0.08	0.10	0.23	0.07	0.17	0.28	NS	*	***	NS	0.02
40.10	37.09	33.45	45.43	42.78	37.19	46.53	45.31	38.64	NS	***	***	NS	0.50
59.90	62.91	66.55	54.57	57.22	62.81	53.47	54.69	61.36	NS	***	***	NS	0.50
28.36	22.46	18.86	41.30	36.73	32.72	44.50	39.93	34.93	NS	***	***	NS	0.90
31.54	40.44	47.68	13.27	20.49	30.09	8.97	14.76	26.44	NS	***	***	NS	1.34
68.46	59.56	52.32	86.73	79.51	69.91	91.03	85.24	73.56	NS	***	***	NS	1.34
0.80	1.12	1.43	0.30	0.50	0.83	0.19	0.33	0.71	NS	***	***	NS	0.04
4.29	5.22	8.62	2.15	3.09	5.99	1.86	2.86	6.34	NS	***	***	NS	0.23
27.25	35.23	39.06	11.13	17.41	24.10	7.11	11.90	20.10	NS	***	***	NS	1.13
6.40	6.77	4.57	5.15	5.56	3.95	3.82	4.17	3.12	**	***	***	+	0.12

Table 3. Fatty acid content (mg/100 g fresh meat) as influenced by the diet, the breed (B) and the muscle location (M).

	Diet		Breed			Muscle		
	Pulp	Barley	BB	LIM	AA	LT	RA	ST
C14:0	19.04	16.42	4.40^{a}	19.65^{b}	29.14^{c}	27.70^{a}	18.31^{b}	7.17^{c}
C14:1	5.49	4.81	1.55^{a}	5.55^{b}	8.36^{c}	4.55^{a}	7.10^{b}	3.82^{a}
C16:0	244.51	213.92	88.69^{a}	251.38^{b}	347.59^{c}	334.52^{a}	222.48^{b}	128.67^{c}
C16:1	23.34	21.06	5.50^{a}	26.51^{ab}	34.24^{b}	32.15^{a}	22.13^{b}	12.31^{c}
C18:0	182.95	169.53	89.10^{a}	186.64^{b}	252.98	286.59^{a}	154.47^{b}	87.66^{c}
C18:1n9/7	366.07	325.04	110.96^{a}	381.25^{b}	544.44^{c}	556.93^{a}	300.10^{b}	179.63^{c}
C18:2n6	102.46^{a}	108.99^{b}	109.83^{a}	105.50^{ab}	101.85^{b}	121.15^{a}	103.94^{b}	92.09^{c}
C18:3n3	13.02	11.83	9.14^{a}	12.14^{b}	16.00^{c}	16.36^{a}	10.50^{b}	10.42^{b}
C20:0	0.91	0.82	0.17^{a}	0.91^{b}	1.51^{c}	1.74^{a}	0.68^{b}	0.17^{c}
C20:1n9	1.24	1.16	0.04^{a}	1.31^{b}	2.25^{c}	2.28^{a}	0.94^{b}	0.38^{c}
C20:2n6	1.38	1.27	0.77^{a}	1.46^{b}	1.75^{b}	1.99^{a}	1.04^{b}	0.94^{b}
C20:3n6	7.26	7.61	8.17^{a}	7.30^{b}	6.85^{b}	7.26^{a}	7.18^{a}	7.87^{b}
C20:4n6	27.7	29.06	31.37^{a}	28.93^{b}	24.84^{c}	25.34^{a}	28.46^{b}	31.34^{c}
C20:5n3	5.86	6.13	5.16^{a}	5.57^{a}	7.26^{b}	5.39^{a}	4.11^{b}	8.47^{c}
C22:0	0.88	0.92	0.58^{a}	0.89^{ab}	1.23^{b}	1.69^{a}	0.45^{a}	0.56^{b}
C22:4n6	2.99	2.94	3.08	3.14	2.68	3.15^{a}	2.96^{b}	2.78^{b}
C22:5n3	12.27	11.92	11.59^{a}	11.83^{a}	12.88^{b}	10.10^{a}	11.35^{b}	14.84^{c}
C22:6n3	1.12	1.21	1.00^{a}	1.06^{a}	1.44^{b}	1.18^{a}	0.81^{b}	1.51^{c}
Total	1018.31	934.52	481.11^{a}	1051.08^{b}	1397.07^{c}	1441.58^{a}	897.01^{b}	590.66^{c}
SFA	448.34	401.62	182.95^{a}	459.54^{b}	632.45^{c}	654.32^{a}	396.39^{b}	224.23^{c}
UFA	569.96	532.91	298.16^{a}	591.53^{b}	764.61^{c}	787.26^{a}	500.61^{b}	366.43^{c}
MUFA	396.14	352.07	118.07^{a}	414.62^{b}	589.62^{c}	595.9^{a}	330.26^{b}	196.15^{c}
PUFA	173.83	180.84	180.1	176.91	174.99	191.36^{a}	170.35^{b}	170.29^{b}
SFA+MUFA	844.48	753.67	301.01^{a}	874.17^{b}	1222.08^{c}	1250.22^{a}	726.65^{b}	420.37^{c}
PUFA/SFA	0.67	0.72	1.12^{a}	0.54^{b}	0.42^{b}	0.44^{a}	0.65^{b}	0.99^{c}
n-3	32.24	30.96	26.88^{a}	30.60^{b}	37.33^{c}	32.56^{a}	26.98^{b}	35.26^{c}
n-6	141.58	149.88	153.22^{a}	146.31^{a}	137.66^{b}	158.79^{a}	143.37^{b}	135.03^{c}

BB = Belgian Blue; LIM = Limousin; AA = Aberdeen Angus; LT = *Longissimus thoracis*; RA = *Rectus abdominis*; ST = *Semitendinosus*; NS = Not Significant (P>0.1); +: P<0.1; *: P<0.05; **: P<0.01; ***: P<0.001; ^{a,b,c}: data within a row and within a group with different superscripts are significantly different (P<0.05).

degradable fibre that induces the production of acetic acid (Van Eenaeme *et al.*, 1990). It was hypothesized that the fat content in the muscle of the bulls offered the cereal based diet would be higher owing to a larger glucose provision for acetyl supply. The lack of effect on the fat content could be due to insufficient differences between the relative proportions of the cereals and the sugar beet pulp. The addition of bicarbonate in the cereal based diet may also have changed the fermentation pattern in the rumen. However, the cereal based diet induced the deposition of more C18:2n-6 than the pulp based diet. This was probably due to maize in which fat content is about 5% in where C18:2n-6 represents about 50% of fatty acids. Similar cereal effects were reported by Enser *et al.* (1998) on fat depots from young steers.

There was a significant breed effect on the fat content of the muscle (Table 2). The present results clearly indicated that BB provided a very lean meat. Similar findings were previously reported by

BB			LIM			AA			Level of significance				
LT	RA	ST	LT	RA	ST	LT	RA	ST	Diet	B	M	B*M	SEM
6.86	4.69	1.64	28.98	20.03	9.95	47.27	30.21	9.93	NS	***	***	***	1.64
0.6	2.54	1.52	4.66	7.37	4.61	8.38	11.37	5.32	NS	***	***	NS	0.47
111.25	85.76	69.06	356.77	245.08	152.3	541.53	336.6	164.64	NS	***	***	***	17.02
7.89	5.25	3.38	35.95	26.5	17.08	52.62	34.64	16.47	NS	***	***	+	1.91
124.61	84.51	58.19	299.9	161.29	98.71	435.27	217.61	106.07	NS	***	***	***	12.37
165.02	98.62	69.25	583.69	332.87	227.21	922.09	468.79	242.45	NS	***	***	***	28.09
116.34	109.86	103.3	121.66	103.52	91.31	125.46	98.42	81.67	*	+	***	*	1.79
9.93	7.46	10.02	15.76	10.5	10.16	23.38	13.54	11.07	+	***	***	***	0.51
0.38	0.14	0	1.88	0.67	0.17	2.96	1.23	0.35	NS	***	***	***	0.1
0.13	0	0	2.56	0.91	0.46	4.13	1.92	0.69	NS	***	***	***	0.14
0.99	0.6	0.71	2.28	1.02	1.07	2.72	1.49	1.05	NS	***	***	*	0.09
7.77	7.96	8.79	7.18	6.88	7.83	6.85	6.7	6.99	NS	*	**	NS	0.14
28.45	32.53	33.14	25.66	28.46	32.66	21.91	24.97	28.21	NS	***	***	NS	0.51
4.42	4.04	7.01	4.84	3.54	8.31	6.91	4.74	10.14	NS	***	***	*	0.23
0.98	0.32	0.44	1.59	0.43	0.6	2.45	0.61	0.63	NS	**	***	+	0.09
3.02	3.02	3.19	3.39	3.28	2.75	3.04	2.58	2.42	NS	NS	***	NS	0.07
9.81	10.73	14.23	9.69	11.14	14.66	10.8	12.19	15.64	NS	*	***	NS	0.24
0.76	0.66	1.58	1.12	0.73	1.33	1.65	1	1.63	NS	*	***	NS	0.07
599.18	458.69	385.45	1507.84	964.23	681.17	2217.73	1268.1	705.37	NS	***	***	***	61.88
244.08	175.43	129.33	689.4	427.5	261.74	1029.48	586.26	281.62	NS	***	***	**	30.81
355.1	283.27	256.11	818.44	536.73	419.43	1188.25	681.84	423.76	NS	***	***	***	31.24
173.63	106.41	74.16	626.86	367.65	249.36	987.22	516.72	264.93	NS	***	***	***	30.4
181.48	176.85	181.96	191.58	169.08	170.07	201.02	165.12	158.83	+	NS	***	*	2.13
417.7	281.84	203.49	1316.26	795.15	511.09	2001.71	1102.98	546.54	NS	***	***	**	61.02
0.8	1.12	1.43	0.3	0.5	0.83	0.22	0.33	0.71	NS	***	***	NS	0.04
24.91	22.88	32.84	31.42	25.92	34.46	41.36	32.15	38.49	NS	***	***	**	0.66
156.56	153.97	149.12	160.16	143.17	135.61	159.66	132.97	120.34	*	**	***	+	2,00

Dufrasne *et al.* (2001) in a comparison between BB, Charolais and LIM on a sugar beet pulp based concentrate diet, the fattest meat being produced by the Charolais bulls. Raes *et al.* (2003) also found lower total fatty acid content in retail beef from BB and LIM compared to Irish and AA.

Although not reported in the present paper, the carcass composition was assessed by dissection of a rib set. Total intramuscular fat content increased with increases in carcass adipose tissue. These results were consistent with those reported by Demeyer and Doreau (1999) in a review paper in which total fatty acids concentrations in the LT increased with increasing carcass fat content. There was also a significant location effect on the fat content, the highest values being observed in the LT and the lowest in the ST (Table 2). Differences in fat and/or total fatty acids contents between muscles were also reported in BB cows by Webb *et al.* (1998) and in BB bulls by Raes *et al.* (2004).

The intramuscular PUFA content expressed as g/100 g fatty acids was significantly higher in the BB than in the AA group (P<0.001) with intermediate values for the LIM bulls. The ranking was opposite for the sum of SFA and monounsaturated fatty acids (MUFA) (Table 2). When expressed in mg/100 g of meat (Table 3), there were no differences in the PUFA content. By contrast, the intramuscular SFA + MUFA content was significantly larger in the AA meat than in the two other breeds and in the LIM as compared to the BB. When the concentrations of PUFA and of SFA + MUFA, expressed in g/100 g fatty acids, were compared between muscle locations, there were also significant differences (P<0.001), the LT muscle being characterised by the lowest PUFA content and the highest SFA + MUFA content. When expressed in mg/100 g of fresh meat, the PUFA content varied to a very small extent while the differences between muscles were much larger for the SFA + MUFA. The fatty acid content - PUFA and SFA + MUFA - expressed in mg/100 g fresh meat were plotted against the fat content in Figure 1a for the LT, in Figure 1b for RA and in Figure 1c for ST. The PUFA content did not change to any extent in the three muscles. By contrast, the SFA + MUFA content increased steadily with the increase in fat content.

Intramuscular lipids are present in 3 different locations. First, triacylglycerols are located in intramuscular adipocytes, associated with the epi- and endomysium, to determine the marbling of the meat. Triacylglycerols can also be present in muscles fibres, as intracellular lipid droplets. Finally, intramuscular lipids are located in the phopholipid membranes of the muscles fibres. PUFA are mostly present in the phospholipids of the cell membranes, while SFA are found in the triacylglycerol of the adipocytes and intracellular lipid droplets. Increasing intramuscular fat content is then not correlated with an increase in the intramuscular PUFA content. These results show that the increase of the intramuscular total fat content is limited to the triacylglycerol fraction. From a nutritional point of view, the PUFA intake by the consumer is similar whatever the meat origin, but much higher amounts of SFA and MUFA are provided with the AA meat. This is also clearly illustrated in Figure 1 in which the PUFA present as phospholipids in the membranes are diluted by SFA + MUFA when the fat content in the muscle is increased.

The C18:3n-3 content was significantly affected by breed and muscle locations (Tables 2 and 3). In fresh meat, the C18:3n-3 content was the highest in the fattest breed (AA) and in the fattest muscle (LT). Similar effects were observed between breeds for the other n-3 fatty acids (C20:5n-3, C22:5n-3, C22:6n-3) but were opposite between muscle location. Within the n-6 fatty acid group, the C18:2n-6 and C20:4n-6 were the predominant fatty acids. Since the PUFA content between the breeds was not affected by the fat content (Figure 1), the increase in total n-3 fatty acids was compensated by a corresponding decrease in n-6 fatty acids. One has to note that the change appeared smaller in n-6 fatty acids (15 mg on an average of 146 mg) than in n-3 fatty acids (10 mg on an average of 32 mg) (Table 3). On the whole, the present results were in line with the data reported by Raes *et al.* (2003) in a comparison between retail meats from 4 breeds and 2 muscles.

The n-6:n-3 PUFA ratio was the highest with the leanest breed. By contrast, a larger ratio was observed for the fattest muscle (Table 2). However, the value for RA was not in line due to the proportionally low n-3 content as previously indicated.

Conclusions

It could be concluded from the present results that factors such as breed and muscle location affect the fatty acid concentration of intramuscular fat by their effects on the fat content of the muscle. Although the n-6:n-3 ratio was higher in the BB than LIM and AA, this breed must be considered as providing the lowest SFA + MUFA levels on the bases of 100 g fresh meat intake.

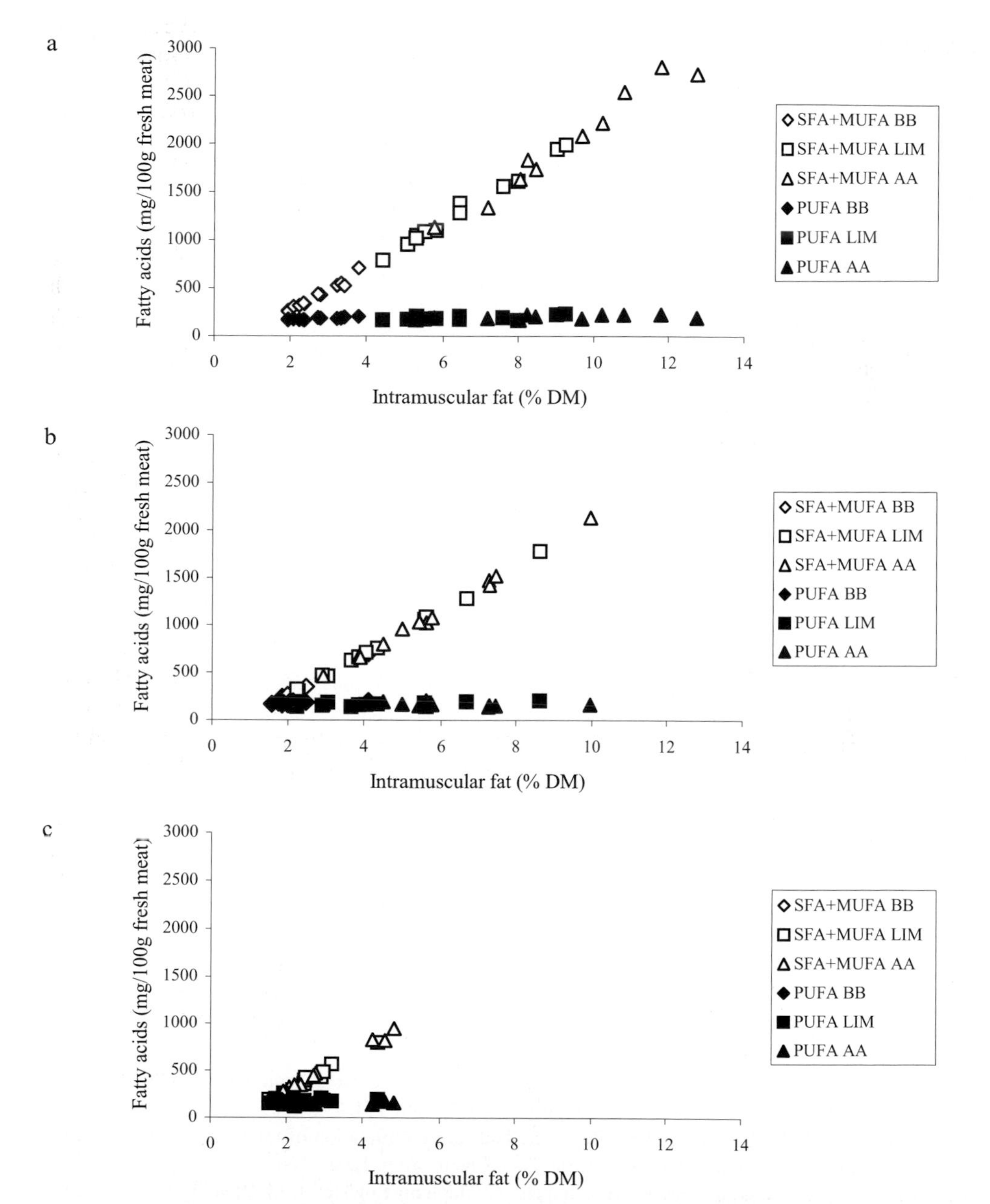

Figure 1. a. Relationship between the PUFA and SFA + MUFA contents and the fat content in the Longissimus thoracis *(LT) of Belgian Blue (BB), Limousin (LIM) and Aberdeen Angus (AA) bulls.*
b. Relationship between the PUFA and SFA + MUFA contents and the fat content in the Rectus abdominis *(RA) of Belgian Blue (BB), Limousin (LIM) and Aberdeen Angus (AA) bulls.*
c. Relationship between the PUFA and SFA + MUFA contents and the fat content in the Semitendinosus *(ST) of Belgian Blue (BB), Limousin (LIM) and Aberdeen Angus (AA) bulls.*

References

Association of Official Analytical Chemists, 1975. Official methods of analysis of the Association of Official Analytical Chemists, 12th Edition. H. Horwitz (editor), Washington, United States of America, 1094 p.

Bhattacharya, A.N. and M. Alulu, 1975. Appetite and insulin-metabolite harmony in portal blood or sheep fed high or low roughage diet with or without intraruminal infusion of VFA. J. Anim. Sci. 41: p. 225-233.

Clinquart, A., J.L. Hornick, C. Van Eenaeme and L. Istasse, 1998. Influence du caractère culard sur la production et la qualité de la viande des bovins Blanc Bleu Belge. INRA Prod. Anim. 11: p. 285-297.

Demeyer, D. and M. Doreau, 1999. Targets and procedures for altering ruminant meat and milk lipids. Proc. Nutr. Soc. 58: p. 593-607.

Dufrasne, I., J.F. Cabaraux, V. de Behr, J.L. Hornick, A. Clinquart and L. Istasse, 2001. Performances zootechniques et la qualité de la viande de taurillons Blanc Bleu Belge, Limousin et Charolais. Renc. Rech. Ruminants, 8: p. 109.

Enser, M., K.G. Hallett, B. Hewett, G.A.J. Fursey, J.D. Wood and G. Harrington, 1998. Fatty acid content and composition of UK beef and lamb muscle in relation to production system and implications for human nutrition. Meat Sci. 49: p. 329-341.

Istasse, L., N.A. Macleod, E.D. Goodall and E.R. Orskow, 1987. Effects on plasma insulin of intermittent infusions of propionic acid, glucose or casein into the alimentary tract of non lactating cows maintained on a liquid diet. Br. J. Nutr. 58: p. 139-148.

Lee, T.H., R.L. Hoover, J.D. Williams, R.I. Sperling, J. Ravalese, B.W. Spur, D.R. Robinson, E.J. Corey, R.A. Lewis and K.F. Austen, 1985. Effect of dietary enrichment with eicosapentaenoic and docosahexaenoic acids on in vitro neutrophil and monocyte leukotriene generation and neutrophil function. New Engl. J. Med. 312: p. 1217-1224.

Mayombo, A.P., I. Dufrasne, P. Baldwin, A. Clinquart and L. Istasse, 1997. Influence de l'incorporation de mélasse à l'ensilage de maïs sur l'ingestion, la digestibilité apparente, les fermentations dans le rumen et les performances zootechniques chez le taurillon à l'engraissement. Ann. Med. Vet. 141: p. 231-238.

Raes, K., A. Balcaen, P. Dirinck, A. De Winne, E. Claeys, D. Demeyer and S. De Smet, 2003. Meat quality, fatty acid composition and flavour analysis in Belgian retail beef. Meat Sci. 65: p. 1237-1246.

Raes, K., L. Haak, A. Balcaen, E. Claeys, D. Demeyer, S. De Smet, 2004. Effect of linseed feeding at similar linoleic acid levels on the fatty acid composition of double-muscled Belgian Blue young bulls. Meat Sci. 66: p. 307-315.

SAS Institute Inc., 1989. SAS/STAT user's guide, Version 6, 4th Ed., Volume 1. SAS Institute Inc., Cary, United States of America, 943 p.

Simopoulos, A.P., 1999. Essential fatty acids in health and chronic disease. Am. J. Clin. Nutr. 70: p. 560S-569S.

Simopoulos, A.P., 2002. The importance of the ratio of omega-6/omega-3 essential fatty acids. Biomed. Pharmacother. 56: p. 365-379.

Smith, S.B. and J.D. Crouse, 1984. Relative contributions of acetate, lactate and glucose to lipogenesis in bovine intramuscular and subcutaneous adipose tissue. J. Nutr. 114: 792-800

Sukhija, P.S. and D.L. Palmquist, 1988. Rapid method for determination of total fatty acid content and composition of feedstuffs and feces. J. Agric. Food Chem. 36: p. 1202-1206.

Van Eenaeme, C., L. Istasse, A. Gabriel, A. Clinquart, G. Maghuin-Rogister, J.-M. Bienfait, 1990. Effects of dietary carbohydrate composition on rumen fermentation, plasma hormones and metabolites in growing-fattening bulls. Anim. Prod. 50: p. 409-416.

Webb, E.C., S. De Smet, C. Van Nevel, B. Martens, D.I. Demeyer, 1998. Effect of anatomical location on the composition of fatty acids in double-muscled Belgian Blue cows. Meat Sci. 50: p. 45-53.

Adipocyte fatty acid-binding protein expression and mitochondrial activity as indicators of intramuscular fat content in young bulls

I. Barnola[1], J.F. Hocquette[1], I. Cassar-Malek[1], C. Jurie[1], G. Gentès[1], J.F. Cabaraux[2], C. Cuvelier[2], L. Istasse[2] and I. Dufrasne[3]*
[1]INRA, Herbivore Research Unit, Theix, France, [2]Nutrition Unit and [3]Experimental Station, Veterinary Faculty, Liège University, Belgium.
**Correspondence : hocquet@clermont.inra.fr*

Summary

Intramuscular fat (IMF) deposition influences many quality attributes of beef meat, especially flavour. Meat from young European bulls contains low amounts of fat. Muscle fat content results from a balance between fat anabolism within intramuscular adipocytes and fat catabolism within mitochondria of myofibers. This work aims to assess the contribution of markers of intramuscular adipocytes and mitochondria to variability in IMF. Belgian Blue (n=12), Limousin (n=12) and Aberdeen Angus (n=12) young bulls fed either a pulp-rich or cereal-rich diet were slaughtered at 18-20 months of age. Samples of *Longissimus thoracis* (LT) muscle were taken at slaughter. Quantification of the mRNA coding for the adipocyte fatty acid-binding protein (A-FABP), a marker of adipocyte differentiation, was performed in quadruplicate using Light Cycler technology relative to a standard curve. Cytochrome-*c* oxidase (COX) activity was determined as a mitochondrial marker. Intramuscular fat (IMF) content, A-FABP mRNA content and COX activity were 3.5, 16.0 and 2.0 fold higher in LT from Angus than from Belgian Blue young bulls, values for Limousin being intermediate. A-FABP mRNA content and COX activity explained 42 and 47% respectively of variability in IMF content between breeds; together, they explained 64% of this variability. However, such relationships were not observed for each individual breed. In conclusion, A-FABP expression and COX activity may be indicators of the ability of bulls to deposit intramuscular fat between breeds, but not within breeds.

Keywords: muscle, intramuscular fat, mitochondrial activity, breed

Introduction

Intramuscular fat (IMF) deposition (marbling) influences many quality attributes of beef meat. Marbling is the primary factor determining the quality grade of beef meat in North American and Asian markets. By contrast, dietary advice in Europe recommends a reduction in the fat content (especially saturated fatty acids) of dietary products, including beef. French (Limousin) and Belgian (Belgian Blue) cattle beef breeds are generally late-maturing and IMF is the last depot to develop. Therefore, beef produced in France generally contains a low amount of fat (about 5% in fresh muscle according to Bas and Sauvant, 2001). This theoretically satisfies the recommendations from nutritionists. The amount of intramuscular fat is even lower in young bulls since it is less than 2.5%.

However, IMF, mainly as neutral lipids, positively influences overall flavour until it reaches a maximum of 4% per g fresh tissue (Goutefongea and Valin, 1978). Therefore, meat from young bulls of late-maturing breeds is not sufficiently tasty for most consumers. Therefore, research has been conducted in order to relate intramuscular fat deposition to muscle metabolic activities. To achieve this goal, it has been proposed to take advantage of the genetic variability in intramuscular fat deposition (Hocquette *et al.*, 2003).

Triacylglycerols (TAG) are the major components of fat in muscles. They are stored to a minor extent within the myofibres and to a major extent within the intramuscular adipocytes. Therefore, the number and the diameter of intramuscular adipocytes may be good predictors of marbling (Cianzo *et al.*, 1985). This is the reason why scientists are looking for markers of adipocyte differentiation. Among them, is the A-FABP (adipocyte fatty acid-binding protein) which was demonstrated in steers as highly related to intramuscular fat content (Hocquette *et al.*, 2003). Fat muscle content also probably results from the balance between the uptake and the synthesis of fatty acids (FA), and degradation of TAG (Hocquette *et al.*, 1998). Therefore, many metabolic pathways in both adipocytes and myofibres could contribute to the variability of intramuscular fat content. It has been suggested in steers that a high mitochondrial activity (assessed by cytochrome-*c* oxidase activity, COX) could favour fatty acid turnover (Hocquette *et al.*, 2003). It has also been suggested that any dietary increase in glucose availability might accelerate lipogenesis within intramuscular adipocytes (Pethick *et al.*, 2004).

The aim of this work was to assess the relevance of A-FABP and COX, two indicators of beef marbling in steers, as indicators of intramuscular fat content in young bulls.

Material and methods

Animals and samples

Two groups of 6 young bulls each from three different genotypes were used: Angus characterized by a marbled meat, Limousin and Belgian Blue which produce only slightly marbled meat. The first group was fed a cereal-rich diet. The second group was fed a sugar beet pulp-based diet. The end products of digestion clearly differ between these two diets: the proportion of acetate being higher with the pulp-rich diet whereas the proportion of neoglucogenic precursors (propionate) being higher with the cereal-rich diet. The calculated energy content was 1.09 and 1.05 UFV/kg DM in the cereals and pulp based diets respectively (UFV = Unité Fourragères Viande, French System of Energy). The protein content was 160 g/kg DM in both diets. The average daily gain was similar in the three genotypes at 1.62 kg/d. Animals were slaughtered at 18-20 months of age.

Soon after slaughter, samples of *Longissimus thoracis* muscle (LT) were taken. Samples were immediately frozen in liquid nitrogen and stored until analyses.

Biochemical measurements

Intramuscular fat (IMF) content was calculated from the individual fatty acid contents obtained by HPLC methods as described in the companion paper by Cuvelier *et al.* Results were expressed per unit muscle dry matter (DM).

Cytochrome-*c* oxydase [COX] activity was determined according to the methods used by Hocquette *et al.* (2003). All results were expressed per g tissue wet weight.

A preliminary experiment indicated that, unlike in steers (Hocquette *et al.*, 2003), A-FABP protein levels were too low to be detected by ELISA. Quantification of the mRNA coding for A-FABP was therefore performed at least in quadruplicate by real-time RT-PCR using the Light Cycler technology relative to a standard curve.

Total RNA was extracted using TRIzol® Reagent (Invitrogen SARL, Cergy-Pontoise, France) and purified using the Nucleospin^R RNA II kit (Macherey Nagel, France). Purity and concentration of RNA was checked with the RNA 600 Nano Assay kit using the Bioanalyser (Agilent

Technologies, Germany). For the assay, cDNA was synthetised by reverse transcription from 2.5 μg of total RNA in a 20 μl total volume using 100U of Superscript^{TM} II reverse transcriptase (Rnase H-) (Invitrogen Life technologies, USA) as described in the manufacturer's protocol, using oligo(dT) for priming. Amplification was performed in a total volume of 20 μl from 2 μl of cDNA with the LightCycler-FastStart DNA Master Hybridization Probes reaction mix according to the manufacturer's protocol. For the PCR, the following primers were used to amplify a fragment of the A-FABP gene; Forward: 5'-GGTACCTGGAAACTTGTCTCC3'; Reverse: -5' CTGATTTAATGGTGACCACAC-3' (0.5 μM final concentration each, MWG-Biotech, Courtabeuf, France). The sequence of the fluorescent probe was 6FAM-ACATGAAAGAAGTGGGCGTGGGCT XT—PH (0.2 μM final concentration; TIB MOLBIOL, Germany). These sequences were determined based on available A-FABP sequences in GenBank database (human: BC003672, NM_001442 ; pig: Y16039 ; rat: U75581). The amplified fragment was sequenced to check its specificity. The cycling conditions of PCR included a first denaturation at 95°C during 8 min and 40 cycles of denaturation at 95°C during 5 sec followed by an hybridation/elongation phase of 60 sec at 60°C. The standard was composed of a pool of cDNA from three different breeds. All the samples (2 μl of RT solution diluted 1/4) and the standard of the PCR (diluted 1, 1/2, 1/4, 1/16, 1/32) were analysed within the same run. Different PCR runs were performed so that, for each sample, 2 RT and 2 PCR per RT at least were performed and analysed relatively to the standard curve. Efficiency of the PCR ranged from 95 to 100%. The results were expressed in arbitrary units per mg total RNA. The technical variability between assays was equal to 30-40% on average. Variability between individuals ranged from 64 to 82% depending on the breed.

Statistical analysis

Differences between breeds and diets were analysed by variance analysis using the GLM procedure of SAS. The effects tested included breed, diet and interaction between diet and breed. For A-FABP mRNA levels, the effects of RT and PCR were also included in the model. Covariance analyses were also made using SAS with COX activity and/or FABP mRNA levels as covariates to explain variability in IMF content.

Results

Biochemical characteristics of *Longissimus thoracis* muscle

Differences in IMF content, COX activity and A-FABP mRNA levels were highly significant between the three breeds ($P < 0.002$). These parameters were the highest in Angus, intermediate in Limousin and the lowest in Belgian Blue (Table 1).

Table 1. Metabolic characteristics of Longissimus thoracis muscle.

	Belgian Blue	Limousin	Angus	SE
Intramuscular fat content (g /100 g dry matter)	2.69 ^{c}	6.51 ^{b}	9.37 ^{a}	0.518
COX activity (μmoles per g fresh tissue)	4.09 ^{c}	6.43 ^{b}	8.12 ^{a}	0.427
A-FABP mRNA levels (arbitrary units per mg total RNA	3.24 ^{b}	21.73 ^{b}	51.88 ^{a}	8.765

Means with different superscripts in the same row are significantly different ($P < 0.05$).

Relative to values in Belgian Blue, IMF content was 2.4 and 3.5-fold higher in Limousin and Angus respectively. COX activity was 1.6 and 2.0-fold higher and A-FABP mRNA levels 6.7 and 16-fold higher in Limousin and Angus respectively.

Intramuscular fat content, COX activity and A-FABP mRNA levels did not significantly differ between the two diets for each breed. However, COX activity tended to be slightly higher (+13%) with the pulp-rich diet although it was not significant (P = 0.12).

Metabolic activities related to intramuscular fat content

Irrespective of breed and diet, covariance analysis indicated that variability in COX activity explained 47% of total variability in IMF content ($P < 0.001$). Similarly, variability in A-FABP mRNA levels explained 42% of total variability in IMF content ($P < 0.001$) (Figure 1). The combination of COX activity and A-FABP mRNA levels in the model accounted for 64% of total variability in IMF content ($P < 0.001$).

Within each breed, there were no relationships between IMF content and COX activity (data not shown) or between IMF content and A-FABP mRNA levels (Figure 1), whereas IMF content was positively related to COX activity or A-FABP mRNA levels when the three breeds were considered together.

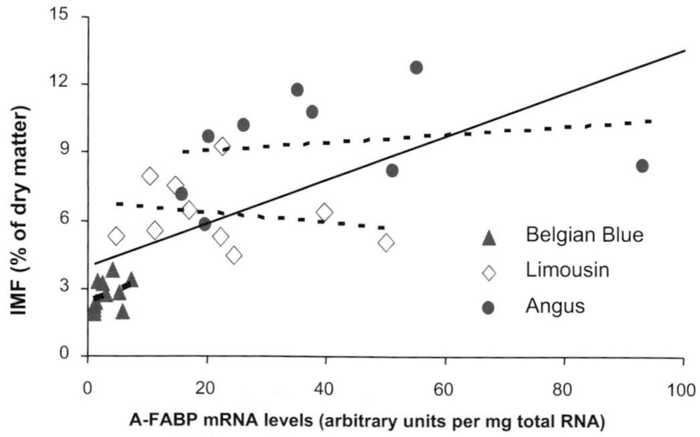

Figure 1. Intramuscular fat (IMF) content plotted against A-FABP mRNA levels. The black line represents the correlation curve of all values for the three breeds (r^2=0.42). The three dotted lines represent the correlation curves for each individual breed.

Discussion

This study was conducted with extreme diets and breeds. The number of individuals allowed a wide range of IMF contents. It was therefore possible to identify some major metabolic indicators which may contribute to the variability of fat deposition within muscle.

The type of diet did not influence IMF content, although a pulp-rich diet slightly increased oxidative muscle activity in Limousin and Angus young bulls only. This confirmed observations in Charolais steers in which the oxidative muscle metabolism was slightly increased with a grass-based diet which provided a high proportion of acetate compared to a cereal-rich diet with a high proportion of propionate (Ortigues-Marty *et al.*, 2002).

Comparison of breeds clearly indicates that IMF content was the highest in Angus, which is characterised by the highest activity of muscle oxidative enzyme (COX) and the highest A-FABP expression. On the contrary, IMF content was lowest in the Belgian Blue breed, which is characterised by the lowest activity of mitochondrial enzyme (COX) and the lowest A-FABP expression. Differences in COX activity confirm previous observations in rabbits which indicated that oxidative muscles deposit more intramuscular fat (Gondret *et al.*, 2001). It is surprising since, theoretically, the more the FA are catabolised, the less they should be deposited. It has been speculated that a high FA turnover which is a characteristic of oxidative muscles would favour fat deposition (Gondret *et al.*, 2001). Differences in A-FABP expression also confirm previous observations in steers from different breeds (Hocquette *et al.*, 2003). This suggests a higher number of adipocytes in muscles from Angus breed, as A-FABP is a late marker of adipocyte differentiation.

The remaining variability in IMF content not explained by the studied metabolic markers may be related to differences in other metabolic activities not explored in this study. In favour of this hypothesis, Bonnet *et al.*, (2003) also demonstrated that another metabolic enzyme involved in the control of fatty acid synthesis (glucose-6-phosphate dehydrogenase) was more active in muscles from Angus than from Limousin steers.

It is of interest to note that the metabolic predictors (COX, A-FABP) of IMF between breeds can not be used within breeds. This suggests that other metabolic pathways implicated in lipogenesis and in lipolysis may have a higher relative importance for intramuscular fat deposition when animals of the same breed are compared.

In conclusion, A-FABP expression and COX activity may be indicators of the ability of young bulls to deposit intramuscular fat, since these indicators differ between breeds which produce lean or fat beef.

Acknowlegdgements

This work was supported by the DGA from Ministry of Agriculture of Region Wallonne-Belgium and the French Ministry of Research (decision P 0405, Nov. 8th, 2001). The authors thank the private company GENOLIFE for the preparation of the bovine A-FABP primers.

References

Bas, P. and D. Sauvant, 2001. Variations de la composition des dépôts lipidiques chez les bovins. INRA Prod. Anim., 14 : p. 311-322.

Bonnet, M., Y. Faulconnier, J.F. Hocquette, C. Leroux, P. Boulesteix, Y. Chilliard and D.W. Pethick, 2003. Lipogenesis in subcutaneaous adipose tissue and in oxidative or glycolytic muscles from Angus, Black Japanese x Angus and Limousin steers. In: Progress in research on energy and protein metabolism, W.B. Souffrant and C.C. Metges (editors), Wageningen Academic Publishers, Wageningen, The Netherlands, p. 469-472.

Cianzo, D.S., D.G. Topel, G.B. Whitehurst, D.C. Beitz and H.L. Self, 1985. Adipose tissue growth and cellularity: changes in bovine adipocyte and number. J. Anim. Sci. 60: p. 970-976.

Gondret, F., J.F. Hocquette and P. Herpin, 2001. Teneur en lipides intramusculaires chez le lapin: contribution relative des différentes voies métaboliques des tissus musculaires. World Rabbit Sci. 9: p. 8.

Goutefongea, R. and C. Valin, 1978. Quality of beef. 2. The comparison of the organol.eptic properties of beef from cows and from young bulls. Ann. Tech. Agric. 27, p. 609-627

Hocquette, J.F., C. Jurie, Y. Ueda, P. Boulesteix, D. Bauchart and D.W. Pethick, 2003. The relationship between muscle metabolic pathways and marbling of beef. In: Progress in research on energy and protein metabolism, W.B. Souffrant and C.C. Metges (editors), Wageningen Academic Publishers, Wageningen, The Netherlands, p. 513-516.

Hocquette, J.F., I. Ortigues-Marty, D.W. Pethick, P. Herpin and X. Fernandez, 1998. Nutritional and hormonal regulation of energy metabolism in skeletal muscles of meat-producing animals. Livest. Prod. Sci. 56: p. 115-143.

Ortigues-Marty, I., C. Jurie, J.F Hocquette, B. Picard, I. Cassar-Malek, A. Listrat, R. Jailler, D. Bauchart, D. Dozias and D. Micol 2002. The use of principal component analysis (PCA) to characterize beef steers. In: Multi-Function grasslands. Quality forages, animal products and landscapes, J.L. Durand, J.C. Emile, C. Huyghe, C. Lemaire (editors). EGF, Volume 7, pages 584-585, Grassland Science in Europe

Pethick, D.W., G.S. Harper and V.H. Oddy, 2004. Growth development and nutritional manipulation of marbling in cattle: a review. Austr. J. Exp. Agric. 44: 705-715.

The fatty acid profiles of meat from calves fed linseed of oily cultivars

M.B. Zymon, J.A. Strzetelski and B. Niwińska*
Research Institute of Animal Production, Department of Animal Nutrition, ul. Sarego 2, 31-047 Krakow, Poland.
**Correspondence: jstrzet@izoo.krakow.pl*

Summary

The effects of supplementing full fat linseeds (*Linum usitatissimum* L. *convar. mediterraneum*) cultivar Opal and ecotype Linola on the fatty acids composition of muscle were investigated with 18 bull calves divided into 3 groups (n=6). Animals of the control group (C) were given a basal concentrate mixture additionally to a whole milk, and animals of the experimental groups were offered the same diet of which the concentrate was supplemented with linseeds from cultivar Opal (LO) rich in C18:3 n-3 (47,4% of total FA) or Linola (LL) rich in C18:2 n-6 (65.6%). Groups were divided into 2 sub-groups differing by the slaughtering age of calves: 42 or 90 days. Samples of *Musculus thoracis (MT)* were taken for fatty acid analysis. In MT muscle, fatty acids from total lipids did not differ between diets except for C18:3 n-3, CLA and C22:6 n-3 which tended to be higher in LL and LO than in group C and for C20:4 n-6 which tended to be higher in the LL group than in C and LO groups, but differences were not significant. Fat of meat of calves slaughtered at 42 days of age was characterized by lower UFA/SFA ratio than fat of meat of calves at 90 days of age.

Keywords: calf, linseed, meat, fatty acids profiles

Introduction

The experimental results showed that composition of fat of young ruminants could be modified by the diets (Jenkins and Kramer, 1990; Wachira *et al.*, 2002). A significant increase in the essential unsaturated fatty acids, especially C18:3 n-3 and conjugated fatty acids in deposited meat tissue fat was found in fattening bulls receiving concentrate mixture containing 19% ground linseed (Strzetelski *et al.*, 2001). Linseeds of oily cultivars are different in chemical composition and fatty acid profiles (Borowiec *et al.*, 2001). As the calves grow intake of solid feed increase and it may much more affect on fatty acids profile of muscle tissue than ingestion of liquid feed. The effects of milk and seeds from traditional and new "high linolenic" or "high linoleic" oily cultivars in calf diets on fatty acid profiles of meat are not fully examined.

The aim of the experiment was to study the effects of dietary supplementation with linseed from oily cultivars Opal and Linola on the fatty acid composition of veal meat.

Material and methods

The experiment was carried out on 18 Black-and-White Lowland bull calves (67.5% HF) randomly assigned at 7 days of age to 3 experimental groups of 6 animals each. Full fat dark-seeded linseed (*Linum usitatissimum* L. *convar. mediterraneum*) cultivar Opal and yellow-seeded ecotype Linola were used as the main source of fat in experimental diets. In the control group C calves were offered the basal concentrate diet without linseed, in group LO they received 11% seeds of cv. Opal and in group LL 10.5% seeds of ecotype Linola. Concentrate diet consisted on a percentage basis of: ground barley (44-50), ground wheat (23-37), soybean meal (12-15) and minerals (4). Groups were divided into 2 sub-groups: slaughtered at 42 and 90 day of age. The samples of *Musculus thoracis* (MT) were taken for analysis.

The requirements of calves, whole milk feeding schedule (from 7 to 56 days of age) and the composition of the concentrate diets were based on feeding standards (IZ-INRA, 2001). The chemical composition of the concentrate diets was determined according to AOAC (1990). The fatty acids (FA) composition of the concentrate diets and MT were determined by gas chromatography (Pye Unicam Sc 104, 30 m column Supelcowax).

The results were statistically analysed using a two-way analysis of variance SAS (1989), taking into account the type of diets and the slaughter age. Differences were declared significant at $P<0.05$.

Results

The chemical composition and nutritive values are given in Table 1. The crude protein (CP) and energy content in all concentrates were similar, but the highest content of PDI and lowest of ether extract (EE) was found in concentrate mixture C.

The average intake of concentrate diets by one calf to 42 days of age was 5.20 ± 1.1 kg and to 90 days of age was 83.1 ± 5.2 kg. The mean daily intake of nutrients by calf for all period of experiment were: 1.36 ± 0.05 kg dry matter, 280 ± 9.5 g crude protein, 225 ± 6.5 g PDI and 1.91±0.06 UFL.

Seeds ecotype Linola were characterized by the higher Cn-6/Cn-3 ratio, higher C16:0, C18:2 n-6 and lower content of C18:3 n3 in fat compared with fat from cv. Opal (Table 2). Concentrate diets LO and LL were characterized by a higher UFA/SFA ratio (1.6 and 1.8 times) compared with concentrate C. The higher content of C18:2 n-6 in FA of LL concentrate diet (62.5%) and of C18:3 n-3 in FA of LO concentrate (38.0%) in comparison with C concentrate (53.2% and 5.5%,

Table 1. Chemical composition and nutritive value of feeds.

Feed	Dry matter %	Content (kg^{-1} dry matter)							% of linseed EE in total E
		Crude protein g	Ether extract g	Crude fibre g	Ash g	UFL[1]	PDIN[1] g	PDIE[1] g	
Linseed var.									
Opal	90.87	285.5	377.2	163.2	39.2	1.47	149	44	
Linola	92.49	216.4	416.9	210.8	42.4	1.48	125	36	
Whole milk	12.56	264.2	318.2	-	63.6	0.24	250.9[2]		
Diets									
C	87.6	178.0	23.88	41.97	28.2	1.14	123	127	
LO	88.2	182.1	55.74	39.12	14.1	1.15	118	113	76,6
LL	87.9	174.8	59.19	44.50	16.1	1.15	115	112	77,5

[1]IZ-INRA (2001): UFL- unit of energy for milk production; PDIN- protein digested in the small intestine dependent on rumen degraded protein; PDIE - protein digested in the small intestine dependent on rumen-fermented organic matter
[2]digestible crude protein of milk for preruminant calves

Table 2. Fatty acid composition of linseed and diets.

Item	Seeds of linseed var		Diets		
	Opal	Linola	C	LO	LL
SFA					
14:0	0.23	0.14	0.31	0.26	0.19
16:0	4.62	7.59	17.98	7.83	10.23
18:0	4.64	4.03	2.13	4.07	3.58
22:0	0.08	0.34	0.20	0.11	0.30
24:0	0.10	0.20	0.19	0.12	0.19
Total SFA	10.54	12.73	21.39	13.18	14.96
UFA					
16:1	0.22	0.16	0.32	0.24	0.19
18:1	21.34	17.93	16.14	20.07	17.42
18:2 n-6	18.81	65.61	53.24	26.35	62.52
18:3 n-3	47.44	2.64	5.52	38.05	3.36
20:1	0.31	0.29	0.66	0.40	0.39
24:1	0.00	0.23	0.12	0.03	0.20
Total UFA	88.33	86.94	76.24	85.35	84.19
UFA/SFA	8.38	6.83	3.56	6.48	5.63
n6/n3	0.40	24.82	9.64	0.69	18.61

respectively) were found. The content of C18:3 n-3 in FA of LL concentrate diet was lower (3.4%) in comparison with LO concentrate (38.0%).

Meat samples of all groups were characterized by an average of 24.7±2% dry matter and a protein to fat ratio was: 8.12 (C), 10.86 (LO) and 11.37 (LL). In MT fat of LO and LL groups there were tendencies for higher level of C18:3 n-3, EPA, DHA and CLA and lower value of n-6/n-3 ratio as compared with group C (Table 3). The lowest content of C16:0 ($P<0.05$) and the highest of C18:0 ($P<0.05$) were found in the LL group as compared with others. The intramuscularly fat of calves slaughtered at 42 days of age was characterized by lower UFA/SFA ratio than that from calves at 90 days of age.

Discussion

Despite the lack of differences in UFA/SFA ratio in FA profile of Musculus **thoracis** fat, the concentrates with linseed could have increased the supply of some UFA to the small intestine. As the rumen developed, part of UFA could have become biohydrogenated. However, due to a higher fat proportion in the concentrate diets with linseeds, the biohydrogenation process could have been restricted and some UFA supply to the duodenum may have been relatively high Wu *et al.* (1991) stated that supplementary fat in the diet linearly increases fatty acids passage into the duodenum, creating the conditions for absorption from the small intestine. The somewhat greater deposition of C18:3 n-3 and C18:2 n-6 in MT of calves in groups LO and LL than in C groups was probably due to the high fat content of linseeds and its level in the concentrates and to FA proportions because feed intake being similar in all groups.

The differences in SFA and UFA in the meat fat of calves before and after weaning indicate that the proportions of UFA and particularly PUFA were mainly influenced by concentrate diets. For the first stage of growth, characterized by low solid feed intake, the meat fat contained more SFA

Table 3. Effect of diet and age of calves on the MT fatty acid composition.

Item	Main effects of							Means	SE	Inter-action
	Diets				Age of calves (days)					
	C[1]	LO[1]	LL[1]	LS =[2]	42	90	LS=[2]			
SFA										
C 12:0	0.43	0.42	0.28	0.21	0.41	0.35	0.51	0.38	0.19	NS
C 14:0	2.85	2.54	2.41	0.19	2.87	2.35	0.06	2.59	0.62	*
C 16:0	22.54[a]	22.63[a]	20.41[b]	0.04	23.62	19.97	0.01	21.80	2.81	**
C 18:0	13.90[b]	14.26[ab]	15.64[a]	0.04	14.22	14.98	0.21	14.60	1.45	NS
C 20:0	0.033	0.040	0.025	0.32	0.031	0.034	0.78	0.03	0.02	NS
C 22:0	0.12	0.003	0.013	0.29	0.010	0.009	0.81	0.009	0.01	NS
Total SFA	39.76	39.89	38.59	0.31	41.16	37.66	0.01	39.42	2.98	*
UFA										
C 16:1	3.14[ab]	3.41[a]	2.49[b]	0.02	3.07	2.96	0.69	3.016	0.64	*
C 18:1	31.79	30.31	29.71	0.21	30.54	30.67	0.70	30.61	2.56	NS
C 18:2 n-6	12.05	12.18	12.94	0.79	11.79	12.32	0.52	12.05	1.69	NS
C 18:3 n-6	0.09	0.11	0.13	0.17	0.9	0.12	0.12	0.11	0.06	NS
C 18:3 n-3	0.66	1.06	0.94	0.20	0.56	1.21	0.02	0.89	0.60	NS
CLA	0.41	0.51	0.47	0.30	0.49	0.41	0.29	0.45	0.16	NS
C 20:4 n-6	1.95	1.98	2.90	0.08	1.93	2.59	0.10	2.26	0.94	*
C 20:5 n-3 EPA	0.32	0.35	0.45	0.07	0.28	0.40	0.96	0.34	0.18	NS
C 22:6 n-3 DHA	0.39	0.50	0.49	0.08	0.47	0.45	0.71	0.46	0.92	NS
Total UFA	50.81	50.27	50.48	0.51	49.21	51.17	0.22	50.19	3.05	NS
UFA/SFA	1.29	1.27	1.31	0.83	1.2	1.37	0.04	1.28	0.16	NS
n-6/n-3	10.98	7.75	8.74	0.11	11.16	7.49	0.02	8.11	2.55	*

[1]C - control diet, LO- with linseed Opal, LL- with Linola,
[2]LS - level of significance;
[a,b] and * -P< 0.05; lack letters or NS-no statistically significant

than UFA, which was mainly due to the FA composition of milk fat and suggests that they could be absorbed similarly as in monogastric animals (Hocquette and Bauchart, 1999). The somewhat higher concentration of linolenic acid in the veal meat of calves receiving linseed cultivar Opal, compared with LL group, could be explained by the higher intake of this acid. The increased proportions of CLA in MT of calves in these groups could result from changes of C18:2 n-6 or C18:3 n-3 (Stasiniewicz *et al.*, 2000). Possibly, this had a beneficial effect on limiting fat synthesis, as evidenced by a greater protein to fat ratio in LO and LL than in control group (Steinhart *et al.* 1998).

Conclusion

Feeding calves with concentrate diets with 10% of linseed Opal or Linola makes it possible to increase the PUFA content of veal meat of calves slaughtered at 90 days of age. Fat of meat obtained from calves at 42 days of age during the milk-feeding period was characterized by lower UFA/SFA ratio than fat of meat obtained from calves at 90 days of age.

References

AOAC, 1990. Association of Official Analytical Chemists. Official Methods of Analysis. 15thEdition, Arlington, VA

Borowiec, F., T. Zając, Z.M. Kowalski, P. Micek and M. Marcin̈ski, 2001. Comparison of nutritive value of new commercial linseed oily cultivars for ruminant. J. Anim. Feed Sci. 10: p. 301-308.

Hocquette, J. F. and D. Bauchart, 1999. Intestinal absorption, blood transport and hepatic and muscle metabolism of fatty acids in preruminant and ruminant animals. Repr. Nutr. Dev. 39: 1, p. 27-48.

IZ-INRA, 2001. Standards for Cattle, Sheep and Goat Nutritional value of Feed for Ruminants (in Polish) Research Institute of Animal Production, Balice, (Poland)

Jenkins K.J. and J.K.G. Kramer, 1990. Effect of dietary corn oil and fish oil concentrate on lipid composition of calf tissues. J. Dairy Sci. 73: p. 2940-2951.

SAS.SAS/STAT User Guide,Version 6.,1989. SAS Inst., Inc., Cary NC.

Stasiniewicz, T., J. Strzetelski, J. Kowalczyk, S. Osięgłowski and H. Pustkowiak, 2000. Performance and meat quality of fattening bulls fed complete feed with rapeseed oil or linseed. J. Anim. Feed Sci. 9: p. 283-296.

Steinhard, H., J. Fritsche and N. Sehat, 1998. Determination of trans-fatty acids and conjugated linoleic acid isomers (CLA) in foods. VS. FDA Washington D.C./VSA. G.I.T. Laboratory J. 2: p. 84-85.

Strzetelski J., J. Kowalczyk, S. Osięgłowski, T. Stasiniewicz, E. Lipiarska and H. Pustkowiak, 2001. Fattening bulls on maize silage and concentrate supplemented with vegetable oils. J. Anim. Feed Sci. 10: p. 259-271.

Wachira, A. M., L.A. Sinclar, R.G. Wilkinson, M. Enser, J.D. Wood, and A.V. Fisher, 2002. Effect of dietary fat source and breed on the carcass composition, n-3 polyunsaturated fatty acids and conjugated linoleic acid content of sheep meat and adipose tissue. Brit. J. Nutr. 88: p. 697-709.

Wu, Z., O.A. Ohajuruka and D.L. Palmquist, 1991. Rumen sythesis, biohydrogenation and digestibility of fatty acids by dairy cows. J. Dairy Sci. 74: p. 3025-3034.

Effects of diets supplemented with oil seeds and vitamin E on specific fatty acids of *rectus abdominis* muscle in Charolais fattening bulls

D. Bauchart[1], C. Gladine[1], D. Gruffat[1], L. Leloutre[2] and D. Durand[1]*
[1]*INRA-Centre de Recherches de Clermont-Ferrand-Theix, Unité de Recherches sur les Herbivores, Equipe Nutriments et Métabolismes, 63122 St-Genès-Champanelle, France,*
[2]*Innovation en Nutrition et Zootechnie, BP 19, 02402 Château-Thierry Cedex, France.*
** Correspondence: bauchart@clermont.inra.fr*

Summary

The effects of lipid (from linseed or rapeseed) and vitamin E supplements on fatty acid (FA) composition of lipids in *rectus abdominis* muscle were determined in 56 fattening Charolais bulls given for 97d either a straw (30%) and concentrate (70%) -based diet (SC) or a corn silage (70%) and concentrate (30%) -based diet (CC). Generally, the health value of beef FA for consumers was better for the main FA considered (except trans vaccenic acid) with the SC diet than with the CC diet. Extruded linseed added to SC and CC diets had a more beneficial effects than extruded rapeseed when added to the basal diets, but vitamin E added to linseed-supplemented diets appeared to play variable effects on the health value of beef FA for humans.

Keywords: Fattening bulls, high-fat diets, oil seeds, fatty acids, muscle

Introduction

Incorporation of vegetable oils into rations is now currently employed in finishing beef cattle since it increases total energy intake (thus reducing the fattening period), but it also facilitates manipulation of muscle fatty acid (FA) composition for a better health value of meat for consumers (Clinquart *et al.*, 1995, Wood *et al.*, 1999). Among factors that influence deposition of FA in muscles (breed, sex, age, diet), oil seeds added to the diet would play an important role by providing large amounts of dietary unsaturated FA, especially n-6 and n-3 polyunsaturated FA (PUFA). However, their effects do not depend only on their relative protection against ruminal biohydrogenation, but also on the composition of the basal diet, especially the proportions and the nature of forage associated with concentrate (Scollan *et al.*, 2005).

The aim of the present experiment was to compare in fattening bulls the impact of two sources of extruded oil seeds (linseed and rapeseed), with or without vitamin E, added to two basal diets which differed by the source and proportion of forage associated to concentrate on the FA profile of total lipids in the *rectus abdominis* (RA) muscle.

Material and methods

Animals and diets

The experiment was performed with 56 Charolais fattening bulls [373 (SD 6) d-old, live weight 531 (SD 6) kg], selected for their live weight and age. The animals were assigned at random to eight rations (n=7 for each diet) for a 97 day feeding study. Four rations were straw (30%) and concentrate (70%) -based (SC) and four rations were corn silage (70%) and concentrate (30%) -based (CC). For each ration type, animals were given the basal diets without any supplements (diets SC and CC) or with extruded linseed alone (SCL and CCL) or plus vitamin E (SCLE and CCLE), or with extruded rapeseed plus vitamin E (SCRE and CCRE) (Table1). The eight diets were calculated to be nearly similar in their energy and nitrogen contents (Table 1) and to allow a mean live weight gain of 1300 g/d. For SC-based diets, oleic (18:1 Δ9c), linoleic (18:2n-6) and linolenic (18:3n-3) acids ingested by bulls (in kg per 97d) amounted to 0.4, 8.8 and 0.7 (SC diet), 7.8, 14.1 and 29.3 (SCL and SCLE diets) and 40.1, 19.6, and 6.5 (SCRE diet) respectively. For CC-based diets, these FA amounted to 1.7, 6.3 and 0.7 (CC diet), 6.5, 10.4 and 25.4 (CCL and CCLE diets), and 34.0, 15.1, and 5.4 (CCRE diet) respectively. Animals were slaughtered at a mean live weight of 682 (SD 12) kg. A sample of RA muscle (100 g) was collected 1d *post-mortem* and stored at -20°C until FA analysis.

Fatty acid analysis

Total lipids of RA muscle were extracted by mixing 5 g of fresh tissue with chloroform/methanol 2/1 (vol./vol.) according to the method of Folch *et al.*, (1957). The FA were extracted from total lipids and then converted into methyl esters (FAME) by transmethylation using borontrifluoride-methanol (14% solution) according to the method of Christie (2001). FAME composition was determined by GLC (Peri 2001, Perichrom, France) using a glass capillary column (length: 100 m; Øi: 0.25mm, H_2 as the carrier gas) coated with CP Sil 88 (oven temperature program : from 70°C to 215°C). Response coefficient of each individual FA was calculated by using the quantitative mix C4-C24 FAME (Supelco, USA).

Statistical analysis

Results were expressed as mean values with their standard errors (SEM). Effects of dietary treatments on individual FA or on a FA family have been analysed by ANOVA according to the GLM procedure of SAS.

Table 1. Ingredient and chemical compositions of the eight experimental diets.

Diets	SC	SCL	SCLE	SCRE	CC	CCL	CCLE	CCRE
Ingredient composition								
Straw (% diet DM)	30	32	32	31	-	-	-	-
Corn silage (% diet DM)	-	-	-	-	63	61	61	62
Concentrate (% diet DM)	70	68	68	69	37	39	39	38
Extruded linseed (% diet DM)	-	14	14	-	-	11	11	-
Extruded rapeseed (% diet DM)	-	-	-	14	-	-	-	11
Vitamin E (IU/kg diet)	10	10	250	250	10	10	250	250
Chemical composition								
Dry matter (g/kg feed)	88	88	88	88	45	45	45	45
Composition of DM (%)								
Crude protein (N x 6.25)	13.4	13.7	13.5	12.8	12.7	13.2	12.9	12
Cellulose	15.2	18.5	18.5	18.9	15.5	19.1	19	19.5
Starch	33.6	23.3	22.8	23.2	30.5	26.1	25.9	26.1
Lipid	2.0	5.8	5.7	7.5	2.5	5.9	5.8	7.0

SC = Straw and Concentrate-based diet; CC = Corn silage and Concentrate-based diet; L = Linseed; R= Rapeseed; E = Vitamin E

Results and discussion

The relative fatty acid composition (% of total fatty acids) of total lipids in RA muscle of Charolais bulls given the eight experimental diets for 97 days is given in Table 2.

Among the main saturated fatty acids (SFA), the proportion of 16:0 was lower in SC than in CC-based diets (P < 0.0001). It decreased in lipid-supplemented diets compared to control diets (P < 0.0001) which was beneficial for the health value of meat since this fatty acid is known to be atherogenic for humans (Grundy and Denke, 1990). Conversely, dietary addition of linseed or rapeseed increased deposition of 18:0 in RA muscle (Table 2), which also improved the health value of meat. This suggested that vegetable oils rich in C18 PUFA (linseed) or C18 monounsaturated FA (MUFA) (rapeseed) would favour production of 18:0 as a result of their extensive biohydrogenation by rumen bacteria. Similar observations were previously reported for cattle given lipid supplements from soybean (Madron *et al.*, 2002) or rapeseed (Flachowski *et al.*, 1994).

Monounsaturated FA were dominated by cis-oleate (18:1 Δ9cis: >75%) neutral or favourable for the health of consumers (Grundy and Denke, 1990). It was higher in CC-based diets (P < 0.001) but did not vary significantly with the addition of oil seeds in diets. The lack of impact of vegetable oils was previously noted with diets supplements with crushed rapeseed (Flachowski *et al.*, 1994)

Table 2. Relative fatty acid composition (% of total fatty acids) of total lipids in rectus abdominis muscle of Charolais bulls assigned to the eight experimental diets (n=7 for each treatment) for 97 days.

	16:0	18:0	18:1Δ9c	18:1Δ11t	18:2n-6	18:3n-3	CLA	PUFA/ SFA	n-6/ n-3
SC	22.0	14.1	27.8	5.3	8.8	1.1	1.02	0.26	4.15
SCL	19.9	17.3	25.9	5.3	8.6	3.6	0.94	0.31	2.30
SCLE	21.7	15.6	27.0	6.2	6.8	3.4	1.03	0.26	2.02
SCRE	20.3	15.3	27.6	6.5	10.3	1.5	0.89	0.31	4.56
CC	25.7	15.8	32.7	2.0	5.8	0.7	0.69	0.15	4.19
CCL	22.4	17.8	28.0	2.5	7.3	3.7	0.78	0.26	1.80
CCLE	22.6	17.0	30.8	3.4	5.9	3.2	1.05	0.22	1.77
CCRE	22.9	17.6	30.4	4.1	6.7	1.3	0.85	0.19	3.27
SEM	0.04	0.04	0.07	0.04	0.04	0.02	0.01	0.01	0.02
Statistical effects									
Basal diet	0.001	0.0035	0.0008	0.0001	0.0001	NS	NS	0.0001	0.003
Supplements	0.0005	0.003	NS	0.065	0.007	0.001	NS	0.0047	0.001
Interaction	NS	NS	NS	NS	NS	NS	NS	NS	NS
Contrasts									
Control vs. lipids	0.0001	0.0017	NS	0.0671	NS	0.0001	NS	0.0053	0.001
L. vs. R.	NS	NS	NS	NS	0.0012	.0001	NS	NS	0.001
L vs. L+vit. E	NS	0.05	NS	NS	0.0081	NS	NS	0.02	NS

NS =P > 0.10; Significant tendency: = 0.10 > P > 0.05; Significant = P < 0.05; SC = Straw and Concentrate-based diet; CC = Corn silage and Concentrate-based diet; L = Linseed; R= Rapeseed; E = Vitamin E; L vs R includes the vitamin E treatment or extruded linseed (Choi *et al.*, 2000). Conversely, trans vaccenic acid (TVA, 18:1Δ11t), is considered to be undesirable for humans since it can increase the cardiovascular disease risk (higher susceptibility of low-density lipoproteins to oxidation and blood pressure) (Lichtenstein, 2000). TVA was two times higher in SC than in CC-based diets ($P < 0.0001$) and tended to increase in groups given lipid supplements, especially in groups given rapeseed (Table 2). Deposition of TVA (produced by partial hydrogenation of PUFA by ruminal bacteria) into muscles, was mainly stimulated by dietary 18:2n-6 (rapeseed in our experiment) ($r = 0.73$, Figure 1c) and not by 18:3n-3 (Table 2), since this n-3 PUFA was mainly converted into 18:0 in the rumen (Demeyer and Doreau, 1999). This would confirmed by the lack of relationships noted between levels of 18:3n-3 and of TVA in lipids of RA muscle (figure not given).

Polyunsaturated FA are generally considered to be favourable for the health of consumers. Compared to dietary advice, the PUFA/SFA ratio of beef (0.12-0.25) is considered to be low, the ideal value for humans averaged 0.40. Moreover, the ratio of n-6/n-3 PUFA of beef (generally higher than 4) should be beneficially decreased up to 2-3 in order to constitute an additional source of n-3 PUFA for humans, since n-3 PUFA are generally low in their diets (Legrand, 2001). For these reasons, linseed was used to supply amounts of linolenic acid (18:3n-3) (29.3 and 25.4 kg per 97d in SC and CC groups, respectively). Such treatments increased the proportion of 18:3n-3 in RA fatty acids by 3.2 and 4.9 times in SC and CC groups, respectively ($P<0.0001$) (Table 2), as noted earlier (Choi *et al.*, 2000). Incorporation of this FA into muscle increased with their dietary intake (Figure 1a). It mainly contributed to the 2-fold decrease of n-6/n-3 ratio in linseed supplemented groups ($P< 0.001$) very beneficial for the health value of muscles. Linoleic acid

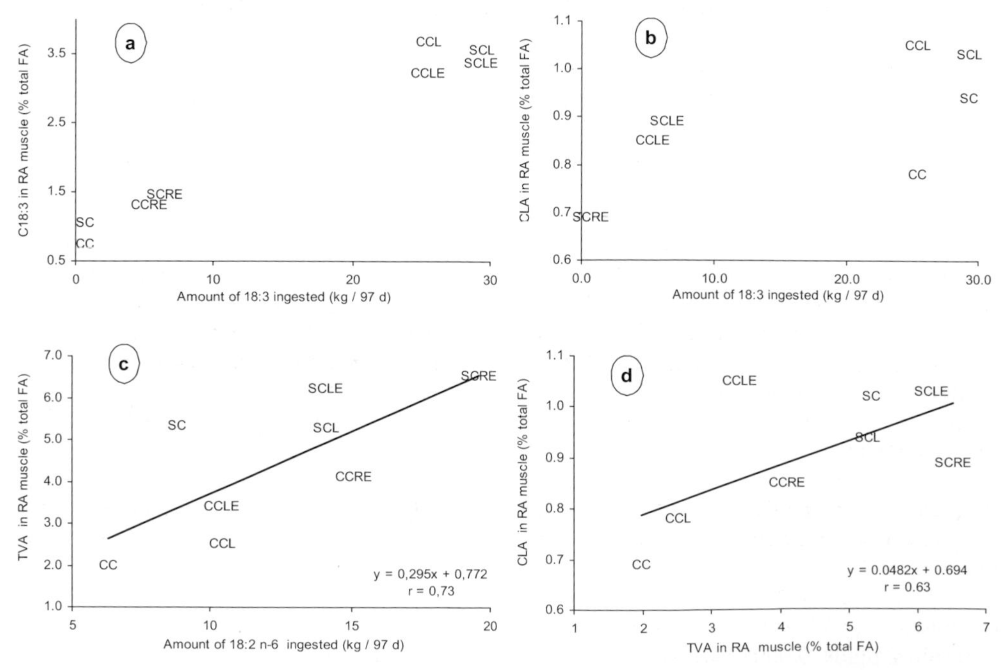

Figure 1. Relationships between the level of 18:3n-3, CLA or TVA in lipids of rectus abdominis *(RA) muscle and the amount of 18:3n-3 (Figure 1a and b) or of 18:2n-6 (Figure 1c) ingested or between RA muscle levels of CLA and TVA (Figure 1d) in Charolais bulls assigned to the eight experimental diets (n=7 for each treatment) for 97 days.*
SC = Straw and Concentrate based diet; CC = Corn silage and Concentrate based diet; L = Linseed; R= Rapeseed; E = Vitamin E .

(18:2n-6) was 40% higher in SC than in CC diets (P<0.0001). It was not affected by lipid supplements compared to control values but was lower in linseed compared to rapeseed supplemented diets (P< 0.0012). Addition of vitamin E in linseed supplemented diets led to a 3-fold increase of vitamin E in muscles (data not given). Unexpectedly, it decreased by 20% the proportion of 18:2n-6 in muscle FA (Table 2). This would be the reflect of an higher biohydrogenation of 18:2n-6 by rumen bacterial, since an higher proportion of TVA was also noted in linseed +vitamin E supplemented diets. Among hypothesis, vitamin E would affect protection of 18:2n-6 against biohydrogenation, possibly by altering its uptake and incorporation by rumen bacteria, especially solid-adherent bacteria (Bauchart *et al.*, 1990).

Conjugated linoleic acid (CLA), which is known to be beneficial for humans by their potential anti carcinogenic and hypocholesterolemic properties (Pariza *et al.*, 2001), is mainly represented by the *9-c,11-t* isomer, but practically devoid of the *10-t,12-c* isomer. The lack of effect of linseed treatment on muscle CLA was unexpected (Table 2). It can be explained by its relatively high amount in RA muscle (0.69 and 1.02%) both in basal SC and CC diets compared to data from the literature (<0.70). Such situation can limit its endogenous synthesis (from TVA) and therefore their subsequent deposition in muscle tissues. Moreover, the large incorporation of 18:3n-3 from dietary linseed into muscle suggested that this FA was relatively well protected against ruminal biohydrogenation, thus limiting production of TVA, the main precursor of *9c,11t* CLA in tissue. A positive and significant correlation was noted between muscle CLA and dietary 18:3n-3 supply (r= 0.66, Figure 1b) and between muscle CLA and muscle TVA (r = 0.63, Figure 1d).

In conclusion, the health value of beef FA for consumers was better for the main FA considered (excepted TVA) with SC diet compared to CC diet. These characteristics were improved by the dietary addition of linseed rather than of rapeseed. Addition of vitamin E in linseed-supplemented diets appeared to play rather a general negative effect on the health value of beef FA for humans since it increased 16:0 and TVA and decreased both n-6 and n-3 PUFA. On the other hand, vitamin E had beneficial effects since it decreased the n-6/n-3 ratio and increased CLA proportion in beef additionally to its protective effects against peroxidation of muscle unsaturated FA.

Further investigations are needed to analyse the long-term effect of linseed supplementation in situation of a high supply of linseed oil. On the other hand, it would be important to precise, in such dietary conditions, the specific effects of vitamin E on the metabolism of n-6 PUFA in the rumen (interaction with bacterial PUFA metabolism) and in muscle tissues (interaction with PUFA cell membranes) in order to understand the negative effect of vitamin E on deposition of 18:2n-6 in beef.

References

Bauchart D., F. Legay-Carmier, M. Doreau and B. Gaillard, 1990. Lipid metabolism of liquid-associated and solid adherent bacteria in rumen contents of dairy cows offered lipid -supplemented diets. Br. J. Nutr. 63: p. 563-578.

Choi N. J., M. Enser, J.D. Wood, and N. D. Scollan, 2000. Effect of breed on the deposition in beef muscle and adipose tissue of dietary n-3 polyunsaturated fatty acids. Anim. Sci. 71: p. 509-519.

Christie W.W. 2001. A practical guide to the analysis of conjugated linoleic acid. Inform 12: p. 147-152.

Clinquart A., D. Micol, C. Brundseaux, I. Dufrasne, and L. Istasse, 1995. Utilisation des matières grasses chez le Bovin (*Use of fats for beef cattle*). INRA Prod. Anim. 8: 29-42

Demeyer D. and M. Doreau, 1999. Targets and procedures for altering ruminant meat and milk lipids. Proc. Nutr. Soc. 58: p. 593-607.

Flachowski G., G.H. Richter, M. Wendemuth, P. Mockel, H. Graf, G. Jahreis and F. Lubbe, 1994. Einfluß von Rapssamen in der Mastrinderernährung auf Fettsäurenmuster, Vitamin E - Gehalt und oxidative Stabilität des Körperfettes. Z. Ernährungswiss (*Influence of rapeseed on fatty acid composition, vitamin E content and stability of body fat against peroxidation in beef cattle)*, 33: p. 277-285.
Folch J., M. Lees and G.H.S. Sloane-Stanley, 1957. A simple method for the isolation and purification of total lipids from animal tissues. J. Biol. Chem. 226: p. 497-509.
Grundy S.M. and M. A. Denke, 1990. Dietary influence on serum lipids and lipoproteins. J. Lipid Res. 31: p. 1149-1172.
Legrand, P. 2001. ANCs for fat. Sci. Aliments 21: p. 348-360.
Lichtenstein A.H., 2000. Trans fatty acids and cardiovascular disease risk. Curr. Opin. Lipid. 11: p. 37-42.
Madron M. S., D. G. Peterson, D. A. Dwyer, B. A. Corl, L. H. Baumgard, D. H. Beermann and N. Bauman, 2002. Effect of extruded full-fat soybeans in conjugated linoleic content of intramuscular, intermuscular and subcutaneous fat in beef steers. J. Anim. Sci. 80: p. 1135-1143.
Pariza M.W., Y. Park and M.E. Cook, 2001. Review : The biologically active isomers of conjugated linoleic acid. Prog. Lipid Res. 40: p. 283-298.
Scollan N.D., R. I. Richardson, S. De Smet, A.P. Moloney, M. Doreau, D. Bauchart and K. Nuernberg, 2005. Enhancing the content of beneficial fatty acids in beef and consequences for meat quality. In: Indicators of milk and beef quality, J.F. Hocquette and S. Gigli (editors), Wageningen Academic Publishers, Wageningen, The Netherlands, this book.
Wood J. D., M. Enser, A. V. Fisher, G.R. Nute, R. I. Richardson and P. R. Sheard, 1999. Manipulating meat quality and composition. Proc. Nutr. Soc. 58: p. 363-370.

Dietary tea catechins and lycopene: effects on meat lipid oxidation

D. Tedesco[1], S. Galletti[1], S. Rossetti[1] and P. Morazzoni[2]*
[1]Department of Veterinary Sciences and Technologies for Food Safety, Via Celoria 10, 20133 Milano, Italy, [2]Indena S.p.A., Viale Ortles 12, 20139 Milano, Italy
**Correspondence: doriana.tedesco@unimi.it*

Summary

Twenty White New Zealand male rabbits aged 45 days and weighing on average 1.360 ± 0.04 kg were equally divided into two groups. A control group was fed a basal commercial diet. The other group was fed the basal diet with 600 mg/kg diet of a mixture of green tea extract (Green Select®, Indena S.p.A., Milan, Italy) and lycopene (Indena S.p.A., Milan, Italy). Feed intake and body weight were monitored weekly. After slaughtering and after 7 days of storage in commercial conditions, muscle brightness and colour indices (lightness, redness, yellowness), pH, and thiobarbituric acid-reactive substances (TBARS) were evaluated on the *M. Longissimus lumborum*.

Feed intake, body weight and feed conversion ratio were not negatively affected by treatment. No differences between groups were found in meat pH and colour indices both at slaughtering time and after 7 days of storage. TBARS were similar in the two groups at slaughtering time. After 7 days of storage TBARS were higher ($P<0.05$) in control group with respect to treated group, indicating an improvement in lipid oxidation status of meat from treated animals.

We conclude that treatment with green tea extract and lycopene have positive effects on oxidative status of rabbit meat and do not negatively affect animal performances.

Keywords: rabbit meat, tea catechins, lycopene, lipid oxidation

Introduction

Lipid oxidation is considered as the main process responsible for the loss of meat quality, leading to oxidative flavors and loss of pigments and vitamins in meat (Monahan *et al.*, 1990). Lipid oxidation can be delayed using antioxidants. Previous researchers have demonstrated that beef from cattle given supplementary vitamin E in the diet was less susceptible to metmyoglobin formation and lipid oxidation than conventionally produced beef was (Arnold *et al.*, 1993), enhancing shelf-life and colour stability. The use of synthetic antioxidants like butylated hydroxytoluene (BHT) and butylated hydroxyanisole (BHA) are not accepted by consumers, then the use of natural antioxidants as alternative has received much attention in recent years (Tang *et al.*, 2000).

Catechins are a predominant group of polyphenols present in green tea leaves, and have recently attracted interest for their antimicrobial activity evaluated on rabbit caecum microflora (Tedesco *et al.*, 2003). Tea catechins have been reported to inhibit lipid oxidation in meat extending the shelf life of muscle food products containing high levels of fat (Buck and Edwards, 1997). Other authors reported that dietary tea catechins exhibited a long-term antioxidative effect on chicken meat during frozen storage (Tang *et al.*, 2001).

Lycopene is an acyclic carotenoid with 11 linearly arranged conjugated double bonds and is found mainly in tomato (*Solanum lycopersicum*) products. Many of the putative biologic effects and health benefits of lycopene, as the other carotenoids, are hypothesized to occur via protection against oxidative damage. Lycopene acts as free radical scavenger and interacts with reactive oxygen species such as hydrogen peroxide and nitrogen dioxide (Bohm *et al.*, 1995; Woodall *et al.*, 1997).

The action sites of the polyphenolic substances are very different from those of carotenoids, therefore the concomitant administration of lipophylic compounds like lycopene and hydrophilic compounds like tea catechins rise to an advantage from the biological point of view. The combination of two antioxidants with different polarity increases the effectiveness of the single compounds (Bombardelli and Morazzoni, 1995).

The objective of this study was to determine the influence of green tea extract and lycopene administration in rabbits diets on meat oxidative stability.

Materials and methods

Animals and diets

A total of 20 forty-five-d-old male New Zealand White rabbits (1.360 ± 0.04 kg BW) were randomly allotted into two groups of 10 animals each. Rabbits were housed in individual cages and adapted for five days prior to treatment. Animals were fed with standard commercial pellet feed diet for a period lasted 5 weeks. Nutrient composition of the diet was crude protein 15.30%, crude fiber 17.40%, ether extract 2.5% and ashes 8.5% as fed. The control group was fed the basal commercial diet, the treated group was fed the basal diet supplemented with 600 mg/kg feed of a mixture of green tea extract (Green Select®, Indena S.p.A., Milan, Italy) and lycopene (Indena S.p.A., Milan, Italy), containing 15.8% of polyphenols and 2.5% of lycopene. Animals were cared on in accordance with the European Directive 86/609/CEE.

Individual feed consumption was measured each day for all animals. Animals were individually weighed every week. The individual feed conversion ratio (FCR) was estimated for each experimental week. Morbidity and mortality were recorded in each group.

Rabbits were sacrificed by cervical dislocation. The *M. Longissumus lumborum* was removed from each carcass and divided in two chops. The chops were randomized for analyses to be performed on day 0 and on day 7 of storage. The d0-chops were immediately analyzed, the d7-chops were placed in plastic trays, wrapped in oxygen permeable PVC wrap and placed in an illuminated (fluorescent light of 616 lx) refrigerated (4°C) display cabinet for 7 days. To assess the effect of dietary treatment on raw meat during refrigerated storage the d0-chops and d7-chops were analysed for the colour, pH and TBARS determination.

Colour and pH measurement

The colour measurements were performed using a tristimulus colourimeter (Minolta Chroma Meter CR-300, Minolta, Osaka, Japan). The L^* (lightness), a^*(redness) and b^*(yellowness) values were measured 3 times on the surface of the chops at day 0 and at day 7. The instrument was calibrated using white calibration plate at the beginning of each session. The pH of meat was measured using a pH meter (HI microcomputer, Hanna Instruments, Milan, Italy).

TBARS values

Lipid oxidation was assessed on the basis of the malondialdehyde formed during refrigerated storage, by the thiobarbituric acid reactive substances (TBARS) method, as described by Du *et al.* (2000). Three g of meat was weighed and homogenized with 15 mL of deionized distilled water, using a Sorvall Omnimixer (DuPont Instruments, Newtown, Conn., U.S.A.), for 10 s at highest speed. One mL of the meat homogenate was transferred to a test tube and butylated hydroxyanisole (BHA, 50 μl) and thiobarbituric acid/tricloroacetic acid (TBA/TCA, 2mL) were added. The mixture was vortexed and then incubated in a boiling water bath for 15 min to develop colour. The sample was then cooled in ice for 10 min, vortexed again and centrifuged for 15 min at 2500 r.p.m. The absorbance of the resulting supernatant solution was determined at 531nm against a blank containing 1 mL of deionized distilled water and 2 mL of TBA/TCA solution. The amounts of TBARS were determined by comparing the standard curve of absorbance at 531 nm for series of malondialdehyde solutions analyzed by the same method, and expressed as mg of malondialdehyde (MDA)/kg of meat.

Statistical analysis

Data were analyzed using the GLM procedure of SAS (SAS, User's Guide, 2001). Significance was declared at P<0.05 and values are presented as least square means (±SEM).

Results and discussion

No morbidity or mortality was recorded in this study. The influence of dietary supplementation with green tea and lycopene on growth performances are presented in Table 1. No significant differences for BW, BW gain, feed intake and FCR were detected for rabbits from control and treated groups. No treatment effect was also observed in carcass weight at slaughter (1.53 ± 0.041 kg *vs* 1.45 ± 0.041 kg in control and treated group respectively).

Values obtained from the evaluation of *M. Longissimus lumborum* at 0 and 7 days of storage are reported in Table 2. No treatment effects on pH, L^*, a^* and b^* were found. Values of pH, L^* and b^* were significantly different with respect to the day of storage (P<0.001). In both groups pH decreased with time of storage. L^* values, a measure of lightness, and b^* values, a measure of yellowness, increased in both groups after 7 days of refrigerated storage. Rabbit meat is paler than beef. Meat color of *M. Longissimus lumborum* shows a greater lightness (L*) and lesser redness (a*) than the same muscle in beef (Realini *et al.*, 2004). Redness a^* decreases with oxidation of myoglobin and in beef lipid and pigment oxidation are related during meat maturation. These antioxidants could be useful in this species to improve appearance of beef cuts extending the retail-display.

The effect of the treatment on TBARS is shown in Figure 1. The TBARS values evaluated at day 0 indicated no differences between control and treated group. The TBARS values of meat from the control group after 7 days of storage in commercial conditions were much higher than that of the day 0, indicating a significant development of lipid oxidation during storage in those meats (0.545 ± 0.045 and 0.832 ± 0.045 mg MDA/kg, at 0 and 7 days respectively; P<0.05). The refrigerated storage-induced lipid oxidation was not observed in meats from animals receiving green tea and lycopene: TBARS values did not present any significant (P>0.05) difference between day 0 and day 7 of storage (Figure 1). The effectiveness of green tea and lycopene as antioxidants during meat storage was evidenced by the significance of treatment effect (P<0.01) found after 7 days of storage, when the TBARS values in chops from treated group were lower with respect to control group (0.528 ± 0.045 mg MDA/kg *vs* 0.832 ± 0.045 mg MDA/kg).

Our results are in accordance with other studies that focused on the single administration of tea polyphenols or lycopene. Tang *et al.* (2001) evidenced that meat from chickens supplemented with dietary tea catechins was less susceptible to lipid oxidation during frozen storage. Botsoglou

Table 1. Effects of the green tea extract and lycopene supplementation on rabbit performances (C=control; T=green tea + lycopene).

	Weeks in treatment														
	1			2			3			4			5		
	C	T	SEM	C	T	SEM	C	T	SEM	C	T	SEM	C	T	SEM
BW (kg)	1.963	1.872	0.047	2.278	2.151	0.058	2.521	2.400	0.069	2.738	2.633	0.065	2.879	2.848	0.053
BW gain (kg/d)	0.601	0.509	0.027	0.315	0.279	0.028	0.242	0.249	0.027	0.217	0.232	0.016	0.141	0.214	0.026
Feed intake (g)	148.1	153.4	9.52	181.3	152.8	9.81	172.5	161.5	10.54	178.7	168.2	7.35	157.3	152.3	9.45
Feed conversion ratio	0.400	0.327	0.129	0.247	0.261	0.221	0.199	0.218	0.221	0.179	0.198	0.012	0.139	0.199	0.237

Table 2. Colour, pH and TBARS values (mg/kg) of M. Longissimus lumborum *after 0 and 7 days of storage*[1]*. (C=control; T=green tea + lycopene).*

	Day 0			Day 7			P		
	C	T	SEM	C	T	SEM	Treatment	Day	Day*Treatment
pH	6.72	6.74	0.11	5.53	5.77	0.07	N.S.	***	N.S.
L	44.89	45.64	0.76	53.67	52.55	1.21	N.S.	***	N.S.
*a**	3.16	3.48	0.35	4.41	3.96	0.44	N.S.	N.S.	N.S.
*b**	2.69	2.89	0.28	8.15	8.09	0.47	N.S.	***	N.S.

[1]Means with different superscript within row differ significantly
*P<0.05, **P<0.01, ***P<0.001

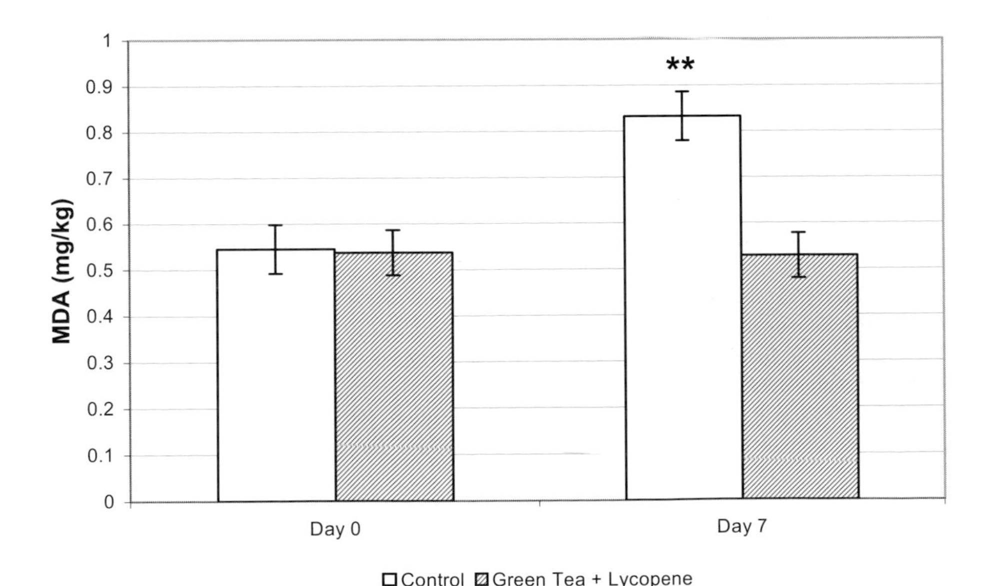

Figure 1. TBARS values (mg/kg) of M. Longissimus lumborum *after 0 and 7 days of storage. Data marked with asterisk differ significantly (**P<0.01).*

et al. (2004) suggested that inclusion of dried tomato pulp in feed at a level of 5%, comparable to the lycopene level of our experimental diet, exerted an antioxidant effect on quail meat, whereas addition at level of 10% exerted prooxidant effect. Considering TBARS values observed in our study after 7 days of storage, the combination of two antioxidants with different polarity inhibited lipid oxidation, reducing the potential risk induced by lipid oxidation products.

Conclusion

In conclusion, the results of this study show an inhibition of lipid oxidation of meat as a result of feeding green tea and lycopene. In addition this study provides evidence that dietary green tea and lycopene do not negatively affect growth performances and feed intake, enabling the safe use of this compounds as feed additives.

References

Arnold, R. N., S. C. Arp, K. K. Scheller, S. N. Williams, and D. M. Schaefer. 1993. Tissue equilibration and subcellular distribution of vitamin E relative to myoglobin and lipid oxidation in displayed beef. J. Anim. Sci. 71: p. 105.

Bohm, F., J.H. Tinkler and T.G. Truscott, 1995. Carotenoids protect against cell membrane damage by the nitrogen dioxide radical. Nat. Med. 1: p. 98-99

Bombardelli, E. and P. Morazzoni, 1995. Formulations containing carotenoids and procarotenoids combined with polyphenols in the prevention of the damages due to an abnormal production of free radicals. Eur. Pat. Appl. 659402.

Botsoglou N., Papageorgiou G., Nikolakakis I., Florou-Paneri P., Giannenas I., Dotas V. and E. Sinapis. 2004. Effect of dietary dried tomato pulp on oxidative stability of japanese quail meat, J. Agric. Food Chem. 52: p. 2982-2988.

Buck, D. F. and M. K. Edwards, 1997. Antioxidants to prolong shelf life. Food Technol. Int. 2: 29-33

Du, M., D. U. Ahn, K. C. Nam and J. L. Sell, 2000. Influence of dietary coniugated linoleic acid on volatiles profile, colour and lipid oxidation of irradiated raw chicken. Meat Sci. 56: p. 387-395.

Monahan, F. J., D. J. Buckley, J. I. Gray, P. A. Morrissey, A. Asghar, T. J. Hanrahan and P. B. Lynch, 1990. Effect of dietary vitamin E on the stability of raw and cooked pork. Meat Sci. 27: p. 99-108.

Realini C.E., S.K. Ducketta, G.W. Brito, M. Dalla Rizza and D. De Mattos, 2004. Effect of pasture vs. concentrate feeding with or without antioxidants on carcass characteristics, fatty acid composition, and quality of Uruguayan beef. Meat Sci. 66: p. 567-577.

SAS® User's Guide: Statistics, 2001. SAS Inst., Inc., Cary, NC.

Tang, S. Z., J. P. Kerry, D. Sheehan, D. J. Buckley and P. A. Morrissey, 2001. Antioxidative effect of dietary tea catechins on lipid oxidation of long-term frozen stored chicken meat. Meat Sci. 57: p. 331-336.

Tang, S. Z., J. P. Kerry, D. Sheehan, D. J. Buckley and P. A. Morrissey, 2000. Dietary tea catechins and iron-induced lipid oxidation in chicken meat, liver and heart. Meat Sci. 56: p. 285-290.

Tedesco, D., S. Stella, S. Galletti and S. Rossetti, 2003. Antimicrobial activity of green tea extract on rabbit caecum microflora. Book of Abstracts, EAAP - 54th Annual Meeting. 9: p. 113.

Woodall, A. A., S. W. Lee, R. J. Weesie, M. J. Jackson and G. Britton, 1997. Oxidation of carotenoids by free radicals: relationship between structure and reactivity. Biochim. Biophys. Acta 1336: p. 33-42.

Lack of correlation between vitamin B12 content in bovine muscles and indicators of the body stores in vitamin B12

I. Ortigues-Marty[1], D. Micol[1], D. Dozias[2] and C.L. Girard[3]*
[1]Unité de Recherches sur les Herbivores, INRA, Theix, 63122 Saint Genès Champanelle, France, [2]Domaine Expérimental INRA du Pin-au-Haras, 61310 Le-Pin-au-Haras, France, [3]Centre de Recherche et de Développement sur le Bovin Laitier et le Porc, Agriculture et Alimentaire Canada, 2000 Route 108 est, C.P. 90, Lennoxville, Québec, Canada.
**Correspondence: ortigues@clermont.inra.fr*

Summary

An important nutritional characteristic of ruminant meat is its high vitamin B12 (Vit B12) content, however its variability and its predictors are not well known. The present experiment aimed first at determining the influence of two muscle types (the more oxidative *Rectus Abdominis*, and the more glycolytic *Semitendinosus*) on their Vit B12 content and second at quantifying the variability of Vit B12 content in muscles. The present work was based on two studies which used 30-32 month-old-Charolais steers and which included differences in nature of the diet and in dietary cobalt allowances. In particular, animals were supplemented in macro and trace minerals according to usual feeding practices in France in order to theoretically avoid any risk of deficiency, without looking for optimum level of supplementation. In this context, it was also tested whether variations in Vit B12 muscle levels correlated with variations in body stores in Vit B12 (evaluated through changes in hepatic and plasma Vit B12 concentrations). Results indicated that: 1) the Vit B12 contents in oxidative type muscle were twice higher than those in glycolytic type muscle (RA 10.8 vs. ST 5.0 ng/g), 2) the variability in Vit B12 content in muscles amounted to 20% for the RA and 45% for the ST and 3) changes in muscle Vit B12 contents were poorly correlated with changes in body stores in Vit B12 as evaluated by plasma and hepatic concentrations in Vit B12.

Keywords: nutritional value, meat, vitamin B12, muscle

Introduction

An important nutritional characteristic of ruminant meat is its high content in vitamin B12 (Vit B12). In ruminants, Vit B12 is synthesized in the rumen by the microbial population using cobalt as the obligatory precursor (Underwood, 1981), before being absorbed and stored in the liver (for about 60%) and in muscles (Le Grusse and Watier, 1993).

No information is available yet on the variability of the Vit B12 content of bovine meat within a given type of animal and of the influence of muscle type. The first objective of the present project was thus to quantify the influence of muscle type on its Vit B12 content and to quantify the variability of those contents.

The second objective of the project was to test whether the variability in muscle Vit B12 concentrations was correlated with changes in Vit B12 body stores using both plasma and hepatic concentrations as indicators of body stores of Vit B12. Plasma Vit B12 concentrations are usually modified by short term changes in Vit B12 absorption while long term changes in body stores are best indicated by hepatic concentrations (Marston, 1970; Underwood, 1981). A clear relationship had been noted by Marston (1970) between intramuscularly injected Vit B12 and hepatic levels in sheep. It is not known whether such a relationship can also be found for 'basal' levels of Vit B12 in muscles.

To answer these objectives, 2 experiments were used, that were carried out on one type of animal only, 2.5-3 year-old-steers, and in which a relatively large variability in Vit B12 status was likely to occur considering the differences in cobalt allowances (Stangl *et al.*, 2000) or in type of diets (Elliott, 1980) provided. In particular, the mineral, including cobalt, supplementation aimed theoretically at avoiding any deficiency and no optimum level of supplementation was looked for, according to usual feeding practices in France. In this context, the allowances of cobalt could vary widely among treatments. This paper presents preliminary results of those studies.

Materials and methods

Two studies were conducted in successive years, using one-year-old Charolais steers. Each study corresponded to a 2 year experimental protocol. In both studies, animals were allotted to 3 treatment groups: 1) Pasture Grazing, 2) Indoors fed Cut Grass and 3) Maize silage. During the winter seasons, animals on Grass were fed a grass silage based diet (85 % rye-grass silage, 12 % hay and 3% concentrate). During the grazing seasons, these animals were either on a rotative grazing of rye-grass pasture or fed indoors with fresh grass which was cut daily from the same plot as that used for pasture grazing. Animals on the maize silage treatment were fed a diet composed of 66% maize silage, 22% wheat straw and 12 % rapeseed meal over the whole duration of the experiment. A total of 24 animals were used in the first study and of 30 animals in the second study.

In both studies, the mineral, including cobalt, supplementation followed usual feeding practices in France for beef cattle in which supplementation is theoretically aimed at avoiding any deficiency. No optimum level of supplementation was looked for. In this context, mineral (cobalt) supplementation differed greatly among treatments. It varied from 100 g of a commercial mineral supplement (10 mg cobalt/kg) per day for animals fed indoors to no supplementation for animals at pasture. Ultimately, the total cobalt content per kg of dry matter intake averaged 308 μg/kg at pasture, 378 μg/kg for indoors grass fed animals and 141 μg/kg for maize silage fed ones. Within each treatment, daily cobalt allowances varied little with time within the last 6 months preceding measurements.

Daily rations were adjusted so that average daily gains would be similar across all treatments within each experiment. Steers were slaughtered at the end of the second grazing seasons, at same final age (30-32 months) and same final live weight. Samples of muscles (Rectus Abdominis, RA, Semi Tendinosus, ST), liver and EDTA plasma were taken at slaughter after an overnight fast.

Feed samples were analysed for their cobalt content by electrothermy atomic absorption spectrometry after wet-ashing the samples in a micro-wave in presence of nitric acid and hydrogen peroxide. All liver and muscle samples were analysed for their content in biologically active 'true' Vit B12 by radioassay (Quantaphase II kit from BioRad) after homogenisation of the sample in an acetate buffer, hydrolysis with papaine and cyanisation with sodium cyanide. Plasma samples were analysed for their true Vit B12 content using the same BioRad kit. Means were compared by Student's t test. The use of plasma and hepatic Vit B12 levels as predictors of muscle Vit B12 contents was tested by linear regression.

Results and discussion

Across both studies, Vit B12 content was twice higher ($P < 0.001$) in the RA (10.76 ± 2.17 ng/g) than in the ST muscle (5.03 ± 2.26 ng/g). To the best of our knowledge, such original finding has not been reported earlier. In muscles, Vit B12 concentrations are known to vary in an inversely proportional manner with the lipid content (Mogens, 1984). However, in the present case, the above mentioned differences in Vit B12 are not the consequences of different lipid contents between muscles, since total lipid content was significantly higher in RA than in ST (Bauchart *et al.*, 2001).There are two main forms of active Vit B12: a methyl form which in humans is preferentially located in the cytosol and an adenosyl form which is found in mitochondria (Markle, 1996). Considering the known structural differences between the more oxidative type (e.g. RA) and the more glycolytic type (e.g. ST) muscles with in particular their difference in mitochondrial content (Hocquette *et al.*, 1998), it may be assumed that differences in Vit B12 contents between oxidative and glycolytic types of muscles might be attributed to the adenosyl form of Vit B12. The metabolic consequences of these differences would be worth investigating as well as the influence of the management factors which are known to modify the orientation of muscle energy metabolism (Hocquette *et al.*, 1998).

The overall variability of Vit B12 concentrations was higher for the ST (coefficient of variation, CV=45%) than for the RA (CV=20%). Within each muscle, no clear treatment differences could be detected. Hepatic Vit B12 levels averaged 658 ± 157.6 ng/g and presented a dispersion (CV=24%) which was moderate considering the range of cobalt allowances. On the other hand, the plasma Vit B12 concentrations which averaged 140 ± 64.8 pg/ml showed a larger dispersion (CV=46%).

It was then tested whether the variability in Vit B12 levels in muscles was correlated to plasma and hepatic levels. Because of the above mentioned differences between muscles and because of the lack of correlation between Vit B12 concentrations in RA and ST ($R^2 = 0.05$, NS), the analysis was carried out for each muscle separately. Muscle Vit B12 concentrations were not correlated to plasma levels at all. Some, albeit low, correlation was noted between muscle and hepatic concentrations (in ng/g; Figure 1):

$[Vit]_{RA} = 7.50 + 0.0049\ [Vit]_{liver}$, RSE=2.04, n=54, $R^2=0.13$ ($P<0.05$)

$[Vit]_{ST} = 0.92 + 0.0063\ [Vit]_{liver}$, RSE=2.06, n=54, $R^2=0.19$ ($P<0.01$)

No correlation whatsoever was noted between plasma and hepatic Vit B12 concentrations nor did the inclusion of both plasma and hepatic Vit B12 concentrations as co-variables improve the prediction of Vit B12 concentrations in muscles.

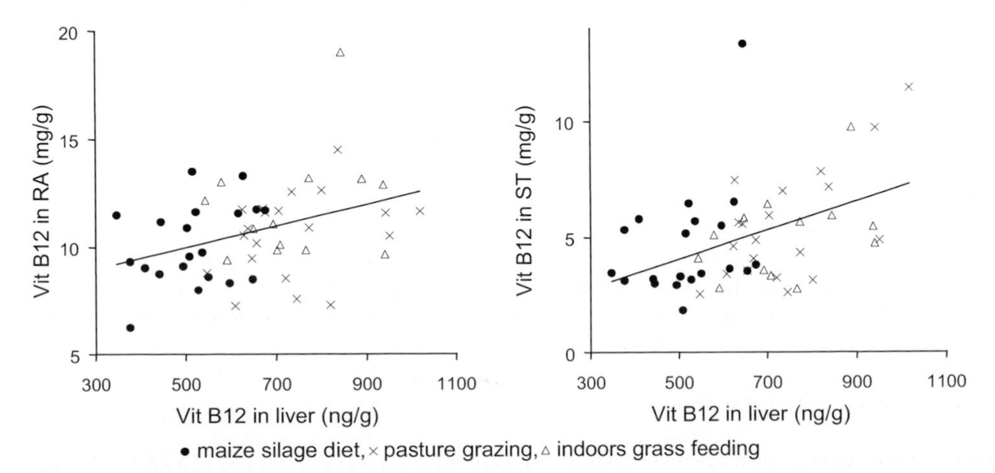

Figure 1. Relationship between Vit B12 concentrations in Rectus Abdominis *(RA) or* Semitendinosus *(ST) muscles and in liver.*

It may thus be concluded that, within a range of cobalt allowances which varied on average from 141 to 378 µg/kg dry matter, the variability of Vit B12 content in muscles from 2.5-3 year-old-steers was moderate and could not be predicted from plasma and hepatic Vit B12 levels with sufficient confidence.

Acknowledgements

Authors gratefully acknowledge the expert contribution of the staff of the INRA experimental station of Le Pin-au-Haras for running the experiments, of the INRA-Theix slaughter house for slaughtering the animals and contributing to tissue sampling, of Chrystiane Plante for the Vit B12 analysis and for Y. Anglaret for the statistical analyses.

References

Bauchart, D., D. Durand, D. Mouty, D. Dozias, I. Ortigues-Marty and D. Micol. 2001. Effets d'un régime à base d'herbe sur la teneur et la composition en acides gras des lipides des muscles et du foie chez le bouvillon à l'engrais. Renc. Rech. Ruminants 8: p. 108.

Elliott, J.M., 1980. Propionate metabolism and vitamin B12. In: Digestive Physiology and Metabolism in Ruminants, Y. Ruckebusch and P. Thivend (editors), Westport CO, USA, p. 485-503.

Hocquette, J.F., I. Ortigues-Marty, D. Pethick, P. Herpin and X. Fernandez, 1998. Nutritional and hormonal regulation of energy metabolism in skeletal muscles of meat-producing animals. Livest. Prod. Sci. 56: p. 115-143.

Le Grusse, J. and B. Watier (editors), 1993. Les Vitamines, Données Biochimiques, Nutritionnelles et Cliniques, CEIV, p. 255-271.

Markle, H.V., 1996. Cobalamin. Clin. Lab. Sci. 33: p. 247-356.

Marston, H.R., 1970. The requirement of sheep for cobalt or for vitamin B12. Brit. J. Nutr. 24: p. 615-633.

Mogens, J. (editor) 1984. The Quality of Frozen Foods. Academic Press, London.

Stangl, G.I., F.J. Schwarz, H. Müller and M. Kirchgessner, 2000. Evaluation of the cobalt requirement of beef cattle based on vitamin B12, folate, homocysteine and methylmalonic acid. Brit. J. Nutr. 84: p. 645-653.

Underwood, E.J. (editor) 1981. The Mineral Nutrition of Livestock. Commonwealth Agricultural Bureaux, Slough, UK, 180 p.

Assessment of the impact of herd management on sensorial quality of Charolais heifer meat

M.P. Oury [1], D. Micol [2], H. Labouré[3], M. Roux[1] and R. Dumont[1]*
[1]ENESAD, 26 bd Petitjean, BP 87999, 21079 Dijon Cedex, France, [2]INRA Clermont-Ferrand Theix, URH, 63122 Saint-Genès-Champanelle, France, [3]UMR ENESAD-INRA, 17 rue Sully, BP 86510, 21065 Dijon Cedex, France.
**Correspondence: mp.oury@enesad.fr*

Summary

An experiment was conducted to explain the variability of meat sensorial qualities by herd management practices and finishing status. One hundred Charolais heifers were selected at the slaughterhouse. Muscle m. *rectus abdominis* of each heifer was sampled, vacuum packed and stored at +4°C for 14 days before freezing. These samples were subsequently assessed by a 16 member taste panel. The management used for each heifer was studied by farm management survey. Carcass data were noted at the slaughterhouse.

Three different meat tenderness classes were constituted on initial and global tenderness: low (L) 4.7 and 4.2 out of 10, medium (M) 5.8 and 5.5 out of 10 and high (H) 7.0 and 6.7 out of 10 respectively ($p<0.0001$). Level H consisted of younger heifers (32 months), which grew more quickly during their whole life (677g/day) than heifers of classes L and M (35 and 33 months; $p=0.07$ and 614 and 671 g/day; $p=0.01$ respectively). There were no consistent variations on concentrate intake and carcass weight between tenderness classes. Heifer carcasses from classes L and M scored higher for bone development than those from H class (62, 63 and 56 on 100 respectively; $p=0.02$). The data suggest that there is a slight possibility of improving beef eating quality by choosing animals with a smaller bone development, by decreasing slaughter age and increasing growth rate.

Keywords: heifer, meat, tenderness, cattle, Charolais

Introduction

The meat industry and some consumers often criticize beef meat because the organoleptic quality and tenderness are highly variable and difficult to predict. So, beef meat consumption could be penalized, by the marketing of an irregular quality product.

There are many livestock rearing methods in France to produce beef meat. Herd management has an impact on meat qualities in most experiments, where the different factors, such as compensatory growth, type of feedstuff, sex, age, exercise, or slaughter characteristics, are tested in controlled situations (Geay *et al.*, 2001).

In order to explain the reasons for tenderness variation in red meat it appears important to make a global study of the effects associated with the management methods. Thus, this is a multifactorial approach to meat quality.

The aim of this study was to analyse and explain the variability of meat tenderness in one type of cattle, Charolais heifers. The originality of this study is that it first analysed tenderness and then tried to explain its variability by herd management practices and finishing status of the animals (age, live weight). This is different to the approaches generally used in the literature which consist of studying the effect of one breeding factor on sensorial quality.

Material and methods

Herd management and carcass characteristics

One hundred Charolais heifers were selected on carcass weight, higher than 330 kg. Heifers were sampled in a commercial slaughterhouse between February 2003 and January 2004. Carcasses were classified according to the EU classification scheme for conformation (E: greatest musculature, U, R, O, P: least musculature).

In order to prevent the effects of technological factors, standardized procedures were applied when slaughtering animals, chilling and storing carcasses.

Livestock rearing methods of each heifer were studied by farm management survey. For the present investigation, herd management data were age at weaning and at slaughter, duration of the different management periods (first winter, finishing) and feeding level (concentrates during animal life, whole life growth rate). The carcass results were recorded directly at the slaughterhouse. The 6th rib joint was removed and dissected in order to assess muscle, fat and bone proportions (Robelin and Geay, 1975). Muscle and bone development were recorded on the farm according to the Institut de l'Elevage method (Bèche *et al.*, 1996).

Sensory evaluations

The muscle *m. rectus abdominis* of each right half-carcass was stored under vacuum for a 14 day period at +4°C. Homogenous slices of 1.5 cm were then cut by a professional butcher and kept at -20°C for the sensorial analyses. Sixteen panellists were recruited and trained for meat assessment. Steaks were grilled on a double-sided grill at 300°C for 1.45 minute, then cut into cubes and served to the jury, in a monadic presentation. Four sensory characteristics (initial tenderness, global tenderness, juiciness and flavour intensity) were described on a 0 to 10 scale (10 for a very tender, juicy and well-flavoured meat).

Statistical analyses

A hierarchical classification, made with Winstat software was used to separate three classes of meat tenderness: low (L), medium (M), high (H). The data were also analysed using StatView 5 (SAS). The effects of the different factors of herd management were tested on the tenderness levels with variance analysis.

Results

Description of the data

The 100 heifers were on average 33 months of age with a carcass weight of 381kg. There were 2 main conformation classes, with 56 heifers in R= and 29 animals in R+, the remaining heifers being quoted in R- (n=9) and U- (n=6). The distribution frequency of the global tenderness mark varied between 3.0 and 7.2, with an average of 5.3 (Figure 1).

Differences in meat tenderness levels

From a hierarchical classification on the initial and global tenderness data, it appeared that three different tenderness classes could be observed in the population (Table 1). Level H consisted of 11 heifers, with initial and global tenderness significantly higher than classes L and M ($p<0.0001$).

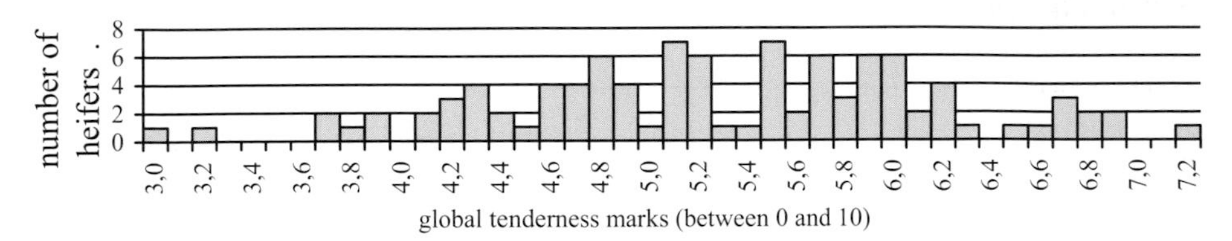

Figure 1. Heifer distribution depending on their global tenderness mark.

Table 1. Average marks of the three levels for each descriptor.

Tenderness level	Number of heifers	Initial tenderness	Global tenderness	Juiciness	Flavour intensity
Low	25	4.7 a	4.2 a	4.7 a	5.6 a
Medium	64	5.8 b	5.5 b	5.4 b	5.7 a
High	11	7.0 c	6.7 c	5.6 b	6.0 b
Statistical significance ($P<F$)		0.0001	0.0001	0.0001	0.04

a, b. Means in the same column with different letters are significantly different at $p<0.05$.

There was also a significant difference between the L class and the two others, with respect to juiciness ($p<0.0001$), even if the classes were only established on tenderness. Thus, meat of animals from class H, which scored better for tenderness than classes L and M, was also more juicy and tended to have a more intense flavour ($p=0.04$).

Relationships between carcass characteristics, herd management and tenderness levels

Analysis of herd management of the heifers in the three tenderness classes showed that H animals were born one month after animals from levels L and M, and were weaned at the same age ($p=0.52$). The first winter (9 to 15 months period) was longer for the class M and H animals, than for the class L animals ($p=0.06$). The analysis of finishing length and concentrate intake did not show any difference between the three tenderness classes ($p=0.79$; $p=0.63$). There was a significant effect of age at slaughter on tenderness levels ($p=0.07$) (Table 2). From the 6th rib joint dissection it appeared that heifers from the different tenderness levels were characterized by the same fat and bone proportions. Moreover, meat tenderness increased when the heifers' bone development mark decreased ($p=0.03$).

Discussion

There were no consistent variations in concentrate intake and carcass weight between tenderness classes. The present findings are in agreement with the works reported on tenderness by Dinius and Cross (1978) and Camfield *et al.* (1997) on young steers. By contrast, according to Harrisson *et al.* (1978), the duration of finishing of crossbred calves influenced tenderness. However, one has to be careful to extend conclusions from veal calves to heifers. The animals from each level received the same concentrate intake, during their whole life, and during the finishing period. The present results may be in agreement with Davis *et al.* (1981), who found in a comparison between a limited or an *ad libitum* allowance of concentrate to steers, that tenderness was not influenced by the amount of concentrates.

Table 2. Average different characteristics of finished heifers classified according to the tenderness levels.

Tenderness level	Low	Medium	High	Statistical significance (P<F)
Slaughter age (months)	35[a]	33[ab]	32[b]	0.07
Average birth date	20/02	10/02	15/03	
Weaning age (days)	228	232	211	0.52
1st winter length/9-15months (day)	164[a]	179[b]	188[b]	0.06
Finishing length (day)	145	151	139	0.79
Total concentrate intake (kg)	1665	1530	1667	0.63
Life growth rate (g/day)	614[a]	671[b]	677[b]	0.01
Carcass weight (kg)	374	384	381	0.43
Fat proportion in 6th rib joint (%)	17.7	17.4	19.1	0.38
Muscle proportion in 6th rib joint (%)	66.0[ab]	66.9[b]	64.4[a]	0.08
Bone proportion in 6th rib joint (%)	16.3	15.7	16.5	0.44
Muscle development (mark out of 100)	61	60	60	0.70
Bone development (mark out of 100)	62[a]	63[a]	56[b]	0.03

[a, b]Means in the same column with different letters are significantly different at $p<0.05$.

Between levels H, M and L, increasing slaughter age (32, 33 and 35 months respectively) leads to a decrease of tenderness (p=0.07). The same conclusion was already seen by Touraille (1982), who reported a negative effect of age on tenderness in males, between 24 and 33 months. This result needs to be emphasized, as the negative effect of age is not often underlined in the literature. Indeed, a lack of effect was noticed in females between 24 and 30 months and between 31 and 35 months of age (Touraille, 1982; Field *et al.*, 1996). Moreover, the average carcass weights of the three classes were not significantly different (p=0.43).

Life growth rate of levels M and H animals was significantly higher than those from level L (p=0.01). We also conclude that a higher growth rate is in favour of greater tenderness, as already reported by Purchas *et al.* (2002). According to Harris *et al.* (1979), Patterson *et al.* (2002) and Trenkle *et al.* (1978), a fast growth was favourable for a higher rate of fat deposit in the carcass and in the muscles. So, meat tenderness may be positively correlated to intramuscular fat. Thus, growth rate could induce improved tenderness by increasing fat deposition.

Conclusion

This experiment allowed firstly to conclude that it was possible within one type of animal, such as the Charolais heifer, to assess three meat tenderness classes. These three tenderness levels may be explained by three characteristics of herd management: slaughter age, growth rate, and bone development. Heifers that produce the most tender meat (level H) were younger, grew more quickly and had less bone development than heifers from levels L and M.

Acknowledgements

The authors gratefully acknowledge the SCICAV Bourgogne Elevage producers group, the Arcadie slaughterhouse and breeders who participated in the present work. The study was carried out with the financial support of the Scientific Council of ENESAD and of the French "Ministère de l'Agriculture de l'Alimentation de la Pêche et des Affaires Rurales".

References

Bèche, J.M., P. Chavatte, P. Berrechet, M. Caillaud, B. Chretien, C. Delahaie, J. Lapray, O. Leudet, E. Rehben and L. Journaux, 1996. Pointage au sevrage des bovines de race à viande. In: CR2495, Institut de l'élevage, 67 p.

Camfield, P.K., A.H. Brown, P.K. Lewis, L.Y. Rakes and Z. B. Johnson, 1997. Effects of frame size and time on feed on carcass characteristics, sensory attributes, and fatty acid profiles of steers. J. Anim. Sci. 75: p. 1837-1844.

Davis, G.W., A.B. Cole, W.R. Backus and S.L. Melton, 1981. Effect of electrical stimulation on carcass quality and meat palatability of beef from forage and grain finished steers. J. Anim. Sci. 53: p. 651-657.

Dinius, D.A. and H.R. Cross, 1978. Feedlot performances, carcass characteristics and meat palatability of steers fed concentrate for short periods. J. Anim. Sci., 47: p. 1109-1113.

Field, R., R. McCormick, V. Balasubramanian, D. Sanson, J. Wise, D. Hixon, M. Riley and W. Russell, 1996. Growth, carcass and tenderness characteristics of virgin, spayed, and single calf heifers. J. Anim. Sci. 74: p. 2178-2186.

Geay, Y., D. Bauchart, J.F. Hocquette and J. Culioli, 2001. Effect of nutritional factors on biochemical, structural and metabolic characteristics of muscles in ruminants, consequences on dietetic value and sensorial qualities of meat. Reprod. Nutr. Dev. 41: 1-26, Erratum, 41: p. 377.

Robelin, J. and Y. Geay 1975. Estimation de la composition des carcasses de jeunes bovins à partir de la composition d'un morceau monocostal prélevé au niveau de la 11ème côte. Ann. Zootech. 24: p. 391-402.

Harris, J.M., E.H. Cash, L.L. Wilson and W.R. Stricklin, 1979. Effects of concentrate level, protein source and growth promotant : growth and carcass traits. J. Anim. Sci. 49: 613-619

Harrisson, A.R., M.E. Smith, D.M. Allen, M.C. Hunt, C.L. Kastner and D.H. Kropf, 1978. Nutritional regime effects on quality and yield characteristics of beef. J. Anim. Sci. 47: p. 383-38.

Patterson, D.C., C.A. Moore, B.W. Moss and D.J. Kilpatrick, 2002. Parity associated changes in slaughter weight and carcass characteristics of 3/4 charolais crossbred cows kept on a lowland grass/grass silage feeding and management system. Anim. Sci. 75: 221-235

Purchas, R.W., D.L. Burnham and S.T. Morris, 2002. Effects of growth potential and growth path on tenderness of beef longissimus muscle from bulls and steers. J. Anim. Sci. 80: p. 3211-3221.

Touraille, C., 1982. Influence du sexe et de l'âge à l'abattage sur les qualités organoleptiques des viandes de bovins limousins abattus entre 16 et 33 mois. CRZV, INRA, 48: p. 83-89.

Trenkle, A., D.L. De Witt and D.G. Topel, 1978. Influence of age, nutrition and genotype on carcass traits and cellular development of the m. longissimus of cattle. J. Anim. Sci. 46: p. 1597-1603.

Comparison of rearing systems at low environmental impact: Quality meat of beef cattle in central Italy

M. Iacurto, A. Gaddini, S. Gigli, G.M. Cerbini, S. Ballico and A. Di Giacomo*
Animal Production Research Institute - Via Salaria, 31, 00016 Monterotondo (Roma), Italy.
**Correspondence: miriam.iacurto@isz.it*

Summary

We analysed physical quality of meat of 30 Chianina young bulls. Animals were reared in three different housing systems: 10 animals on slatted floor (SF), 12 in a feed-lot (FL) and 8 on pasture (PA). Carcasses were anatomically dissected after 8 days from slaughter and meat samples were drawn from 5 different muscles: *Longissimus thoracis* (LT), *Gluteobiceps* (GB), *Semimembranosus* (SM), *Semitendinosus* (ST) and *Caput longum Triceps Brachiis* (CloTB). The Warner-Bratzler shear force was not different between groups on raw meat, while on the cooked one it was the highest in PA (11.15 kg vs. less than 10.2 kg); drip losses were the lowest in FL (1,34% vs. more than 1.68%); PA had the lowest cooking losses (27.00% vs. more than 29%). SF had the highest Lightness (42.91 vs. less than 41.3), no difference was found in Chroma, while Hue was different between the three production systems (from 28.29 to 39.52). The comparison between muscles showed that ST was significantly tougher (20.63 kg on raw meat, 12.65 kg on cooked meat); ST showed the highest drip losses (2.79% vs. less than 1.51%) and the highest cooking losses (31.42%). Lightness and Hue were also the highest in ST (46.00 vs. less than 40.60 for lightness; 38.45 vs. less than 33.91 for Hue) while Chroma was the lowest in LT (25.84 vs. more than 28.42). In conclusion, the meat from PA animals was, as expected, the toughest, the darkest and showed less cooking losses. Among muscles, ST was the toughest, with the highest lightness and the highest drip and water losses; these results being also expected.

Keywords: rearing systems, quality meat, Chianina bulls

Introduction

The recurring food safety crises in Europe, involving mainly meat, strengthened the demand for traditional products, deemed as coming from production systems more natural and healthy for the consumer. Moreover, a production chain distant from industrial systems is also seen as animal-friendly and with a lower environmental impact (Issanchou, 1996; Gibon *et al.*, 1999). In Italy, dry summers and winters allow to feed cattle on pasture only in spring and part of the summer (Gigli and Iacurto, 1995; Gigli *et al.*, 1997), while, in the dry seasons, satisfying daily gains are obtainable only by an integration, usually consisting in grains and hay. Chianina is one of the most important Italian traditional beef breeds, perfectly fit to the environment and renown for the quality of its meat and for the *Fiorentina* steak. The *Fiorentina* steak is a particular cut, obtained from the lumbar region, 5 cm high, and weighing minimum 0.5 kg; it should be quickly roasted on embers.

In this work, we analysed cattle reared in a sustainable system based on pasture, compared to an intensive and a semi-intensive system. In a previous work (Iacurto *et al.*, 2003), *in vivo* and slaughtering performances were examined and the authors concluded that Chianina breed, reared at pasture, keeps the good conformation, but does not reach an excellent state of fattening, which is a problem for the Italian market because these carcasses are not appreciated by the consumers, therefore not accepted by the butchers. Thus, this work aimed to analyse some objective quality traits of beef from Chianina young bulls reared in different conditions.

Materials and methods

We analysed physical quality of meat of 30 Chianina young bulls. Animals were reared in three different housing systems: 10 animals on slatted floor (SF), 12 in a feed-lot (FL) and 8 on pasture (PA).

Housing system

The animals were all born in the experimental farm from a herd at pasture, and were taken back from rangeland and weaned at 6 months of age; the trial then started at 11-12 months (average initial live weight of 270 kg).

The SF bulls were fed maize silage *ad libitum*, maize grain (1.0 kg for100 kg body weight); soybean (1.1 kg) and mineral-vitamin integration. On average, diet was of 9.27 kg/d of dry matter (DM) at about 9.0% of crude protein (CP) and 7.5 meat forage unit (MFU) per day of energy.

The FL feed was the same of SF group until 14 months of age, successively the maize silage was substituted with hay (10.0% CP), to follow the prescriptions of a labelling given by the European Union: the PGI (Protected Geographical Indication Reg. UE 134/981998), maize grain (1.0 kg for100 kg body weight); soybean (1.7 kg), mineral-vitamin integration; with a ratio hay/concentrate of 1, average daily intake was of 11 kg DM at about 14.0% of CP and 13 MFU.

The PA group had a grazing land surface of 7 ha, mainly of natural meadow (12% of CP), and along all the rearing period, only an integration with maize grains (1.0 kg/100 kg body weight) was administered (Iacurto *et al.*, 2000a; 2000b).

Physical meat quality

The animals were slaughtered at commercial maturity (Iacurto *et al.*, 2003) which is calculated at about the 75% of the maturity weight that is typical of the breed. Carcasses were anatomically dissected after 8 days from slaughter and meat samples were drawn from 5 different muscles: *Longissimus thoracis* (LT), *Gluteobiceps* (GB), *Semimembranosus* (SM), *Semitendinosus* (ST) and *Caput longum Triceps Brachiis* (CloTB).

The following parameters were measured on the samples: Shear force on raw and cooked meat were measured with a Warner-Bratzler Shear on Instron 1011, on 4 cores of 2 inch section. Water losses were determined as drip losses and cooking losses. Drip losses were determined as a percentage after 48 h storage at 2°±2°C and cooked losses as a percentage in a waterbath until internal temperature 75°C (ASPA 1996). Colour was measured by a Minolta CM-2600d colorimeter-spectrometer, with CieLAB system and a C illuminant (6774°K); Chroma ($C=(a^2+b^2)^{1/2}$) that indicates the percentage of pure colour (for C=0 is grey) and Hue (H=arctg b/a) that indicates colour tonality (for H=0 is purple red) were calculated.

Statistical analysis

Statistical analysis was performed with SAS (2001), using the General Linear Model procedures with the following model: $Y_{ijk}=\mu+A_i+Nu_j(Ai)+B_k+(AB)_{ik}+\varepsilon_{ijkl}$ where: μ=Means; A_i=Housing type (i=1, ..., 3); $Nuj(A_i)$ = animal number nested with the housing type effect, B_k =Muscle type (j=1, ..., 5); ε_{ij}=Residual variance. Because muscles have to be considered as repeated measurements on the same animals, the housing type factor was tested against the animal number factor. The residual mean square was used as the error term for the muscle effect. The interactions were not significant.

Results and discussion

Housing system

Comparing housing systems about hardness (Table 1), we found significant differences only on cooked meat. Higher toughness in animals reared on pasture was made evident (11.15 kg vs. 10.10 kg in average). The higher values of hardness in pastured animals are largely expected, by the more intense muscular activity which characterizes this production system; but these animals were slaughtered at higher age compared to the other groups (Table 2) in fact cooking losses were lower (27.00 %) although intramuscular fat content of the carcass was lower (Table 2). However, the shear force values on raw meat, even if not significantly different, were higher compared to the standard values of Chianina meat (Iacurto *et al.*, 1994) probably because the fatness score of these animals was lower (Table 2).

The drip losses (Table 1) were higher in SF and PA groups (+0.47 point percent (p.p.) in average). In the first case it is probably due to the feedstuff composed of maize silage even if the literature is contradictory (Iacurto and Gigli, 1996), while, in the second case, the lower value is due to lower internal fat content in the carcass (Table 2). In the FL group, the value of drip losses was intermediate (1.34%). About the cooking losses, the lower values were found in the PA group, as previously mentioned; this was probably due to the higher age of the animals (approximately 3.5 months compared to the other groups).

The PA group showed meat of dark and lower lightness colour (Table 1); in fact they had higher Hue values compared to the other groups (+28.4% vs. FL group and +9.1% vs. SF group) and

Table 1. Physical quality parameters according to production systems and muscles.

	Shear force (kg)		Water losses (%)		Colour		
	Raw	cooked	drip	cooking	Lightness	Chroma	Hue
Housing system							
SF	16.08	10.02 ^{b}	1.94^{a}	29.94^{a}	42.91^{a}	28.68	35.91^{b}
FL	16.27	10.17^{b}	1.34^{b}	29.41^{a}	41.24^{b}	28.40	28.29^{c}
PA	15.49	11.15^{a}	1.68^{a}	27.00^{b}	41.29^{b}	29.10	39.52^{a}
Muscles type							
CloTB	17.76^{b}	11.19^{b}	1.34^{b}	30.24^{ab}	41.60^{b}	29.67^{a}	33.87^{b}
GB	13.08^{c}	8.18^{d}	1.30^{b}	27.02^{c}	40.15^{c}	29.48^{a}	33.08^{b}
LT	11.89^{c}	9.57^{c}	1.33^{b}	25.98^{c}	40.12^{c}	25.84^{b}	33.55^{b}
SM	16.37^{b}	10.63^{b}	1.51^{b}	29.25^{b}	41.20^{bc}	30.21^{a}	33.91^{b}
ST	20.63^{a}	12.65^{a}	2.79^{a}	31.42^{a}	46.00^{a}	28.42^{a}	38.45^{a}
Means	16.00	10.38	1.63	28.94	41.81	28.68	33.82
RMSE(*)	2.915	1.561	1.379	3.376	1.943	3.678	2.278

^{a,b}Means with different superscripts differ significantly at P≤0.05

SF=slatted floor; FL=Feed lot; PA=Pasture; CloTB=*Caput Longum Triceps Brachiis*; GB=*Gluteo Biceps*; LT=*Longissimus thoracis*; SM=*Semimembranosus*; ST=*Semitendinosus*.

(*) RMSE= Root Means Standard Error

Table 2. Slaughtering and dissection data (Iacurto et al., 2003).

	Slatted fl.	Feed-lot	Pasture	means	RMSE**
Number (total: 33)	10	12	8		
Average daily gain (g/d)	1.42 [a]	1.18 [ab]	1.07 [b]	1.22	0.323
Age at slaughtering (d)	556.6 [b]	579.8 [b]	654.8 [a]	593.3	39.58
Live weight (kg)	606.6 [b]	602.3 [b]	663.1 [a]	620.2	33.97
Carcass weight (kg)	388.1 [a]	359.7 [b]	403.8 [a]	380.3	24.95
Net dressing percentage (%)	68.16 [a]	64.66 [b]	65.89 [ab]	66.06	3.32
Conformation score *	10.88 (U)	10.45 (U-)	10.19 (U-)	10.51	1.14
Fatness score *	5.40 (2) [a]	3.50 (2-) [b]	3.02 (1+) [b]	3.95	1.02
Meat (%)	70.89 [b]	72.03 [b]	74.50 [a]	72.36	2.08
Intramuscular fat (%)	6.89 [a]	4.77 [b]	3.11 [c]	4.96	1.07
Subcutaneous fat (%)	4.00 [a]	2.46 [b]	2.03 b	2.81	0.71

* correspondence with EUROP scale: Conformation score: 1=P-,..., 15=E+; Fatness score: 1=1-,..., 15=5+.
** RMSE= Root Means Standard Error

lower lightness than SF group (-3.8%). The clearest meat was found in the FS group (Hue=28.29), even if lightness was low. Chroma values were not different between rearing systems, with an average value of 28.68. Ender *et al.*, (1997), on Black Pied steers, found similar values between animals permanently housed, finished indoors and permanently on pasture.

Physical meat quality

When comparison was made between muscles (Table 1), ST was significantly tougher on raw (+28.3% in average) and cooked meat (+21.8% in average); CloTB and SM were intermediate (17.76 and 16.37 kg on raw, 11.19 and 10.63 on cooked meat), and LT and GB were the most tender (11.89 and 13.08 kg on raw and 9.57 and 8.18 kg on cooked meat). Torrescano *et al.* (2003) on Brown Swiss young bulls analysed shear force on raw meat in 14 muscles and found the same order that we found, with the same differences, but lower absolute values; the same was found by Crouse *et al.* (1984) on Hereford x Angus heifers. Also Johnson *et al.*(1988) analysed 34 muscles of Angus steers forequarters, cooked in a waterbath, and found in LT a lower shear force value than in CloTB; while Shackelford *et al.* (1995) on crossbred steers of several breeds (*Bos taurus* and *Bos indicus*) did not find any difference on shear force on cooked meat between ST, SM and LT.

We found in ST a higher drip loss (2.79%) compared to the average of the other four muscles (1.37%) not differing from one to another; also cooking losses were higher in ST (31.42%) than in the other muscles (28.12% in average), even if not significantly different from CloTB (30.24%), which in its turn did not differ from SM (29.25%); LT and GB had lower losses, respectively 25.98% and 27.02%. Crouse *et al.*, (1984) analysing cooking losses, found similar values, ST having higher losses with respect to SM and LT that was lower.

About the color measurement, ST showed also a higher lightness, with a value of 46.00, while the other muscles had markedly less lightness (40.77 in average), Chroma was lower in LT (25.84) with respect to the others muscles (in average 29.45) which did not differ between each other's and Hue was significantly higher in ST (+12.6% in average) compared to the other muscles that not differ between them. Torrescano *et al.* (2003) found that the meat of ST was the lightest, the one of SM the intermediate and the one of LT the less light, confirming our values.

Conclusion

The comparison between muscles made evident that ST was the toughest, the lightest and had the lowest water holding capacity; CloTB and SM were intermediate and generally did not differ between them; GB and LT were the most tender, the lightest and had the lowest cooking losses; confirming the individuality of the quality of the muscles.

This research confirmed that in Italy, in spite of the seasonal allowance of pasture, it is possible to rear beef cattle in an extensive way; nevertheless the extensification of production is linked to integration with grains and to hay supplying in periods when pastures are not disposable.

However, animals reach slaughter maturity at too elevated age for the Italian market; the meat is hard, darker and with lower water losses compared to the intensive and semi-intensive systems. Nevertheless, an extensive system can provide a type of meat that can be recognized by the consumers as safe, healthy, with low environmental impact and with attention to animal welfare.

References

ASPA. Associazione Scientifica di Produzione Animale 1996. Metodiche per la determinazione della caratteristiche qualitative della carne. Ed. Università degli Studi di Perugia.

Crouse, J.D., H.R. Cross and S.C. Seideman, 1984. Effects of a grass or grain diet on the quality of three beef muscles. J. Anim. Sci. 58: 619-625

Ender, K., H.-J. Papstein,, K. Nürnberg and J. Wegner, 1997. Muscle and fat related characteristics of grazing steers and lambs in extensive systems. Proceedings of the Workshop "Effects of Extensification on Animal Performance, Carcass Composition and Product Quality", Melle-Gontrode, Belgium, May 16-17: p. 229-237.

European Commission Regulation No 134/98 of 20 January 1998, "Vitellone Bianco dell'Appennino Centrale".

Gibon, A., A.R. Sibbald and C. Thomas, 1999. Improved sustainability in livestock systems, a challenge for animal production science. Livest. Prod. Sci. 61: p. 107-110.

Gigli, S., and M. Iacurto, 1995 Present and future cattle and sheep meat production systems in Italy. Proceedings of the Workshop "Extensification of Beef and Sheep Production on Grassland", Paris, November 22-24, p. 253-267.

Gigli, S., M. Iacurto, U. Francia and A. Carretta. 1997. Land and animals stock situation and their use in Italy. Effect of extensive feeding systems on Maremmana young bulls. Proceedings of the Workshop "Effects of Extensification on Animal Performance, Carcass Composition and Product Quality", Melle-Gontrode, Belgium, May 16-17: 133-156.

Iacurto, M., S. Gigli, S. Failla and M. Mormile, 1994. Caratteristiche chimico-fisiche della carne (Longissimus Dorsi) di vitelloni di razza Chianina iscritti al Consorzio 5R. VII International Congress of Chianina Cattle "Italian Beef Cattle Contest". Perugia, 16-18 September.

Iacurto M., D. Settineri, S. Gigli, M. Mormile, A. Di Giacomo, 2000a. Daily gain and dressing of Maremmana Young bull crossbreeds: use of specific techniques (Geographic Identification Scheme). 35° Simposio Internazionale di Zootecnia, Ragusa 25 Maggio 2000.

Iacurto M., S. Gigli, S. Failla, M. Mormile, A. Di Giacomo, 2000b. Comparison of susteinable farming system: beef cattle in Central Italy. Atti Convegno Nazionale "Parliamo di ... allevamenti del 3° millennio". Fossano (Cuneo) 12-13 Ottobre 2000.

Iacurto, M. and S. Gigli, 1996. Physical Characteristics variations in Chianina young bulls meat according to maize sillage use. Atti Convegno Nazionale "Parliamo di....." Fossano (Cuneo) 17-18 Ottobre.

Iacurto, M., A. Gaddini, S. Gigli, M. Mormile andF. Vincenti, 2003. Comparison of rearing systems at low environmental impact: the beef cattle of central Italy. 54th Annual Meeting of European Association for Animal Production, Rome, August 31st - September 3rd.

Issanchou, S. 1996. Consumer expectations and perception of meat and meat product quality. Meat Sci. 43: p. S5-S19.

Johnson, R.C., C.M. Chen , T.S. Muller, W.J. Costello, J.R. Romans and K.W. Jones , 1988. Characterization of the muscles within the beef forequarter. J. Food Sci. 53: p. 1247-1250.

SAS Institute Inc. 1999-2001, Release 8.02 for Windows

Shackelford, S.D., T.L. Wheeler and M. Koohmaraie 1995. Relationship between shear force and trained sensory panel tenderness ratings of 10 major muscles from Bos indicus and Bos taurus cattle. J. Anim. Sci. 73: p. 3333-3340.

Torrescano, G., A. Sánchez-Escalante, B. Giménez, P. Roncalés and J.A. Beltrán, 2003. Shear values of raw samples of 14 bovine muscles and their relation to muscle collagen characteristics. Meat Sci. 64: p. 85-91.

Authors index

Keyword index